Abbreviations for Units

A	ampere	H	henry	nm	nanometer (10^{-9} m)
Å	angstrom (10^{-10} m)	h	hour	pt	pint
atm	atmosphere	Hz	hertz	qt	quart
Btu	British thermal unit	in	inch	rev	revolution
Bq	becquerel	J	joule	R	roentgen
C	coulomb	K	kelvin	Sv	seivert
°C	degree Celsius	kg	kilogram	s	second
cal	calorie	km	kilometer	T	tesla
Ci	curie	keV	kilo-electron volt	u	unified mass unit
cm	centimeter	lb	pound	V	volt
dyn	dyne	L	liter	W	watt
eV	electron volt	m	meter	Wb	weber
°F	degree Fahrenheit	MeV	mega-electron volt	y	year
fm	femtometer, fermi (10^{-15} m)	Mm	megameter (10^6 m)	yd	yard
ft	foot	mi	mile	μm	micrometer (10^{-6} m)
Gm	gigameter (10^9 m)	min	minute	μs	microsecond
G	gauss	mm	millimeter	μC	microcoulomb
Gy	gray	ms	millisecond	Ω	ohm
g	gram	N	newton		

Some Conversion Factors

Length

1 m = 39.37 in = 3.281 ft = 1.094 yd

1 m = 10^{15} fm = 10^{10} Å = 10^9 nm

1 km = 0.6215 mi

1 mi = 5280 ft = 1.609 km

1 lightyear = 1 c·y = 9.461×10^{15} m

1 in = 2.540 cm

Volume

1 L = 10^3 cm^3 = 10^{-3} m^3 = 1.057 qt

Time

1 h = 3600 s = 3.6 ks

1 y = 365.24 d = 3.156×10^7 s

Speed

1 km/h = 0.278 m/s = 0.6215 mi/h

1 ft/s = 0.3048 m/s = 0.6818 mi/h

Angle–angular speed

1 rev = 2π rad = 360°

1 rad = 57.30°

1 rev/min = 0.1047 rad/s

Force–pressure

1 N = 10^5 dyn = 0.2248 lb

1 lb = 4.448 N

1 atm = 101.3 kPa = 1.013 bar = 76.00 cmHg = 14.70 lb/in^2

Mass

1 u = $[(10^{-3}\ \text{mol}^{-1})/N_A]$ kg = 1.661×10^{-27} kg

1 tonne = 10^3 kg = 1 Mg

1 slug = 14.59 kg

1 kg weighs about 2.205 lb

Energy–power

1 J = 10^7 erg = 0.7373 ft·lb = 9.869×10^{-3} L·atm

1 kW·h = 3.6 MJ

1 cal = 4.184 J = 4.129×10^{-2} L·atm

1 L·atm = 101.325 J = 24.22 cal

1 eV = 1.602×10^{-19} J

1 Btu = 778 ft·lb = 252 cal = 1054 J

1 horsepower = 550 ft·lb/s = 746 W

Thermal conductivity

1 W/(m·K) = 6.938 Btu·in/(h·ft^2·F°)

Magnetic field

1 T = 10^4 G

Viscosity

1 Pa·s = 10 poise

fifth edition

PHYSICS

FOR SCIENTISTS AND ENGINEERS

Volume 1B
Oscillations and Waves;
Thermodynamics

W. H. Freeman and Company
New York

PT: For Claudia

GM: For Vivian

Publisher: Susan Finnemore Brennan
Senior Development Editor: Kathleen Civetta/Jennifer Van Hove
Assistant Editors: Rebecca Pearce/Amanda McCorquodale/Eileen McGinnis
Marketing Manager: Mark Santee
Project Editors: Georgia L. Hadler/Cathy Townsend, PreMediaONE, A Black Dot Group Company
Cover and Text Designers: Marcia Cohen/Blake Logan
Illustrations: Network Graphics/PreMediaONE, A Black Dot Group Company
Photo Editors: Patricia Marx/Dena Betz
Production Manager: Julia DeRosa
Media and Supplements Editors: Brian Donnellan
Composition: PreMediaONE, A Black Dot Group Company
Manufacturing: RR Donnelley & Sons Company

Cover image: Digital Vision

Library of Congress Cataloging-in-Publication Data
Physics for Scientists and Engineers. - 5th ed.
p. cm.
By Paul A. Tipler and Gene Mosca
Includes index.
ISBN: 0-7167-0809-4 (Vol. 1 Hardback Ch. 1-20, R)
ISBN: 0-7167-0900-7 (Vol. 1A Softcover Ch. 1-13, R)
ISBN: 0-7167-0903-1 (Vol. 1B Softcover Ch. 14-20)
ISBN: 0-7167-0810-8 (Vol. 2 Hardback Ch. 21-41)
ISBN: 0-7167-0902-3 (Vol. 2A Softcover Ch. 21-25)
ISBN: 0-7167-0901-5 (Vol. 2B Softcover Ch. 26-33)
ISBN: 0-7167-0906-6 (Vol. 2C Softcover Ch. 34-41)
ISBN: 0-7167-8339-8 (Standard Hardback Ch. 1-33, R)
ISBN: 0-7167-4389-2 (Extended Hardback Ch. 1-41)

Printed in the United States of America

First printing 2003

CONTENTS

VOLUME 1

CHAPTER 5

APPLICATIONS OF NEWTON'S LAWS / 117

CHAPTER 6

WORK AND ENERGY / 151

CHAPTER 7

CONSERVATION OF ENERGY / 183

CHAPTER 8

SYSTEMS OF PARTICLES AND CONSERVATION OF LINEAR MOMENTUM / 217

CHAPTER 9

ROTATION / 267

PART III THERMODYNAMICS /532

CHAPTER 17

TEMPERATURE AND THE KINETIC THEORY OF GASES / 532

CHAPTER 18

HEAT AND THE FIRST LAW OF THERMODYNAMICS / 558

CHAPTER 19

THE SECOND LAW OF THERMODYNAMICS / 595

CHAPTER 20*

THERMAL PROPERTIES AND PROCESSES / 628

APPENDIX A

APPENDIX B

APPENDIX C

APPENDIX D

VOLUME 2

PART IV ELECTRICITY AND MAGNETISM/651

CHAPTER 21

CHAPTER 22

CHAPTER 26

THE MAGNETIC FIELD / 829

CHAPTER 27

SOURCES OF THE MAGNETIC FIELD / 856

CHAPTER 28

MAGNETIC INDUCTION / 897

CHAPTER 29

ALTERNATING-CURRENT CIRCUITS / 934

CHAPTER 30

MAXWELL'S EQUATIONS AND ELECTROMAGNETIC WAVES / 971

PART V LIGHT/997

CHAPTER 31

PROPERTIES OF LIGHT / 997

CHAPTER 32

OPTICAL IMAGES / 1038

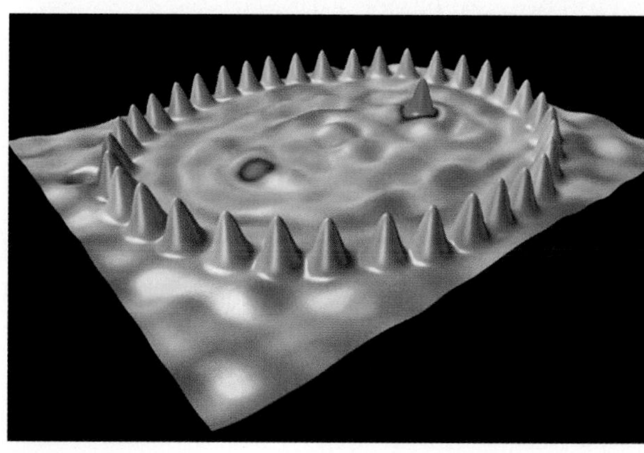

CHAPTER 36

ATOMS

CHAPTER 37

MOLECULES

CHAPTER 38

SOLIDS AND THE THEORY OF CONDUCTION

CHAPTER 39

RELATIVITY

CHAPTER 40

NUCLEAR PHYSICS

CHAPTER 41

ELEMENTARY PARTICLES AND THE BEGINNING OF THE UNIVERSE

PART II OSCILLATIONS AND WAVES

CHAPTER 14

Oscillations

THIS BOAT IS RISING AND FALLING ON THE SWELLS OF THE SEA. ITS MOTION IS AN EXAMPLE OF OSCILLATORY MOTION. THE MAXIMUM CHANGE IN THE VERTICAL POSITION OF THE BOAT CAN BE READILY MEASURED, AS CAN THE TIME FOR THE BOAT TO COMPLETE ONE CYCLE OF THIS UP AND DOWN MOTION.

? How can the vertical position of the boat be expressed as a function of time? (See Example 14-1.)

Oscillation occurs when a system is disturbed from a position of stable equilibrium. There are many familiar examples: boats bob up and down, clock pendulums swing back and forth, and the strings and reeds of musical instruments vibrate. Other, less familiar examples are the oscillations of air molecules in a sound wave and the oscillations of electric currents in radios and television sets.

➢ **In this chapter, we deal mostly with simple harmonic motion—the most basic type of oscillatory motion. Applying the kinematics and dynamics of simple harmonic motion provides the analysis of the oscillatory motion of a variety of interesting systems. In some situations dissipative forces dampen the oscillatory motion, but in other situations driving forces sustain the motion by compensating for the damping.**

14-1 Simple Harmonic Motion

A common, very important, and very basic kind of oscillatory motion is **simple harmonic motion** such as the motion of an object attached to a spring (Figure 14-1). In equilibrium, the spring exerts no force on the object. When the object is displaced an amount x from its equilibrium position, the spring exerts a force $-kx$, as given by Hooke's law†:

$$F_x = -kx \tag{14-1}$$

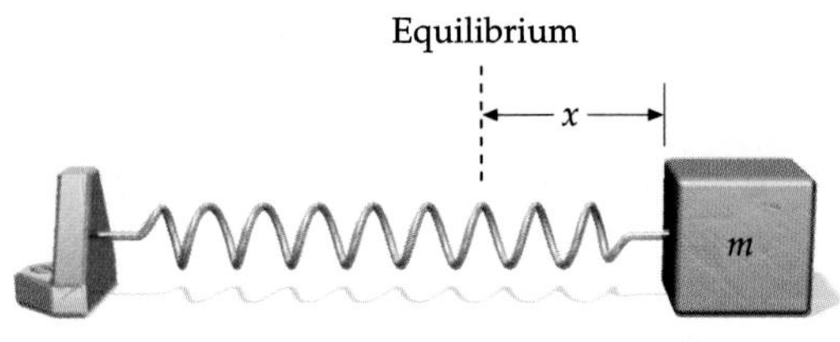

FIGURE 14-1 An object and spring on a frictionless surface. The displacement x, measured from the equilibrium position, is positive if the spring is stretched and negative if the spring is compressed.

where k is the force constant of the spring, a measure of the spring's stiffness. The minus sign indicates that the force is a restoring force; that is, it is opposite to the direction of the displacement from the equilibrium position. Combining Equation 14-1 with Newton's second law ($F_x = ma_x$), we have

$$-kx = ma_x$$

or

$$a_x = -\frac{k}{m}x \quad \left(\text{or} \quad \frac{d^2x}{dt^2} = -\frac{k}{m}x\right) \tag{14-2}$$

The acceleration is proportional to the displacement and is oppositely directed. This is the defining characteristic of simple harmonic motion and can be used to identify systems that will exhibit it:

> Whenever the acceleration of an object is proportional to its displacement and is oppositely directed, the object will move with simple harmonic motion.

CONDITIONS FOR SIMPLE HARMONIC MOTION IN TERMS OF ACCELERATION

Because the acceleration is proportional to the net force, whenever the net force on an object is proportional to its displacement and is oppositely directed, the object will move with simple harmonic motion.

The time it takes for a displaced object to execute a complete cycle of oscillatory motion—from one extreme to the other extreme and back—is called the **period** T. The reciprocal of the period is the **frequency** f, which is the number of cycles per second:

$$f = \frac{1}{T} \tag{14-3}$$

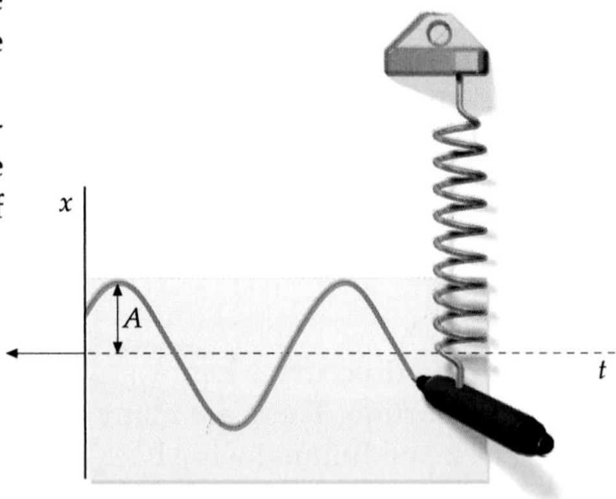

FIGURE 14-2 A marking pen is attached to a mass on a spring, and the paper is pulled to the left. As the paper moves with constant speed, the pen traces out the displacement x as a function of time t. (Here we have chosen x to be positive when the spring is compressed.)

The unit of frequency is the cycle per second (cy/s), which is called a **hertz** (Hz). For example, if the time for one complete cycle of oscillation is 0.25 s, the frequency is 4 Hz.

Figure 14-2 shows how we can experimentally obtain x versus t for a mass on a spring. The general equation for such a curve is

$$x = A\cos(\omega t + \delta) \tag{14-4}$$

POSITION IN SIMPLE HARMONIC MOTION

† Hooke's law is introduced in Chapter 4, Section 4.

where A, ω, and δ are constants. The maximum displacement x_{max} from equilibrium is called the **amplitude** A. The argument of the cosine function, $\omega t + \delta$, is called the **phase** of the motion, and the constant δ is called the **phase constant,** which is the phase at $t = 0$. (Note that $\cos(\omega t + \delta) = \sin(\omega t + \delta + \pi/2)$; thus, whether the equation is expressed as a cosine function or a sine function simply depends on the phase of the oscillation at the moment we designate to be $t = 0$.) If we have just one oscillating system, we can always choose $t = 0$ at which $\delta = 0$. If we have two systems oscillating with the same amplitude and frequency but different phase, we can choose $\delta = 0$ for one of them. The equations for the two systems are then

$$x_1 = A\cos(\omega t)$$

and

$$x_2 = A\cos(\omega t + \delta)$$

If the phase difference δ is 0 or an integer times 2π, then $x_2 = x_1$ and the systems are said to be in phase. If the phase difference δ is π or an odd integer times π, then $x_2 = -x_1$, and the systems are said to be 180° out of phase.

The swaying of the Citicorp Building in New York City during high winds is reduced by this tuned-mass damper mounted on an upper floor. It consists of a 400-ton sliding block connected to the building by a spring. The spring constant is chosen so that the natural frequency of the spring–block system is the same as the natural sway frequency of the building. Set into motion by winds, the building and damper oscillate 180° out of phase with each other, thereby significantly reducing the swaying.

We can show that Equation 14-4 is a solution of Equation 14-2 by differentiating x twice with respect to time. The first derivative of x gives the velocity v:

$$v = \frac{dx}{dt} = -\omega A\sin(\omega t + \delta) \qquad 14\text{-}5$$

VELOCITY IN SIMPLE HARMONIC MOTION

Differentiating velocity with respect to time gives the acceleration:

$$a = \frac{dv}{dt} = \frac{d^2x}{dt^2} = -\omega^2 A\cos(\omega t + \delta) \qquad 14\text{-}6$$

Substituting x for $A\cos(\omega t + \delta)$ (see Equation 14-4) gives

$$a = -\omega^2 x \qquad 14\text{-}7$$

ACCELERATION IN SIMPLE HARMONIC MOTION

Comparing $a = -\omega^2 x$ (Equation 14-7) with $a = -(k/m)x$ (Equation 14-2), we see that $x = A\cos(\omega t + \delta)$ is a solution of Equation 14-2 (which can be expressed $d^2x/dt^2 = -(k/m)x$) if

$$\omega = \sqrt{\frac{k}{m}} \qquad 14\text{-}8$$

The amplitude A and the phase constant δ can be determined from the initial position x_0 and the initial velocity v_0 of the system. Setting $t = 0$ in $x = A\cos(\omega t + \delta)$ gives

$$x_0 = A\cos\delta \qquad 14\text{-}9$$

Similarly, setting $t = 0$ in $v = dx/dt = -A\omega\sin(\omega t + \delta)$ gives

$$v_0 = -A\omega\sin\delta \qquad 14\text{-}10$$

These equations can be solved for A and δ in terms of x_0 and v_0.

The period T is the shortest time satisfying the relation

$$x(t) = x(t + T)$$

for all t. Substituting into this relation using Equation 14-4 gives

$$\begin{aligned} A\cos(\omega t + \delta) &= A\cos[\omega(t + T) + \delta] \\ &= A\cos(\omega t + \delta + \omega T) \end{aligned}$$

The cosine (and sine) function repeats in value when the phase increases by 2π, so

$$\omega T = 2\pi \qquad \left(\text{or} \quad \omega = \frac{2\pi}{T}\right)$$

The constant ω is called the **angular frequency.** It has units of radians per second and dimensions of inverse time, the same as angular speed, which is also designated by ω. Substituting $2\pi/T$ for ω in Equation 14-4 gives

$$x = A\cos\left(2\pi\frac{t}{T} + \delta\right)$$

We can see by inspection that each time t increases by T, the phase increases by 2π and one cycle of the motion is completed.

The frequency is the reciprocal of the period:

$$f = \frac{1}{T} = \frac{\omega}{2\pi} \qquad 14\text{-}11$$

DEFINITION—FREQUENCY, PERIOD, AND ANGULAR FREQUENCY

Because $\omega = \sqrt{k/m}$, the frequency and period of an object on a spring are related to the force constant k and the mass m by

$$f = \frac{1}{T} = \frac{1}{2\pi}\sqrt{\frac{k}{m}} \qquad \text{14-12}$$

FREQUENCY AND PERIOD FOR AN OBJECT ON A SPRING

The frequency increases with increasing k (spring stiffness) and decreases with increasing mass.

Astronaut Alan L. Bean measures his body mass during the second Skylab mission by sitting in a seat attached to a spring and oscillating back and forth. The total mass of the astronaut plus the seat is related to his frequency of vibration by Equation 14-12.

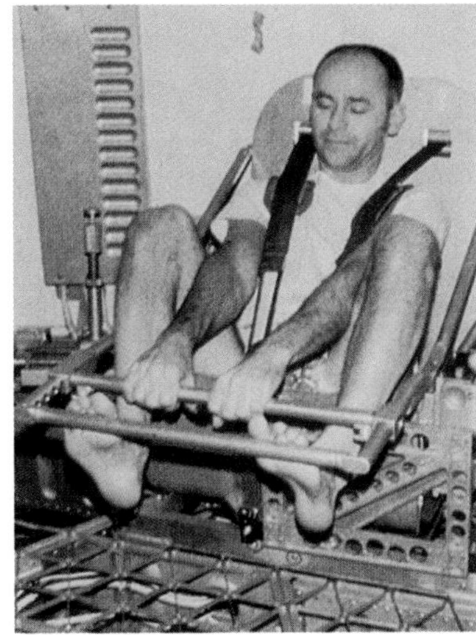

RIDING THE WAVES — EXAMPLE 14-1

You are on a boat, which is bobbing up and down. The boat's vertical displacement y is given by

$$y = (1.2\text{ m})\cos\left(\frac{1}{2\text{ s}}t + \frac{\pi}{6}\right)$$

(*a*) Find the amplitude, angular frequency, phase constant, frequency, and period of the motion. (*b*) Where is the boat at $t = 1$ s? (*c*) Find the velocity and acceleration as functions of time t. (*d*) Find the initial values of the position, velocity, and acceleration of the boat.

PICTURE THE PROBLEM We find the quantities asked for in (*a*) by comparing the equation of motion

$$y = (1.2\text{ m})\cos\left(\frac{1}{2\text{ s}}t + \frac{\pi}{6}\right)$$

with the standard equation for simple harmonic motion, Equation 14-4. The velocity and acceleration are found by differentiating $y(t)$.

(*a*) 1. Compare this equation with Equation 14-4, $y = A\cos(\omega t + \delta)$, to get A, ω, and δ:

$$y = (1.2\text{ m})\cos\left(\frac{1}{2\text{ s}}t + \frac{\pi}{6}\right)$$

$$A = \boxed{1.2\text{ m}} \qquad \omega = \boxed{1/2\text{ rad/s}} \qquad \delta = \boxed{\pi/6\text{ rad}}$$

2. The frequency and period are found from ω:

$$f = \frac{\omega}{2\pi} = \boxed{0.0796\text{ Hz}} \qquad T = \frac{1}{f} = \boxed{12.6\text{ s}}$$

(*b*) Set $t = 1$ s to find the boat's position:

$$y = (1.2\text{ m})\cos\left[\frac{1}{2\text{ s}}(1\text{ s}) + \frac{\pi}{6}\right] = \boxed{0.624\text{ m}}$$

(*c*) The velocity and acceleration are obtained from the position by differentiation with respect to time:

$$v_y = \frac{dy}{dt} = \frac{d}{dt}[A\cos(\omega t + \delta)]$$

$$= -\omega A\sin(\omega t + \delta)$$

$$= -\frac{1}{2\text{ s}}(1.2\text{ m})\sin\left(\frac{1}{2\text{ s}}t + \frac{\pi}{6}\right)$$

$$= \boxed{-(0.6\text{ m/s})\sin\left(\frac{1}{2\text{ s}}t + \frac{\pi}{6}\right)}$$

$$a_y = \frac{dv_y}{dt} = \frac{d}{dt}[-\omega A\sin(\omega t + \delta)]$$

$$= -\omega^2 A\cos(\omega t + \delta)$$

$$= -\left(\frac{1}{2\text{ s}}\right)^2 (1.2\text{ m}) \cos\left(\frac{1}{2\text{ s}}t + \frac{\pi}{6}\right)$$

$$= \boxed{-(0.3\text{ m/s}^2) \cos\left(\frac{1}{2\text{ s}}t + \frac{\pi}{6}\right)}$$

(*d*) Set $t = 0$ to find y_0, v_{y0}, and a_{y0}:

$$y_0 = (1.2\text{ m}) \cos\frac{\pi}{6} = \boxed{1.04\text{ m}}$$

$$v_{y0} = -(0.6\text{ m/s}) \sin\frac{\pi}{6} = \boxed{-0.300\text{ m/s}}$$

$$a_{y0} = -(0.3\text{ m/s}^2) \cos\frac{\pi}{6} = \boxed{-0.260\text{ m/s}^2}$$

EXERCISE A 0.8-kg object is attached to a spring of force constant $k = 400$ N/m. Find the frequency and period of motion of the object when it is displaced from equilibrium. (*Answer* $f = 3.56$ Hz, $T = 0.281$ s)

Figure 14-3 shows two identical masses attached to identical springs and resting on a frictionless surface. One spring is stretched 10 cm and the other 5 cm. If they are released at the same time, which object reaches the equilibrium position first?

According to Equation 14-12, the period depends only on k and m and not on the amplitude. Since k and m are the same for both systems, the periods are the same. Thus, the objects reach the equilibrium position at the same time. The second object has twice as far to go to reach equilibrium, but it will also have twice the speed at any given instant. Figure 14-4 shows a sketch of the position functions for the two objects. This illustrates an important general property of simple harmonic motion:

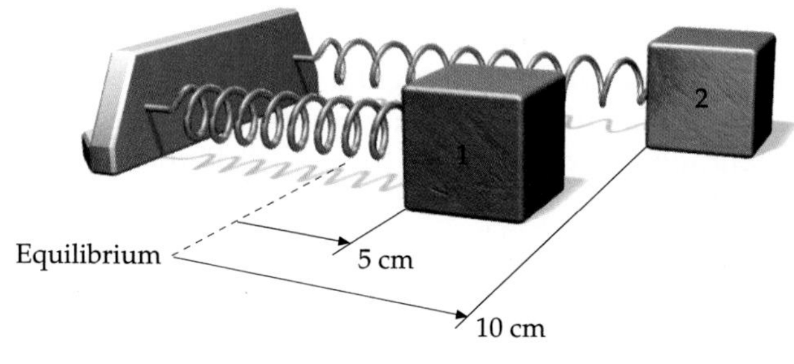

FIGURE 14-3 Two identical mass–spring systems.

> The frequency and period of simple harmonic motion are independent of the amplitude.

The fact that the frequency in simple harmonic motion is independent of the amplitude has important consequences in many fields. In music, for example, it means that when a note is struck on the piano, the pitch (which corresponds to the frequency) does not depend on how loudly the note is played (which corresponds to the amplitude).† If changes in amplitude had a large effect on the frequency, then musical instruments would be unplayable.

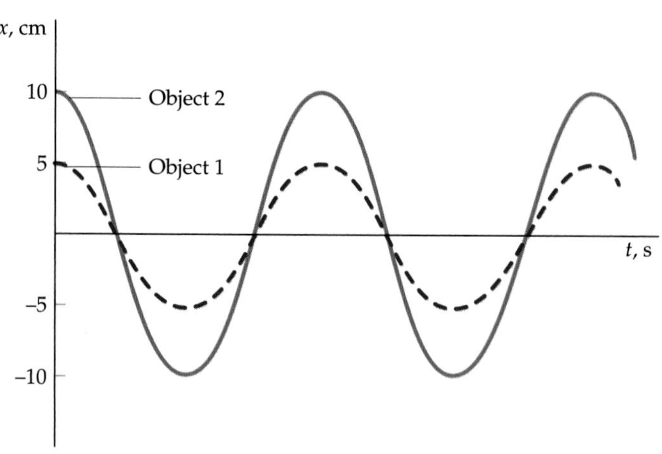

FIGURE 14-4 Plots of x versus t for the systems in Figure 14-3. Both reach their equilibrium positions at the same time.

† For many musical instruments, there is a slight dependence of frequency on amplitude. The vibration of an oboe reed, for example, is not exactly simple harmonic, thus its pitch depends slightly on how hard it is blown. This effect can be corrected for by a skilled musician.

AN OSCILLATING OBJECT **EXAMPLE 14-2**

An object oscillates with angular frequency $\omega = 8.0$ rad/s. At $t = 0$, the object is at $x = 4$ cm with an initial velocity $v = -25$ cm/s. (*a*) Find the amplitude and phase constant for the motion. (*b*) Write x as a function of time.

PICTURE THE PROBLEM The initial position and velocity give us two equations from which to determine the amplitude A and the phase constant δ.

(*a*) 1. The initial position and velocity are related to the amplitude and phase constant. The position is given by Equation 14-4. The velocity is found by taking the derivative with respect to time:

$$x = A\cos(\omega t + \delta)$$

and

$$v = \frac{dx}{dt} = -\omega A\sin(\omega t + \delta)$$

2. At $t = 0$ the position and velocity are:

$$x_0 = A\cos\delta \quad \text{and} \quad v_0 = -\omega A\sin\delta$$

3. Divide these equations to eliminate A:

$$\frac{v_0}{x_0} = \frac{-\omega A\sin\delta}{A\cos\delta} = -\omega\tan\delta$$

4. Substituting numerical values yields δ:

$$\tan\delta = -\frac{v_0}{\omega x_0}$$

so

$$\delta = \tan^{-1}\left(-\frac{v_0}{\omega x_0}\right)$$

$$= \tan^{-1}\left[-\frac{-25\text{ cm/s}}{(8.0\text{ rad/s})(4\text{ cm})}\right]$$

$$= \boxed{0.663\text{ rad}}$$

5. The amplitude can be found using either the x_0 or v_0 equation. Here we use x_0:

$$A = \frac{x_0}{\cos\delta} = \frac{4\text{ cm}}{\cos 0.663} = \boxed{5.08\text{ cm}}$$

(*b*) Comparing with Equation 14-4 yields x:

$$x = \boxed{(5.08\text{ cm})\cos[(8.0\text{ s}^{-1})t + 0.663]}$$

When the phase constant is $\delta = 0$, Equations 14-4, 14-5, and 14-6 then become

$$x = A\cos\omega t \qquad \text{14-13}a$$

$$v = -\omega A\sin\omega t \qquad \text{14-13}b$$

and

$$a = -\omega^2 A\cos\omega t \qquad \text{14-13}c$$

These functions are plotted in Figure 14-5.

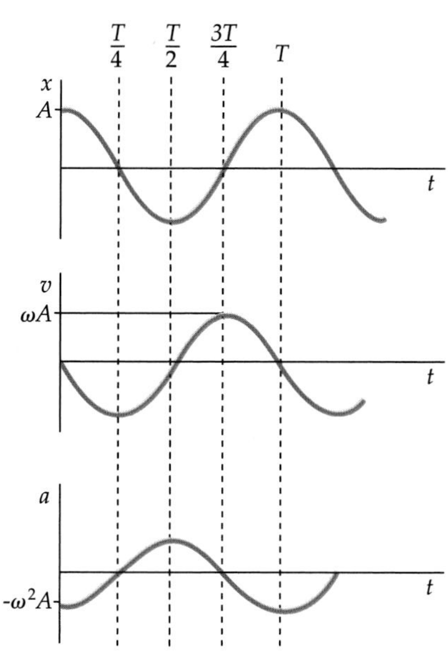

FIGURE 14-5 Plots of x, v, and a as functions of time t for $\delta = 0$. At $t = 0$, the displacement is maximum, the velocity is zero, and the acceleration is negative and equal to $-\omega^2 A$. The velocity becomes negative as the object moves back toward its equilibrium position. After one quarter-period ($t = T/4$), the object is at equilibrium, $x = 0$, $a = 0$, and the speed has its maximum value of ωA. At $t = T/2$, the displacement is $-A$, the velocity is again zero, and the acceleration is $+\omega^2 A$. At $t = 3T/4$, $x = 0$, $a = 0$, and $v = +\omega A$.

A BLOCK ON A SPRING — EXAMPLE 14-3 Try It Yourself

A 2-kg block is attached to a spring as in Figure 14-1. The force constant of the spring is $k = 196$ N/m. The block is held a distance 5 cm from the equilibrium position and is released at $t = 0$. (*a*) Find the angular frequency ω, the frequency f, and the period T. (*b*) Write x as a function of time.

Cover the column to the right and try these on your own before looking at the answers.

Steps	Answers
(*a*) 1. Calculate ω from $\omega = \sqrt{k/m}$.	$\omega = \boxed{9.90\ \text{rad/s}}$
2. Use your result to find f and T.	$f = \boxed{1.58\ \text{Hz}}$ $T = \boxed{0.635\ \text{s}}$
3. Find A and δ from the initial conditions.	$A = 5$ cm $\delta = 0$
(*b*) Write $x(t)$ using your results for A, ω, and δ.	$x = \boxed{(5\ \text{cm})\cos[(9.90\ \text{s}^{-1})t]}$

SPEED AND ACCELERATION OF AN OBJECT ON A SPRING — EXAMPLE 14-4

Consider an object on a spring whose position is given by $x = (5\ \text{cm})\cos(9.90\ \text{s}^{-1}\ t)$. (*a*) What is the maximum speed of the object? (*b*) When does this maximum speed first occur? (*c*) What is the maximum of the acceleration of the object? (*d*) When does maximum acceleration first occur after $t = 0$?

PICTURE THE PROBLEM Because the object is released from rest, $\delta = 0$, and the position, velocity, and acceleration are given by Equations 14-13*a*, *b*, and *c*.

(*a*) 1. Equation 14-13*a*, with $\delta = 0$, gives the position. We get the velocity by taking the derivative with respect to time:

$$x = A\cos\omega t$$

so

$$v = \frac{dx}{dt} = -\omega A\sin\omega t$$

2. Maximum speed occurs when $|\sin\omega t| = 1$:

$$|v| = \omega A\,|\sin\omega t|$$

so

$$|v|_{\text{max}} = \omega A = (9.90\ \text{rad/s})(5\ \text{cm})$$

$$= \boxed{49.5\ \text{cm/s}}$$

(*b*) 1. $|\sin\omega t| = 1$ first occurs when $\omega t = \pi/2$:

$$|\sin\omega t| = 1 \Rightarrow \omega t = \frac{\pi}{2}, \frac{3\pi}{2}, \frac{5\pi}{2}, \ldots$$

2. Solve for t when $\omega t = \pi/2$:

$$t = \frac{\pi}{2\omega} = \frac{\pi}{2(9.90\ \text{s}^{-1})} = \boxed{0.159\ \text{s}}$$

(*c*) 1. We find the acceleration by taking the derivative of the velocity, obtained in step 1 of Part (*a*):

$$a = \frac{dv}{dt} = -\omega^2 A\cos\omega t$$

2. Maximum acceleration corresponds to $\cos \omega t = -1$.

$$a_{max} = \omega^2 A = (9.90\text{ rad/s})^2\,(5\text{ cm})$$
$$= \boxed{490\text{ cm/s}^2 \approx \tfrac{1}{2}g}$$

(*d*) The maximum acceleration occurs when $|\cos \omega t| = 1$, which is when $\omega t = 0, \pi, 2\pi, \ldots$:

$$\omega t = \pi$$

so

$$t = \frac{\pi}{\omega} = \frac{\pi}{9.90\text{ s}^{-1}} = \boxed{0.317\text{ s}}$$

REMARKS The maximum speed first occurs after one quarter-period

$$t = \frac{\pi}{2\omega} = \frac{\pi}{2(2\pi/T)} = \frac{1}{4}T$$

The maxima of the magnitude of the acceleration occur when $\omega t = 0, \pi, 2\pi, \ldots$. These correspond to $t = 0, \tfrac{1}{2}T, \tfrac{2}{2}T, \tfrac{3}{2}T, \ldots$.

Simple Harmonic Motion and Circular Motion

There is a relation between simple harmonic motion and circular motion with constant speed. Imagine a particle moving with constant speed v in a circle of radius A (Figure 14-6*a*). Its angular displacement relative to the x axis is

$$\theta = \omega t + \delta \qquad \text{14-14}$$

where δ is the angular displacement at time $t = 0$ and $\omega = v/A$ is the angular speed of the particle. The x component of the particle's position (Figure 14-6*b*) is

$$x = A\cos\theta = A\cos(\omega t + \delta)$$

which is the same as Equation 14-4 for simple harmonic motion.

When a particle moves with constant speed in a circle, its projection onto a diameter of the circle moves with simple harmonic motion (Figure 14-6).

The speed of a particle moving in a circle is $r\omega$, where r is the radius. For the particle in Figure 14-6*b*, $r = A$, so its speed is $A\omega$. The projection of the velocity vector onto the x axis gives $v_x = -v\sin\theta$. Substituting for v and θ gives

$$v_x = -\omega A\sin\theta = -\omega A\sin(\omega t + \delta)$$

which is the same as Equation 14-5 for simple harmonic motion.

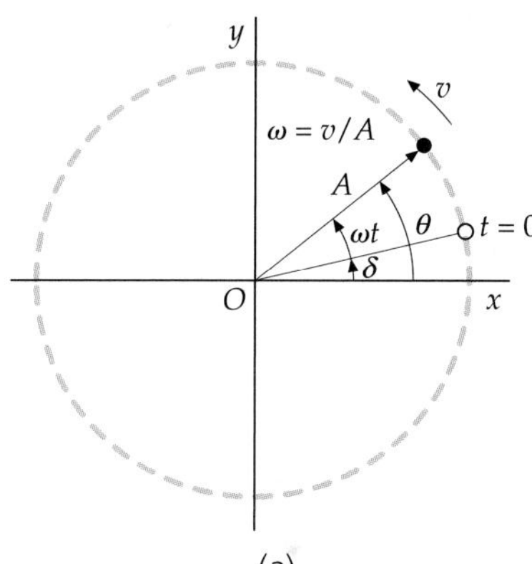

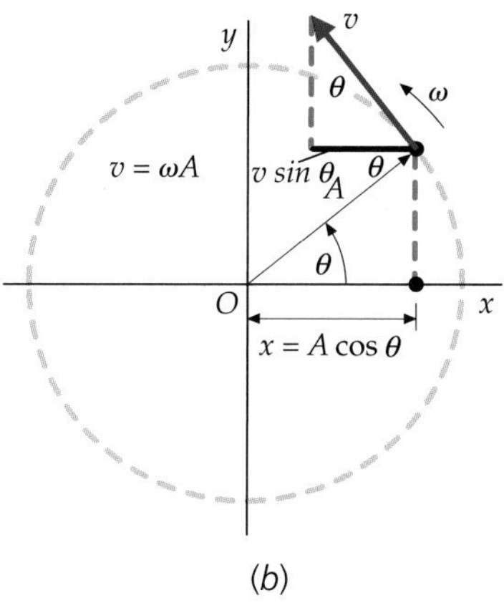

FIGURE 14-6 A particle moves in a circular path with constant speed. (*a*) Its *x* component of position describes simple harmonic motion, and (*b*) its *x* component of velocity describes the velocity of the simple harmonic motion.

Bubbles foaming off the edge of a rotating propeller that is moving through water produce a sinusoidal pattern.

14-2 Energy in Simple Harmonic Motion

When an object on a spring undergoes simple harmonic motion, the system's potential energy and kinetic energy vary with time. Their sum, the total mechanical energy $E = K + U$, is constant. Consider an object a distance x from equilibrium, acted on by a restoring force $-kx$. The system's potential energy is

$$U = \tfrac{1}{2}kx^2$$

This is Equation 6-21. For simple harmonic motion, $x = A\cos(\omega t + \delta)$. Substituting gives

$$U = \tfrac{1}{2}kA^2\cos^2(\omega t + \delta) \qquad \text{14-15}$$

POTENTIAL ENERGY IN SIMPLE HARMONIC MOTION

The kinetic energy of the system is

$$K = \tfrac{1}{2}mv^2$$

where m is the object's mass and v is its speed. For simple harmonic motion, $v_x = -\omega A\sin(\omega t + \delta)$. Substituting gives

$$K = \tfrac{1}{2}m\omega^2A^2\sin^2(\omega t + \delta)$$

Then using $\omega^2 = k/m$,

$$K = \tfrac{1}{2}kA^2\sin^2(\omega t + \delta) \qquad \text{14-16}$$

KINETIC ENERGY IN SIMPLE HARMONIC MOTION

The total mechanical energy is the sum of the potential and kinetic energies:

$$\begin{aligned} E_{\text{total}} &= U + K = \tfrac{1}{2}kA^2\cos^2(\omega t + \delta) + \tfrac{1}{2}kA\sin^2(\omega t + \delta) \\ &= \tfrac{1}{2}kA^2[\cos^2(\omega t + \delta) + \sin^2(\omega t + \delta)] \end{aligned}$$

Since $\sin^2(\omega t + \delta) + \cos^2(\omega t + \delta) = 1$,

$$E_{\text{total}} = \tfrac{1}{2}kA^2 \qquad \text{14-17}$$

TOTAL MECHANICAL ENERGY IN SIMPLE HARMONIC MOTION

This equation reveals an important general property of simple harmonic motion:

> The total mechanical energy in simple harmonic motion is proportional to the square of the amplitude.

For an object at its maximum displacement, the total energy is all potential energy. As the object moves toward its equilibrium position, the kinetic energy of the system increases and its potential energy decreases. As it moves through its equilibrium position, the kinetic energy of the object is maximum, the potential energy of the system is zero, and the total energy is kinetic.

As the object moves past the equilibrium point, its kinetic energy begins to decrease, and the potential energy of the system increases until the object again stops momentarily at its maximum displacement (now in the other direction). At all times, the sum of the potential and kinetic energies is constant. Figures 14-7*b* and *c* show plots of *U* and *K* versus time. These curves have the same shape except that one is zero when the other is maximum. Their average values over one or more cycles are equal, and because $U + K = E$, their average values are given by

$$U_{av} = K_{av} = \tfrac{1}{2}E \qquad \text{14-18}$$

In Figure 14-8, the potential energy *U* is graphed as a function of *x*. The total energy E_{total} is constant and is therefore plotted as a horizontal line. This line intersects the potential-energy curve at $x = A$ and $x = -A$. These are the points at which oscillating objects reverse direction and head back toward the equilibrium position, and are called the **turning points.** Because $U \leq E_{total}$, the motion is restricted to $A \leq x \leq +A$.

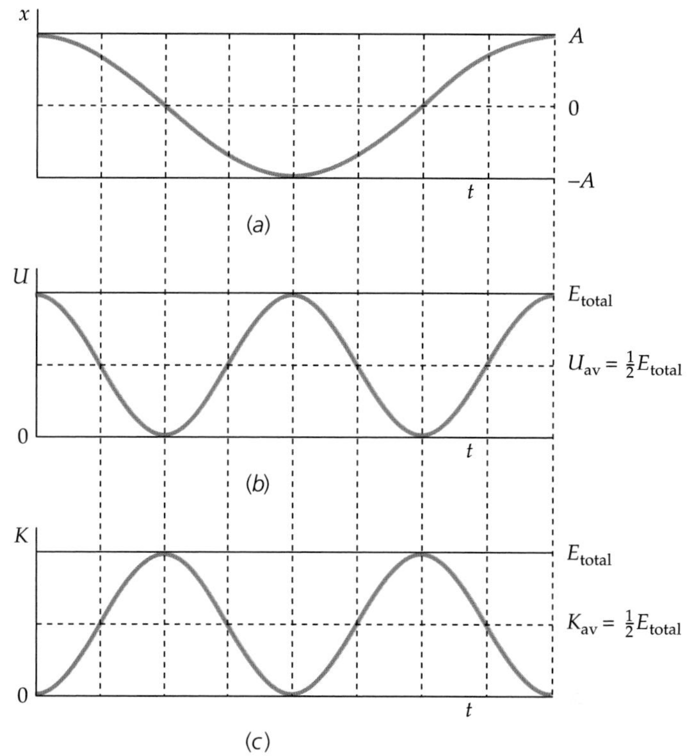

FIGURE 14-7 Plots of *x*, *U*, and *K* versus *t*.

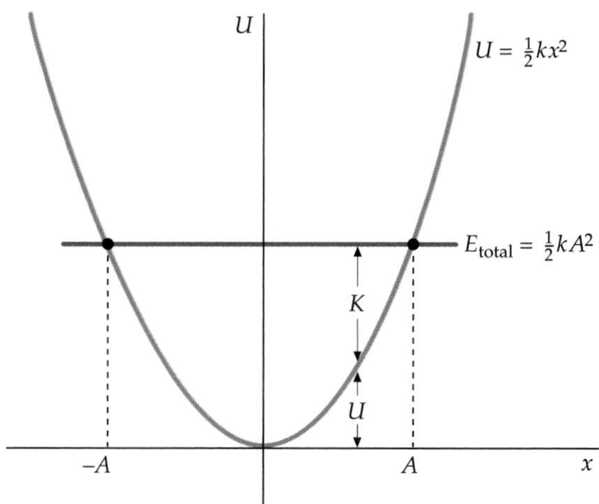

FIGURE 14-8 The potential-energy function $U = \frac{1}{2}kx^2$ for an object of mechanical mass *m* on a (massless) spring of force constant *k*. The horizontal blue line represents the total mechanical energy E_{total} for an amplitude of *A*. The kinetic energy *K* is represented by the vertical distance $K = E_{total} - U$. Since $E_{total} \geq U$, the motion is restricted to $-A \leq x \leq +A$.

ENERGY AND SPEED OF AN OSCILLATING OBJECT — EXAMPLE 14-5

A 3-kg object attached to a spring oscillates with an amplitude of 4 cm and a period of 2 s. (*a*) What is the total energy? (*b*) What is the maximum speed of the object? (*c*) At what position x_1 is the speed equal to half its maximum value?

PICTURE THE PROBLEM (*a*) The total energy can be found from the amplitude and force constant, and the force constant can be found from the mass and period. (*b*) The maximum speed occurs when the kinetic energy equals the total energy. (*c*) We can relate the position to the speed by using conservation of energy.

(*a*) 1. Write the total energy *E* in terms of the force constant *k* and amplitude *A*:

$$E = \tfrac{1}{2}kA^2$$

2. The force constant is related to the period and mass:

$$k = m\omega^2 = m\left(\frac{2\pi}{T}\right)^2$$

3. Substitute the given values to find *E*:

$$E = \frac{1}{2}kA^2 = \frac{1}{2}m\left(\frac{2\pi}{T}\right)^2 A^2$$

$$= \frac{1}{2}(3\text{ kg})\left(\frac{2\pi}{2\text{ s}}\right)^2 (0.04\text{ m})^2 = \boxed{2.37 \times 10^{-2}\text{ J}}$$

(*b*) To find v_{max}, set the kinetic energy equal to the total energy and solve for v:

$$\frac{1}{2}mv_{max}^2 = E$$

so

$$v_{max} = \sqrt{\frac{2E}{m}} = \sqrt{\frac{2(2.37 \times 10^{-2}\text{ J})}{3\text{ kg}}}$$

$$= \boxed{0.126\text{ m/s}}$$

(*c*) 1. Conservation of energy relates the position x to the speed v:

$$E = \frac{1}{2}mv^2 + \frac{1}{2}kx^2$$

2. Substitute $v = \frac{1}{2}v_{max}$ and solve for x_1. It is convenient to find x in terms of E and then write $E = \frac{1}{2}kA^2$ to obtain an expression for x in terms of A:

$$E = \frac{1}{2}m(\tfrac{1}{2}v_{max})^2 + \frac{1}{2}kx_1^2$$

$$= \frac{1}{4}(\tfrac{1}{2}mv_{max}^2) + \frac{1}{2}kx_1^2 = \frac{1}{4}E + \frac{1}{2}kx_1^2$$

so

$$\frac{1}{2}kx_1^2 = E - \frac{1}{4}E = \frac{3}{4}E$$

and

$$x_1 = \sqrt{\frac{3E}{2k}} = \sqrt{\frac{3}{2k}\left(\frac{1}{2}kA^2\right)} = \frac{\sqrt{3}}{2}A$$

$$= \frac{\sqrt{3}}{2}(4\text{ cm}) = \boxed{3.46\text{ cm}}$$

EXERCISE Calculate ω for this example and find v_{max} from $v_{max} = \omega A$. (*Answer* $\omega = 3.14$ rad/s, $v_{max} = 0.126$ m/s)

EXERCISE An object of mass 2 kg is attached to a spring of force constant 40 N/m. The object is moving at 25 cm/s when it is at its equilibrium position. (*a*) What is the total energy of the object? (*b*) What is the amplitude of the motion? (*Answer* (*a*) $E_{total} = \frac{1}{2}mv_{max}^2 = 0.0625$ J (*b*) $A = \sqrt{2E_{total}/k} = 5.59$ cm)

*General Motion Near Equilibrium

Simple harmonic motion typically occurs when a particle is displaced slightly from a position of stable equilibrium. Figure 14-9 is a graph of the potential energy U versus x for a force that has a position of stable equilibrium and a position of unstable equilibrium. As discussed in Chapter 6, the maximum at x_2 on Figure 14-9 corresponds to unstable equilibrium, whereas the minimum at x_1 corresponds to stable equilibrium. Many smooth curves with a minimum as in Figure 14-9 can be approximated near the minimum by a parabola. The dashed curve in this figure is a parabolic curve that approximately fits U near the stable equilibrium point. The general equation for a parabola that has a minimum at point x_1 can be written

$$U = A + B(x - x_1)^2 \tag{14-19}$$

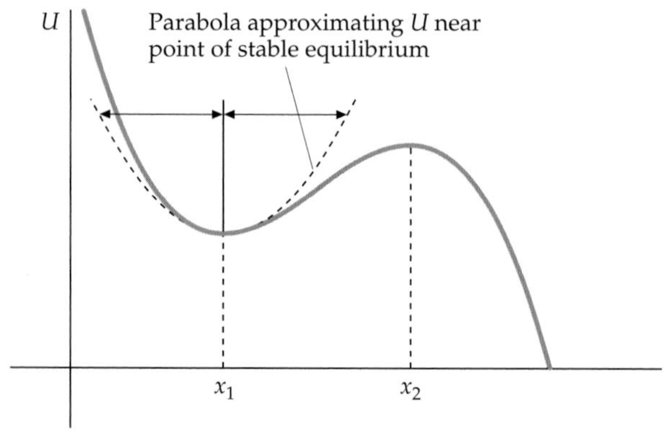

FIGURE 14-9 Plot of U versus x for a force that has a position of stable equilibrium (x_1) and a position of unstable equilibrium (x_2).

where A and B are constants. The constant A is the value of U at the equilibrium position $x = x_1$. The force is related to the potential energy curve by $F_x = -dU/dx$. Then

$$F_x = -\frac{dU}{dx} = -2B(x - x_1)$$

If we set $2B = k$, this equation reduces to

$$F_x = -\frac{dU}{dx} = -k(x - x_1) \qquad \text{14-20}$$

According to Equation 14-20, the force is proportional to the displacement and oppositely directed, so the motion will be simple harmonic. Figure 14-9 shows a graph of this system's potential energy function $U(x)$, which has a position of stable equilibrium at $x = x_1$. Figure 14-10 shows a potential energy function that has a position of stable equilibrium at $x = 0$. The system for this function is a small particle of mass m oscillating back and forth at the bottom of a frictionless spherical bowl.

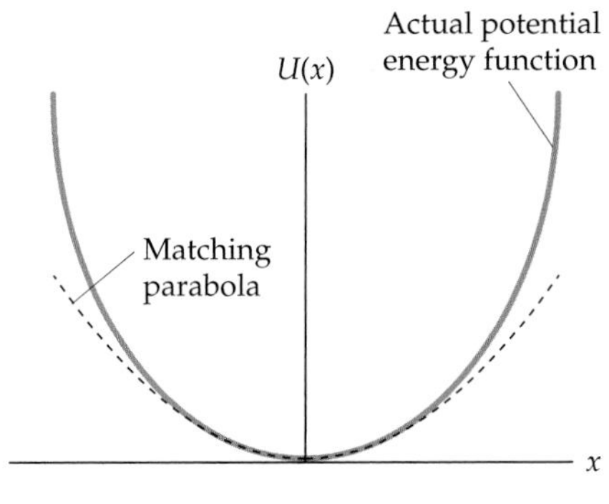

FIGURE 14-10 Plot of U versus x for a small particle oscillating back and forth at the bottom of a spherical bowl.

14-3 Some Oscillating Systems

Object on a Vertical Spring

When an object hangs from a vertical spring, there is a downward force mg in addition to the force of the spring (Figure 14-11). If we choose downward as the positive y direction, then the spring's force on the object is $-ky$, where y is the extension of the spring. The net force on the object is then

$$\sum F_y = -ky + mg \qquad \text{14-21}$$

We can simplify this equation by changing to a new variable $y' = y - y_0$, where $y_0 = mg/k$ is the amount the spring is stretched when the object is in equilibrium. Substituting $y = y' + y_0$ gives

$$\sum F_y = -k(y' + y_0) + mg$$

But $ky_0 = mg$, so

$$\sum F_y = -ky' \qquad \text{14-22}$$

Newton's second law ($\Sigma F_y = ma_y$) gives

$$-ky' = m\frac{d^2y}{dt^2}$$

However, $y = y' + y_0$, where $y_0 = mg/k$ is a constant. Thus $d^2y/dt^2 = d^2y'/dt^2$, so

$$-ky' = m\frac{d^2y'}{dt^2}$$

Rearranging gives

$$\frac{d^2y'}{dt^2} = -\frac{k}{m}y'$$

which is the same as Equation 14-2 with y' replacing x. It has the now familiar solution

$$y' = A\cos(\omega t + \delta)$$

where $\omega = \sqrt{k/m}$.

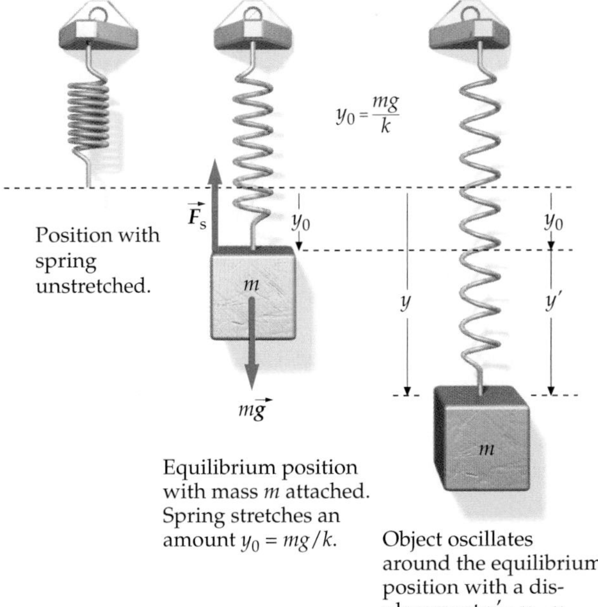

FIGURE 14-11 The problem of a mass on a vertical spring is simplified if the displacement (y') is measured from the equilibrium position of the spring with the mass attached.

Thus the effect of the gravitational force mg is merely to shift the equilibrium position from $y = 0$ to $y' = 0$. When the object is displaced from this equilibrium position by the amount y', the net force is $-ky'$. The object oscillates about this equilibrium position with an angular frequency $\omega = \sqrt{k/m}$, the same angular frequency as that for an object on a horizontal spring.

A force is conservative if the work done by it is independent of path. Both the force of the spring and the force of gravity are conservative, and the sum of these forces (Equations 14-21 and 14-22) also is conservative. The potential energy function U associated with the sum of these forces is the negative of the work done plus an arbitrary integration constant. That is,

$$U = -\int -ky'\,dy' = \tfrac{1}{2}ky'^2 + U_0$$

where the integration constant U_0 is the value of U at the equilibrium position ($y' = 0$). Thus,

$$U = \tfrac{1}{2}ky'^2 + U_0 \qquad 14\text{-}23$$

PAPER SPRINGS **EXAMPLE 14-6** **Put It in Context**

You are showing your nieces how to make paper party decorations using paper springs. One niece makes a paper spring and, with a single sheet of colored paper suspended from it, the spring is stretched 8 cm. You want the decorations to bounce at approximately 1 cy/s. How many sheets of colored paper should be used for the decoration on that spring if it is to bounce at 1 cy/s?

PICTURE THE PROBLEM The frequency depends on the ratio of the spring constant to the suspended mass (Equation 14-12), and you do not know either the spring constant or the mass. However, Hooke's law (Equation 14-1) can be used to find the required ratio from the information given.

1. Write the frequency in terms of the force constant k and the mass M (Equation 14-12), where M is the mass of N sheets. We need to find N:

$$f = \frac{\omega}{2\pi} = \frac{1}{2\pi}\sqrt{\frac{k}{M}}$$

2. The spring stretches a distance of $y_0 = 8$ cm when a single sheet of mass m is suspended:

$$ky_0 = mg \quad \text{so} \quad \frac{k}{m} = \frac{g}{y_0}$$

3. The mass of N sheets equals N times the mass of a single sheet:

$$M = Nm$$

4. Using the steps 2 and 3 results, solve for k/M:

$$\frac{k}{M} = \frac{k}{Nm} = \frac{1}{N}\frac{g}{y_0}$$

5. Substitute the step 4 result into the step 1 result and solve for N:

$$f = \frac{1}{2\pi}\sqrt{\frac{k}{M}} = \frac{1}{2\pi}\sqrt{\frac{1}{N}\frac{g}{y_0}}$$

so

$$N = \frac{g}{(2\pi f)^2 y_0} = \frac{9.81\ \text{m/s}^2}{4\pi^2\,(1\ \text{Hz})^2\,(0.08\ \text{m})} = 3.11$$

Three sheets are needed.

REMARKS Note that in this example we didn't need to use the value of m or k because the frequency depends on the ratio k/m, which equals g/y_0. Also, the units work out since 1 Hz = 1 cy/s, and a cycle is a dimensionless unit.

EXERCISE How much is the paper spring stretched when a decoration made from three sheets of paper is suspended from it and the paper is in equilibrium? (*Answer* 24 cm)

A Bead on a Block — EXAMPLE 14-7

A block attached to a spring oscillates vertically with a frequency of 4 Hz and an amplitude of 7 cm. A tiny bead is placed on top of the oscillating block just as it reaches its lowest point. Assume that the bead's mass is so small that its effect on the motion of the block is negligible. At what distance from the block's equilibrium position does the bead lose contact with the block?

PICTURE THE PROBLEM The forces on the bead are its weight mg downward and the upward normal force exerted by the block. The magnitude of this normal force changes as the acceleration changes. As the block moves upward *from equilibrium,* its acceleration and that of the bead is *downward* and increasing in magnitude. When the acceleration reaches g downward, the normal force will be zero. If the block's downward acceleration becomes even slightly larger, the bead will leave the block.

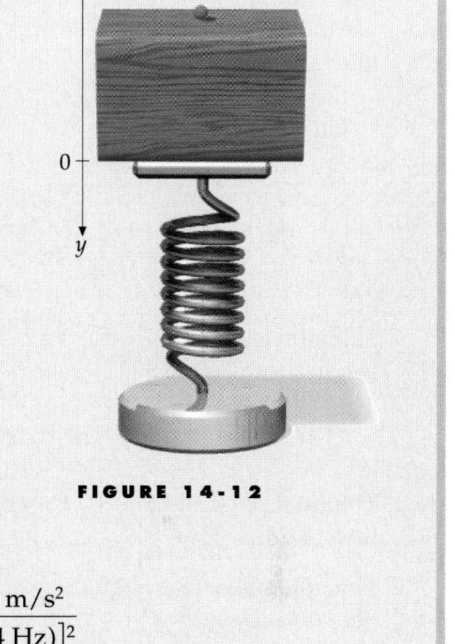

FIGURE 14-12

1. Draw a sketch of the system (Figure 14-12). Include a y coordinate axis with its origin at the equilibrium position and with down as the positive direction:

2. We are looking for the value of y when the acceleration is g downward. Use Equation 14-7:
$$a_y = -\omega^2 y$$
$$g = -\omega^2 y$$

3. Substitute $2\pi f$ for ω and solve for y:
$$g = -(2\pi f)^2\, y$$
so
$$y = -\frac{g}{(2\pi f)^2} = -\frac{9.81\ \text{m/s}^2}{[2\pi(4\ \text{Hz})]^2}$$
$$= -0.0155\ \text{m} = \boxed{-1.55\ \text{cm}}$$

PLAUSIBILITY CHECK The bead leaves the block when y is negative, which is when the bead is above the equilibrium position, as expected.

Potential Energy of the Spring–Earth System — EXAMPLE 14-8

A 3-kg object stretches a spring 16 cm when it hangs vertically in equilibrium. The spring is then stretched an additional 5 cm and the object is released. Let U be the total potential energy of the spring-object-planet system. When the mass is at its maximum displacement from equilibrium find U (*a*) with $U = 0$ at the equilibrium position and (*b*) with $U = 0$ when the spring is unstretched.

PICTURE THE PROBLEM (*a*) With $U = 0$ at the equilibrium position the total potential energy U is $\frac{1}{2}ky'^2$, where y' is the displacement from the equilibrium position. (*b*) With $U = 0$ when the spring is unstretched the total potential energy is the potential energy of the spring plus the gravitational potential energy.

(*a*) 1. Make three sketches of the system, one with the spring unstretched, one with it stretched 16 cm, and a third with it stretched 21 cm (Figure 14-13).

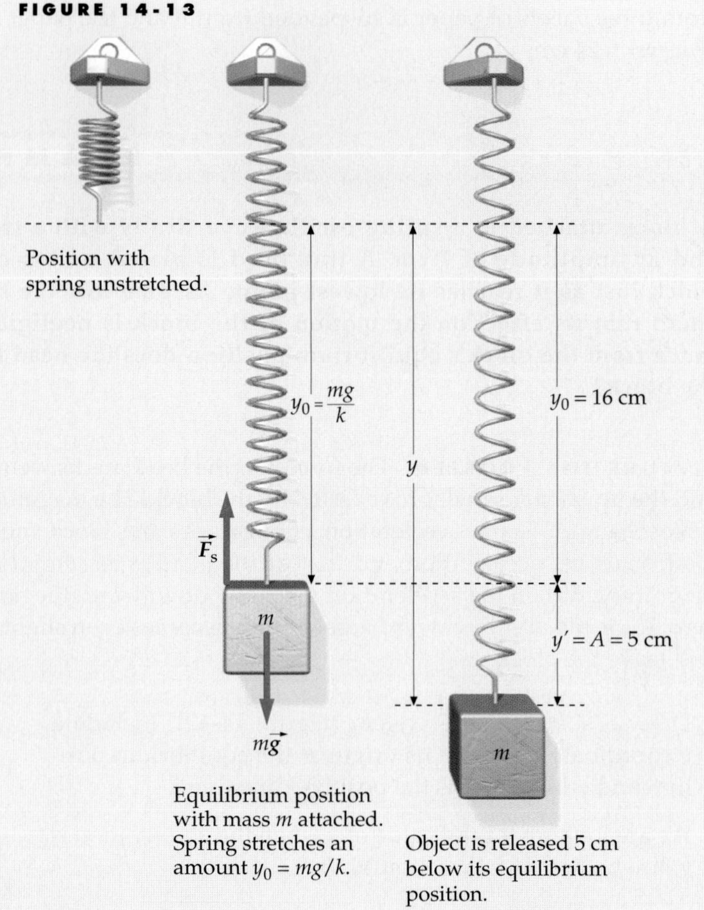

FIGURE 14-13

2. Let the positive y' direction be downward and let $y' = 0$ at the equilibrium position. The total potential energy function (Equation 14-23) is:

$$U = \tfrac{1}{2}ky'^2$$

3. To determine U we first need to find the spring constant k. At equilibrium the upward force of the spring equals the downward force of gravity. Use this to calculate the value of k:

$$ky_0 = mg$$

so

$$k = \frac{mg}{y_0} = \frac{(3\text{ kg})(9.81\text{ m/s}^2)}{0.16\text{ m}} = 184\text{ N/m}$$

4. Substituting for k in the step 1 result and solving for U gives:

$$U = \tfrac{1}{2}ky'^2 = \tfrac{1}{2}(184\text{ N/m})(0.05\text{ m})^2 = \boxed{0.230\text{ J}}$$

(*b*) 1. The total potential energy function is given by Equation 14-23:

$$U = \tfrac{1}{2}ky'^2 + U_0$$

2. The potential energy equals zero at $y' = y'_{\text{ref}} = -16$ cm. Set $U = 0$, set $y' = y'_{\text{ref}} = -16$ cm, and solve for U_0:

$$0 = \tfrac{1}{2}ky'^2_{\text{ref}} + U_0$$

so

$$U_0 = -\tfrac{1}{2}ky'^2_{\text{ref}} = -\tfrac{1}{2}(184\text{ N/m})(-0.16\text{ m})^2$$
$$= -2.35\text{ J}$$

3. Substituting for U_0 in the Part (*b*) step 1 result gives:

$$U = \tfrac{1}{2}ky'^2 + U_0$$
$$= \tfrac{1}{2}(184\text{ N/m})(0.05\text{ m})^2 - 2.35\text{ J} = \boxed{-2.12\text{ J}}$$

PLAUSIBILITY CHECK The potential energy calculated in Part (*b*) should equal the sum of the spring's potential energy U_s at $y = 21$ cm plus the gravitational potential energy U_g at $y = 21$ cm, where each of these potential energies is zero if the spring is unstretched, the positive y direction is downward, and $y = 0$ if the spring is unstretched. $U_s = \tfrac{1}{2}ky^2 = \tfrac{1}{2}(184\text{ N/m})(0.21\text{ m})^2 = 4.06$ J and $U_g = mg(-y) = (3\text{ kg})(9.81\text{ N/kg})(-0.21\text{ m}) = -6.18$ J. Adding these gives $4.06\text{ J} - 6.18\text{ J} = -2.12$ J, which agrees with the Part (*b*) result.

The Simple Pendulum

A simple pendulum consists of a string of length L and a bob of mass m. When the bob is released from an initial angle ϕ_0 with the vertical, it swings back and forth with some period T.

EXERCISE IN DIMENSIONAL ANALYSIS We might expect the period of a simple pendulum to depend on the mass m of a pendulum bob, the length L of the pendulum, the acceleration due to gravity g, and the initial angle ϕ_0. Find a simple combination of some or all of these quantities that gives the correct dimensions for the period. (*Answer* $\sqrt{L/g}$)

REMARKS The units of length, mass, and g, are m, kg, and m/s^2, respectively. If we divide L by g, the meters cancel and we are left with seconds squared, suggesting the form $\sqrt{L/g}$. If the formula for the period contains the mass, then the unit kg must be canceled by some other quantity. But there is no combination of L and g that can cancel mass units. So the period cannot depend on the mass of the bob. Since the initial angle ϕ_0 is dimensionless, we cannot tell whether or not it is a factor in the period. We will see below that for small ϕ_0, the period is given by $T = 2\pi\sqrt{L/g}$.

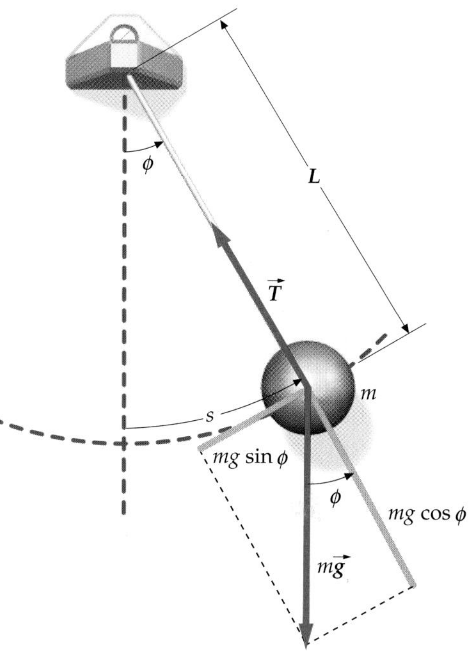

FIGURE 14-14 Forces on a pendulum bob.

The forces on the bob are its weight $m\vec{g}$ and the string tension $\vec{T}$(Figure 14-14). At an angle ϕ with the vertical, the weight has components $mg \cos \phi$ along the string and $mg \sin \phi$ tangential to the circular arc in the direction of decreasing ϕ. Using tangential components, Newton's second law ($\Sigma F_t = m\, a_t$) gives

$$-mg \sin \phi = m\frac{d^2s}{dt^2} \qquad \text{14-24}$$

where the arc length s is related to the angle ϕ by $s = L\phi$. Repeatedly differentiating both sides of $s = L\phi$ gives

$$\frac{d^2s}{dt^2} = L\frac{d^2\phi}{dt^2}$$

Substituting $L\, d^2\phi/dt^2$ into Equation 14-24 for d^2s/dt^2 and rearranging gives

$$\frac{d^2\phi}{dt^2} = -\frac{g}{L}\sin \phi \qquad \text{14-25}$$

Note that the mass m does not appear in Equation 14-25—the motion of a pendulum does not depend on its mass. For small ϕ, $\sin \phi \approx \phi$, and

$$\frac{d^2\phi}{dt^2} \approx -\frac{g}{L}\phi \qquad \text{14-26}$$

Equation 14-26 is of the same form as Equation 14-2 for an object on a spring. Thus, the motion of a pendulum approximates simple harmonic motion for small angular displacements.

Equation 14-26 can be written

$$\frac{d^2\phi}{dt^2} = -\omega^2\phi, \qquad \text{where} \quad \omega^2 = \frac{g}{L} \qquad \text{14-27}$$

and ω is the angular frequency—not the angular speed—of the motion of the pendulum. The period of the motion is thus

$$T = \frac{2\pi}{\omega} = 2\pi\sqrt{\frac{L}{g}} \qquad \text{14-28}$$

PERIOD OF A SIMPLE PENDULUM

The solution of Equation 14-27 is

$$\phi = \phi_0 \cos(\omega t + \delta)$$

where ϕ_0 is the maximum angular displacement.

According to Equation 14-28, the greater the length of a pendulum, the greater the period, which is consistent with experimental observation. The period, and therefore the frequency, are independent of the amplitude of oscillation (as long as the amplitude is small), a general feature of simple harmonic motion.

EXERCISE Find the period of a simple pendulum of length 1 m. (*Answer* 2.01 s)

The acceleration due to gravity can be measured using a simple pendulum. We need only measure the length L and period T of the pendulum, and using Equation 14-28, solve for g. (When finding T, we usually measure the time for n oscillations and then divide by n, which minimizes measurement error.)

All mechanical clocks keep time because the period of the oscillating part of the mechanism remains constant. The period of any pendulum changes with changes in amplitude. However, the driving mechanism of a pendulum clock maintains the amplitude at a constant value.

Pendulum in an Accelerated Reference Frame Figure 14-15*a* shows a simple pendulum suspended from the ceiling of a boxcar that has acceleration $\vec{a}_0$ relative to the ground, to the right, and $\vec{a}$ is the acceleration of the bob relative to the ground. Applying Newton's second law to the bob gives

$$\sum \vec{F} = \vec{T} + m\vec{g} = m\vec{a} \qquad 14\text{-}29$$

If the bob remains at rest relative to the boxcar, then $\vec{a} = \vec{a}_0$ and

$$\sum F_x = T \sin \theta_0 = ma_0$$

$$\sum F_y = T \cos \theta_0 - mg = 0$$

where θ_0 is the equilibrium angle. θ_0 is thus given by $\tan \theta_0 = a_0/g$. If the bob is moving relative to the boxcar, then $\vec{a}' = \vec{a} - \vec{a}_0$, where $\vec{a}'$ is the acceleration of the bob relative to the boxcar. Substituting for $\vec{a}$ in Equation 14-29 gives

$$\sum \vec{F} = \vec{T} + m\vec{g} = m(\vec{a}' + \vec{a}_0)$$

Subtracting $m\vec{a}_0$ from both sides of this equation and rearranging terms gives

$$\vec{T} + m\vec{g}' = m\vec{a}'$$

where $\vec{g}' = \vec{g} - \vec{a}_0$. Thus by replacing $\vec{g}$ by $\vec{g}'$ and $\vec{a}$ by $\vec{a}'$ in Equation 14-29 we can solve for the motion of the bob relative to the boxcar. The vectors $\vec{T}$ and $m\vec{g}'$ are shown in Figure 14-15*b*. If the string breaks so that $\vec{T} = 0$, then our equation gives $\vec{a}' = \vec{g}'$, which means that $\vec{g}'$ is the free-fall acceleration in the reference frame of the boxcar. If the bob is displaced slightly from equilibrium, it will oscillate with a period T given by Equation 14-28 with g replaced by g'.

EXERCISE A simple pendulum of length 1 m is in a boxcar that is accelerating horizontally with acceleration $a_0 = 3\ \text{m/s}^2$. Find g' and the period T. (*Answer* $g' = 10.26\ \text{m/s}^2$, $T = 1.96$ s)

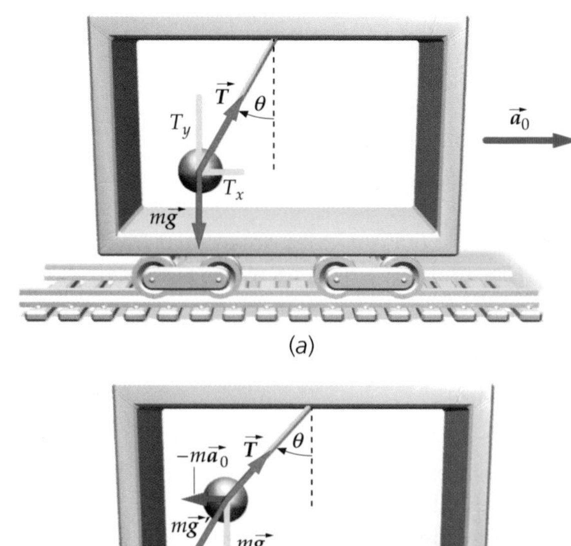

FIGURE 14-15 (*a*) Simple pendulum in apparent equilibrium in an accelerating boxcar. Forces are those as seen from a separate stationary frame. (*b*) Forces on the bob as seen in the accelerated frame. Adding the pseudoforce $-m\vec{a}_0$ is equivalent to replacing $\vec{g}$ by $\vec{g}'$.

Large-Amplitude Oscillations When the amplitude of a pendulum's oscillation becomes large, its motion continues to be periodic, but it is no longer simple harmonic. A slight dependence on the amplitude

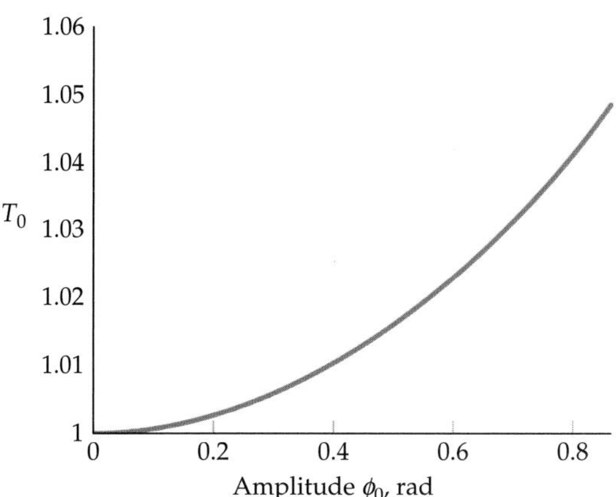

FIGURE 14-16 Note that the ordinate values range from 1 to 1.06. Over a range of ϕ from 0 to 0.8 rad (46°), the period varies by about 5 percent.

must be accounted for when determining the period. For a general angular amplitude ϕ_0, the period can be shown to be

$$T = T_0\left[1 + \frac{1}{2^2}\sin^2\frac{1}{2}\phi_0 + \frac{1}{2^2}\left(\frac{3}{4}\right)^2\sin^4\frac{1}{2}\phi_0 + \cdots\right] \qquad 14\text{-}30$$

PERIOD FOR LARGE-AMPLITUDE OSCILLATIONS

where $T_0 = 2\pi\sqrt{L/g}$ is the period for very small amplitudes. Figure 14-16 shows T/T_0 as a function of amplitude ϕ_0.

A Pendulum Clock — EXAMPLE 14-9 Try It Yourself

A simple pendulum clock is calibrated to keep accurate time at an angular amplitude of $\phi_0 = 10°$. When the amplitude has decreased to the point where it is very small, does the clock gain or lose time? How much time will the clock gain or lose in one day if the amplitude remains very small.

Cover the column to the right and try these on your own before looking at the answers.

Steps	Answers
1. Answer the first question by finding out if the period increases or decreases.	T decreases as ϕ decreases, so **the clock gains time.**
2. Use Equation 14-30 to find the percentage change $[(T - T_0)/T] \times 100\%$ for $\phi = 10°$. Use only the first correction term.	0.190%
3. Find the number of minutes in a day.	There are 1440 minutes in a day.
4. Combine the steps 2 and 3 to find the change in the number of minutes in a day.	The gain is **2.73 min/d**

REMARKS To avoid this gain, pendulum-clock mechanisms are designed to keep the amplitude fairly constant.

*The Physical Pendulum

A rigid object free to rotate about a horizontal axis that is not through its center of mass will oscillate when displaced from equilibrium. Such a system is called a **physical pendulum.** Consider a plane figure with a rotation axis a distance D from the figure's center of mass and displaced from equilibrium by the angle ϕ (Figure 14-17). The torque about the axis has a magnitude $MgD \sin \phi$ and tends to decrease $|\phi|$. Newton's second law applied to rotation is

$$\tau = I\alpha$$

where α is the angular acceleration and I is the moment of inertia about the axis. Substituting $-MgD \sin \phi$ for the net torque and $d^2\phi/dt^2$ for α, we have

$$-MgD \sin \phi = I\frac{d^2\phi}{dt^2}$$

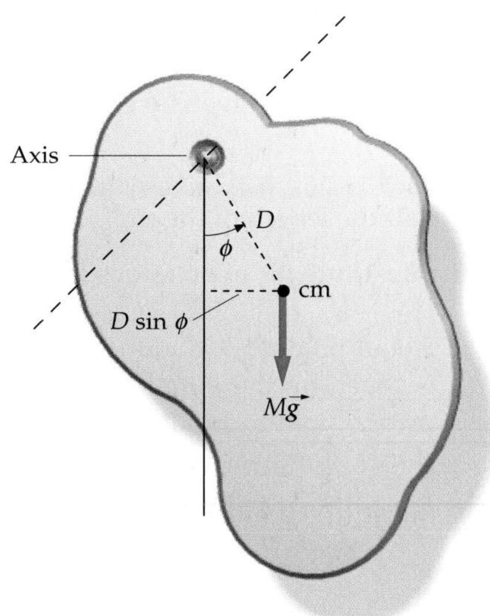

FIGURE 14-17 A physical pendulum.

or

$$\frac{d^2\phi}{dt^2} = -\frac{MgD}{I}\sin\phi \qquad \text{14-31}$$

As with the simple pendulum, the motion is approximately simple harmonic if the angular displacements are small, so $\sin\phi \approx \phi$. In this case, we have

$$\frac{d^2\phi}{dt^2} \approx -\frac{MgD}{I}\phi = -\omega^2\phi \qquad \text{14-32}$$

where $\omega = \sqrt{MgD/I}$ is the angular frequency—not the angular speed—of the motion. The period is therefore

$$T = \frac{2\pi}{\omega} = 2\pi\sqrt{\frac{I}{MgD}} \qquad \text{14-33}$$

PERIOD OF A PHYSICAL PENDULUM

For large amplitudes, the period is given by Equation 14-30, with T_0 given by Equation 14-33. For a simple pendulum of length L, the moment of inertia is $I = ML^2$ and $D = L$. Then Equation 14-33 gives $T = 2\pi\sqrt{ML^2/(MgL)} = 2\pi\sqrt{L/g}$, the same as Equation 14-28.

A Rotating Rod — EXAMPLE 14-10

A uniform rod of mass M and length L is free to rotate about a horizontal axis perpendicular to the rod and through one end. (*a*) Find the period of oscillation for small angular displacements. (*b*) Find the period of oscillation if the rotation axis is a distance x from the center of mass.

FIGURE 14-18 Plot of the period versus the distance from the pivot to the center of mass. For $x > 0.5$ m the pivot is beyond the end of the rod.

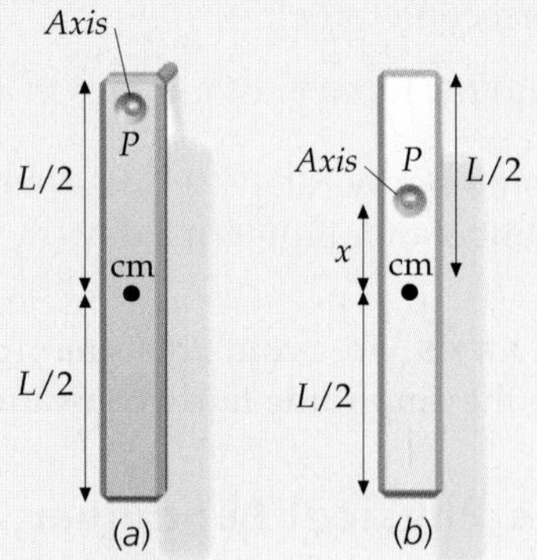

PICTURE THE PROBLEM (*a*) The period is given by Equation 14-33. The center of mass is at the center of the rod, so the distance from the center of mass to the rotation axis is half the length of the rod (Figure 14-18*a*). The moment of inertia of a uniform rod can be found in Table 9-1. (*b*) For rotations around an axis through point *P* (Figure 14-18*b*), the moment of inertia can be found from the parallel-axis theorem $I = I_{cm} + MD^2$ (Equation 9-44), where I_{cm} can be found in Table 9-1.

(*a*) 1. The period is given by Equation 14-33: $T = 2\pi\sqrt{\frac{I}{MgD}}$

2. I about the end is found in Table 9-1 and D is half the length of the rod: $I = \frac{1}{3}ML^2; \quad D = \frac{1}{2}L$

3. Substitute the expressions for I and D to find T: $T = 2\pi\sqrt{\frac{\frac{1}{3}ML^2}{Mg(\frac{1}{2}L)}} = \boxed{2\pi\sqrt{\frac{2L}{3g}}}$

(*b*) 1. About point *P*, $D = x$, and the moment of inertia is given by the parallel-axis theorem. The moment of inertia about a parallel axis through the center of mass is found in Table 9-1: $D = x$

$I = I_{cm} + MD^2 = \frac{1}{12}ML^2 + Mx^2$

2. Substitute these values to find T: $T = 2\pi\sqrt{\frac{I}{MgD}} = 2\pi\sqrt{\frac{(\frac{1}{12}ML^2 + Mx^2)}{Mgx}}$

$$= \boxed{2\pi\sqrt{\frac{(\frac{1}{12}L^2 + x^2)}{gx}}}$$

PLAUSIBILITY CHECK As $x \to 0$, $T \to \infty$ as expected. (If the rotation axis of the rod passes through its center of mass, we do not expect gravity to exert a restoring torque.) Also, if $x = L/2$, we get the same result as found in Part (*a*), and if $x >> L$, the expression for the period approaches $T = 2\pi\sqrt{x/g}$, which is the expression for the period of a simple pendulum of length x (Equation 14-28).

EXERCISE What is the period of oscillation for small angular displacements of a 1-m-long uniform rod about an axis through one end? (*Answer* $T = 1.64$ s) Note that this is a smaller period than for a simple pendulum of length $L = 1$ m. The period of the simple pendulum is greater because the ratio of its moment of inertia to the restoring torque is greater.

EXERCISE Show that when $x = L/6$, the period is the same as when $x = L/2$.

REMARKS The period T versus distance x from the center of mass for a rod of length 1 m is shown in Figure 14-19.

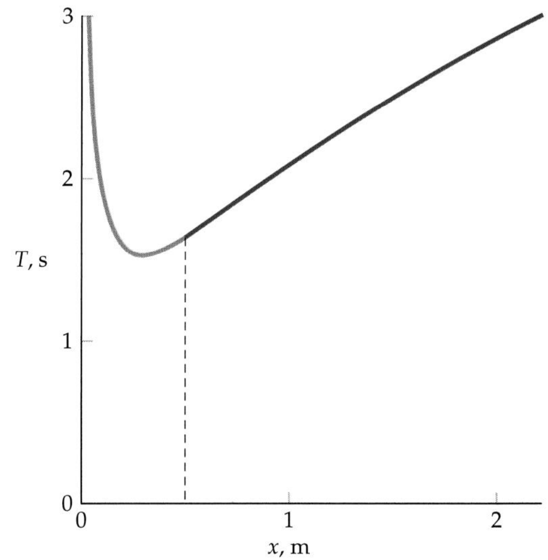

FIGURE 14-19

THE ROTATING ROD REVISITED **EXAMPLE 14-11** **Try It Yourself**

Find the value of x in Example 14-10 for which the period is a minimum.

PICTURE THE PROBLEM At the value of x for which T is a minimum, $dT/dx = 0$.

Cover the column to the right and try these on your own before looking at the answers.

Steps	Answers
1. The period, given by the Example 14-10 Part (*b*) result, equals $T = 2\pi\sqrt{Z/g}$, where $Z = (\frac{1}{12}L^2 + x^2)/x$. Find the period both as x approaches zero and as x approaches infinity.	As $x \to 0$, $Z \to \infty$, and $T \to \infty$. As $x \to \infty$, $Z \to \infty$, and $T \to \infty$.
2. The period goes to infinity as x approaches zero and as x approaches infinity. Somewhere in the range $0 < x < \infty$ the period is a minimum. To find the minimum, evaluate dT/dx, set it equal to zero, and solve for x.	$\frac{dT}{dx} = \frac{dT}{dZ}\frac{dZ}{dx} = \frac{\pi}{\sqrt{g}}Z^{-1/2}\frac{dZ}{dx}$ $Z > 0$ throughout the range $0 < x < \infty$, so $\frac{dT}{dx} = 0 \Rightarrow \frac{dZ}{dx} = 0$. $\frac{dZ}{dx} = 0 \Rightarrow x = \boxed{\frac{L}{\sqrt{12}} = 0.289L}$

14-4 Damped Oscillations

Left to itself, a spring or a pendulum eventually stops oscillating because the mechanical energy is dissipated by frictional forces. Such motion is said to be **damped.** If the damping is large enough, as, for example, a pendulum submerged in molasses, the oscillator fails to complete even one cycle of oscillation. Instead it just moves toward the equilibrium position with a speed that approaches zero as the object approaches the equilibrium position. This type of motion is referred to as **overdamped.** If the damping is small enough that the system oscillates with an amplitude that decreases slowly with time—like a child on a playground swing when Mom stops providing a push each cycle—the motion is said to be **underdamped.** Motion with the minimum damping for nonoscillatory motion is said to be **critically damped.** (With any less damping, the motion would be underdamped.)

Underdamped Motion The damping force exerted on an oscillator such as the one shown in Figure 14-20*a* can be represented by the empirical expression

$$\vec{F}_d = -b\vec{v}$$

(*a*)

(*b*)

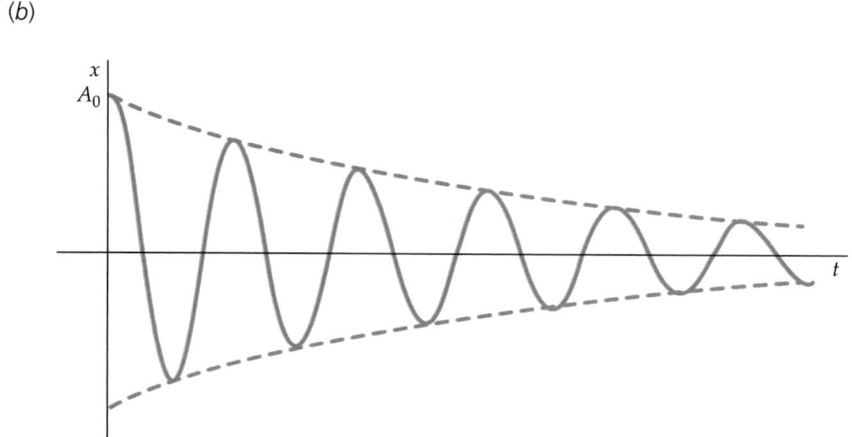

FIGURE 14-20 (*a*) A damped oscillator. The motion is damped by the plunger immersed in the liquid. (*b*) Damped oscillation curve.

where b is a constant. Such a system is said to be linearly damped. The discussion here is for linearly damped motion. Because the damping force is opposite to the direction of motion, it does negative work and causes the mechanical energy of the system to decrease. This energy is proportional to the square of the amplitude (Equation 14-17), and the square of the amplitude decreases exponentially with increasing time. That is,

$$A^2 = A_0^2 e^{-t/\tau} \qquad 14\text{-}34$$

DEFINITION—TIME CONSTANT

where A is the amplitude, A_0 is the amplitude at $t = 0$, and τ is the **decay time** or **time constant.** The time constant is the time for the energy to decrease by a factor of e.

The motion of a damped system can be obtained from Newton's second law. For an object of mass m on a spring of force constant k, the net force is $-kx - b(dx/dt)$. Setting the net force equal to the mass times the acceleration d^2x/dt^2, we obtain

$$-kx - b\frac{dx}{dt} = m\frac{d^2x}{dt^2} \qquad 14\text{-}35$$

DIFFERENTIAL EQUATION FOR A DAMPED OSCILLATOR

The exact solution of this equation can be found using standard methods for solving differential equations. The solution for the underdamped case is

$$x = A_0 e^{-(b/2m)t} \cos(\omega' t + \delta) \qquad 14\text{-}36$$

where A_0 is the initial amplitude. The frequency ω' is given by

$$\omega' = \omega_0 \sqrt{1 - \left(\frac{b}{2m\omega_0}\right)^2} \qquad 14\text{-}37$$

where ω_0 is the frequency with no damping ($\omega_0 = \sqrt{k/m}$ for a mass on a spring). For weak damping, $b/(2m\omega_0) << 1$ and ω' is nearly equal to ω_0. The dashed curves in Figure 14-20*b* correspond to $x = A$ and $x = -A$, where A is given by

$$A = A_0 e^{-(b/2m)t} \qquad 14\text{-}38$$

By squaring both sides of this equation and comparing the results with Equation 14-34 we have

$$\tau = \frac{m}{b} \qquad 14\text{-}39$$

If the damping constant b is gradually increased, the angular frequency ω' decreases until it becomes zero at the critical value

$$b_c = 2m\omega_0 \qquad 14\text{-}40$$

When b is greater than or equal to b_c, the system does not oscillate. If $b > b_c$, the system is overdamped. The smaller b is, the more rapidly the object returns to equilibrium. If $b = b_c$, the system is said to be critically damped and the object returns to equilibrium (without oscillation) most rapidly. Figure 14-21 shows plots of the displacement versus time for a critically damped and an overdamped oscillator. We often use critical damping when we want a system to avoid oscillations and yet return to equilibrium quickly. For example, shock absorbers are used to damp the oscillations of an automobile on its springs. You can test the damping of a car's shock absorbers by pushing down on one fender of the car and then releasing it. If the car returns to equilibrium with no oscillation, then the system is critically damped or overdamped. (You will usually observe one or two oscillations for an unoccupied vehicle, indicating that the damping constant is just under the critical value.)

Because the energy of an oscillator is proportional to the square of its amplitude, the energy of an underdamped oscillator (averaged over a cycle) also decreases exponentially with time:

$$E = \tfrac{1}{2}m\omega^2 A^2 = \tfrac{1}{2}m\omega^2 (A_0 e^{-(b/2m)t})^2 = \tfrac{1}{2}m\omega^2 A_0^2 e^{-(b/m)t} = E_0 e^{-t/\tau} \qquad 14\text{-}41$$

where $E_0 = \tfrac{1}{2}m\omega^2 A_0^2$ and

$$\tau = \frac{m}{b} \qquad 14\text{-}42$$

A damped oscillator is often described by its Q factor (for quality factor),

$$Q = \omega_0 \tau \qquad 14\text{-}43$$

DEFINITION—Q FACTOR

Shock absorbers (yellow cylinders) are used to damp the oscillations of this truck.

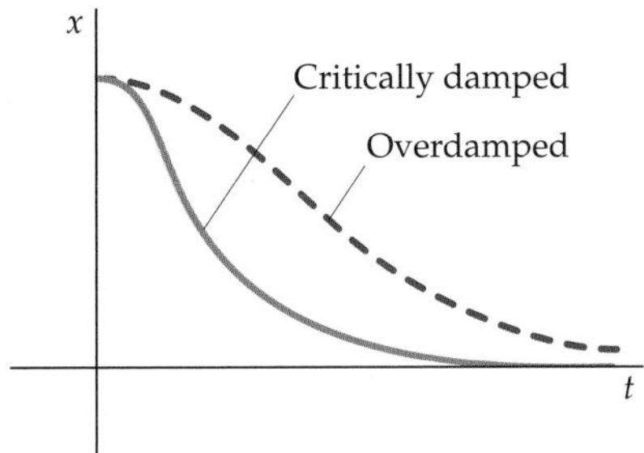

FIGURE 14-21 **Plots of displacement versus time for a critically damped and an overdamped oscillator.**

Weights are placed in automobile wheels when the wheels are "balanced." The purpose of balancing the wheels is to prevent vibrations that will drive oscillations of the wheel assembly.

The Q factor is dimensionless. (Since ω_0 has dimensions of reciprocal time, $\omega_0\tau$ is without dimension.) We can relate Q to the fractional energy loss per cycle. Differentiating Equation 14-41 gives

$$dE = -(1/\tau)E_0e^{-t/\tau}\,dt = -(1/\tau)E\,dt$$

If the damping is weak so that the energy loss per cycle is small, we can replace dE by ΔE and dt by the period T. Then $|\Delta E|/E$ in one cycle (one period) is given by

$$\left(\frac{|\Delta E|}{E}\right)_{\text{cycle}} = \frac{T}{\tau} = \frac{2\pi}{\omega_0\tau} = \frac{2\pi}{Q} \qquad \text{14-44}$$

by

$$Q = \frac{2\pi}{(|\Delta E|/E)_{\text{cycle}}}, \qquad \frac{|\Delta E|}{E} << 1 \qquad \text{14-45}$$

PHYSICAL INTERPRETATION OF Q FOR WEAK DAMPING

Q is thus inversely proportional to the fractional energy loss per cycle.

MAKING MUSIC **EXAMPLE 14-12**

When middle C on a piano (frequency 262 Hz) is struck, it loses half its energy after 4 s. (*a*) What is the decay time τ? (*b*) What is the Q factor for this piano wire? (*c*) What is the fractional energy loss per cycle?

PICTURE THE PROBLEM (*a*) We use $E = E_0e^{-t/\tau}$ and set E equal to $\frac{1}{2}E_0$. (*b*) The Q value can then be found from the decay time and the frequency.

(*a*) 1. Set the energy at time $t = 4$ s equal to half the original energy:

$$E = E_0e^{-t/\tau} \quad \text{so} \quad \tfrac{1}{2}E_0 = E_0e^{-4\,\text{s}/\tau}$$

$$\tfrac{1}{2} = e^{-4\,\text{s}/\tau}$$

2. Solve for the time τ by taking the natural log of both sides:

$$\ln\frac{1}{2} = -\frac{4\text{ s}}{\tau}$$

so

$$\tau = \frac{4\text{ s}}{\ln 2} = \boxed{5.77\text{ s}}$$

(*b*) Calculate Q from τ and ω_0:

$$Q = \omega_0\tau = 2\pi f\tau$$

$$= 2\pi(262\text{ Hz})(5.77\text{ s}) = \boxed{9.50 \times 10^3}$$

(*c*) The fractional energy loss in a cycle is given by Equation 14-44 and the frequency $f = 1/T$:

$$\left(\frac{|\Delta E|}{E}\right)_{\text{cycle}} = \frac{T}{\tau} = \frac{1}{f\tau} = \frac{1}{(262\text{ Hz})(5.77\text{ s})} = \boxed{6.61 \times 10^{-4}}$$

PLAUSIBILITY CHECK Q can also be calculated from $Q = 2\pi/(\Delta E/E)_{\text{cycle}} = 2\pi/(6.61 \times 10^{-4}) = 9.50 \times 10^3$. Note that the fractional energy loss after 4 s is not just the number of cycles (4×262) times the fractional energy loss per cycle, because the energy decrease is exponential, not constant.

REMARKS Figure 14-22 shows the relative amplitude A/A_0 versus time and the relative energy E/E_0 versus time for the oscillation of a piano string after middle C is struck. After 4 s, the amplitude has decreased to about 0.7 times its initial value, and the energy, which is proportional to the amplitude squared, drops to about half its initial value.

Note that Q is quite large. You can estimate τ and Q of various oscillating systems. Tap a crystal wine glass and see how long it rings. The longer it rings, the greater the value of τ and Q and the lower the damping. Glass beakers from the laboratory may also have a high Q. Try tapping a plastic cup. How does the damping compare to that of the beaker?

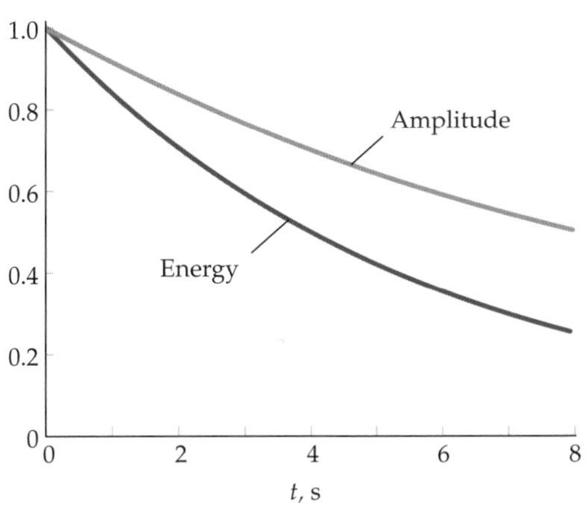

FIGURE 14-22 Plots of A/A_0 and E/E_0 for a struck piano string.

In terms of Q, the exact frequency of an underdamped oscillator is

$$\omega' = \omega_0\sqrt{1 - \left(\frac{b}{2m\omega_0}\right)^2} = \omega_0\sqrt{1 - \frac{1}{4Q^2}} \qquad 14\text{-}46$$

Because b is quite small (and Q is quite large) for a weakly damped oscillator (Example 14-12), we see that ω' is nearly equal to ω_0.

We can understand much of the behavior of a weakly damped oscillator by considering its energy. The power dissipated by the damping force equals the instantaneous rate of change of the total mechanical energy

$$P = \frac{dE}{dt} = \vec{F}_d \cdot \vec{v} = -b\vec{v} \cdot \vec{v} = -bv^2 \qquad 14\text{-}47$$

For a weakly damped oscillator, the total mechanical energy decreases slowly with time. The average kinetic energy per cycle equals half the total energy

$$\left(\frac{1}{2}mv^2\right)_{av} = \frac{1}{2}E \qquad \text{or} \qquad (v^2)_{av} = \frac{E}{m}$$

If we substitute $(v^2)_{av} = E/m$ for v^2 in Equation 14-47, we have

$$\frac{dE}{dt} = -bv^2 \approx -b(v^2)_{av} = -\frac{b}{m}E \qquad 14\text{-}48$$

Rearranging Equation 14-48 gives

$$\frac{dE}{E} = -\frac{b}{m}dt$$

which upon integration gives

$$E = E_0e^{-(b/m)t} = E_0e^{-t/\tau}$$

which is Equation 14-41.

14-5 Driven Oscillations and Resonance

By pumping the swing, she is transferring her internal energy into the mechanical energy of the oscillator.

To keep a damped system going, mechanical energy must be put into the system. When this is done, the oscillator is said to be driven or forced. When you keep a swing going by "pumping," that is, by moving your body and legs, you are driving an oscillator. If you put mechanical energy into the system faster than it is dissipated, the mechanical energy increases with time, and the amplitude increases. If you put mechanical energy in at the same rate it is being dissipated, the amplitude remains constant over time. The motion of the oscillator is then said to be in steady state.

Figure 14-23 shows a system consisting of an object on a spring that is being driven by moving the point of support up and down with simple harmonic motion of frequency ω. At first the motion is complicated, but eventually steady-state motion is reached in which the system oscillates with the same frequency as that of the driver and with a constant amplitude and, therefore, at constant energy. In the steady state, the energy put into the system per cycle by the driving force equals the energy dissipated per cycle due to the damping.

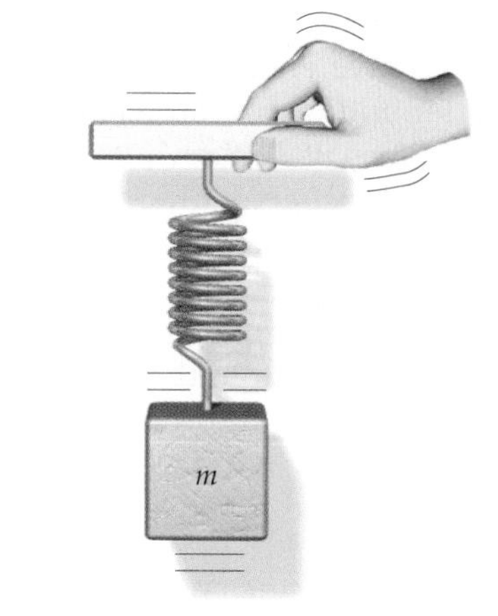

FIGURE 14-23 An object on a vertical spring can be driven by moving the support up and down.

The amplitude, and therefore the energy, of a system in the steady state depends not only on the amplitude of the driving force, but also on its frequency. The **natural frequency** of an oscillator, ω_0, is its frequency when no driving or damping forces are present. (In the case of a spring, for example, $\omega_0 = \sqrt{k/m}$.) If the driving frequency is approximately equal to the natural frequency of the system, the system will oscillate with a relatively large amplitude. For example, if the support in Figure 14-23 oscillates at a frequency close to the natural frequency of the mass–spring system, the mass will oscillate with a much greater amplitude than it would if the support oscillates at higher or lower frequencies. This phenomenon is called **resonance.** When the driving frequency equals the natural frequency of the oscillator, the energy per cycle transferred to the oscillator is maximum. The natural frequency of the system is thus called the **resonance frequency.** (Mathematically, the angular frequency ω is more convenient to use than the frequency $f = \omega/(2\pi)$. Because ω and f are proportional, most statements concerning angular frequency also hold for frequency. In verbal descriptions, we usually omit the word angular when the omission will not cause confusion.) Figure 14-24 shows plots of the average power delivered to an oscillator as a function of the driving frequency for two different values of damping. These curves are called **resonance curves.** When the damping is weak (large Q), the width of the peak of the resonance curve is correspondingly narrow, and we speak of the resonance as being sharp. For strong damping, the resonance curve is broad. The width of each resonance curve $\Delta\omega$, indicated in the figure, is the width at half the maximum height. For weak damping, the ratio of the width of the resonance to the resonant frequency can be shown to equal the reciprocal of the Q factor (see Problem 126):

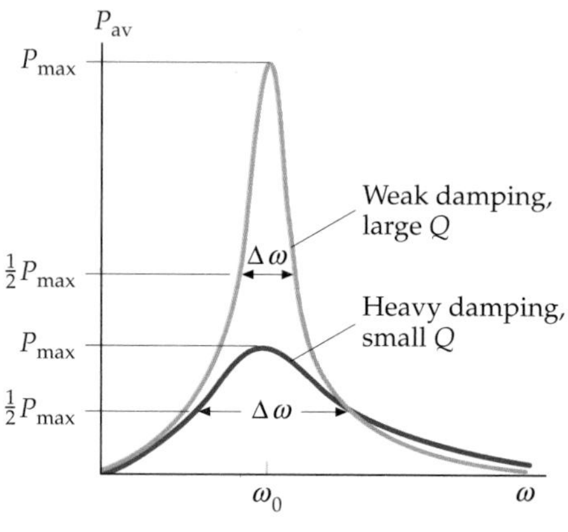

FIGURE 14-24 Resonance for an oscillator. The width $\Delta\omega$ of the resonance peak for a high-Q oscillator is small compared to the natural frequency of ω_0.

$$\frac{\Delta\omega}{\omega_0} = \frac{1}{Q} \tag{14-49}$$

RESONANCE WIDTH FOR WEAK DAMPING

Thus, the Q factor is a direct measure of the sharpness of resonance.

You can do a simple experiment to demonstrate resonance. Hold a meterstick at one end between two fingers so that it acts like a pendulum. (If a meterstick is not available, use whatever is convenient. A golf club works fine.) Release the stick from some initial angular displacement and observe the natural frequency of its motion. Then move your hand back and forth horizontally, driving it at its natural frequency. Even if the amplitude of the motion of your hand is small, the stick will oscillate with a substantial amplitude. Now move your hand back and forth at a frequency two or three times the natural frequency and note the decrease in amplitude of the oscillating stick.

There are many familiar examples of resonance. When you sit on a swing, you learn intuitively to pump with the same frequency as the natural frequency of the swing. Many machines vibrate because they have rotating parts that are not in perfect balance. (Observe a washing machine in the spin cycle for an example.) If such a machine is attached to a structure that can vibrate, the structure becomes a driven oscillatory system that is set in motion by the machine. Engineers pay great attention to balancing the rotary parts of such machines, damping their vibrations, and isolating them from building supports.

A glass with low damping can be broken by an intense sound wave at a frequency equal to or very nearly equal to the natural frequency of vibration of the glass. This is often done in physics demonstrations using an audio oscillator and an amplifier.

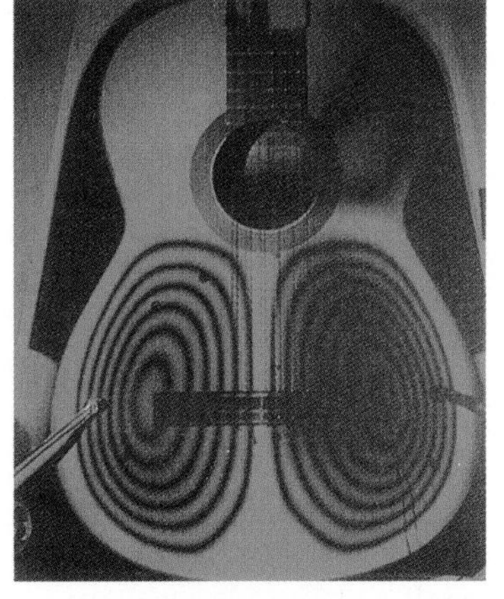

268 Hz ($Q = 52$)

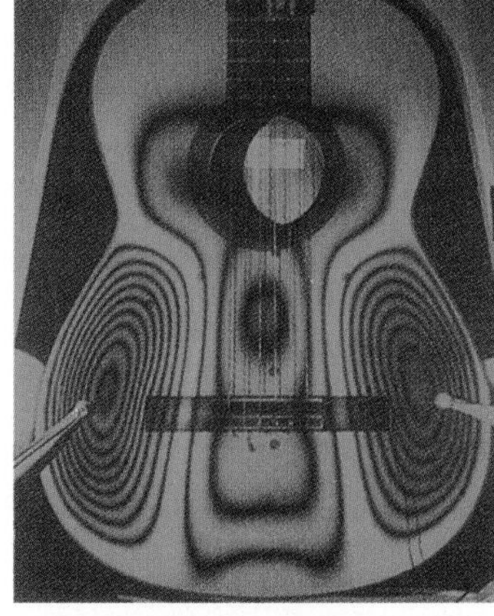

553 Hz ($Q = 66$)

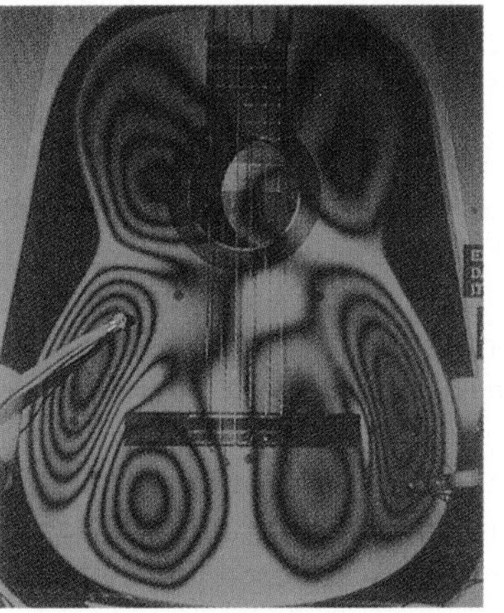

672 Hz ($Q = 61$)

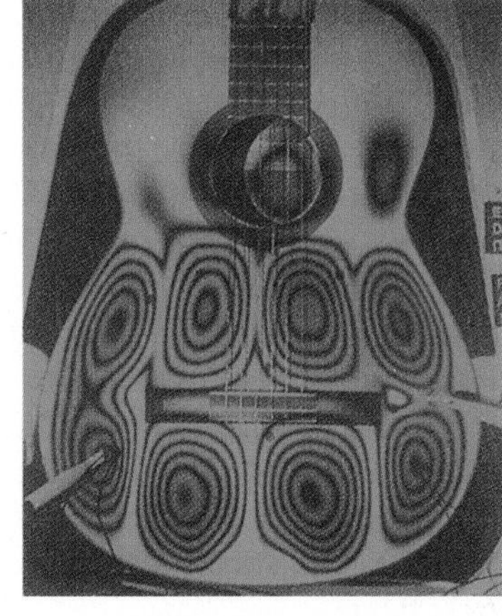

1010 Hz ($Q = 80$)

Extended objects have more than one resonance frequency. When plucked, a guitar string transmits its energy to the body of the guitar. The body's oscillations, coupled to those of the air mass it encloses, produce the resonance patterns shown.

*Mathematical Treatment of Resonance

We can treat a driven oscillator mathematically by assuming that, in addition to the restoring force and a damping force, the oscillator is subject to an external driving force that varies harmonically with time:

$$F_{\text{ext}} = F_0 \cos \omega t \qquad \text{14-50}$$

where F_0 and ω are the amplitude and angular frequency of the driving force. This frequency is generally not related to the natural angular frequency of the system ω_0.

Newton's second law applied to an object of mass m attached to a spring of force constant k and subject to a damping force $-bv_x$ and an external force $F_0 \cos \omega t$ gives

$$\sum F_x = ma_x$$

$$-kx - bv_x + F_0 \cos \omega t = m\frac{d^2x}{dt^2}$$

where we have used $a_x = d^2x/dt^2$. Substituting $m\omega_0^2$ for k (Equation 14-8) and rearranging gives

$$m\frac{d^2x}{dt^2} + b\frac{dx}{dt} + m\omega_0^2 x = F_0 \cos \omega t \qquad \text{14-51}$$

DIFFERENTIAL EQUATION FOR A DRIVEN OSCILLATOR

We will discuss the general solution of Equation 14-51 qualitatively. It consists of two parts, the **transient solution** and the **steady-state solution.** The transient part of the solution is identical to that for a damped oscillator given in Equation 14-36. The constants in this part of the solution depend on the initial conditions. Over time, this part of the solution becomes negligible because of the exponential decrease of the amplitude. We are then left with the steady-state solution, which can be written as

$$x = A \cos(\omega t - \delta) \qquad \text{14-52}$$

POSITION FOR A DRIVEN OSCILLATOR

where the angular frequency ω is the same as that of the driving force. The amplitude A is given by

$$A = \frac{F_0}{\sqrt{m^2(\omega_0^2 - \omega^2)^2 + b^2\omega^2}} \qquad 14\text{-}53$$

AMPLITUDE FOR A DRIVEN OSCILLATOR

and the phase constant δ is given by

$$\tan \delta = \frac{b\omega}{m(\omega_0^2 - \omega^2)} \qquad 14\text{-}54$$

PHASE CONSTANT FOR A DRIVEN OSCILLATOR

Comparing Equations 14-50 and 14-52, we can see that the displacement and the driving force oscillate with the same frequency, but they differ in phase by δ. When the driving frequency ω is much less than the natural frequency ω_0, $\delta \approx 0$, as can be seen from Equation 14-54. At resonance, $\omega = \omega_0$, $\delta = \pi/2$, and when ω is much greater than ω_0, $\delta \approx \pi$. At the beginning of this chapter the displacement of a particle undergoing simple harmonic motion is written $x = A\cos(\omega t + \delta)$ (Equation 14-4). This equation is identical with Equation 14-52 except for the sign preceding the phase constant δ. The phase of a driven oscillator always lags behind the phase of the driving force. The negative sign in Equation 14-52 ensures that δ is always positive (rather than always negative).

In your simple experiment to drive a meterstick by moving your hand back and forth, you should note that at resonance the oscillation of your hand is neither in phase nor 180° out of phase with the oscillation of the stick. If you move your hand back and forth at a frequency several times the natural frequency of the stick, the stick's steady state motion will be almost 180° out of phase with your hand.

The velocity of the object in the steady state is obtained by differentiating x with respect to t:

$$v = \frac{dx}{dt} = -\omega A \sin(\omega t - \delta)$$

At resonance, $\delta = \pi/2$, and the velocity is in phase with the driving force:

$$v = -\omega A \sin\left(\omega t - \frac{\pi}{2}\right) = +\omega A \cos \omega t$$

Thus, at resonance the object is always moving in the direction of the driving force, as would be expected for maximum power input. The velocity amplitude ωA is maximum at $\omega = \omega_0$.

EXPLORING

Is there a numerical solution to the equations for damped and driven oscillators? Find out this, and more, at www.whfreeman.com/tipler5e.

AN OBJECT ON A SPRING — EXAMPLE 14-13 — Try It Yourself

An object of mass 1.5 kg on a spring of force constant 600 N/m loses 3 percent of its energy in each cycle. The same system is driven by a sinusoidal force with a maximum value of F_0 = 0.5 N. (*a*) What is Q for this system? (*b*) What is the resonance (angular) frequency? (*c*) If the driving frequency is slowly varied through resonance, what is the width $\Delta\omega$ of the resonance? (*d*) What is the amplitude at resonance? (*e*) What is the amplitude if the driving frequency is ω = 19 rad/s?

PICTURE THE PROBLEM The energy loss per cycle is only 3 percent, so the damping is weak. We can find Q from $Q = 2\pi/ (\Delta E/E)_{\text{cycle}}$ (Equation 14-45) and then use this result and $\Delta\omega/\omega_0 = 1/Q$ (Equation 14-49) to find the width $\Delta\omega$ of the resonance. The resonance frequency is the natural frequency. The amplitude both at resonance and off resonance can be found from Equation 14-53, with the damping constant calculated from Q using the definition of Q (Equation 14-43) $Q = \omega_0\tau = \omega_0 m/b$.

Cover the column to the right and try these on your own before looking at the answers.

Steps	Answers
(*a*) The damping is weak. Relate Q to the fractional energy loss using Equation 14-45 and solve for Q.	$Q \approx \frac{2\pi}{(\lvert\Delta E\rvert/E)_{\text{cycle}}} = \frac{2\pi}{0.03} = \boxed{209}$
(*b*) Relate the resonance frequency to the natural frequency of the system.	$\omega_0 = \sqrt{\frac{k}{m}} = \boxed{20\text{ rad/s}}$
(*c*) Relate the width of the resonance $\Delta\omega$ to Q.	$\Delta\omega = \frac{\omega_0}{Q} = \boxed{0.0957\text{ rad/s}}$
(*d*) 1. Write an expression for the amplitude A for any driving frequency ω.	$A(\omega) = \frac{F_0}{\sqrt{m^2(\omega_0^2 - \omega^2)^2 + b^2\omega^2}}$
2. Substitute $\omega = \omega_0$ to calculate A at resonance.	$A(\omega_0) = \frac{F_0}{b\omega_0}$
3. Use Equation 14-43 to relate the damping constant b to Q.	$b = \frac{m\omega_0}{Q} = 0.144\text{ kg/s}$
4. Use the results of the previous two steps to calculate the amplitude at resonance.	$A(\omega_0) = \frac{F_0}{b\omega_0} = \boxed{17.4\text{ cm}}$
(*e*) Calculate the amplitude for $\omega = 19$ rad/s. (We can omit the units to simplify the equation. Since all quantities are in SI units, A will be in meters.)	$A(19\text{ s}^{-1}) = \frac{0.5}{\sqrt{1.5^2(20^2 - 19^2)^2 + 0.144^2(19)^2}} = \boxed{0.854\text{ cm}}$

REMARKS At just 1 rad/s off resonance, the amplitude drops by a factor of 20. This is not surprising, because the width $\Delta\omega$ of the resonance is only 0.0957 rad/s. Note that off resonance the term $b^2\omega^2$ is negligible compared with the other term in the denominator of the expression for A. When $\omega - \omega_0$ is more than several times the half width $\Delta\omega$, as it was in this example, we can neglect the $b^2\omega^2$ term and calculate A from $A \approx F_0/[m(\omega_0^2 - \omega^2)]$. Figure 14-25 shows the amplitude versus driving frequency ω. Note that the horizontal scale is over a small range of ω.

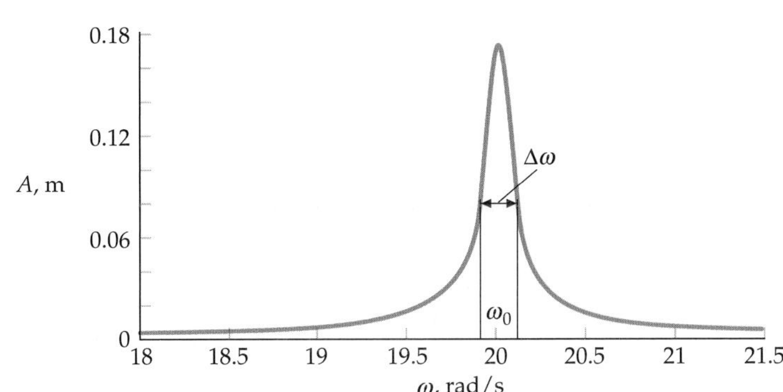

FIGURE 14-25

SUMMARY

1. Simple harmonic motion occurs whenever the restoring force is proportional to the displacement from equilibrium. It has wide application in the study of oscillations, waves, electrical circuits, and molecular dynamics.
2. Resonance is an important phenomenon in many areas of physics. It occurs when the frequency of the driving force is close to the natural frequency of the oscillating system.

Topic	Relevant Equations and Remarks	
1. Simple Harmonic Motion	In simple harmonic motion, the net force and acceleration are both proportional to the displacement and oppositely directed.	
Position function	$x = A\cos(\omega t + \delta)$	**14-4**
Velocity	$v = -\omega A\sin(\omega t + \delta)$	**14-5**
Acceleration	$a_x = -\omega^2 A\cos(\omega t + \delta)$	**14-6**
	$a_x = -\omega^2 x$	**14-7**
Angular frequency	$\omega = 2\pi f = \dfrac{2\pi}{T}$	**14-11**
Total energy	$E_{\text{total}} = K + U = \frac{1}{2}kA^2$	**14-17**
Average kinetic or potential energy	$K_{\text{av}} = U_{\text{av}} = \frac{1}{2}E_{\text{total}}$	**14-18**
Circular motion	If a particle moves in a circle with constant speed, the projection of the particle onto a diameter of the circle moves in simple harmonic motion.	
General motion near equilibrium	If an object is given a small displacement from a position of stable equilibrium, it typically oscillates about this position with simple harmonic motion.	
2. Angular Frequencies for Various Systems		
Mass on a spring	$\omega = \sqrt{\dfrac{k}{m}}$	**14-8**
Simple pendulum	$\omega = \sqrt{\dfrac{g}{L}}$	**14-27**
*Physical pendulum	$\omega = \sqrt{\dfrac{MgD}{I}}$	**14-32**
	where D is the distance from the center of mass to the rotation axis and I is the moment of inertia about the rotation axis.	
3. Damped Oscillations	In the oscillations of real systems, the motion is damped because of dissipative forces. If the damping is greater than some critical value, the system does not oscillate when disturbed but merely returns to its equilibrium position. The motion of a weakly damped system is nearly simple harmonic with an amplitude that decreases exponentially with time.	

Frequency	$\omega' = \omega_0\sqrt{1 - \frac{1}{4Q^2}}$	14-46
Amplitude	$A = A_0 e^{-(b/2m)t}$	14-38
Energy	$E = E_0 e^{-t/\tau}$	14-41
Decay time	$\tau = \frac{m}{b}$	14-42
Q factor defined	$Q = \omega_0 \tau$	14-43
Q factor for weak damping	$Q = \frac{2\pi}{(\lvert\Delta E\rvert/E)_{\text{cycle}}}, \quad \frac{\lvert\Delta E\rvert}{E} << 1$	14-45
4. Driven Oscillations	When an underdamped ($b < b_c$) system is driven by an external sinusoidal force $F_{\text{ext}} = F_0 \cos \omega t$, the system oscillates with a frequency ω equal to the driving frequency and an amplitude A that depends on the driving frequency.	
Resonance frequency	$\omega = \omega_0$	
Resonance width for weak damping	$\frac{\Delta\omega}{\omega_0} = \frac{1}{Q}$	14-49
*Position function	$x = A\cos(\omega t - \delta)$	14-52
*Amplitude	$A = \frac{F_0}{\sqrt{m^2(\omega_0^2 - \omega^2)^2 + b^2\omega^2}}$	14-53
*Phase constant	$\tan\delta = \frac{b\omega}{m(\omega_0^2 - \omega^2)}$	14-54

PROBLEMS

In a few problems, you are given more data than you actually need; in a few other problems, you are required to supply data from your general knowledge, outside sources, or informed estimates.

- • Single-concept, single-step, relatively easy
- •• Intermediate-level, may require synthesis of concepts
- ••• Challenging
- SSM Solution is in the *Student Solutions Manual*
- iSOLVE Problems available on iSOLVE online homework service
- iSOLVE ✓ These "Checkpoint" online homework service problems ask students additional questions about their confidence level, and how they arrived at their answer

Conceptual Problems

1 • What is the magnitude of the acceleration of an oscillator of amplitude A and frequency f when its speed is maximum? When its displacement from equilibrium is maximum?

2 • Are the acceleration and the displacement (from equilibrium) of a simple harmonic oscillator ever in the same direction? The acceleration and the velocity? The velocity and the displacement? Explain.

3 • True or false:

(*a*) For a simple harmonic oscillator, the period is proportional to the square of the amplitude.
(*b*) For a simple harmonic oscillator, the frequency does not depend on the amplitude.
(*c*) If the acceleration of a particle undergoing 1-dimensional motion is proportional to the displacement from equilibrium and oppositely directed, the motion is simple harmonic.

4 • SSM If the amplitude of a simple harmonic oscillator is tripled, by what factor is the energy changed?

5 •• An object attached to a spring has simple harmonic motion with an amplitude of 4.0 cm. When the object is 2.0 cm from the equilibrium position, what fraction of its total energy is potential energy? (*a*) One-quarter. (*b*) One-third. (*c*) One-half. (*d*) Two-thirds. (*e*) Three-quarters.

6 • True or false:

(*a*) For a given object on a given spring, the period is the same whether the spring is vertical or horizontal.
(*b*) For a given object oscillating with amplitude *A* on a given spring, the maximum speed is the same whether the spring is vertical or horizontal.

7 • True or false: The motion of a simple pendulum is simple harmonic for any initial angular displacement.

8 • True or false: The motion of a simple pendulum is periodic for any initial angular displacement.

9 •• SSM Two identical carts on a frictionless air track are attached by a spring. One is suddenly struck a blow that sends it moving away from the other cart. The motion of the carts is seen to be very jerky—first one cart moves, then stops as the other cart moves and stops in its turn. Explain the motion in a qualitative way.

10 •• SSM The length of the string or wire supporting a pendulum bob increases slightly when its temperature is raised. How would this affect a clock operated by a simple pendulum?

11 • True or false: The mechanical energy of a damped, undriven oscillator decreases exponentially with time.

12 • True or false:

(*a*) Resonance occurs when the driving frequency equals the natural frequency.
(*b*) If the *Q* value is high, the resonance is sharp.

13 • Give some examples of common systems that can be considered to be driven oscillators.

14 • A crystal wineglass shattered by an intense sound is an example of (*a*) resonance, (*b*) critical damping, (*c*) an exponential decrease in energy, (*d*) overdamping.

15 • SSM The effect of the mass of a spring on the motion of an object attached to it is usually neglected. Describe qualitatively its effect when it is not neglected.

16 •• A lamp hanging from the ceiling of the club car in a train oscillates with period T_0 when the train is at rest. The period will be (match left and right columns)

1. greater than T_0 when	A. the train moves horizontally with constant velocity.
2. less than T_0 when	B. the train rounds a curve of radius *R* with speed *v*.
3. equal to T_0 when	C. the train climbs a hill of inclination θ at constant speed.
	D. the train goes over the crest of a hill of radius of curvature *R* with constant speed.

17 •• Two mass–spring systems oscillate at frequencies f_A and f_B. If $f_A = 2f_B$ and the spring constants of the two springs are equal, it follows that the masses are related by (*a*) $M_A = 4M_B$, (*b*) $M_A = M_B/\sqrt{2}$, (*c*) $M_A = M_B/2$, (*d*) $M_A = M_B/4$.

18 •• Two mass–spring systems A and B oscillate so that their energies are equal. If $M_A = 2M_B$, which formula relates the amplitudes of oscillation? (*a*) $A_A = A_B/4$. (*b*) $A_A = A_B/\sqrt{2}$. (*c*) $A_A = A_B$. (*d*) Not enough information is given to determine the ratio of the amplitudes.

19 •• Two mass–spring systems A and B oscillate so that their energies are equal. If $k_A = 2k_B$, then which formula relates the amplitudes of oscillation? (*a*) $A_A = A_B/4$. (*b*) $A_A = A_B/\sqrt{2}$. (*c*) $A_A = A_B$. (*d*) Not enough information is given to determine the ratio of the amplitudes.

20 •• Pendulum A has a bob of mass M_A and a length L_A; pendulum B has a bob of mass M_B and a length L_B. If the period of A is twice that of B, then (*a*) $L_A = 2L_B$ and $M_A = 2M_B$, (*b*) $L_A = 4L_B$ and $M_A = M_B$, (*c*) $L_A = 4L_B$ whatever the ratio M_A/M_B, (*d*) $L_A = \sqrt{2}\,L_B$ whatever the ratio M_A/M_B.

Estimation and Approximation

21 •• For a child on a swing, the amplitude drops by a factor of $1/e$ in about eight periods if no mechanical energy is fed in. Estimate the *Q* factor for this system.

22 •• SSM (*a*) Estimate the natural period of oscillation for swinging your arms as you walk, with your hands empty. (*b*) Now estimate it when carrying a heavy briefcase. Look around at other people as they walk by—do these two estimates seem on target?

Simple Harmonic Motion

23 • ISOLVE The position of a particle is given by $x = (7\text{ cm})\cos 6\pi t$, where *t* is in seconds. What are (*a*) the frequency, (*b*) the period, and (*c*) the amplitude of the particle's motion? (*d*) What is the first time after $t = 0$ that the particle is at its equilibrium position? In what direction is it moving at that time?

24 • What is the phase constant δ in Equation 14-4 if the position of the oscillating particle at time $t = 0$ is (*a*) 0, (*b*) $-A$, (*c*) *A*, (*d*) $A/2$?

25 • SSM A particle of mass *m* begins at rest from $x = +25$ cm and oscillates about its equilibrium position at $x = 0$ with a period of 1.5 s. Write equations for (*a*) the position *x* as a function of *t*, (*b*) the velocity *v* as a function of *t*, and (*c*) the acceleration *a* as a function of *t*.

26 • Find (*a*) the maximum speed and (*b*) the maximum acceleration of the particle in Problem 23. (*c*) What is the first time that the particle is at $x = 0$ and moving to the right?

27 •• Work Problem 25 with the particle initially at $x = 25$ cm and moving with velocity $v_0 = +50$ cm/s.

28 •• The period of an oscillating particle is 8 s and its amplitude is 12 cm. At $t = 0$ it is at its equilibrium position. Find the distance traveled during the intervals (*a*) $t = 0$ to

$t = 2$ s, (*b*) $t = 2$ s to $t = 4$ s, (*c*) $t = 0$ to $t = 1$ s, and (*d*) $t = 1$ s to $t = 2$ s.

29 •• The period of an oscillating particle is 8 s. At $t = 0$, the particle is at rest at $x = A = 10$ cm. (*a*) Sketch x as a function of t. (*b*) Find the distance traveled in the first, second, third, and fourth second after $t = 0$.

30 •• SSM ISOLVE Military specifications often call for electronic devices to be able to withstand accelerations of $10g = 98.1\ \text{m/s}^2$. To make sure that their products meet this specification, manufacturers test them using a shaking table that can vibrate a device at various specified frequencies and amplitudes. If a device is given a vibration of amplitude 1.5 cm, what should its frequency be in order to test for compliance with the $10g$ military specification?

31 •• ISOLVE✓ The position of a particle is given by $x = 2.5 \cos \pi t$, where x is in meters and t is in seconds. (*a*) Find the maximum speed and maximum acceleration of the particle. (*b*) Find the speed and acceleration of the particle when $x = 1.5$ m.

32 •• SSM (*a*) Show that $A_0 \cos(\omega t + \delta)$ can be written as $A_s \sin(\omega t) + A_c \cos(\omega t)$, and determine A_s and A_c in terms of A_0 and δ. (*b*) Relate A_c and A_s to the initial position and velocity of a particle undergoing simple harmonic motion.

Simple Harmonic Motion and Circular Motion

33 • ISOLVE A particle moves in a circle of radius 40 cm with a constant speed of 80 cm/s. Find (*a*) the frequency of the motion and (*b*) the period of the motion. (*c*) Write an equation for the x component of the position of the particle as a function of time t, assuming that the particle is on the positive x axis at time $t = 0$.

34 • SSM A particle moves in a circle of radius 15 cm, making 1 revolution every 3 s. (*a*) What is the speed of the particle? (*b*) What is its angular velocity ω? (*c*) Write an equation for the x component of the position of the particle as a function of time t, assuming that the particle is on the positive x axis at time $t = 0$.

Energy in Simple Harmonic Motion

35 • A 2.4-kg object is attached to a horizontal spring of force constant $k = 4.5$ kN/m. The spring is stretched 10 cm from equilibrium and released. Find its total energy.

36 • Find the total energy of a 3-kg object oscillating on a horizontal spring with an amplitude of 10 cm and a frequency of 2.4 Hz.

37 • A 1.5-kg object oscillates with simple harmonic motion on a spring of force constant $k = 500$ N/m. Its maximum speed is 70 cm/s. (*a*) What is the total mechanical energy? (*b*) What is the amplitude of the oscillation?

38 • A 3-kg object oscillating on a spring of force constant 2 kN/m has a total energy of 0.9 J. (*a*) What is the amplitude of the motion? (*b*) What is the maximum speed?

39 • An object oscillates on a spring with an amplitude of 4.5 cm. Its total energy is 1.4 J. What is the force constant of the spring?

40 •• SSM ISOLVE✓ A 3-kg object oscillates on a spring with an amplitude of 8 cm. Its maximum acceleration is $3.50\ \text{m/s}^2$. Find the total energy.

Springs

41 • A 2.4-kg object is attached to a horizontal spring of force constant $k = 4.5$ kN/m. The spring is stretched 10 cm from equilibrium and released. What are (*a*) the frequency of the motion, (*b*) the period, (*c*) the amplitude, (*d*) the maximum speed, and (*e*) the maximum acceleration? (*f*) When does the object first reach its equilibrium position? What is its acceleration at this time?

42 • Answer the questions in Problem 41 for a 5-kg object attached to a spring of force constant $k = 700$ N/m when the spring is initially stretched 8 cm from equilibrium.

43 • A 3-kg object attached to a horizontal spring oscillates with an amplitude $A = 10$ cm and a frequency $f = 2.4$ Hz. (*a*) What is the force constant of the spring? (*b*) What is the period of the motion? (*c*) What is the maximum speed of the object? (*d*) What is the maximum acceleration of the object?

44 • SSM ISOLVE✓ An 85-kg person steps into a car of mass 2400 kg, causing it to sink 2.35 cm on its springs. Assuming no damping, with what frequency will the car and passenger vibrate on the springs?

45 • A 4.5-kg object oscillates on a horizontal spring with an amplitude of 3.8 cm. Its maximum acceleration is $26\ \text{m/s}^2$. Find (*a*) the force constant k, (*b*) the frequency, and (*c*) the period of the motion.

46 • An object oscillates with an amplitude of 5.8 cm on a horizontal spring of force constant 1.8 kN/m. Its maximum speed is 2.20 m/s. Find (*a*) the mass of the object, (*b*) the frequency of the motion, and (*c*) the period of the motion.

47 •• ISOLVE✓ A 0.4-kg block attached to a spring of force constant 12 N/m oscillates with an amplitude of 8 cm. Find (*a*) the maximum speed of the block, (*b*) the speed and acceleration of the block when it is at $x = 4$ cm from the equilibrium position, and (*c*) the time it takes the block to move from $x = 0$ to $x = 4$ cm.

48 •• SSM An object of mass m is supported by a vertical spring of force constant 1800 N/m. When pulled down 2.5 cm from equilibrium and released from rest, the object oscillates at 5.5 Hz. (*a*) Find m. (*b*) Find the amount the spring is stretched from its natural length when the object is in equilibrium. (*c*) Write expressions for the displacement x, the velocity v, and the acceleration a as functions of time t.

49 •• ISOLVE✓ An object of unknown mass is hung on the end of an unstretched spring and is released from rest. If the object falls 3.42 cm before first coming to rest, find the period of the motion.

50 •• A spring of force constant $k = 250$ N/m is suspended from a rigid support. An object of mass 1 kg is attached to the unstretched spring and the object is released from rest. (*a*) How far below the starting point is the equilibrium position for the object? (*b*) How far down does the object move before it starts up again? (*c*) What is the period of oscillation? (*d*) What is the speed of the object when it first reaches its equilibrium position? (*e*) When does it first reach its equilibrium position?

51 •• ISOLVE The St. Louis Arch has a height of 192 m. Suppose that a stunt woman of mass 60 kg jumps off the top of the arch with an elastic band attached to her feet. She reaches the ground at zero speed. Find her kinetic energy K after 2.00 s of the flight. (Assume that the elastic band obeys Hooke's law, and neglect its length when relaxed.)

52 •• SSM A suitcase of mass 20 kg is hung from two bungie cords, as shown in Figure 14-26. Each cord is stretched 5 cm when the suitcase is in equilibrium. If the suitcase is pulled down a little and released, what will be its oscillation frequency?

FIGURE 14-26 Problem 52

53 •• A 0.12-kg block is suspended from a spring. When a small stone of mass 30 g is placed on the block, the spring stretches an additional 5 cm. With the stone on the block, the spring oscillates with an amplitude of 12 cm. (*a*) What is the frequency of the motion? (*b*) How long does the block take to travel from its lowest point to its highest point? (*c*) What is the net force on the stone when it is at the point of maximum upward displacement?

54 •• In Problem 53, find the maximum amplitude of oscillation at which the stone will remain in contact with the block.

55 •• An object of mass 2.0 kg is attached to the top of a vertical spring that is anchored to the floor. The uncompressed length of the spring is 8.0 cm and the length of the spring when the object is in equilibrium is 5.0 cm. When the object is resting at its equilibrium position, it is given a downward impulse with a hammer so that its initial speed is 0.3 m/s. (*a*) To what maximum height above the floor does the object eventually rise? (*b*) How long does it take for the object to reach its maximum height for the first time? (*c*) Does the spring ever become uncompressed? What minimum initial velocity must be given to the object for the spring to be uncompressed at some time?

56 •• SSM A winch cable has a cross-sectional area of 1.5 cm^2 and a length of 2.5 m. Young's modulus for the cable is 150 GN/m^2. A 950-kg engine block is hung from the end of the cable. (*a*) By what length does the cable stretch? (*b*) Treating the cable as a simple spring, what is the oscillation frequency of the engine block at the end of the cable?

Energy of an Object on a Vertical Spring

57 •• A 2.5-kg object hanging from a vertical spring of force constant 600 N/m oscillates with an amplitude of 3 cm. When the object is at its maximum downward displacement, find (*a*) the total energy of the system, (*b*) the gravitational potential energy, and (*c*) the potential energy in the spring. (*d*) What is the maximum kinetic energy of the object? (Choose $U = 0$ when the object is in equilibrium.)

58 •• ISOLVE A 1.5-kg object that stretches a spring 2.8 cm from its natural length when hanging at rest oscillates with an amplitude of 2.2 cm. (*a*) Find the total energy of the system. (*b*) Find the gravitational potential energy at maximum downward displacement. (*c*) Find the potential energy in the spring at maximum downward displacement. (*d*) What is the maximum kinetic energy of the object?

59 •• SSM A 1.2-kg object hanging from a spring of force constant 300 N/m oscillates with a maximum speed of 30 cm/s. (*a*) What is its maximum displacement? When the object is at its maximum displacement, find (*b*) the total energy of the system, (*c*) the gravitational potential energy, and (*d*) the potential energy in the spring.

Simple Pendulums

60 • Find the length of a simple pendulum if the period is 5 s at a location where $g = 9.81$ m/s^2.

61 • What would be the period of the pendulum in Problem 60 if the pendulum were on the moon, where the acceleration due to gravity is one-sixth that on earth?

62 • ISOLVE✓ If the period of a pendulum 70 cm long is 1.68 s, what is the value of g at the location of the pendulum?

63 • SSM ISOLVE A pendulum set up in the stairwell of a 10-story building consists of a heavy weight suspended on a 34.0-m wire. If $g = 9.81$ m/s^2, what is the period of oscillation?

64 •• Show that the total energy of a simple pendulum undergoing oscillations of small amplitude ϕ_0 is approximately $E \approx \frac{1}{2}mgL\phi_0^2$. *Hint: Use the approximation* $\cos\phi \approx 1 - \frac{1}{2}\phi^2$ *for small* ϕ.

65 •• A simple pendulum of length L is attached to a massive cart that slides without friction down a plane inclined at angle θ with the horizontal as shown in Figure 14-27. Find the period of oscillation of the pendulum on the sliding cart.

FIGURE 14-27 Problem 65

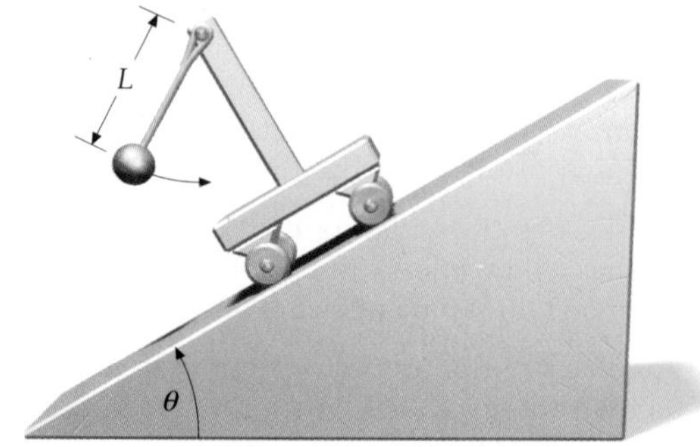

66 •• A simple pendulum of length L is released from rest from an angle ϕ_0. (*a*) Assuming that the pendulum undergoes simple harmonic motion, find its speed as it passes through $\phi = 0$. (*b*) Using the conservation of energy, find this speed exactly. (*c*) Show that your results for (*a*) and (*b*) are the same when ϕ_0 is small. (*d*) Find the difference in your results for $\phi_0 = 0.20$ rad and $L = 1$ m.

*Physical Pendulums

67 • ISOLVE ✓ A thin disk of mass 5 kg and radius 20 cm is suspended by a horizontal axis perpendicular to the disk through its rim. The disk is displaced slightly from equilibrium and released. Find the period of the subsequent simple harmonic motion.

68 • A circular hoop of radius 50 cm is hung on a narrow horizontal rod and allowed to swing in the plane of the hoop. What is the period of its oscillation, assuming that the amplitude is small?

69 • A 3-kg plane figure is suspended at a point 10 cm from its center of mass. When it is oscillating with small amplitude, the period of oscillation is 2.6 s. Find the moment of inertia I about an axis perpendicular to the plane of the figure through the pivot point.

70 •• The pendulum bob of a large town-hall clock has a length of 4 m. (*a*) What is its period of oscillation? Treat it as a simple pendulum with small amplitude oscillations. (*b*) To regulate the period of the pendulum there is a tray attached to its shaft, halfway up. The tray holds a stack of coins. To change the period by a little bit, coins are added or removed from the tray. Explain in detail why this works. Will adding coins increase or decrease the period of the pendulum?

71 •• Figure 14-28 shows a dumbbell with two equal masses, to be considered as point masses attached to a thin massless rod of length L. (*a*) Show that the period of this pendulum is a minimum when the pivot point P is at one of the masses. (*b*) Find the period of this physical pendulum if the distance between P and the upper mass is $L/4$.

FIGURE 14-28 Problem 71

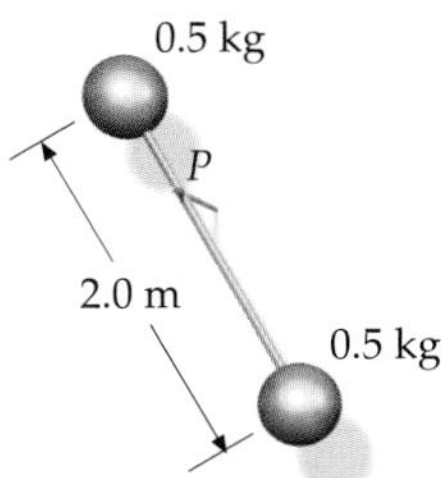

72 •• Suppose the rod in Problem 71 has a mass of $2m$ (Figure 14-29). Determine the distance between the upper mass and the pivot point P when the period of this physical pendulum is a minimum.

FIGURE 14-29 Problem 72

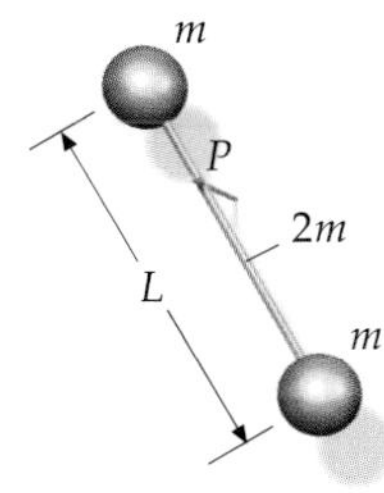

73 •• SSM You are given a meterstick and asked to drill a narrow hole in it so that, when the stick is pivoted about the hole, the period of the pendulum will be a minimum. Where should you drill the hole?

74 •• ISOLVE Figure 14-30 shows a uniform disk of radius $R = 0.8$ m and 6-kg mass with a small hole a distance d from the disk's center that can serve as a pivot point. (*a*) What should be the distance d so that the period of this physical pendulum is 2.5 s? (*b*) What should be the distance d so that this physical pendulum will have the shortest possible period? What is this period?

FIGURE 14-30 Problem 74

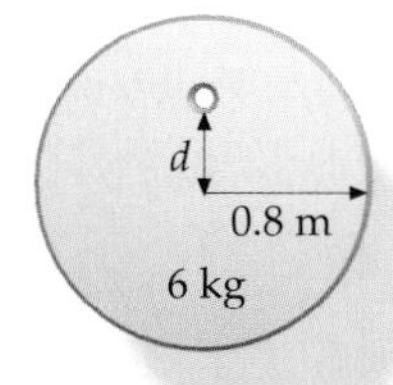

75 ••• A plane object has a moment of inertia I about its center of mass. When pivoted at point P_1, as shown in Figure 14-31, it oscillates about the pivot with a period T. There is a second point P_2 on the opposite side of the center of mass about which the object can be pivoted so that the period of oscillation is also T. Show that $h_1 + h_2 = gT^2/(4\pi^2)$.

FIGURE 14-31 Problem 75

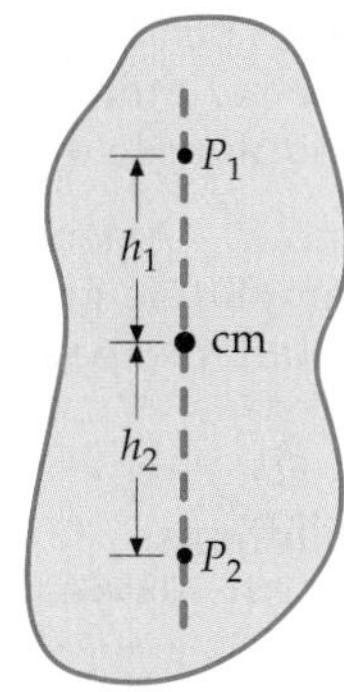

76 ••• A physical pendulum consists of a spherical bob of radius r and mass m suspended from a string (Figure 14-32). The distance from the center of the sphere to the point of support is L. When r is much less than L, such a pendulum is often treated as a simple pendulum of length L. (*a*) Show that the period for small oscillations is given by

FIGURE 14-32 Problem 76

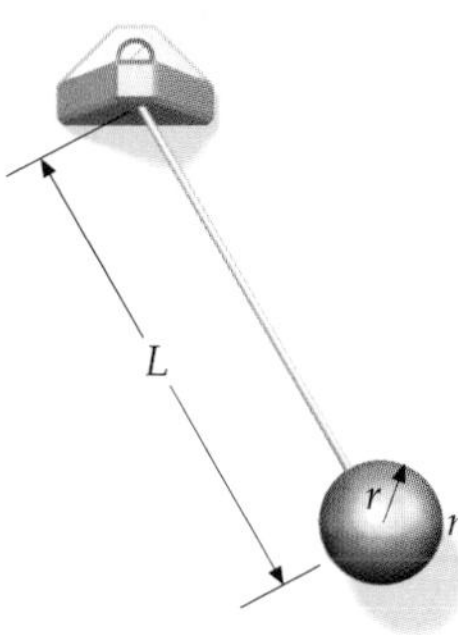

$$T = T_0\sqrt{1 + \frac{2r^2}{5L^2}}$$

where $T_0 = 2\pi\sqrt{L/g}$ is the period of a simple pendulum of length L. (*b*) Show that when r is much smaller than L, the period is approximately $T \approx T_0(1 + r^2/5L^2)$. (*c*) If $L = 1$ m and

$r = 2$ cm, find the error when the approximation $T = T_0$ is used for this pendulum. How large must be the radius of the bob for the error to be 1 percent?

77 ••• Figure 14-33 shows the pendulum of a clock. The uniform rod of length $L = 2.0$ m has a mass $m = 0.8$ kg. Attached to the rod is a uniform disk of mass $M = 1.2$ kg and radius 0.15 m. The clock is constructed to keep perfect time if the period of the pendulum is exactly 3.50 s. (*a*) What should be the distance d so that the period of this pendulum is 2.50 s? (*b*) Suppose that the pendulum clock loses 5.0 min/d. How far and in what direction should the disk be moved to ensure that the clock will keep perfect time?

FIGURE 14-33 Problem 77

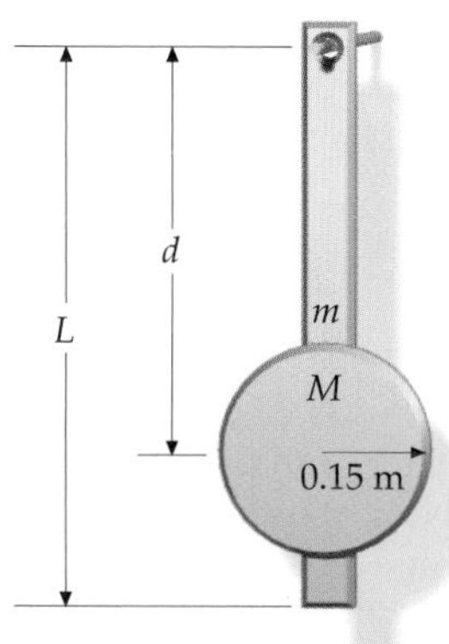

78 •• SSM A pendulum clock loses 48 s/d when the amplitude of the pendulum is 8.4°. What should be the amplitude of the pendulum so that the clock keeps perfect time?

79 •• ISOLVE ✓ A pendulum clock that has run down to a very small amplitude gains 5 min each day. What angular amplitude should the pendulum have to keep the correct time?

Damped Oscillations

80 • Show that the damping constant b has units of kg/s.

81 • An oscillator has a Q factor of 200. By what percentage does its energy decrease during one period?

82 • ISOLVE A 2-kg object oscillates with an initial amplitude of 3 cm on a spring of force constant $k = 400$ N/m. Find (*a*) the period and (*b*) the total initial energy. (*c*) If the energy decreases by 1 percent per period, find the damping constant b and the Q factor.

83 •• Show that the ratio of the amplitudes for two successive oscillations is constant for a damped oscillator.

84 •• An oscillator has a period of 3 s. Its amplitude decreases by 5 percent during each cycle. (*a*) By how much does its energy decrease during each cycle? (*b*) What is the time constant τ? (*c*) What is the Q factor?

85 •• An oscillator has a Q factor of 20. (*a*) By what fraction does the energy decrease during each cycle? (*b*) Use Equation 14-37 to find the percentage difference between ω' and ω_0. *Hint: Use the approximation* $(1 + x)^{\frac{1}{2}} \approx 1 + \frac{1}{2}x$ *for small x.*

86 •• ISOLVE A damped mass–spring system oscillates at 200 Hz. The time constant of the system is 2.0 s. At $t = 0$ the amplitude of oscillation is 6.0 cm and the energy of the oscillating system is 60 J. (*a*) What are the amplitudes of oscillation at $t = 2.0$ s and $t = 4.0$ s? (*b*) How much energy is dissipated in the first 2-s interval and in the second 2-s interval?

87 •• SSM ISOLVE It has been stated that the vibrating earth has a resonance period of 54 min and a Q factor of about 400 and that after a large earthquake, the earth "rings" (continues to vibrate) for about 2 months. (*a*) Find the percentage of the energy of vibration lost to damping forces during each cycle. (*b*) Show that after n periods the energy is $E_n = (0.984)^n E_0$, where E_0 is the original energy. (*c*) If the original energy of vibration of an earthquake is E_0, what is the energy after 2 d?

88 •• A compact pendulum used in a physics experiment has a mass of 15 g; the length of the pendulum is 75 cm. To start the bob oscillating, a physics student puts a fan next to it that blows a horizontal stream of air on the bob. With the fan on, the bob is in equilibrium when the pendulum is displaced by an angle of 5° from the vertical. The speed of the air from the fan is 7 m/s. The fan is then turned off, and the pendulum is allowed to oscillate. (*a*) If we assume that the drag force due to the air is of the form $-bv$, what is the decay time constant τ for the oscillations of the pendulum? (*b*) How long will it take for the amplitude of oscillation of the pendulum to reach 1°?

Driven Oscillations and Resonance

89 • Find the resonance frequency for each of the three systems shown in Figure 14-34.

FIGURE 14-34 Problem 89

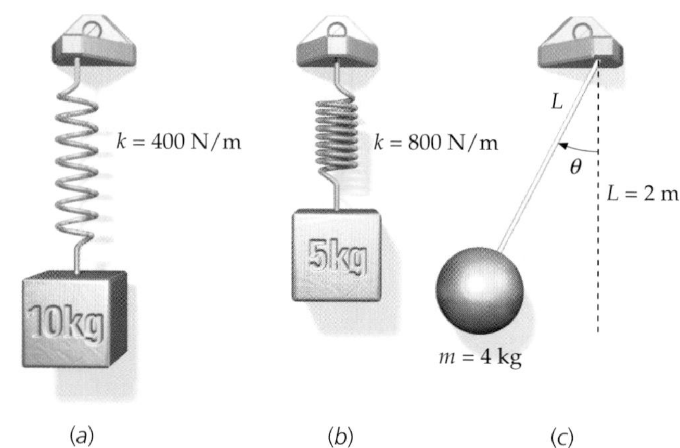

90 • A damped oscillator loses 2 percent of its energy during each cycle. (*a*) What is its Q factor? (*b*) If its resonance frequency is 300 Hz, what is the width of the resonance curve $\Delta\omega$ when the oscillator is driven?

91 •• A 2-kg object oscillates on a spring of force constant $k = 400$ N/m. The damping constant has a value of $b = 2.00$ kg/s. The system is driven by a sinusoidal force of maximum value 10 N and angular frequency $\omega = 10$ rad/s. (*a*) What is the amplitude of the oscillations? (*b*) If the driving frequency is varied, at what frequency will resonance occur? (*c*) What is the amplitude of oscillation at resonance? (*d*) What is the width of the resonance curve $\Delta\omega$?

92 •• A damped oscillator loses 3.5 percent of its energy during each cycle. (*a*) How many cycles elapse before half of its original energy is dissipated? (*b*) What is its Q factor? (*c*) If the natural frequency is 100 Hz, what is the width of the resonance curve when the oscillator is driven?

Collisions

93 ••• Figure 14-35 shows a vibrating mass–spring system supported on a frictionless surface and a second, equal mass that is moving toward the vibrating mass with velocity v. The motion of the vibrating mass is given by $x(t) = (0.1\text{ m})\cos(40\text{ s}^{-1}\,t)$, where x is the displacement of the mass from its equilibrium position. The two masses collide elastically just as the vibrating mass passes through its equilibrium position traveling to the right. (*a*) What should be the velocity v of the second mass so that the mass–spring system is at rest following the elastic collision? (*b*) What is the velocity of the second mass after the elastic collision?

FIGURE 14-35 Problem 93

94 ••• Following the elastic collision in Problem 93, the kinetic energy of the recoiling mass is 8.0 J. Find the masses m and the spring constant k.

95 ••• An object of mass 2 kg resting on a frictionless horizontal surface is attached to a spring of force constant 600 N/m. A second object of mass 1 kg slides along the surface toward the first object at 6 m/s. (*a*) Find the amplitude of oscillation if the objects make a perfectly inelastic collision and remain together on the spring. What is the period of oscillation? (*b*) Find the amplitude and period of oscillation if the collision is elastic. (*c*) For each type of collision, write an expression for the position x as a function of time t for the object attached to the spring, assuming that the collision occurs at time $t = 0$.

General Problems

96 • A particle has a displacement $x = 0.4\cos(3t + \pi/4)$, where x is in meters and t is in seconds. (*a*) Find the frequency f and period T of the motion. (*b*) Where is the particle at $t = 0$? (*c*) Where is the particle at $t = 0.5$ s?

97 • (*a*) Find an expression for the velocity of the particle whose position is given in Problem 96. (*b*) What is the velocity at time $t = 0$? (*c*) What is the maximum velocity? (*d*) At what time after $t = 0$ does this maximum velocity first occur?

98 • An object on a horizontal spring oscillates with a period of 4.5 s. If the object is suspended from the spring vertically, by how much is the spring stretched from its natural length when the object is in equilibrium?

99 •• SSM A small particle of mass m slides without friction in a spherical bowl of radius r. (*a*) Show that the motion of the particle is the same as if it were attached to a string of length r. (*b*) Figure 14-36 shows a particle of mass m_1 that is displaced a small distance s_1 from the bottom of the bowl, where s_1 is much smaller than r. A second particle of mass m_2 is displaced in the opposite direction a distance $s_2 = 3s_1$, where s_2 is also much smaller than r. If the particles are released at the same time, where do they meet? Explain.

FIGURE 14-36 Problems 99, 100

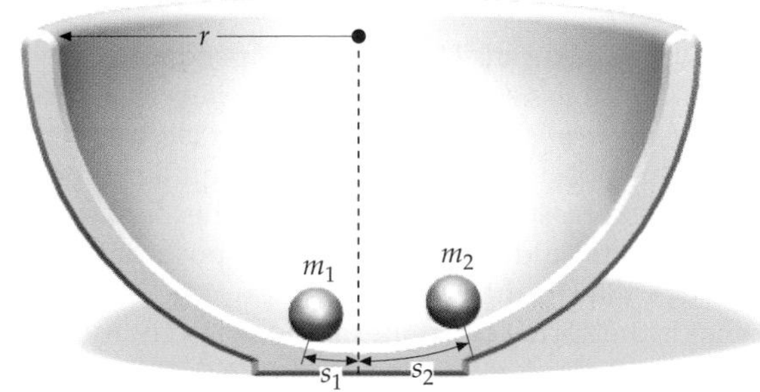

100 •• Now consider a very small uniform ball of mass m and radius R rolling without slipping near the bottom of the bowl in Figure 14-36. (*a*) Write an expression for the total energy of the ball in terms of its velocity and the distance (assumed small) from the center of the bowl. (*b*) By comparing this expression with that for the total energy of a frictionless ball of mass m sliding down the side of the bowl, determine the oscillation frequency of the ball about the center of the bowl.

101 •• ISOLVE✓ As your jet plane speeds down the runway on takeoff, you measure its acceleration by suspending your yo-yo as a simple pendulum and noting that when the bob (mass 40 g) is at rest relative to you, the string (length 70 cm) makes an angle of 22° with the vertical. Find the period T for small oscillations of this pendulum.

102 •• If a wire is twisted, there will be a restoring torque $\tau = -\kappa\theta$, where κ is a torsional spring constant and θ is the total twist angle. A torsion balance consists of an object with moment of inertia I hung at the end of a wire. If the object is given a twist, show that the frequency of small torsional oscillations is $\omega = \sqrt{\kappa/I}$.

103 •• A simple torsion balance (see Problem 102) used in a variety of physics experiments is shown in Figure 14-37. There is a cross-arm of negligible mass at the end of the wire with identical particles attached at each end. If each particle has a mass of 50 g, the length of the cross-arm is 5.0 cm, and the oscillation period of the balance is 80 s, what is the wire's torsion constant κ?

FIGURE 14-37 Problem 103

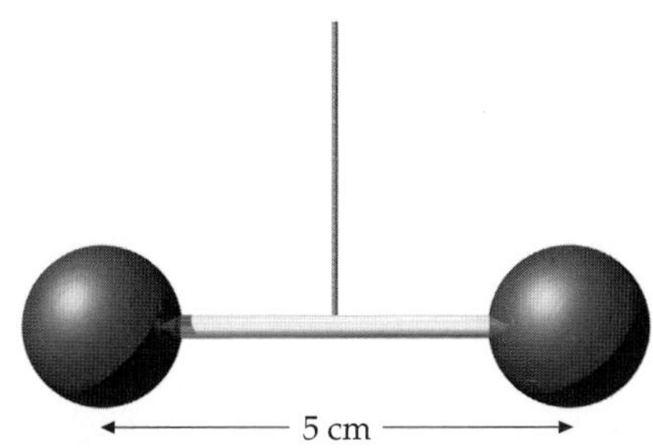

104 •• SSM ISOLVE A wooden cube with edge length a and mass m floats in water with one of its faces parallel to the water surface. The density of the water is ρ. Find the period of oscillation in the vertical direction if the cube is pushed down slightly.

105 •• A clock with a pendulum keeps perfect time on the earth's surface. In which case will the error be greater: if the clock is placed in a mine of depth h or if the clock is elevated to a height h? Assume that $h << R_E$.

106 •• ISOLVE Figure 14-38 shows a pendulum of length L with a bob of mass M. The bob is attached to a spring of spring constant k as shown. When the bob is directly below the pendulum support, the spring is at its equilibrium length. (*a*) Derive an expression for the period of this oscillating system for small amplitude vibrations. (*b*) Suppose that $M = 1$ kg and L is such that in the absence of the spring the period is 2.0 s. What is the spring constant k if the period of the oscillating system is 1.0 s?

FIGURE 14-38 Problem 106

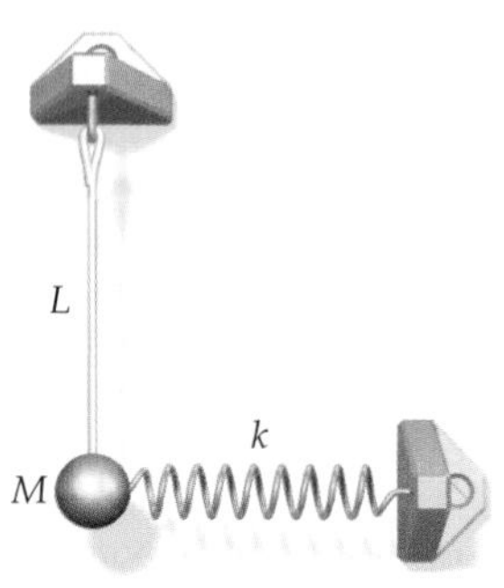

107 •• An object of mass m_1 sliding on a frictionless horizontal surface is attached to a spring of force constant k and oscillates with an amplitude A. When the spring is at its greatest extension and the mass is instantaneously at rest, a second object of mass m_2 is placed on top of it. (*a*) What is the smallest value for the coefficient of static friction μ_s such that the second object does not slip on the first? (*b*) Explain how the total energy E, the amplitude A, the angular frequency ω, and the period T of the system are changed by placing m_2 on m_1, assuming that the friction is great enough so that there is no slippage.

108 •• A box with a mass of 100 kg hangs from the ceiling of a room by a spring with a spring constant of 500 N/m. The uncompressed length of the spring is 0.5 m. (*a*) Find the equilibrium position of the box. (*b*) An identical spring is stretched and attached to the ceiling and box in parallel with the first spring. Find the frequency of the oscillations when the box is released. (*c*) What is the new equilibrium position of the box once it comes to rest?

109 •• The acceleration due to gravity g varies with geographical location because of the earth's rotation and because the earth is not exactly spherical. This was first discovered in the seventeenth century, when it was noted that a pendulum clock carefully adjusted to keep correct time in Paris lost about 90 s/d near the equator. (*a*) Show that a small change in the acceleration of gravity Δg produces a small change in the period ΔT of a pendulum given by $\Delta T/T \approx -\frac{1}{2}\Delta g/g$. (Use differentials to approximate ΔT and Δg.) (*b*) How great a change in g is needed to account for a change in the period of 90 s/d?

110 •• Figure 14-39 shows two 0.6-kg blocks glued to each other and connected to a spring of spring constant $k = 240$ N/m. The blocks, which rest on a frictionless horizontal surface, are displaced 0.6 m from their equilibrium position and released. Before they are released, a few drops of solvent are deposited on the glue. (*a*) Find the frequency of vibration and total energy of the vibrating system before the glue has dissolved. (*b*) Find the frequency, amplitude of vibration, and energy of the vibrating system if the glue dissolves when the spring is (1) at maximum compression and (2) at maximum extension.

FIGURE 14-39 Problem 110

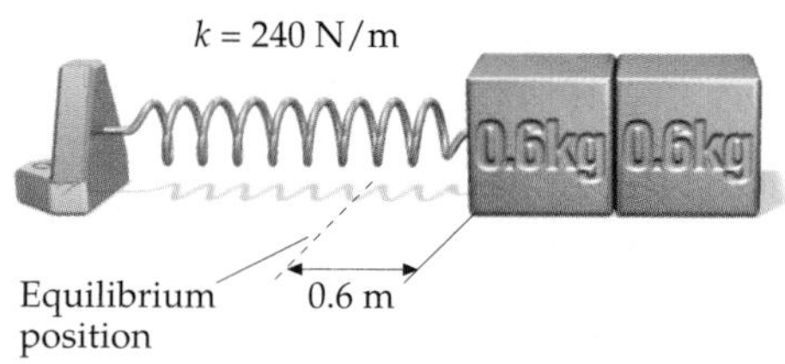

111 •• Show that for the situations in Figure 14-40(*a*) and (*b*), the object oscillates with a frequency $f = (1/2\pi)\sqrt{k_{eff}/m}$, where k_{eff} is given by (*a*) $k_{eff} = k_1 + k_2$ and (*b*) $1/k_{eff} = 1/k_1 + 1/k_2$. *Hint: Find the net force* F *on the object for a small displacement x and write* $F = -k_{eff}x$. *Note that in (b) the springs stretch by different amounts, the sum of which is x.*

FIGURE 14-40 Problem 111

(*a*)

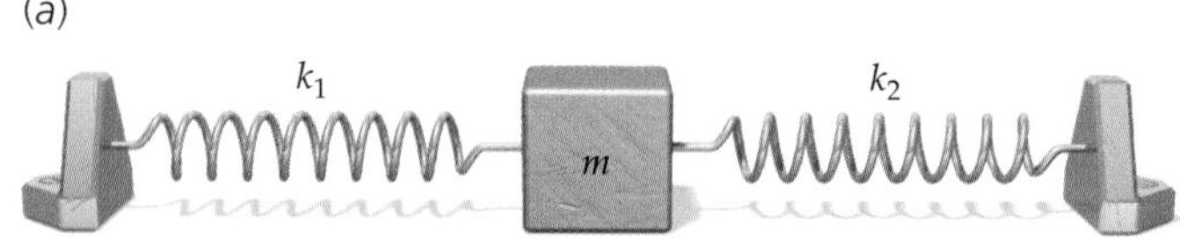

(*b*)

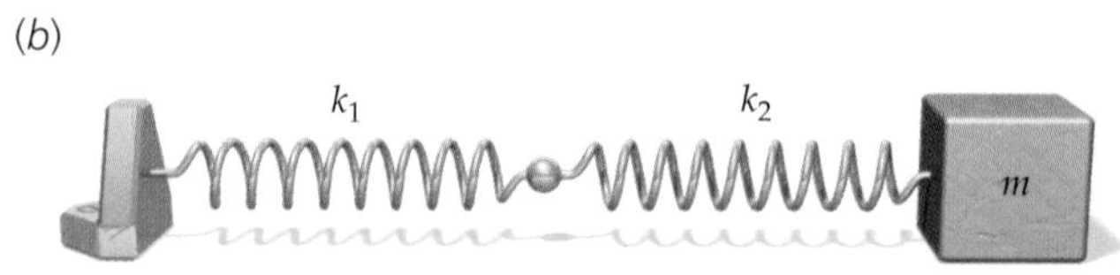

112 •• SSM A small block of mass m_1 rests on a piston that is vibrating vertically with simple harmonic motion given by $y = A \sin \omega t$. (*a*) Show that the block will leave the piston if $\omega^2 A > g$. (*b*) If $\omega^2 A = 3g$ and $A = 15$ cm, at what time will the block leave the piston?

113 •• ISOLVE The plunger of a pinball machine has mass m_p and is attached to a spring of force constant k (Figure 14-41). The spring is compressed a distance x_0 from its equilibrium position $x = 0$ and released. A ball of mass m_b is next to the plunger. (*a*) Where does the ball leave the plunger? (*b*) What is the speed v_s of the ball when it separates? (*c*) At what distance x_f does the plunger come to rest momentarily? (Assume that the surface is horizontal and frictionless so that the ball slides rather than rolls.)

FIGURE 14-41 Problem 113

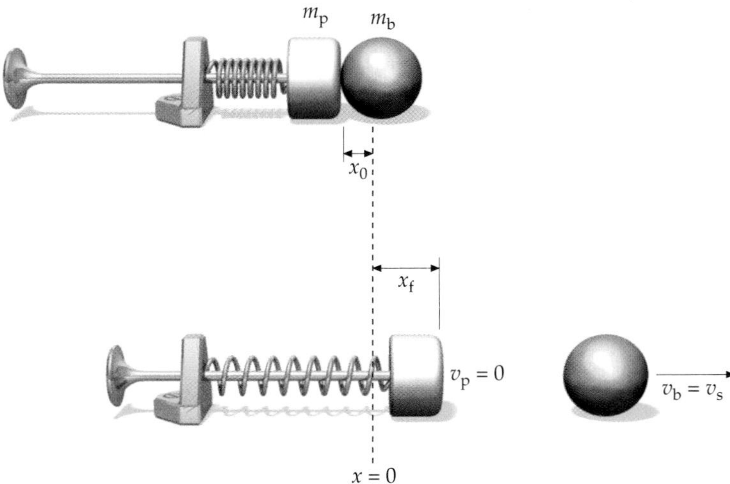

114 •• A level platform vibrates horizontally with simple harmonic motion with a period of 0.8 s. (*a*) A box on the platform starts to slide when the amplitude of vibration reaches 40 cm; what is the coefficient of static friction between the box and the platform? (*b*) If the coefficient of friction between the box and platform were 0.40, what would be the maximum amplitude of vibration before the box would slip?

115 ••• The potential energy of a particle of mass m as a function of position is given by $U(x) = U_0(\alpha + 1/\alpha)$, where $\alpha = x/a$ and a is a constant. (*a*) Plot $U(x)$ versus x for $0.1a < x < 3a$. (*b*) Find the value of $x = x_0$ at stable equilibrium. (*c*) Write the potential energy $U(x)$ for $x = x_0 + \varepsilon$, where ε is a small displacement from the equilibrium position x_0. (*d*) Approximate the $1/x$ term using the binomial expansion

$$(1 + r)^n = 1 + nr + \frac{n(n-1)}{(2)(1)}r^2 + \frac{n(n-1)(n-2)}{(3)(2)(1)}r^3 + \cdots,$$

with $r = \varepsilon/x_0 << 1$ and discarding all terms of power greater than r^2. (*e*) Compare your result with the potential for a simple harmonic oscillator. Show that the mass will undergo simple harmonic motion for small displacements from equilibrium and determine the frequency of this motion.

116 ••• A solid cylindrical drum of mass 6.0 kg and diameter 0.06 m rolls without slipping on a horizontal surface (Figure 14-42). The axle of the drum is attached to a spring of spring constant $k = 4000$ N/m as shown. (*a*) Determine the frequency of oscillation of this system for small displacements from equilibrium. (*b*) What is the minimum value of the coefficient of static friction so that the drum will not slip when the vibrational energy is 5.0 J?

FIGURE 14-42 Problem 116

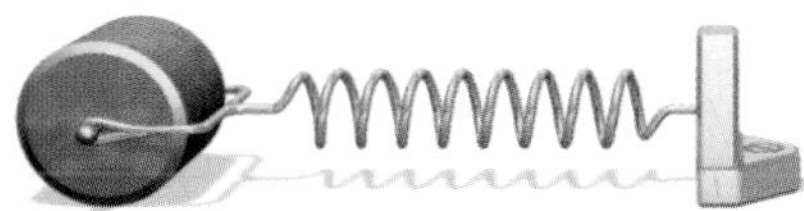

117 ••• SSM If we attach two blocks of masses m_1 and m_2, to either end of a spring of spring constant k and set them into oscillation, show that the oscillation frequency $\omega = (k/\mu)^{1/2}$, where $\mu = m_1m_2/(m_1 + m_2)$ is the reduced mass of the system.

118 •• One of the vibrational modes of the HCl molecule has a frequency of $8.969 \times 10^{13}\ \text{s}^{-1}$. Using the relation stated in Problem 117, find the "spring constant" for the HCl molecule.

119 •• In Problem 118, if we were to replace the hydrogen atom in HCl by a deuterium atom, what would be the new vibration frequency of the molecule? Deuterium consists of 1 proton and 1 neutron.

120 ••• A block of mass m on a horizontal table is attached to a spring of force constant k as shown in Figure 14-43. The coefficient of kinetic friction between the block and the table is μ_k. The spring is unstretched if the block is at the origin ($x = 0$). The block is released with the spring stretched a distance A, where $kA > \mu_k mg$. (*a*) Apply Newton's second law to the block to obtain an equation for its acceleration d^2x/dt^2 for the first half-cycle, during which the block is moving to the left. Show that the resulting equation can be written $d^2x'/dt^2 = -\omega^2x'$, where $\omega = \sqrt{k/m}$ and $x' = x - x_0$, with $x_0 = \mu_k mg/k = \mu_k g/\omega^2$. (*b*) Repeat Part (*a*) for the second half-cycle as the block moves to the right, and show that $d^2x''/dt^2 = -\omega^2x''$, where $x'' = x + x_0$ and x_0 has the same value. (*c*) Use a spreadsheet program to graph the first 5 half-cycles for $A = 10x_0$. Describe the motion, if any, after the fifth half-cycle.

FIGURE 14-43
Problem 120

121 ••• Figure 14-44 shows a uniform solid half-cylinder of mass M and radius R resting on a horizontal surface. If one side of this cylinder is pushed down slightly and then released, the object will oscillate about its equilibrium position. Determine the period of this oscillation.

FIGURE 14-44
Problem 121

122 ••• SSM A straight tunnel is dug through the earth as shown in Figure 14-45. Assume that the walls of the tunnel are frictionless. (*a*) The gravitational force exerted by the earth on a particle of mass m at a distance r from the center of the earth when $r < R_E$ is $F_r = -(GmM_E/R_E^3)r$, where M_E is the mass of the earth and R_E is its radius. Show that the net force on a particle of mass m at a distance x from the middle of the tunnel is given by $F_x = -(GmM_E/R_E^3)x$, and that the motion of the particle is therefore simple harmonic motion. (*b*) Show that the period of the motion is given by $T = 2\pi\sqrt{R_E/g}$ and find its value in minutes. (This is the same period as that of a satellite orbiting near the surface of the earth and is independent of the length of the tunnel.)

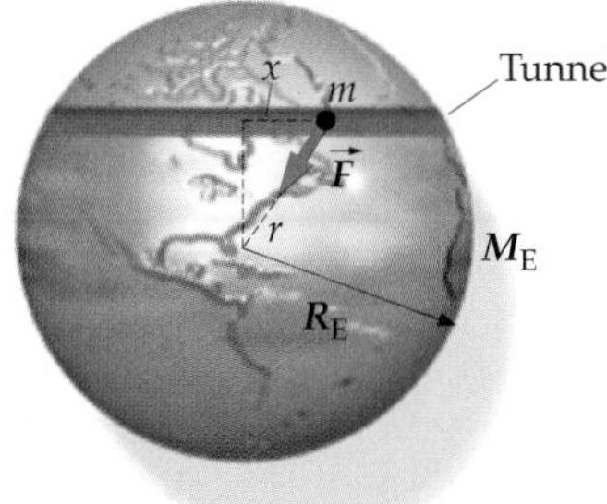

FIGURE 14-45
Problem 122

123 ••• A damped oscillator has a frequency ω' that is 10 percent less than its undamped frequency. (*a*) By what factor is the amplitude of the oscillator decreased during each oscillation? (*b*) By what factor is its energy reduced during each oscillation?

124 ••• Show by direct substitution that Equation 14-52 is a solution of Equation 14-51.

125 ••• SSM In this problem, you will derive the expression for the average power delivered by a driving force to a driven oscillator (Figure 14-24, page 450).

(*a*) Show that the instantaneous power input of the driving force is given by

$$P = Fv = -A\omega F_0 \cos \omega t \sin(\omega t - \delta).$$

(*b*) Use the trigonometric identity $\sin(\theta_1 - \theta_2) = \sin\theta_1 \cos\theta_2 - \cos\theta_1 \sin\theta_2$ to show that the equation in (*a*) can be written

$$P = A\omega F_0 \sin\delta \cos^2 \omega t - A\omega F_0 \cos\delta \cos\omega t \sin\omega t.$$

(*c*) Show that the average value of the second term in your result for (*b*) over one or more periods is zero and that therefore

$$P_{av} = \tfrac{1}{2} A\omega F_0 \sin\delta.$$

(*d*) From Equation 14-54 for $\tan\delta$, construct a right triangle in which the side opposite the angle δ is $b\omega$ and the side adjacent is $m(\omega_0^2 - \omega^2)$, and use this triangle to show that

$$\sin\delta = \frac{b\omega}{\sqrt{m^2(\omega_0^2 - \omega^2)^2 + b^2\omega^2}} = \frac{b\omega A}{F_0}$$

(*e*) Use your result for (*d*) to eliminate ωA from your result for (*c*) so that the average power input can be written

$$P_{av} = \frac{1}{2}\frac{F_0^2}{b}\sin^2\delta = \frac{1}{2}\left[\frac{b\omega^2 F_0^2}{m^2(\omega_0^2 - \omega^2)^2 + b^2\omega^2}\right] \qquad \text{14-55}$$

126 ••• In this problem, you are to use the result of Problem 125 to derive Equation 14-49, which relates the width of the resonance curve to the Q value when the resonance is sharp. At resonance, the denominator of the fraction in brackets in Equation 14-55 is $b^2\omega_0^2$ and P_{av} has its maximum value. (Equation 14-55 can be found in Problem 125.) For a sharp resonance, the variation in ω in the numerator in Equation 14-55 can be neglected. Then the power input will be half its maximum value at the values of ω, for which the denominator is $2b^2\omega_0^2$.

(*a*) Show that ω then satisfies $m^2(\omega - \omega_0)^2(\omega + \omega_0)^2 \approx b^2\omega_0^2$.
(*b*) Using the approximation $\omega + \omega_0 \approx 2\omega_0$, show that $\omega - \omega_0 \approx \pm b/2m$.
(*c*) Express b in terms of Q.
(*d*) Combine the results of (*b*) and (*c*) to show that there are two values of ω for which the power input is half that at resonance and that they are given by

$$\omega_1 = \omega_0 - \frac{\omega_0}{2Q} \quad \text{and} \quad \omega_2 = \omega_0 + \frac{\omega_0}{2Q}$$

Therefore, $\omega_2 - \omega_1 = \Delta\omega = \omega_0/Q$, which is equivalent to Equation 14-49.

127 ••• The Morse potential, which is often used to model interatomic forces, can be written in the form $U(r) = D(1 - e^{-\beta(r-r_0)})^2$, where r is the distance between the two atomic nuclei. (*a*) Using a spreadsheet program or graphing calculator, make a graph of the Morse potential using $D = 5$ eV, $\beta = 0.2\ \text{nm}^{-1}$, and $r_0 = 0.75$ nm. (*b*) Determine the equilibrium separation and "spring constant" for small displacements from equilibrium for the Morse potential. (*c*) Determine a formula for the oscillation frequency for a homonuclear diatomic molecule (that is, two of the same atoms), where the atoms have mass m.

CHAPTER 15

Traveling Waves

THIS POLICE RADAR UNIT SENDS OUT ELECTROMAGNETIC WAVES THAT TRAVEL AT THE SPEED OF LIGHT AND REFLECT FROM THE MOVING CAR.

? How does the police officer determine the speed of the car? (See Example 15-12.)

Waves transport energy and momentum through space without transporting matter. As a water wave moves across a pond, for example, the molecules of water oscillate up and down, but do not cross the pond with the wave. Energy and momentum are transported by the wave, but matter is not. A rowboat will bob up and down on the waves but will not be moved by them across the pond. Water waves, waves on a stretched guitar string, and sound waves all involve oscillation.

➢ **In this chapter we continue the study of oscillatory motion that we began in Chapter 14 by examining periodic waves, particularly harmonic waves. We will see that mechanical waves occur when there is a disturbance in a medium, such as air or water, while electromagnetic waves exist without a material medium.**

15-1 Simple Wave Motion

Transverse and Longitudinal Waves

A mechanical wave is caused by a disturbance in a medium. For example, when a taut string is plucked, the pulses produced travel down the string as waves. The disturbance in this case is the change in shape of the string from its equilibrium shape. Its propagation arises from the interaction of each string segment with the adjacent segments. The segments of the string (the medium) move in the direction perpendicular to the string as the pulses propagate up and down the string. Waves such as these, in which the disturbance is perpendicular to the direction of propagation, are called **transverse** (Figure 15-1). Waves in which the disturbance is parallel to the direction of propagation are called **longitudinal** (Figure 15-2). Sound waves are examples of longitudinal waves—the molecules of a gas, liquid, or solid, through which sound travels oscillate (move back and forth) along the line of propagation, alternately compressing and rarefying (expanding) the medium.

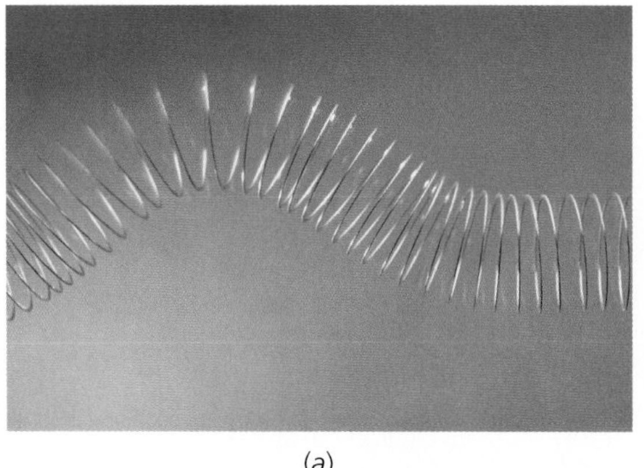

(a)

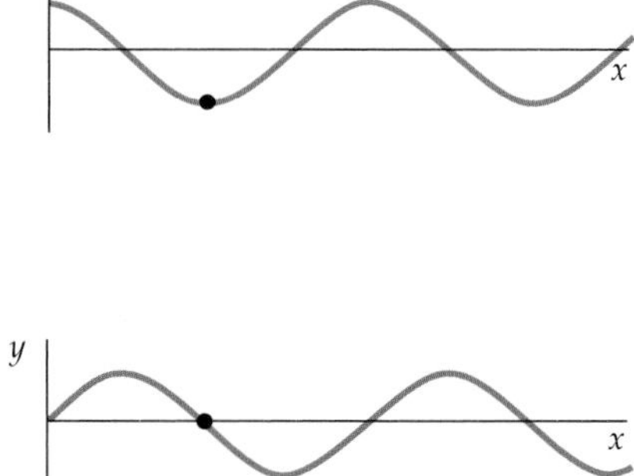

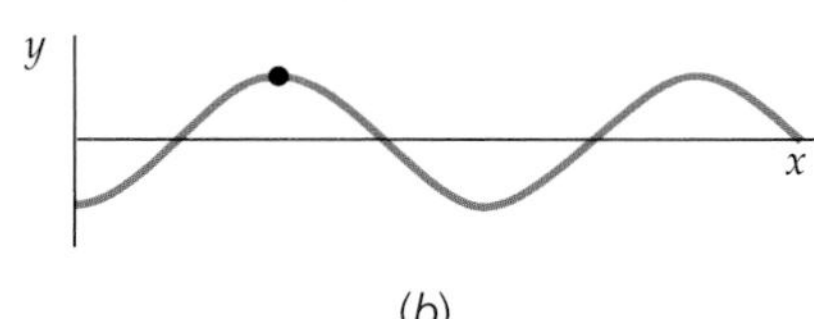

(b)

FIGURE 15-1 (*a*) Transverse wave pulse on a spring. The disturbance is perpendicular to the direction of the motion of the wave. (*b*) Three successive drawings of a transverse wave on a string traveling to the right. An element of the string moves up and down.

FIGURE 15-2 Longitudinal wave pulse on a spring. The disturbance is in the direction of the motion of the wave.

Wave Pulses

Figure 15-3*a* shows a pulse on a string at time $t = 0$. The shape of the string at this time can be represented by some function $y = f(x)$. At some later time (Figure 15-3*b*), the pulse is farther down the string. In a new coordinate system with origin O' that moves with the speed of the pulse, the pulse is stationary. The string is described in this frame by $f(x')$ for all times. The x coordinates of the two reference frames are related by

$$x' = x - vt$$

so $f(x') = f(x - vt)$.

Thus, the shape of the string in the original frame is

$$y = f(x - vt), \qquad \text{wave moving in the positive } x \text{ direction} \tag{15-1}$$

The same line of reasoning for a pulse moving to the left leads to

$$y = f(x + vt), \qquad \text{wave moving in the negative } x \text{ direction} \tag{15-2}$$

In both expressions, v is the speed of propagation of the wave. The function $y = f(x - vt)$ is called a **wave function.** For waves on a string, the wave function represents the transverse displacement of the string. For sound waves in air, the wave function can be the longitudinal displacement of the air molecules, or the pressure of the air. These wave functions are solutions of a differential equation called the wave equation, which can be derived from Newton's laws.

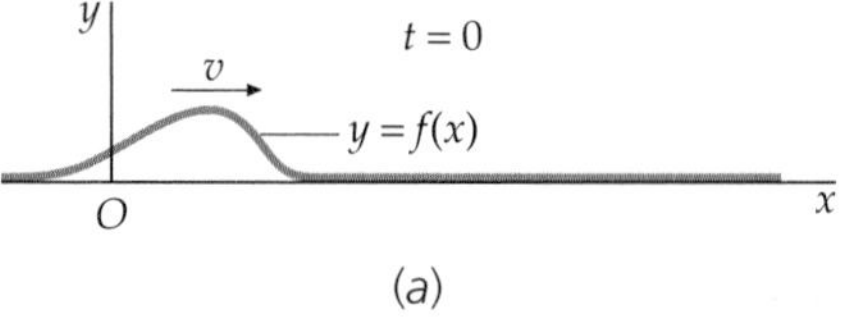

(a)

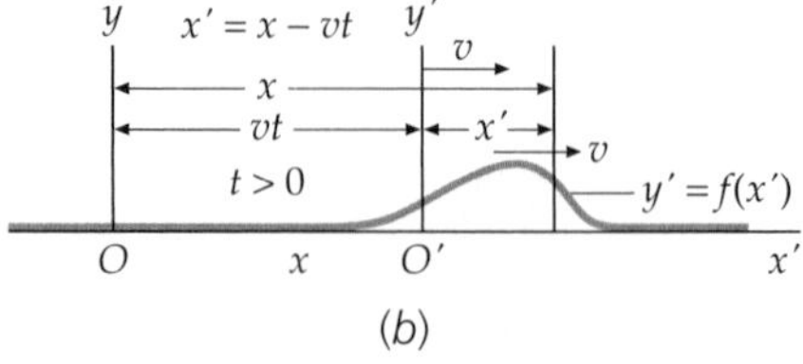

(b)

FIGURE 15-3

Speed of Waves

A general property of waves is that their speed relative to the medium depends on the properties of the medium but is independent of the motion of the source of the waves. For example, the speed of a sound from a car horn depends only on the properties of air and not on the motion of the car.

For wave pulses on a rope, we can easily demonstrate that the greater the tension, the faster the propagation of the waves. Furthermore, waves propagate faster in a light rope than in a heavy rope under the same tension. If F_T is the tension (we use F_T for tension because we use T for the period) and μ is the linear mass density (mass per unit length), then the wave speed is

$$v = \sqrt{\frac{F_T}{\mu}} \qquad 15\text{-}3$$

SPEED OF WAVES ON A STRING

INCHY RUNS FOR HIS LIFE

EXAMPLE 15-1 Put It in Context

Inchy, an inchworm, is inching along a cotton clothesline. The 25-m-long clothesline has a mass of 0.25 kg and is kept taut by a hanging object of mass 10 kg as shown in Figure 15-4. Vivian is hanging up her swimsuit 5 m from one end when she sees Inchy 2.5 cm from the opposite end. She plucks the line sending a terrifying 3-cm-high pulse toward Inchy. If Inchy crawls at 1 in./s, will he get to the end of the clothesline before the pulse reaches him?

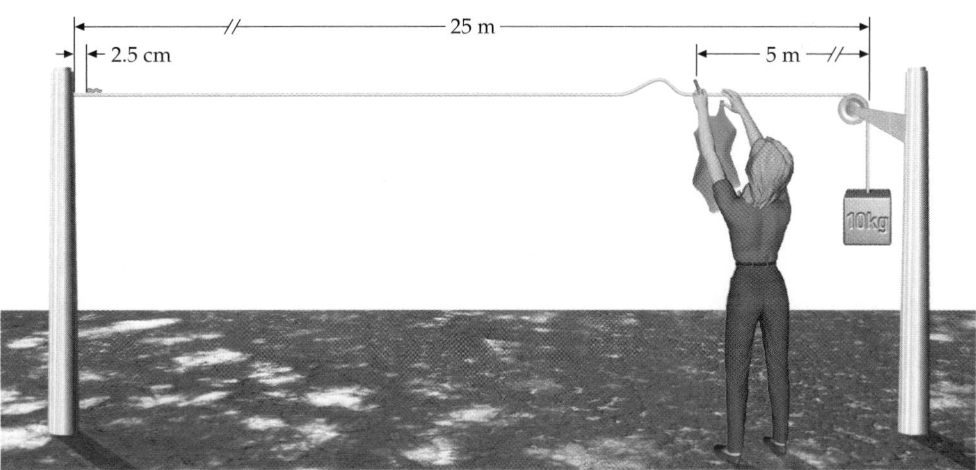

FIGURE 15-4

PICTURE THE PROBLEM We need to know how fast the wave travels. To find the wave speed we use the formula $v = \sqrt{F_T/\mu}$.

1. The speed of the pulse is related to the tension F_T and mass density μ:
$$v = \sqrt{\frac{F_T}{\mu}}$$

2. Express the mass density and tension in terms of the given parameters:
$$\mu = \frac{m_s}{L} \quad \text{and} \quad F_T = mg$$

3. Substitute these values to calculate the speed:
$$v = \sqrt{\frac{F_T}{\mu}} = \sqrt{\frac{mgL}{m_s}} = \sqrt{\frac{(10\text{ kg})(9.81\text{ m/s}^2)(25\text{ m})}{0.25\text{ kg}}}$$
$$= 99.0\text{ m/s}$$

4. Use this speed to find the time for the pulse to travel the 20 m to the far end.
$$\Delta t = \frac{\Delta x}{v} = \frac{20\text{ m}}{99.0\text{ s}} = 0.202\text{ s}$$

5. Find the time it takes Inchy to travel the 2.5 cm to the end traveling at 1 in./s:
$$\Delta t' = \frac{\Delta x'}{v'} = \frac{2.5\text{ cm}}{1\text{ in./s}} \times \frac{1\text{ in.}}{2.54\text{ cm}} = 0.984\text{ s}$$

Inchy does not beat the pulse.

EXERCISE If the 10-kg mass is replaced with a 20-kg mass, what is the speed of waves on the clothesline? (*Answer* 140 m/s)

EXERCISE Show that the units of $\sqrt{F_T/\mu}$ are m/s when F_T is in newtons and μ is in kg/m.

For sound waves in a fluid such as air or water, the speed v is given by

$$v = \sqrt{\frac{B}{\rho}} \qquad 15\text{-}4$$

where ρ is the equilibrium density of the medium and B is the bulk modulus[†] (Equation 13-6). Comparing Equations 15-3 and 15-4, we can see that, in general, the speed of waves depends on an elastic property of the medium (the tension for string waves and the bulk modulus for sound waves) and on an inertial property of the medium (the linear mass density or the volume mass density).

For sound waves in a gas such as air, the bulk modulus[‡] is proportional to the pressure, which in turn is proportional to the density ρ and to the absolute temperature T of the gas. The ratio B/ρ is thus independent of density and is merely proportional to the absolute temperature T. In Chapter 17 we will show that, in this case, Equation 15-4 is equivalent to

$$v = \sqrt{\frac{\gamma RT}{M}} \qquad 15\text{-}5$$

SPEED OF SOUND IN A GAS

In this equation T is the absolute temperature measured in kelvins (K), which is related to the Celsius temperature t_C by

$$T = t_C + 273 \qquad 15\text{-}6$$

The constant γ depends on the kind of gas. For diatomic molecules, such as O_2 and N_2, γ has the value 1.4, and since O_2 and N_2 comprise 98 percent of the atmosphere, that is the value for air. (For monatomic molecules such as He, γ has the value 1.67.) The constant R is the universal gas constant

$$R = 8.314 \text{ J/(mol}\cdot\text{K)} \qquad 15\text{-}7$$

and M is the molar mass of the gas (that is, the mass of 1 mol of the gas), which for air is

$$M = 29 \times 10^{-3} \text{ kg/mol}$$

---

† The bulk modulus is the negative ratio of the pressure change to the fractional change in volume (Chapter 13):

$$B = -\frac{\Delta P}{\Delta V/V}$$

‡ The **isothermal bulk modulus,** which describes changes that occur at constant temperature, differs from the **adiabatic bulk modulus,** which describes changes with no heat transfer. For sound waves at audible frequencies the changes occur too rapidly for appreciable heat flow, so the appropriate bulk modulus is the adiabatic bulk modulus.

SPEED OF SOUND IN AIR **EXAMPLE 15-2** Try It Yourself

Calculate the speed of sound in air at (*a*) 0°C and (*b*) 20°C.

Cover the column to the right and try these on your own before looking at the answers.

Steps	**Answers**
(*a*) 1. Write Equation 15-5.	$v_a = \sqrt{\frac{\gamma R T_a}{M}}$
2. Enter the given values into the equation and solve for the speed. (Be sure to convert the temperature to kelvins.)	$v_a =$ 331 m/s
(*b*) 1. From Equation 15-5 we can see that v is proportional to $\sqrt{T}$. Use this to express the ratio of the speed at 293 K to the speed at 273 K.	$\frac{v_b}{v_a} = \sqrt{\frac{T_b}{T_a}}$
2. Calculate v at 293 K.	$v_b =$ 343 m/s

REMARKS We see from this example that the speed of sound in air is about 340 m/s at normal temperatures.

EXERCISE For helium, $M = 4 \times 10^{-3}$ kg/mol and $\gamma = 1.67$. What is the speed of sound waves in helium at 20°C? (*Answer* 1.01 km/s)

Derivation of v for Waves on a String Equation 15-3 can be obtained from Newton's laws. Consider a pulse traveling along a string with a speed v to the right (Figure 15-5*a*). If the amplitude of the pulse is small compared to the length of the string, then the tension F_T will be approximately constant along the string. In a reference frame moving with speed v to the right, the pulse is stationary and the string moves with a speed v to the left. Figure 15-5*b* shows a small segment of the string of length Δs at the top of the pulse. The segment forms part of a circular arc of radius R. Instantaneously it is moving with speed v in a circular path, so it has an acceleration v^2/R in the centripetal direction. The magnitudes of the forces acting on the segment are the tension F_T at each end. The horizontal components of these forces are equal and opposite and thus cancel. The vertical components of these forces point radially inward toward the center of the circular arc for sufficiently small Δs. These radial forces provide the centripetal acceleration.

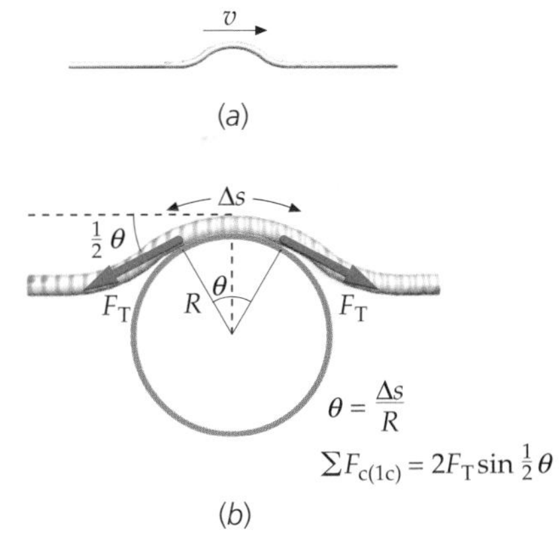

FIGURE 15-5 (*a*) Wave pulse moving with a speed v along a string. (*b*) In a frame in which the wave pulse of (*a*) is at rest, the string is moving with a speed v to the left. A small segment of the string of length Δs is moving in a circular arc of radius R. The centripetal acceleration of the segment is provided by the radial components of the tension.

Let the angle subtended by the string be θ. The centripetal component of the net force on the segment is

$$\sum F_c = 2F_T \sin\tfrac{1}{2}\theta \approx 2F_T(\tfrac{1}{2}\theta) = F_T\theta$$

where we have used the small angle approximation $\sin\frac{1}{2}\theta \approx \frac{1}{2}\theta$. If μ is the mass per unit length of the string, the mass of a segment of length Δs is $m = \mu\,\Delta s$. The angle θ is related to Δs by

$$\theta = \frac{\Delta s}{R}$$

The mass of the element is thus

$$m = \mu\,\Delta s = \mu R\theta$$

Newton's second law ($\Sigma F_c = ma_c$) gives

$$F_T\theta = \mu R\theta\frac{v^2}{R}$$

Solving for v, we obtain $v = \sqrt{F_T/\mu}$.

In the original frame, the string is fixed, and the pulse moves with speed $v = \sqrt{F_T/\mu}$, which is Equation 15-3. Since v is independent of R and θ, this result holds for the peak of any pulse. In the following discussion we will show that this result is true not only for the peak but for other parts of the pulse.

*The Wave Equation

We can apply Newton's laws to a segment of the string to derive a differential equation known as the wave equation, which relates the spatial derivatives of $y(x,t)$ to its time derivatives. Figure 15-6 shows one segment of a string. We consider only small angles θ_1 and θ_2. Then the length of the segment is approximately Δx and its mass is $m = \mu\,\Delta x$, where μ is the string's mass per unit length. First we will show that, for small vertical displacements, the net horizontal force on a segment is zero and the tension is uniform and constant. The net force in the horizontal direction is zero. That is,

$$\sum F_x = F_{T2}\cos\theta_2 - F_{T1}\cos\theta_1 = 0$$

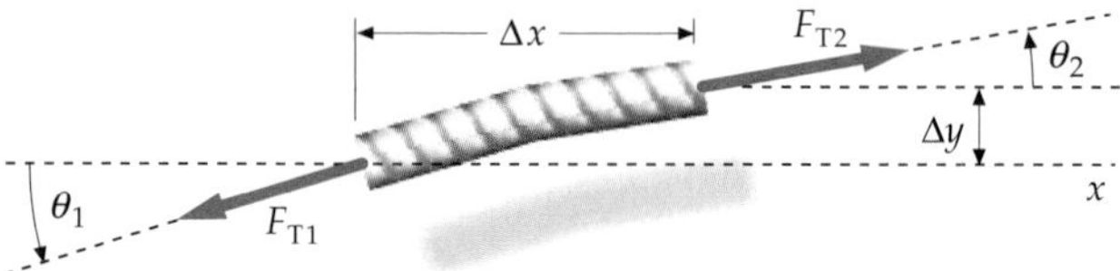

FIGURE 15-6 Segment of a stretched string used for the derivation of the wave equation. The net vertical force on the segment is $F_{T2}\sin\theta_2 - F_{T1}\sin\theta_1$, where F is the tension in the string. The wave equation is derived by applying Newton's second law to the segment.

where θ_2 and θ_1 are the angles shown and F_T is the tension in the string. Since the angles are assumed to be small, we may approximate $\cos\theta$ by 1 for each angle. Then the net horizontal force on the segment can be written

$$\sum F_x = F_{T2} - F_{T1} = 0$$

Thus

$$F_{T2} = F_{T1} = F_T$$

The segment moves vertically, and the net force in this direction is

$$\sum F_y = F_T\sin\theta_2 - F_T\sin\theta_1$$

Since the angles are assumed to be small, we may approximate $\sin\theta$ by $\tan\theta$ for each angle. Then the net vertical force on the string segment can be written

$$\sum F_y = F_T(\sin\theta_2 - \sin\theta_1) \approx F_T(\tan\theta_2 - \tan\theta_1)$$

The tangent of the angle made by the string with the horizontal is the slope of the curve formed by the string. The slope S is the first derivative of $y(x,t)$ with respect to x for constant t. A derivative of a function of two variables with respect to one of the variables with the other held constant is called a **partial derivative.** The partial derivative of y with respect to x is written $\partial y/\partial x$. Thus we have

$$S = \tan\theta = \frac{\partial y}{\partial x}$$

Then

$$\sum F_y = F_T(S_2 - S_1) = F_T\,\Delta S$$

where S_1 and S_2 are the slopes of either end of the string segment and ΔS is the change in the slope. Setting this net force equal to the mass $\mu\,\Delta x$ times the acceleration $\partial^2 y/\partial t^2$ gives

$$F_T\,\Delta S = \mu\,\Delta x\,\frac{\partial^2 y}{\partial t^2}$$

or

$$F_T\,\frac{\Delta S}{\Delta x} = \mu\,\frac{\partial^2 y}{\partial t^2} \qquad \text{15-8}$$

In the limit $\Delta x \to 0$, we have

$$\lim_{\Delta x \to 0} \frac{\Delta S}{\Delta x} = \frac{\partial S}{\partial x} = \frac{\partial}{\partial x}\frac{\partial y}{\partial x} = \frac{\partial^2 y}{\partial x^2}$$

Thus Equation 15-8 becomes

$$\frac{\partial^2 y}{\partial x^2} = \frac{\mu}{F_T}\frac{\partial^2 y}{\partial t^2} \qquad \text{15-9}a$$

Equation 15-9*a* is the **wave equation** for a stretched string.

We now show that the wave equation is satisfied by any function $x - vt$. Let $\alpha = x - vt$ and consider any wave function

$$y = y(x - vt) = y(\alpha)$$

We will use y' for the derivative of y with respect to α. Then, by the chain rule for derivatives,

$$\frac{\partial y}{\partial x} = \frac{\partial y}{\partial \alpha}\frac{\partial \alpha}{\partial x} = y'\,\frac{\partial \alpha}{\partial x}$$

and

$$\frac{\partial y}{\partial t} = \frac{\partial y}{\partial \alpha}\frac{\partial \alpha}{\partial t} = y'\,\frac{\partial \alpha}{\partial t}$$

Since

$$\frac{\partial \alpha}{\partial x} = \frac{\partial(x - vt)}{\partial x} = 1 \qquad \text{and} \qquad \frac{\partial \alpha}{\partial t} = \frac{\partial(x - vt)}{\partial t} = -v$$

we have

$$\frac{\partial y}{\partial x} = y' \qquad \text{and} \qquad \frac{\partial y}{\partial t} = -vy'$$

Taking the second derivatives, we obtain

$$\frac{\partial^2 y}{\partial x^2} = y''$$

and

$$\frac{\partial^2 y}{\partial t^2} = -v\frac{\partial y'}{\partial t} = -v\frac{\partial y'}{\partial \alpha}\frac{\partial \alpha}{\partial t} = +v^2 y''$$

Thus

$$\frac{\partial^2 y}{\partial x^2} = \frac{1}{v^2}\frac{\partial^2 y}{\partial t^2} \qquad \text{15-9}b$$

WAVE EQUATION

The same result can be obtained for any function of $x + vt$. Comparing Equations 15-9*a* and 15-9*b*, we see that the speed of propagation of the wave is $v = \sqrt{F_T/\mu}$, which is Equation 15-3.

HARMONIC WAVE FUNCTION — **EXAMPLE 15-3**

In the following section harmonic waves are defined by the wave function $y(x,t) = A\sin(kx - \omega t)$, where $v = \omega/k$. Show by explicitly calculating the derivatives that this wave function satisfies Equation 15-9*b*.

1. Calculate the second derivative of y with respect to x:

$$\frac{\partial y}{\partial x} = \frac{\partial}{\partial x}[A\sin(kx - \omega t)] = A\cos(kx - \omega t)\frac{\partial(kx - \omega t)}{\partial x}$$

$$= kA\cos(kx - \omega t)$$

$$\frac{\partial^2 y}{\partial x^2} = \frac{\partial}{\partial x}\frac{\partial y}{\partial x} = \frac{\partial}{\partial x}kA\cos(kx - \omega t)$$

$$= -kA\sin(kx - \omega t)\frac{\partial(kx - \omega t)}{\partial x}$$

$$= -k^2 A\sin(kx - \omega t)$$

2. Similarly, the two partial derivatives with respect to t are:

$$\frac{\partial y}{\partial t} = \frac{\partial}{\partial t}[A\sin(kx - \omega t)] = A\cos(kx - \omega t)\frac{\partial(kx - \omega t)}{\partial t}$$

$$= -\omega A\cos(kx - \omega t)$$

$$\frac{\partial^2 y}{\partial t^2} = \omega A\sin(kx - \omega t)\frac{\partial(kx - \omega t)}{\partial t} = -\omega^2 A\sin(kx - \omega t)$$

3. Substituting these results in Equation 15-9*b* gives:

$$-k^2 A\sin(kx - \omega t) = \frac{1}{v^2}[-\omega^2 A\sin(kx - \omega t)]$$

or

$$A\sin(kx - \omega t) = \frac{\omega^2/k^2}{v^2}A\sin(kx - \omega t)$$

4. Substituting for k using $k = \omega/v$ gives:

$$A\sin(kx - \omega t) = \frac{v^2}{v^2}A\sin(kx - \omega t) = \boxed{A\sin(kx - \omega t)}$$

REMARKS We have shown that the function $y = A\sin(kx - \omega t)$ is a solution to the wave equation provided $v = \omega/k$.

EXERCISE Show that any function $y(x + vt)$ satisfies Equation 15-9*b*.

A wave equation for sound waves can also be derived using Newton's laws. In one dimension, this equation is

$$\frac{\partial^2 s}{\partial x^2} = \frac{1}{v_s^2} \frac{\partial^2 s}{\partial t^2}$$

where s is the displacement of the medium in the x direction and v_s is the speed of sound.

15-2 Periodic Waves

If one end of a long taut string is shaken back and forth in periodic motion, then a **periodic wave** is generated. If a periodic wave is traveling along a taut string or any other medium, each point along the medium oscillates with the same period.

Harmonic Waves

Harmonic waves are the most basic type of periodic waves. All waves, whether they are periodic or not, can be modeled as a superposition of harmonic waves. Consequently, an understanding of harmonic wave motion can be generalized to form an understanding of any type of wave motion. If a **harmonic wave** is traveling through a medium, each point of the medium oscillates in simple harmonic motion.

If one end of a string is attached to a vibrating tuning fork that is moving up and down with simple harmonic motion, a sinusoidal wave train propagates along the string. This wave train is a harmonic wave. As shown in Figure 15-7, the shape of the string is that of a sine function. The minimum distance after which the wave repeats (the distance between crests, for example) in this figure is called the **wavelength** λ.

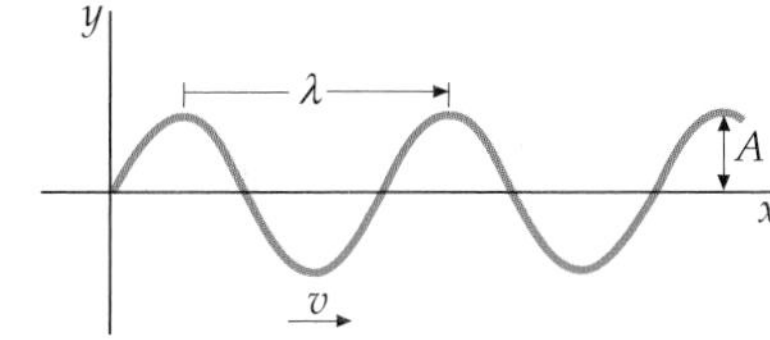

FIGURE 15-7 Harmonic wave at some instant in time. A is amplitude and λ is the wavelength. For a wave on a string, this figure can be obtained by taking a photographic snapshot of the string.

As the wave propagates along the string, each point on the string moves up and down—perpendicular to the direction of propagation—in simple harmonic motion with the frequency f of the tuning fork. During one period $T = 1/f$, the wave moves a distance of one wavelength, so its speed is given by

$$v = \frac{\lambda}{T} = f\lambda \qquad 15\text{-}10$$

Since this relation arises only from the definitions of wavelength and frequency, it applies to all harmonic waves.

The sine function that describes the displacements in Figure 15-7 is

$$y(x) = A \sin\left(2\pi \frac{x}{\lambda} + \delta\right)$$

where A is the amplitude, λ is the wavelength, and δ is a phase constant that depends on the choice of the origin $x = 0$. This equation is expressed more simply as

$$y(x) = A \sin(kx + \delta) \qquad 15\text{-}11$$

where k, called the **wave number,** is given by

$$k = \frac{2\pi}{\lambda} \qquad 15\text{-}12$$

Note that k has dimensions of m^{-1}. (Because the angle must be in radians, we sometimes write the units of k as rad/m.) When dealing with a single harmonic wave we usually choose the location of the origin so that $\delta = 0$.

For a wave traveling in the direction of increasing x with speed v, replace x in Equation 15-11 with $x - vt$ (see "Wave Pulses" in Section 15-1). With δ chosen to be zero, this gives

$$y(x,t) = A \sin k(x - vt) = A \sin(kx - kvt)$$

or

$$y(x,t) = A \sin(kx - \omega t) \qquad 15\text{-}13$$

HARMONIC WAVE FUNCTION

where

$$\omega = kv \qquad 15\text{-}14$$

is the angular frequency, and the argument of the sine function, $(kx - wt)$, is called the **phase.** The angular frequency is related to the frequency f and period T by

$$\omega = 2\pi f = \frac{2\pi}{T} \qquad 15\text{-}15$$

Substituting $\omega = 2\pi f$ into Equation 15-14 and using $k = 2\pi/\lambda$, we obtain

$$2\pi f = kv = \frac{2\pi}{\lambda} v$$

or $v = f\lambda$, which is Equation 15-10.

If a harmonic wave traveling along a string is described by $y(x,t) = A \sin(kx - \omega t)$, the velocity of a point on the string at a fixed value of x is

$$v_y = \frac{\partial y}{\partial t} = \frac{\partial}{\partial t}[A \sin(kx - \omega t)] = -\omega A \cos(kx - \omega t) \qquad 15\text{-}16$$

TRANSVERSE VELOCITY

The acceleration of this point is given by $\partial^2 y/\partial t^2$.

A HARMONIC WAVE ON A STRING — EXAMPLE 15-4

The wave function for a harmonic wave on a string is $y(x,t) = (0.03\text{ m}) \times \sin[(2.2\text{ m}^{-1})x - (3.5\text{ s}^{-1})t]$. (*a*) In what direction does this wave travel and what is its speed? (*b*) Find the wavelength, frequency, and period of this wave. (*c*) What is the maximum displacement of any string segment? (*d*) What is the maximum speed of any short string segment?

PICTURE THE PROBLEM (*a*) To find the direction of travel, express $y(x,t)$ as either a function of $(x - vt)$ or as a function of $(x + vt)$ and use Equations 15-1 and 15-2. To find the wave speed use $\omega = kv$ (Equation 15-14). (*b*) The wavelength, frequency, and period can be found from the wave number k and the angular frequency ω. (*c*) The maximum displacement of a point on the string is the amplitude A. (*d*) The velocity of any short string segment is $\partial y/\partial t$.

(*a*) 1. The given wave function is of the form $y(x,t) = A\sin(kx - \omega t)$. Using $\omega = kv$ (Equation 15-14), write the wave function as a function of $x - vt$. Then use Equations 15-1 and 15-2 to find the direction of travel:

$y(x,t) = A\sin(kx - \omega t)$ and $\omega = kv$

so

$y(x,t) = A\sin(kx - kvt) = A\sin[k(x - vt)]$

The wave travels in the $\boxed{+x \text{ direction}}$

2. Since the form is $y = A\sin(kx - \omega t)$, we know A as well as both ω and k. Use these to calculate the speed:

$$v = \frac{\lambda}{T} = \frac{\lambda}{2\pi}\frac{2\pi}{T} = \frac{\omega}{k} = \frac{3.5\ \text{s}^{-1}}{2.2\ \text{m}^{-1}} = \boxed{1.59\ \text{m/s}}$$

(*b*) The wavelength λ is related to the wave number k, and the period T and frequency f are related to ω:

$$\lambda = \frac{2\pi}{k} = \frac{2\pi}{2.2\ \text{m}^{-1}} = \boxed{2.86\ \text{m}}$$

$$T = \frac{2\pi}{\omega} = \frac{2\pi}{3.5\ \text{s}^{-1}} = \boxed{1.80\ \text{s}}$$

$$f = \frac{1}{T} = \frac{1}{1.80\ \text{s}} = \boxed{0.557\ \text{Hz}}$$

(*c*) The maximum displacement of a string segment is the amplitude A:

$$A = \boxed{0.03\ \text{m}}$$

(*d*) 1. Compute $\partial y/\partial t$ to find the velocity of a point on the string:

$$v_y = \frac{\partial y}{\partial t} = (0.03\ \text{m})\frac{\partial[\sin(2.2\ \text{m}^{-1}x - 3.5\ \text{s}^{-1}t)]}{\partial t}$$
$$= (0.03\ \text{m})(-3.5\text{s}^{-1})\cos(2.2\ \text{m}^{-1}x - 3.5\ \text{s}^{-1}t)$$
$$= -(0.105\ \text{m/s})\cos(2.2\ \text{m}^{-1}x - 3.5\ \text{s}^{-1}t)$$

2. The maximum transverse speed occurs when the cosine function has the value of ± 1:

$$v_{y,\text{max}} = \boxed{0.105\ \text{m/s}}$$

REMARKS We have included the units explicitly to show how they work out. Often we will omit the units for simplicity.

Energy Transfer via Waves on a String Consider again a string attached to a tuning fork. As the fork vibrates, it transfers energy to the segment of the string attached to it. For example, as the fork moves through its equilibrium position, it stretches the adjacent string segment slightly, increasing its potential energy, and the fork imparts a transverse speed to it, increasing its kinetic energy. As a wave moves along the string, energy is transferred to the other segments of the string.

Power is the rate of energy transfer. We can calculate the power by considering work done by the force that one segment of the string exerts on a neighboring segment. The rate of work done by this force is the power. Figure 15-8 shows a harmonic wave moving to the right along a string segment. The tension force $\vec{F}_T$ on the left end of the segment is directed tangent to the string as shown. To calculate the power transferred by this force we use the formula $P = \vec{F}_T \cdot \vec{v}_t$ (Equation 6-16), where F_T is the tension and $\vec{v}_t$, the transverse velocity, is the velocity of the end of the segment. To obtain an expression for the power we first express the vectors in component form. That is, $\vec{F}_T = F_{Tx}\hat{i} + F_{Ty}\hat{j}$ and $\vec{v}_t = v_y\hat{j}$, so $P = F_{Ty}v_y$. We obtain v_y from Equation 15-16. From the figure we see that $F_{Ty} = -F_T \sin\theta \approx -F_T \tan\theta$, where we have used the small angle approximation $\sin\theta \approx \tan\theta$. Since $\tan\theta$ is the slope of the string, we have $\tan\theta = \partial y/\partial x$. Thus

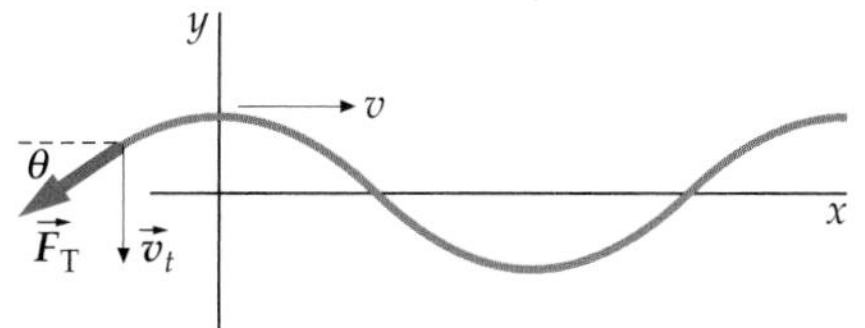

FIGURE 15-8

$$P = F_{Ty}v_y \approx -F_T \tan\theta v_y = -F_T \frac{\partial y}{\partial x}\frac{\partial y}{\partial t}$$
$$= -F_T[kA\cos(kx - \omega t)][-A\omega\cos(kx - \omega t)]$$

Using Equations 15-3 and 15-14, we substitute for F_T and the leading k to obtain

$$P = \mu v \omega^2 A^2 \cos^2 (kx - \omega t) \qquad 15\text{-}17$$

where v is the wave speed. The average power is

$$P_{av} = \tfrac{1}{2} \mu v \omega^2 A^2 \qquad 15\text{-}18$$

since the average value of $\cos^2(kx - \omega t)$ is $\frac{1}{2}$ if the average is taken over an entire period of the motion and x remains constant.

The energy travels along a string at the wave speed v, so the average energy $(\Delta E)_{av}$ flowing past point P_1 during time Δt (Figures 15-9*a* and 15-9*b*) is

$$(\Delta E)_{av} = P_{av}\, \Delta t = \tfrac{1}{2} \mu v \omega^2 A^2\, \Delta t$$

This energy is distributed over a length $\Delta x = v\, \Delta t$ so the average energy in length Δx is

$$(\Delta E)_{av} = \tfrac{1}{2} \mu \omega^2 A^2\, \Delta x \qquad 15\text{-}19$$

Note that both the average energy and the power transmitted are proportional to the square of the amplitude of the wave.

FIGURE 15-9

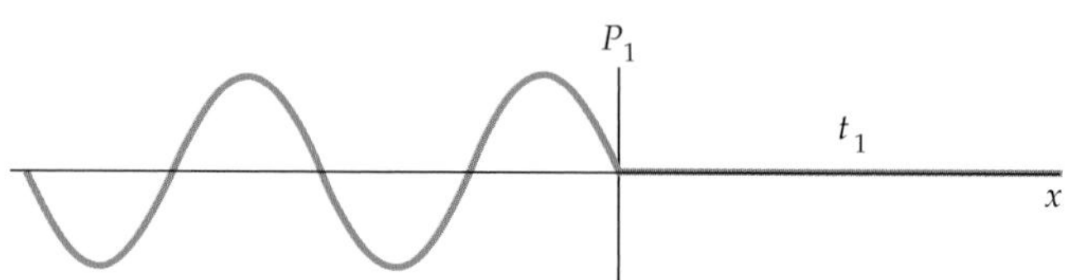

(*a*)

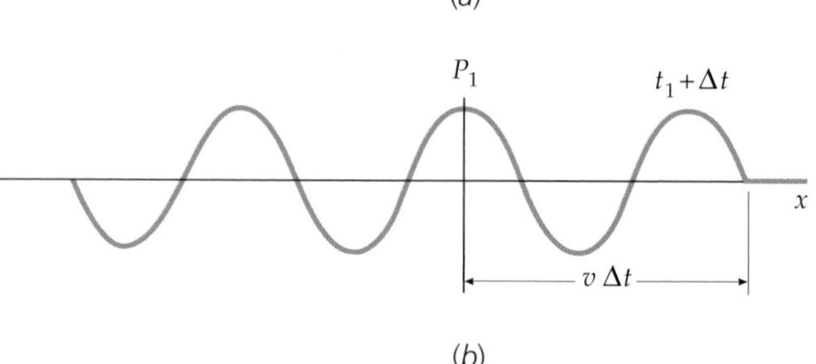

(*b*)

AVERAGE TOTAL ENERGY OF A WAVE ON A STRING **EXAMPLE 15-5**

A harmonic wave of wavelength 25 cm and amplitude 1.2 cm moves along a 15-m-long segment of a 60-m-long string that has a mass of 320 g and a tension of 12 N. (*a*) What is the speed and angular frequency of the wave? (*b*) What is the average total energy of the wave?

PICTURE THE PROBLEM The speed of the waves is $v = \sqrt{F_T/\mu}$, where F_T is given and $\mu = m/L$. We find ω from $\omega = 2\pi f$, where $f = v/\lambda$. The energy is found using Equation 15-19.

(*a*) 1. The speed is related to the tension and mass density: $v = \sqrt{\dfrac{F_T}{\mu}}$

2. Calculate the linear mass density: $\mu = \dfrac{m}{L}$

so

$$v = \sqrt{\frac{F_T}{\mu}} = \sqrt{\frac{F_T L}{m}} = \sqrt{\frac{(12\text{ N})(60\text{ m})}{(0.32\text{ kg})}} = \boxed{47.4\text{ m/s}}$$

3. The angular frequency is found from the frequency, which is found from the speed and wavelength: $\omega = 2\pi f = 2\pi \dfrac{v}{\lambda} = 2\pi \dfrac{47.4\text{ m/s}}{0.25\text{ m}} = \boxed{1190\text{ rad/s}}$

(*b*) The average total energy of waves on the string is given by Equation 15-19 with $\mu\, \Delta x = m = 80$ g:

$$(\Delta E)_{av} = \frac{1}{2}\mu\omega^2 A^2 \Delta x = \frac{1}{2}\frac{m}{L}\omega^2 A^2 \Delta x$$

$$= \frac{1}{2}\frac{0.32\text{ kg}}{60\text{m}}(1190\text{ s}^{-1})^2\,(0.012\text{ m})^2\,(15\text{ m})$$

$$= \boxed{8.19\text{ J}}$$

EXERCISE Calculate the rate at which energy is transmitted along the string. (*Answer* 25.9 W)

Harmonic Sound Waves

Harmonic sound waves can be generated by a tuning fork or loudspeaker that is vibrating with simple harmonic motion. The vibrating source causes the air molecules next to it to oscillate with simple harmonic motion about their equilibrium positions. These molecules collide with neighboring molecules, causing them to oscillate, thereby propagating the sound wave. Equation 15-13 describes a harmonic sound wave if the wave function $y(x,t)$ is replaced by $s(x,t)$, the displacement of the molecules from equilibrium:

$$s(x, t) = s_0 \sin(kx - \omega t) \qquad \text{15-20}$$

These displacements are along the direction of the motion of the wave, and lead to variations in the density and pressure of the air. Figure 15-10 shows the displacement of air molecules and the density changes caused by a sound wave at some fixed time. Because the pressure in a gas is proportional to its density, the change in pressure is maximum where the change in density is maximum. We see from this figure that the pressure or density wave is 90° out of phase with the displacement wave. (In the arguments of sine or cosine functions we will always express phase angles in radians. However, in verbal descriptions we usually say that "two waves are 90° out of phase" rather than "two waves are out of phase by $\pi/2$ rad.") When the displacement is zero, the pressure and density changes are either maximum or minimum. When the displacement is a maximum or minimum, the pressure and density changes are zero. A displacement wave given by Equation 15-20 thus implies a pressure wave given by

$$p = p_0 \sin\left(kx - \omega t - \frac{\pi}{2}\right) \qquad \text{15-21}$$

where p stands for the *change* in pressure from the equilibrium pressure and p_0, the maximum value of this change, is called the pressure amplitude. It can be shown that the pressure amplitude p_0 is related to the displacement amplitude s_0 by

$$p_0 = \rho\omega v s_0 \qquad \text{15-22}$$

where v is the speed of propagation and ρ is the equilibrium density of the gas. Thus, as a sound wave moves in time, the displacement of air molecules, the pressure, and the density all vary sinusoidally with the frequency of the vibrating source.

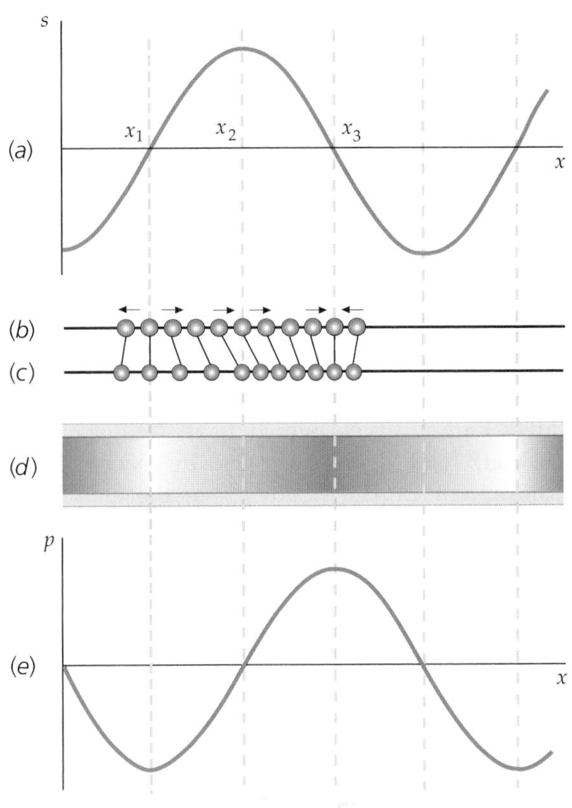

FIGURE 15-10 (*a*) Displacement from equilibrium of air molecules in a harmonic sound wave versus position at some instant. Points x_1 and x_3 are points of zero displacement. (*b*) Some representative molecules equally spaced at their equilibrium positions 1/4 cycle earlier. The arrows indicate the directions of their velocities at that instant. (*c*) Molecules near points x_1, x_2, and x_3 after the sound wave arrives. Just to the left of x_1, the displacement is negative, indicating that the gas molecules are displaced to the left, away from point x_1, at this time. Just to the right of x_1, the displacement is positive, indicating that the molecules are displaced to the right, which is again away from point x_1. So at point x_1, the density is a minimum because the gas molecules on both sides are displaced away from that point. At point x_3, the density is a maximum because the molecules on both sides of that point are displaced toward point x_3. At point x_2, the density does not change because the gas molecules on both sides of that point have equal displacements in the same direction. (*d*) Density of the air at this instant. The density is maximum at x_3 and minimum at x_1, which are both points of zero displacement. It is equal to the equilibrium value at point x_2, which is a maximum in displacement. (*e*) Pressure change, which is proportional to the density change, versus position. The pressure change and displacement (position change) are 90° out of phase.

EXERCISE We can hear sound of frequencies from about 20 Hz to about 20,000 Hz (although many people have rather limited hearing above 15,000 Hz). If the speed of sound in air is 340 m/s, what are the wavelengths that correspond to these extreme frequencies? (*Answer* λ = 17 m at 20 Hz, 1.7 cm at 20,000 Hz)

Energy of Sound Waves The average energy of a harmonic sound wave in a volume element ΔV is given by Equation 15-19 with A replaced by s_0 and $\mu\ \Delta x$ replaced by $\rho\ \Delta V$, where ρ is the average density of the medium.

$$(\Delta E)_{\text{av}} = \tfrac{1}{2}\rho\omega^2 s_0^2\, \Delta V \qquad \text{15-23}$$

The energy per unit volume is the average **energy density** η_{av}:

$$\eta_{av} = \frac{\Delta E_{av}}{\Delta V} = \frac{1}{2}\rho\omega^2 s_0^2 \qquad 15\text{-}24$$

Electromagnetic Waves

Electromagnetic waves include light, radio waves, X rays, gamma rays, and microwaves, among others. The various types of electromagnetic waves differ only in wavelength and frequency. Unlike mechanical waves, electromagnetic waves do not require a medium for propagation. They travel through a vacuum with speed c, which is a universal constant, $c \approx 3 \times 10^8$ m/s. The wave function for electromagnetic waves is an electric field associated with the wave, $\vec{E}(x,t)$. (Electric fields are discussed in Chapter 21. A wave equation, similar to those for string waves and sound waves, is derived from the laws of electricity and magnetism in Chapter 30.) The electric field is perpendicular to the direction of propagation, so electromagnetic waves are transverse waves.

Electromagnetic waves are produced when free electric charges accelerate or when electrons bound to atoms and molecules make transitions to lower energy states. Radio waves, which have frequencies of about 1 MHz for AM and 100 MHz for FM, are produced by macroscopic electric currents oscillating in radio antennas. The frequency of the emitted waves equals the frequency of oscillation of the charges. Light waves, which have frequencies of the order of 10^{14} Hz, are generally produced by atomic or molecular transitions involving bound electrons. The spectrum of electromagnetic waves is discussed in Chapter 31.

15-3 Waves in Three Dimensions

Figure 15-11 shows two-dimensional circular waves on the surface of water in a ripple tank. These waves are generated by a point source moving up and down with simple harmonic motion. The wavelength is the distance between successive wave crests, which in this case are concentric circles. These circles are called **wavefronts.** For a point source of sound, the waves move out in three dimensions, and the wavefronts are concentric spherical surfaces.

The motion of any set of wavefronts can be indicated by **rays,** which are directed lines perpendicular to the wavefronts (Figure 15-12). For circular or spherical waves, the rays are radial lines.

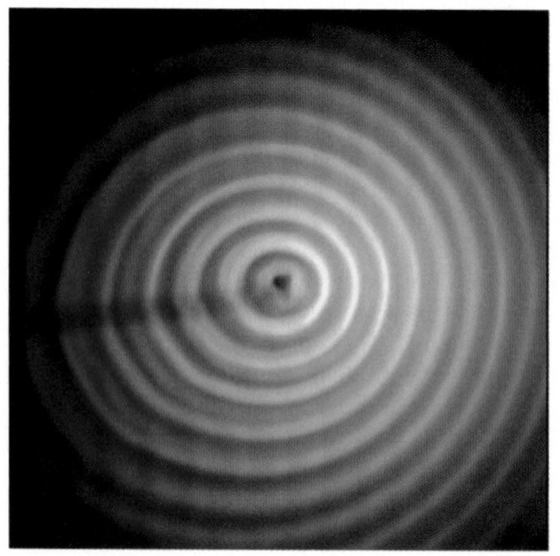

FIGURE 15-11 Circular wavefronts diverging from a point source in a ripple tank.

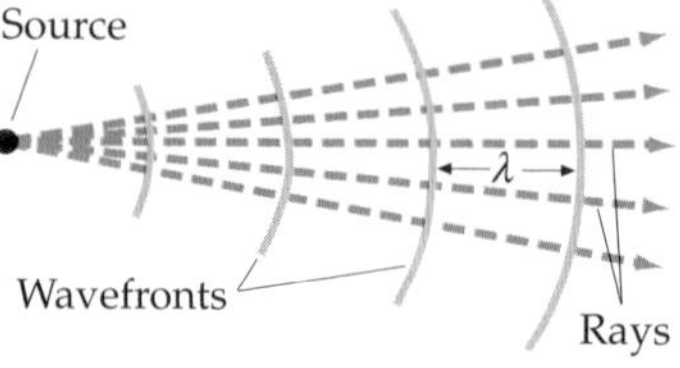

FIGURE 15-12 The motion of wavefronts can be represented by rays drawn perpendicular to the wavefronts. For a point source, the rays are radial lines diverging from the source.

In a homogeneous medium, such as air at constant density, a wave travels in straight lines in the direction of the rays, much like a beam of particles. At a great distance from a point source, a small part of the wavefront can be approximated by a plane, and the rays are approximately parallel lines; such a wave is called a **plane wave** (Figure 15-13). The two-dimensional analog of a plane wave is a line wave, which is a small part of a circular wavefront at a great distance from the source. Such waves can also be produced in a ripple tank by a line source, as in Figure 15-14.

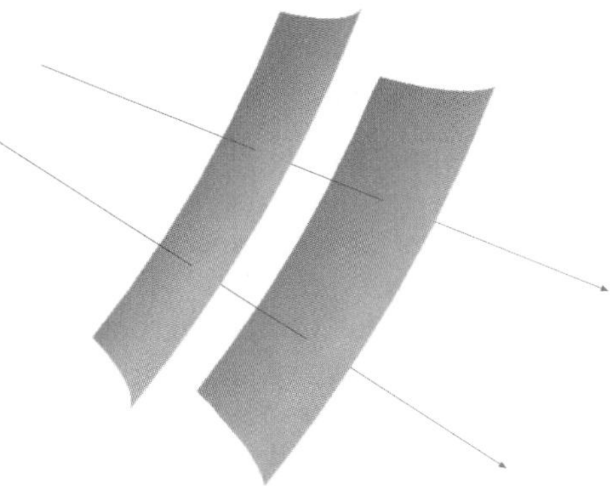

FIGURE 15-13 Plane waves. At great distances from a point source, the wavefronts are approximately parallel planes, and the rays are approximately parallel lines perpendicular to the wavefronts.

Wave Intensity

If a point source emits waves uniformly in all directions, then the energy at a distance r from the source is distributed uniformly on a spherical surface of radius r and area $A = 4\pi r^2$. If P is the power emitted by the source, then the power per unit area at a distance r from the source is $P/(4\pi r^2)$. The average power per unit area that is incident perpendicular to the direction of propagation is called the **intensity:**

$$I = \frac{P_{av}}{A} \qquad 15\text{-}25$$

INTENSITY DEFINED

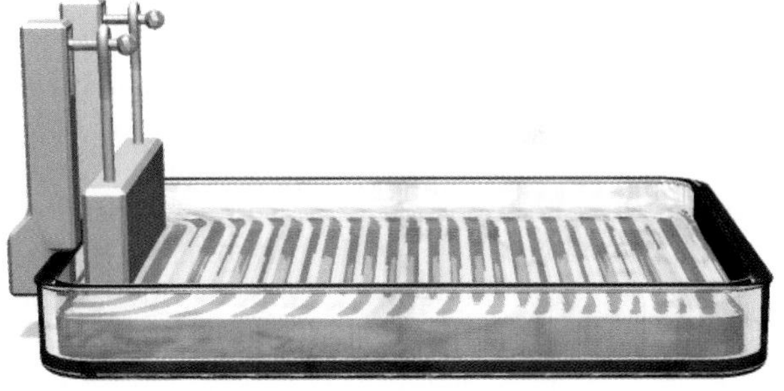

FIGURE 15-14 A two-dimensional analog of a plane wave can be generated in a ripple tank by a flat board that oscillates up and down in the water to produce the wavefronts, which are straight lines.

The units of intensity are watts per square meter. At a distance r from a point source, the intensity is

$$I = \frac{P_{av}}{4\pi r^2} \qquad 15\text{-}26$$

INTENSITY DUE TO A POINT SOURCE

The intensity of a three-dimensional wave varies inversely with the square of the distance from a point source.

There is a simple relation between the intensity of a wave and the energy density in the medium through which it propagates. Figure 15-15 shows a spherical wave that has just reached the radius r_1. The volume inside the radius r_1 contains energy because the particles in that region are oscillating with simple harmonic motion. The region outside r_1 contains no energy because the wave has not yet reached it. After a short time Δt, the wave moves out a short distance $\Delta r = v\,\Delta t$ past r_1. The average energy in the spherical shell of surface area A, thickness $v\,\Delta t$, and volume $\Delta V = A\,\Delta r = Av\,\Delta t$ is

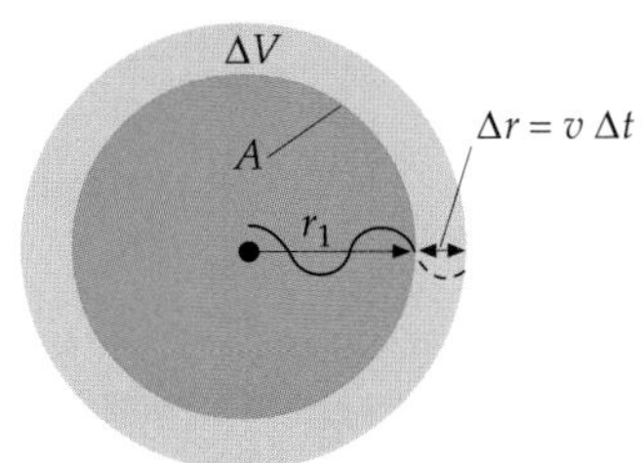

Volume of shell = $\Delta V = A\,\Delta r = Av\,\Delta t$

FIGURE 15-15

$$(\Delta E)_{av} = \eta_{av}\,\Delta V = \eta_{av}\,Av\,\Delta t$$

The rate of transfer of energy is the power passing into the shell. The average incident power is

$$P_{av} = \frac{(\Delta E)_{av}}{\Delta t} = \eta_{av} Av$$

and the intensity of the wave is

$$I = \frac{P_{av}}{A} = \eta_{av} v \qquad 15\text{-}27$$

Thus the intensity equals the product of the wave speed v and the average energy density η_{av}. Substituting $\eta_{av} = \frac{1}{2}\rho\omega^2 s_0^2$ from Equation 15-24 for the energy density in a sound wave, we obtain

$$I = \eta_{av}v = \frac{1}{2}\rho\omega^2 s_0^2 v = \frac{1}{2}\frac{p_0^2}{\rho v} \qquad 15\text{-}28$$

where we have used $s_0 = p_0/(\rho\omega v)$ from Equation 15-22. This result—that the intensity of a sound wave is proportional to the square of the amplitude—is a general property of harmonic waves.

The human ear can accommodate a large range of sound-wave intensities, from about 10^{-12} W/m^2 (which is usually taken to be the threshold of hearing) to about 1 W/m^2 (an intensity great enough to stimulate pain in most people). The pressure variations that correspond to these extreme intensities are about 3×10^{-5} Pa for the hearing threshold and 30 Pa for the pain threshold. (Recall that a pascal is a newton per square meter.) These very small pressure variations add to or subtract from the normal atmospheric pressure of about 101 kPa.

Sound waves from a telephone handset spreading out in the air. The waves have been made visible by sweeping out the space in front of the handset with a light source whose brightness is controlled by a microphone.

A Loudspeaker **EXAMPLE 15-6**

A loudspeaker diaphragm 30 cm in diameter is vibrating at 1 kHz with an amplitude of 0.020 mm. Assuming that the air molecules in the vicinity have the same amplitude of vibration, find (*a*) the pressure amplitude immediately in front of the diaphragm, (*b*) the sound intensity in front of the diaphragm, and (*c*) the acoustic power being radiated. (*d*) If the sound is radiated uniformly into the forward hemisphere, find the intensity at 5 m from the loudspeaker.

PICTURE THE PROBLEM (*a*) and (*b*) The pressure amplitude is calculated directly from $p_0 = \rho\omega v s_0$ (Equation 15-22), and the intensity from $I = \frac{1}{2}\rho\omega^2 s_0^2 v$ (Equation 15-28). (*c*) The power radiated is the intensity times the area of the diaphragm. (*d*) The area of a hemisphere of radius r is $2\pi r^2$. We can use Equation 15-25 with $A = 2\pi r^2$.

(*a*) Equation 15-22 relates the pressure amplitude to the displacement amplitude, frequency, wave velocity, and air density:

$$p_0 = \rho\omega v s_0 = (1.29\text{ kg/m}^3)2\pi(10^3\text{ Hz})(340\text{ m/s})(2 \times 10^{-5}\text{ m}) = \boxed{55.1\text{ N/m}^2}$$

(*b*) Equation 15-28 relates the intensity to these same known quantities:

$$I = \tfrac{1}{2}\rho\omega^2 s_0^2 v = \tfrac{1}{2}(1.29\text{ kg/m}^3)[2\pi(10^3\text{ Hz})]^2(2 \times 10^{-5}\text{ m})^2(340\text{ m/s}) = \boxed{3.46\text{ W/m}^2}$$

(*c*) The power is the intensity times the area of the diaphragm:

$$P = IA = (3.46\text{ W/m}^2)\pi(0.15\text{ m})^2 = \boxed{0.245\text{ W}}$$

(*d*) Calculate the intensity at $r = 5$ m, assuming uniform radiation into the forward hemisphere:

$$I = \frac{P_{av}}{A} = \frac{0.245\text{ W}}{2\pi(5\text{ m})^2} = \boxed{1.56 \times 10^{-3}\text{ W/m}^2}$$

REMARKS The assumption of uniform radiation in the forward hemisphere is not a very good one because the wavelength in this case [$\lambda = v/f = (340\text{ m/s})/(1000\text{s}^{-1}) = 34$ cm] is not large compared with the speaker diameter. There is also some radiation in the backward direction, as can be observed if you stand behind a loudspeaker.

Loudspeakers at a rock concert may put out more than 100 times as much power as the speaker in this example.

***Intensity Level and Loudness** Our perception of loudness is not proportional to the intensity but varies logarithmically. We therefore use a logarithmic scale to describe the **intensity level** β of a sound wave, which is measured in **decibels** (dB) and defined by

$$\beta = 10 \log \frac{I}{I_0} \tag{15-29}$$

DEFINITION—INTENSITY LEVEL IN DB

Here I is the intensity of the sound and I_0 is a reference level, which usually is taken to be the threshold of hearing:

$$I_0 = 10^{-12}\ \mathrm{W/m^2} \tag{15-30}$$

On this scale, the threshold of hearing is $\beta = 10 \log(I/I_0) = 0$ dB and the pain threshold ($I = 1\ \mathrm{W/m^2}$) is $\beta = 10 \log(1/10^{-12}) = 10 \log 10^{12} = 120$ dB. Thus, the range of sound intensities from $10^{-12}\ \mathrm{W/m^2}$ to $1\ \mathrm{W/m^2}$ corresponds to intensity levels from 0 dB to 120 dB. Table 15-1 lists the intensity levels of some common sounds.

SOUNDPROOFING **EXAMPLE 15-7**

A sound absorber attenuates the sound level by 30 dB. By what factor is the intensity decreased?

From Table 15-1, we can see that for every 10-dB drop in the intensity level, the intensity decreases by a factor of 10. Thus, if the sound level drops 30 db then the intensity drops by a factor of $10^3 = \boxed{1000}$.

TABLE 15-1

Intensity and Intensity Level of Some Common Sounds ($I_0 = 10^{-12}\ \mathrm{W/m^2}$)

Source	I/I_0	dB	Description
	10^0	0	Hearing threshold
Normal breathing	10^1	10	Barely audible
Rustling leaves	10^2	20	
Soft whisper (at 5 m)	10^3	30	Very quiet
Library	10^4	40	
Quiet office	10^5	50	Quiet
Normal conversation (at 1 m)	10^6	60	
Busy traffic	10^7	70	
Noisy office with machines; average factory	10^8	80	
Heavy truck (at 15 m); Niagara Falls	10^9	90	Constant exposure endangers hearing
Old subway train	10^{10}	100	
Construction noise (at 3 m)	10^{11}	110	
Rock concert with amplifiers (at 2 m); jet takeoff (at 60 m)	10^{12}	120	Pain threshold
Pneumatic riveter; machine gun	10^{13}	130	
Jet takeoff (nearby)	10^{15}	150	
Large rocket engine (nearby)	10^{18}	180	

The sensation of loudness depends on the frequency as well as the intensity of a sound. Figure 15-16 is a plot of intensity level versus frequency for sounds of equal loudness to the human ear. (In this figure, the frequency is plotted on a logarithmic scale to display the wide range of frequencies from 20 Hz to 10 kHz.) We note from this figure that the human ear is most sensitive at about 4 kHz for all intensity levels.

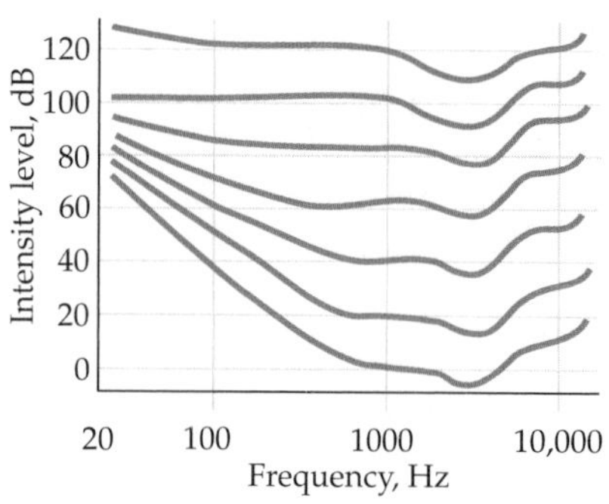

FIGURE 15-16 Intensity level versus frequency for sounds perceived to be of equal loudness. The lowest curve is below the threshold for hearing of all but about 1 percent of the population. The second lowest curve is approximately the hearing threshold for about 50 percent of the population.

BARKING DOGS **EXAMPLE 15-8**

A barking dog delivers about 1 mW of power. (*a*) If this power is uniformly distributed in all directions, what is the sound intensity level at a distance of 5 m? (*b*) What would be the intensity level of two dogs barking at the same time if each delivered 1 mW of power?

PICTURE THE PROBLEM The intensity level is found from the intensity, which is found from $I = P/(4\pi r^2)$. For two dogs, the intensities are added.

(*a*) 1. The intensity level is related to the intensity:

$$\beta = 10 \log \frac{I}{I_0}$$

2. Calculate the intensity at $r = 5$ m:

$$I_1 = \frac{P_1}{4\pi r^2} = \frac{10^{-3}\text{ W}}{4\pi(5\text{ m})^2} = 3.18 \times 10^{-6}\text{ W/m}^2$$

3. Use your result to find the intensity level at 5 m:

$$\beta_1 = 10 \log \frac{I_1}{I_0} = 10 \log \frac{3.18 \times 10^{-6}}{10^{-12}} = \boxed{65.0\text{ dB}}$$

(*b*) If I_1 is the intensity for one dog, the intensity for two dogs is $I_2 = 2I_1$:

$$\beta_2 = 10 \log \frac{I_2}{I_0} = 10 \log \frac{2I_1}{I_0} = 10 \log 2 + 10 \log \frac{I_1}{I_0}$$

$$= 10 \log 2 + \beta_1 = 3.01 + 65.0 = \boxed{68.0\text{ dB}}$$

REMARKS We can see from this example that whenever the intensity is doubled, the intensity level increases by 3 dB.

15-4 Waves Encountering Barriers

Reflection and Refraction

When a wave is incident on a boundary that separates two regions of differing wave speed, part of the wave is reflected and part is transmitted. Figure 15-17*a* shows a pulse on a light string that is attached to a heavier string. In this case, the pulse reflected at the boundary is inverted. If the second string is lighter than the first (Figure 15-17*b*), then the reflected pulse is not inverted. In either case,

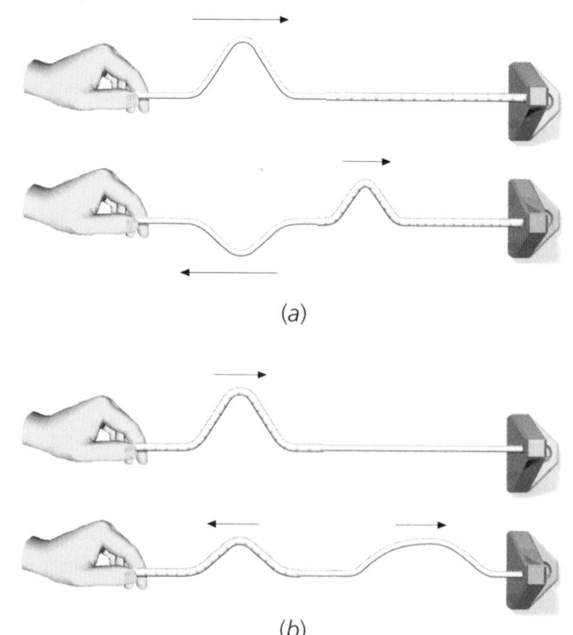

FIGURE 15-17 (*a*) A wave pulse traveling on a light string attached to a heavier string in which the wave speed is smaller. The reflected pulse is inverted, whereas the transmitted pulse is not. (*b*) A wave pulse traveling on a heavy string attached to a light string in which the wave speed is greater. In this case, the reflected pulse is not inverted.

the transmitted pulse is not inverted. If the string is tied to a fixed point, then the pulse is reflected and inverted. If it is tied to a string of negligible mass, then the pulse is reflected, but not inverted.

TWO SOLDERED WIRES **EXAMPLE 15-9**

Two wires of different linear mass densities are soldered together end to end and then stretched under a tension F_T (the tension is the same in both wires). The wave speed in the first wire is twice that in the second. If a harmonic wave traveling in the first wire is incident on the junction of the wires, the amplitude of the reflected wave is half the amplitude of the transmitted wave. (*a*) If the amplitude of the incident wave is *A*, what are the amplitudes of the reflected and transmitted waves? (*b*) What fraction of the incident power is reflected at the junction and what fraction is transmitted?

PICTURE THE PROBLEM By conservation of energy, the power incident on the junction equals the power reflected plus the power transmitted. Each power is expressed in Equation 15-18 as a function of the density μ, amplitude A, frequency ω, and wave speed v (Figure 15-18). The angular frequencies of all the waves are equal. Since the reflected wave and incident wave are in the same medium, they have the same wave speed v_1. We are given that the speed in the second wire is $v_2 = \frac{1}{2}v_1$.

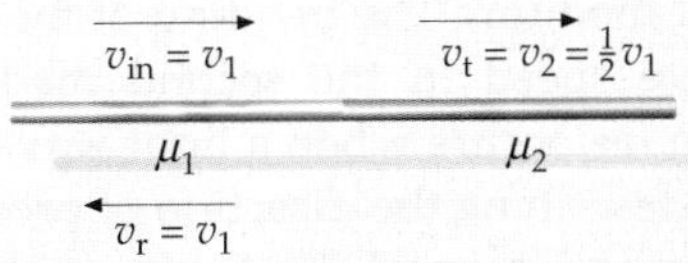

$v_{in} = v_1$ $\quad v_t = v_2 = \frac{1}{2}v_1$ $\quad \mu_1$ $\quad \mu_2$ $\quad v_r = v_1$

FIGURE 15-18

(*a*) 1. By conservation of energy, the incident power equals the transmitted power plus the reflected power:

$$P_{in} = P_t + P_r$$

2. Write Equation 15-18:

$$P_{av} = \tfrac{1}{2}\mu v \omega^2 A^2$$

3. Substitute into the step 1 result and simplify: The angular frequency is the same for all three waves.

$$\tfrac{1}{2}\mu_1\omega^2 A_{in}^2 v_1 = \tfrac{1}{2}\mu_2\omega^2 A_t^2 v_2 + \tfrac{1}{2}\mu_1\omega^2 A_r^2 v_1$$

$$\mu_1 A_{in}^2 v_1 = \mu_2 A_t^2 v_2 + \mu_1 A_r^2 v_1$$

4. Using the relation $v = \sqrt{F_T/\mu}$ (Equation 15-3), substitute for μ_1 and μ_2 and simplify. F_T is the same on either side of the junction:

$$\frac{F_T}{v_1^2}A_{in}^2 v_1 = \frac{F_T}{v_2^2}A_t^2 v_2 + \frac{F_T}{v_1^2}A_r^2 v_1$$

$$\frac{A_{in}^2}{v_1} = \frac{A_t^2}{v_2} + \frac{A_r^2}{v_1}$$

5. Using the given relations $v_2 = \frac{1}{2}v_1$ and $A_r = \frac{1}{2}A_t$, substitute and solve for the amplitudes:

$$\frac{A_{in}^2}{v_1} = \frac{A_t^2}{\frac{1}{2}v_1} + \frac{(\frac{1}{2}A_t)^2}{v_1} = \frac{9}{4}\frac{A_t^2}{v_1}$$

so

$$A_t = \boxed{\tfrac{2}{3}A_{in}} \quad \text{and} \quad A_r = \boxed{\tfrac{1}{3}A_{in}}$$

(*b*) 1. In Part (*a*) steps 1–4 it was shown that the power is proportional to A^2/v. Express each of the three powers, using b as the proportionality constant:

$$P_{in} = b\frac{A_{in}^2}{v_1} \qquad P_t = b\frac{A_t^2}{v_2} \qquad P_r = b\frac{A_r^2}{v_1}$$

2. Using the Part (*a*) step 5 results, eliminate v_2, A_t, and A_r from the expressions for P_t and P_r:

$$P_t = b\frac{(\frac{2}{3}A_{in})^2}{\frac{1}{2}v_1} = \frac{8}{9}b\frac{A_{in}^2}{v_1} = \boxed{\frac{8}{9}P_{in}}$$

$$P_r = b\frac{(\frac{1}{3}A_{in})^2}{v_1} = \frac{1}{9}b\frac{A_{in}^2}{v_1} = \boxed{\frac{1}{9}P_{in}}$$

REMARKS The reflected wave is inverted relative to the incident wave, so it is 180° out of phase with it. When the displacement of the wire just to the left of the junction would be y_1 due only to the incident wave, it would be $-(y_1/3)$ due only to the reflected wave. These add (according to the principle of superposition to be studied in the next chapter), giving a resultant displacement of $2y_1/3$, which is also the displacement that occurs just to the right of the junction due to the transmitted wave. It can be shown that, given the ratio of the wave speeds, the amplitudes of the transmitted and reflected waves can be determined by requiring that the displacement and the slope be continuous at the junction.

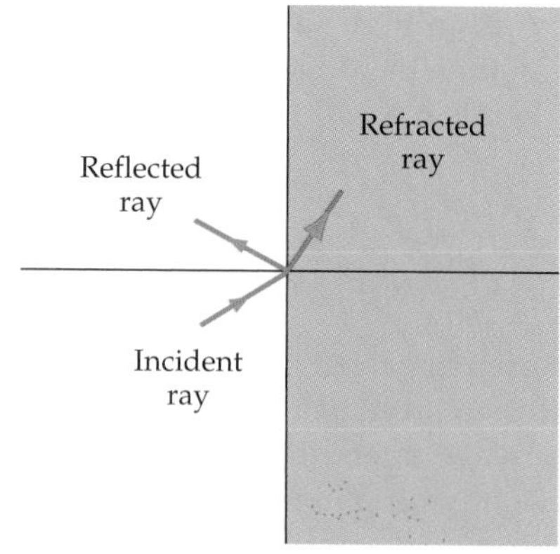

FIGURE 15-19 A wave striking a boundary surface between two media in which the wave speed differs. Part of the wave is reflected and part is transmitted. The change in direction of the transmitted (refracted) ray is called refraction.

In three dimensions, a boundary between two regions of differing wave speed is a surface. Figure 15-19 shows a ray incident on such a boundary surface. This example could be a sound wave in air striking a solid or liquid surface. The reflected ray makes an angle with the normal to the surface equal to that of the incident ray, as shown.

The transmitted ray is bent toward or away from the normal—depending on whether the wave speed in the second medium is less or greater than that in the incident medium. The bending of the transmitted ray is called **refraction.** When the wave speed in the second medium is greater than that in the incident medium (as occurs when a light wave in glass or water is refracted into the air), the ray describing the direction of propagation is bent away from the normal, as shown in Figure 15-20. As the angle of incidence is increased, the angle of refraction increases, until a critical angle of incidence is reached for which the angle of refraction is 90°. For incident angles greater than the critical angle, there is no refracted ray, a phenomenon known as **total internal reflection.**

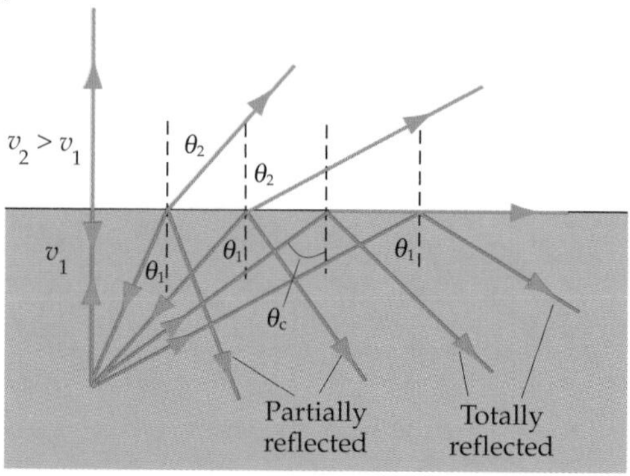

FIGURE 15-20 Light from a source in the water is bent away from the normal when it enters the air. For angles of incidence above a critical angle, there is no transmitted ray, a condition known as total internal reflection.

The amount of energy reflected from a surface depends on the surface. Flat walls, floors, and ceilings make good reflectors for sound waves, whereas porous and less rigid materials, such as cloth in draperies and furniture coverings, absorb much of the incident sound. The reflection of sound waves plays an important role in the design of a lecture hall, a library, or a music auditorium. If a lecture hall has many flat reflecting surfaces, speech is difficult to understand because of the many echoes that arrive at different times at the listener's ear. Absorbent material is often placed on the walls and ceiling to reduce such reflections. In a concert hall, a reflecting shell is placed behind the orchestra, and reflecting panels are hung from the ceiling to reflect and direct the sound back toward the listeners.

Diffraction

If a wavefront is partially blocked by an obstacle, the unblocked part of the wavefront bends behind the obstacle. This bending of the wavefronts is called **diffraction.** Almost all of the diffraction occurs for that part of the wavefront that passes within a few wavelengths of the edge of the obstacle. For the parts of the wavefront that pass farther than a few wavelengths from the edge, diffraction is negligible and the wave propagates in straight lines in the direction of the incident rays. When wavefronts encounter a barrier with an aperture (hole) only a few wavelengths across, the part of the wavefronts passing through the aperture all pass within a few wavelengths of an edge. Thus, flat

wavefronts bend and spread out and become spherical or circular (Figure 15-21). In contrast, for a beam of *particles* falling upon a barrier with an aperture, the part of the beam passing through the aperture does so with no change in the direction of the particles (Figure 15-22). Diffraction is one of the key characteristics that distinguishes waves from particles. We will discuss how diffraction arises when we study the interference and diffraction of light in Chapter 35.

Though waves passing through an aperture always bend, or diffract, to some extent, the amount of diffraction depends on whether the wavelength is small or large relative to the size of the aperture. If the wavelength is large relative to the aperture, as in Figure 15-23, the diffraction effects are large, and the waves spread out as they pass through the aperture—as if the waves were originating from a point source. On the other hand, if the wavelength is small relative to the aperture, the effect of diffraction is small, as shown in Figure 15-21. Near the edges of the aperture the wavefronts are distorted and the waves appear to bend slightly. For the most part, however, the wavefronts are not affected and the waves propagate in straight lines, much like a beam of particles. The approximation that waves propagate in straight lines in the direction of the rays with no diffraction is known as the **ray approximation.** Wavefronts are distorted *near* the edges of any obstacle blocking part of the wavefronts. By *near* we mean within a few wavelengths of the edges.

Because the wavelengths of audible sound (which range from a few centimeters to several meters) are generally large compared with apertures and obstacles (doors or windows, and people, for example), diffraction of sound waves is a phenomenon that is often observed. On the other hand, the wavelengths of visible light (4×10^{-7} to 7×10^{-7} m) are so small compared with the size of ordinary objects and apertures that the diffraction of light is not easily noticed; light appears to travel in straight lines. Nevertheless, the diffraction of light is an important phenomena, one we will study in detail in Chapter 35.

Diffraction places a limitation on how accurately small objects can be located by reflecting waves off them and on how well details of the objects can be resolved. Waves are not reflected appreciably from objects smaller than the wavelength, so detail cannot be observed on a scale smaller than the wavelength used. If waves of wavelength λ are used to locate an object, then its position can be known only to within an uncertainty of one wavelength.

FIGURE 15-21 Plane waves in a ripple tank meeting a barrier with an opening that is only a few wavelengths wide. Beyond the barrier are circular waves that are concentric about the opening, just as if there were a point source at the opening.

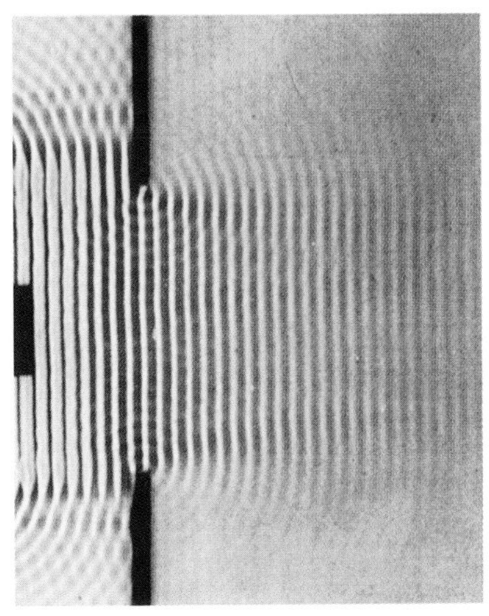

FIGURE 15-23 Plane waves in a ripple tank meeting a barrier with an opening width that is large compared to λ. The wave continues in the forward direction, with only a small amount of spreading into the regions to either side of the opening.

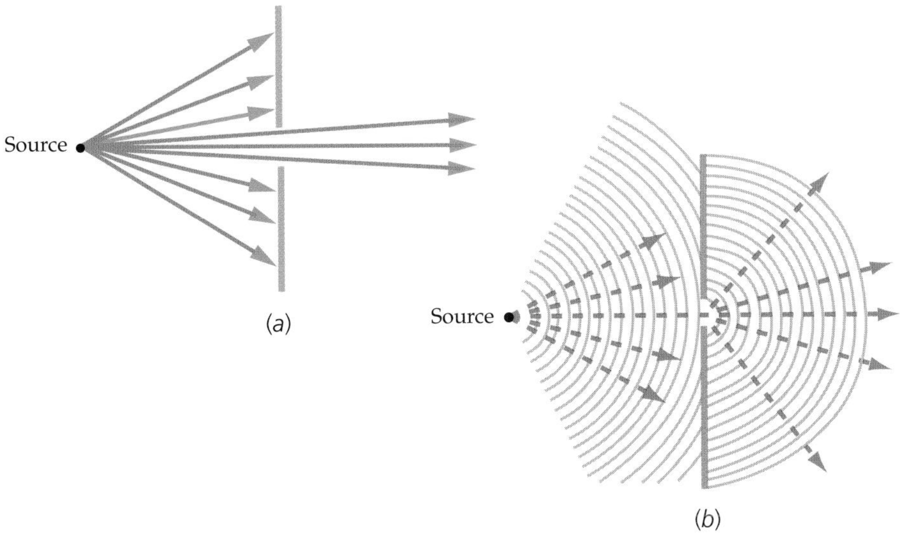

FIGURE 15-22 Comparison of particles and waves passing through a narrow opening in a barrier. (*a*) Transmitted particles are confined to a narrow-angle beam. (*b*) Transmitted waves spread out (radiate widely) from the aperture, which acts like a point source of circular waves.

Sound waves with frequencies above 20,000 Hz are called **ultrasonic waves.** Because of their very small wavelengths, narrow beams of ultrasonic waves can be sent out and reflected from small objects. Bats can emit and detect frequencies up to about 120,000 Hz, corresponding to a wavelength of 2.8 mm, which they use to locate small prey such as moths. Echolocation systems, called sonar (from *so*und and *na*vigation *r*anging), are used to detect the outlines of submerged objects with sound waves. The frequency used by commercially available fish finders ranges from about 25 to 200 kHz, and porpoises produce echolocation clicks in the same frequency range. In medicine, ultrasonic waves are used for diagnostic purposes. Ultrasonic waves are passed through the human body and information about the frequency and intensity of the transmitted and reflected waves is processed to construct a three-dimensional picture of the body's interior, called a sonogram.

15-5 The Doppler Effect

If a wave source and a receiver are moving relative to each other, the received frequency is not the same as the frequency of the source. If they are moving closer together, the received frequency is greater than the source frequency; and if they are moving farther apart, the received frequency is less than the source frequency. This is called the **Doppler effect.** A familiar example is the drop in pitch of the sound of the horn of an approaching car as the car passes by—and then recedes.

In the following discussion, all motions are relative to the medium. Consider the source moving with speed u_s, shown in Figure 15-24*a* and *b*, and a stationary receiver. The source has frequency f_s (and period $T_s = 1/f_s$). The received frequency f_r, the number of wave crests passing the receiver per unit time, is

$$f_r = \frac{v}{\lambda} \quad \text{(stationary receiver)} \qquad 15\text{-}31$$

where v is the wave speed and λ is the wavelength (the distance between successive crests). To find f_r we first need to find λ. Consider event 1—a wave crest leaves the source—and event two—the next wave crest leaves the source—as shown in Figure 15-24*c*. The time between these two events is T_s, and between these events the crest leaving the source first travels a distance vT_s while the source itself travels a distance u_sT_s. Consequently, at the time of the second event, the distance between the source and the crest leaving first equals the wavelength λ.

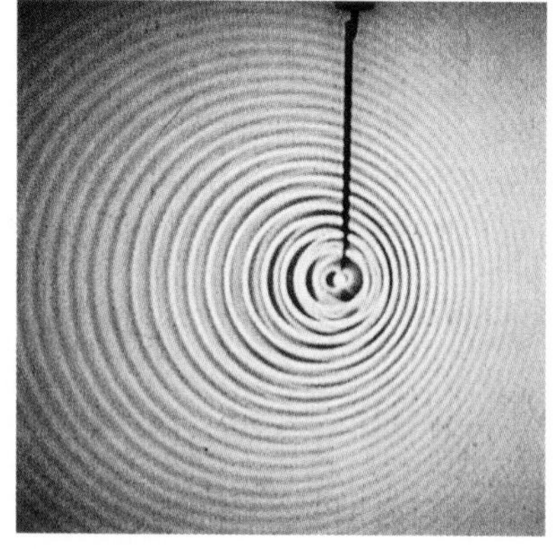

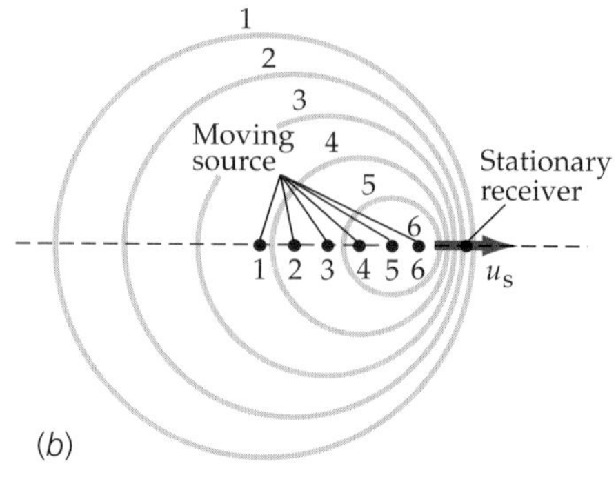

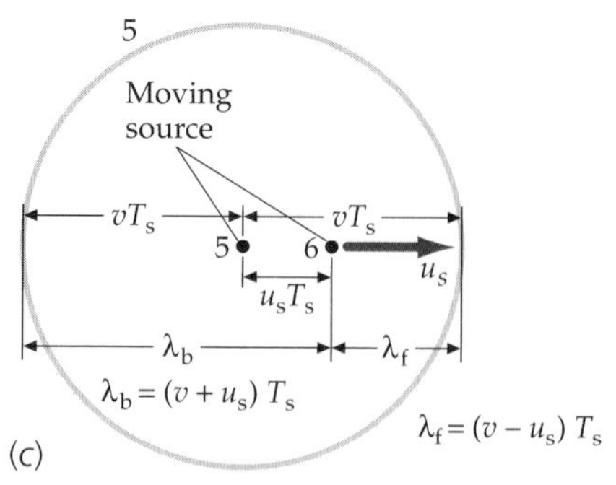

FIGURE 15-24 (*a*) Waves in a ripple tank produced by a point source moving to the right. The wavefronts are closer together in front of the source and farther apart behind the source. (*b*) Successive wavefronts emitted by a point source moving with speed u_s to the right. The numbers of the wavefronts correspond to the positions of the source when the wave was emitted. (*c*) The source vibrates one cycle in time T_s. During time T_s the source moves a distance u_sT_s and the 5th wavefront travels a distance vT_s. In front of the source the wavelength $\lambda_f = (v - u_s)T_s$, while behind the source $\lambda_b = (v + u_s)T_s$.

Behind the source $\lambda = \lambda_b = (v + u_s)T_s$, and in front of the source $\lambda = \lambda_f = (v - u_s)T_s$, provided $u_s < v$. (If $u_s \geq v$, no wavefronts reach the region ahead of the source.) We can express both these wavelengths as

$$\lambda = (v \pm u_s)T_s = \frac{v \pm u_s}{f_s} \qquad 15\text{-}32$$

where we have substituted $1/f_s$ for T_s. In front of the source the wavelength is shortest, so the minus sign applies. Behind the source the plus sign applies. Substituting for λ in Equation 15-31 gives

$$f_r = \frac{v}{\lambda} = \frac{v}{v \pm u_s} f \qquad \text{(stationary receiver)} \qquad 15\text{-}33$$

When the receiver moves relative to the medium, the received frequency is different simply because the receiver moves past more or fewer wave crests in a given time. For a receiver moving with speed u_r, let T_r denote the time between arrivals of successive crests. Then, during the time between the arrivals of two successive crests, each crest will have traveled a distance vT_r, and during the same time the receiver will have traveled a distance u_rT_r. If the receiver moves in the direction opposite to that of the wave (Figure 15-25), then during time T_r the distance each crest moves plus the distance the receiver moves equals the wavelength. That is, $vT_r + u_rT_r = \lambda$, or $T_r = \lambda/(v + u_r)$. [If the receiver moves in the same direction as the wave, then $vT_r - \lambda = u_rT_r$, so $T_r = \lambda/(v - u_r)$]. Since $f_r = 1/T_r$ we have

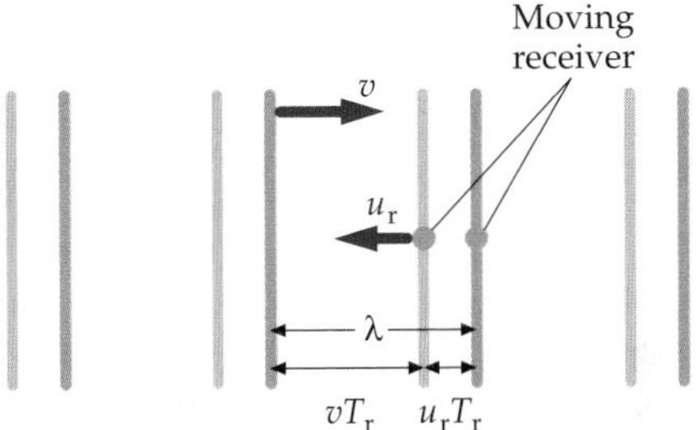

FIGURE 15-25

$$f_r = \frac{1}{T_r} = \frac{v \pm u_r}{\lambda} \qquad 15\text{-}34$$

where, if the receiver moves in the same direction as the wave, the received frequency is lower, so we choose the negative sign. If the receiver moves in the direction opposite to that of the wave, the frequency is higher, so we choose the positive sign. Substituting for λ from Equation 15-32 we obtain

$$f_r = \frac{v \pm u_r}{v \pm u_s} f_s \qquad 15\text{-}35a$$

The correct choices for the plus or minus signs are most easily determined by remembering that the frequency tends to increase both when the source moves toward the receiver and when the receiver moves toward the source. For example, if the receiver is moving toward the source the plus sign is selected in the numerator, which tends to increase the received frequency, and if the source is moving away from the receiver the plus sign is selected in the denominator, which tends to decrease the received frequency. Equation 15-35*a* appears more symmetric, and thus is easier to remember, if expressed in the form

$$\frac{f_r}{v \pm u_r} = \frac{f_s}{v \pm u_s} \qquad 15\text{-}35b$$

It can be shown (see Problem 89) that if both u_s and u_r are much smaller than the wave speed v, then the shift in frequency $\Delta f = f_r - f_s$ is given approximately by

$$\frac{\Delta f}{f_s} \approx \pm\frac{u}{v} \qquad (u << v) \qquad 15\text{-}36$$

where $u = u_s \pm u_r$ is the speed of the source relative to the receiver.

Equations 15-31 through 15-36 are valid *only* in the reference frame of the medium. In a reference frame in which the medium is moving (for example, the reference frame of the ground if air is the medium and there is a wind blowing), the wave speed v is replaced by $v' = v \pm u_w$, where u_w is the speed of the wind relative to the ground.

SOUNDING THE HORN — EXAMPLE 15-10

The frequency of a car horn is 400 Hz. If the horn is honked as the car moves with a speed u_s = 34 m/s (about 122 km/h) through still air toward a stationary receiver, find (*a*) the wavelength of the sound passing the receiver and (*b*) the frequency received. Take the speed of sound in air to be 340 m/s. (*c*) Find the wavelength of the sound passing the receiver and find the frequency received if the car is stationary as the horn is honked and a receiver moves with a speed u_r = 34 m/s toward the car.

PICTURE THE PROBLEM (*a*) The waves in front of the source are compressed, so we use the minus sign in Equation 15-32. (*b*) We calculate the received frequency from Equation 15-35*a*. (*c*) For a moving receiver we use the same equations as in Parts (*a*) and (*b*).

(*a*) Using Equation 15-32, calculate the wavelength in front of the car. In front of the source the wavelength is shorter, so choose the sign accordingly:

$$\lambda = \frac{v - u_s}{f_s} = \frac{340\text{ m/s} - 34\text{ m/s}}{400\text{ Hz}} = \boxed{0.765\text{ m}}$$

(*b*) Using Equation 15-35*a*, solve for the received frequency:

$$f_r = \frac{v \pm u_r}{v \pm u_s} f_s = \frac{v + 0}{v - u_s} f_s$$

$$= \left(\frac{340}{340 - 34}\right)(400\text{ Hz}) = \boxed{444\text{ Hz}}$$

(*c*) 1. Using Equation 15-32, calculate the wavelength in the vicinity of the receiver:

$$\lambda = \frac{v \pm u_s}{f_s} = \frac{340\text{ m/s} \pm 0}{400\text{ Hz}} = \boxed{0.850\text{ m}}$$

2. The received frequency is given by Equation 15-35*a*. The source is approaching the receiver so the frequency is shifted upward. Choose the sign accordingly:

$$f_r = \frac{v \pm u_r}{v \pm u_s} f_s = \frac{v + u_r}{v \pm 0} f_s = \left(1 + \frac{u_r}{v}\right) f_s$$

$$= \left(1 + \frac{34}{340}\right)(400\text{ Hz}) = \boxed{440\text{ Hz}}$$

REMARKS The frequency f_r can also be obtained using Equation 15-34.

EXERCISE As a train moving at 90 km/h is approaching a stationary listener, it blows its horn, which has a frequency of 630 Hz. There is no wind. (*a*) What is the wavelength of the sound waves in front of the train? (*b*) What frequency is heard by the listener? (Use 340 m/s for the speed of sound.) (*Answer* (*a*) λ = 0.5 m (*b*) f_r = 680 Hz)

ANOTHER CAR HORN — EXAMPLE 15-11 — Try It Yourself

The ratio of the frequency of a note to the frequency of the semitone above it on the diatonic scale is about 15:16. How fast is a car going if its horn drops a semitone as it passes you? There is no wind and you are standing next to the road. Take the speed of sound in air to be 340 m/s.

PICTURE THE PROBLEM Let u_s be the speed of the car and f_s be the frequency of the horn. The frequency received as the car approaches, f_r, is greater than f_s and the frequency received as the car recedes, f_r', is less than f_s. Set the ratio f_r'/f_r = 15/16 and solve for u_s.

Cover the column to the right and try these on your own before looking at the answers.

Steps	Answers
1. Write the frequency received as the car approaches in terms of f_s.	$f_r = \frac{v \pm u_r}{v \pm u_s} f_s = \frac{v}{v - u_s} f_s$
2. Write the frequency received as the car recedes in terms of f_s.	$f_r' = \frac{v \pm u_r}{v \pm u_s} f_s = \frac{v}{v + u_s} f_s$
3. Set the ratio f_r'/f_r equal to 15/16.	$\frac{f_r'}{f_r} = \frac{v - u_s}{v + u_s} = \frac{15}{16}$
4. Solve for u_s.	$u_s = 0.0323v = \boxed{39.5\ \text{km/h} = 24.5\ \text{mi/h}}$

REMARKS The wavelength of the sound behind the car is longer than the wavelength of the sound in front of the car. Also, 1 m/s = 3.6 km/h.

Another familiar example of the Doppler effect is the radar used by police to measure the speed of a car. Electromagnetic waves emitted by the radar transmitter strike the moving car. The car acts as both a moving receiver and a moving source as the waves reflect off it back to the radar receiver. Since electromagnetic waves travel at the speed of light, $v = c = 3 \times 10^8$ m/s, the condition $u << v$ is certainly met and Equation 15-36 can be used to calculate the Doppler shift.

POLICE RADAR — EXAMPLE 15-12 Try It Yourself

The radar unit in a police car sends out electromagnetic waves that travel at the speed of light c. The electric current in the antenna of the radar unit oscillates at frequency f_s. The waves reflect from a speeding car moving away from the police car at speed u relative to the police car. There is a frequency difference of Δf between f_s and f_r', the frequency received at the police car. Find u in terms of f_s and Δf.

PICTURE THE PROBLEM The radar wave strikes the speeding car at frequency f_r. This frequency is less than f_s because the car is moving away from the source. The frequency shift is given by Equation 15-36. The car then acts as a moving source emitting waves of frequency f_r. The police unit detects waves of frequency $f_r' < f_r$ because the source (the speeding car) is moving away from the police car. The frequency difference is $f_r' - f_s$.

Cover the column to the right and try these on your own before looking at the answers.

Steps	Answers
1. The radar unit must be able to determine the speed based only on what it transmits and what it detects.	The radar unit must determine u in terms of f_s and f_r'. Because of the way Equation 15-36 is written, we will solve for u in terms of f_s and $\Delta f = f_r' - f_s$.
2. The frequency difference Δf is the frequency difference $\Delta f_1 = f_r - f_s$ plus the frequency difference $\Delta f_2 = f_r' - f_r$.	$\Delta f = \Delta f_1 + \Delta f_2$
3. Using Equation 15-36, substitute for the frequency differences in step 2.	$\Delta f = -\frac{u}{c} f_s - \frac{u}{c} f_r = -\frac{u}{c}(f_s + f_r)$

4. Again using Equation 15-36, solve for f_r in terms of f_s. $\frac{\Delta f_1}{f_s} = -\frac{u}{c}$ so $f_r = \left(1 - \frac{u}{c}\right)f_s$

5. Substitute your step 4 result into your step 3 result and simplify. $\Delta f = -\frac{u}{c}\left(2 - \frac{u}{c}\right)f_s$

6. u/c is negligible compared to 2. Use this to simplify the step 5 result and solve for u in terms of Δf and f_s. $\Delta f \approx -2f_s\frac{u}{c}$ so $u = -\frac{\Delta f}{2f_s}c = \boxed{\frac{|\Delta f|}{2f_s}c}$

REMARKS The difference in frequency between two waves of nearly equal frequency is easy to detect because the two waves interfere to produce a wave whose amplitude oscillates with frequency $|\Delta f|$, which is called the beat frequency. Interference and beats are discussed in Chapter 16.

EXERCISE Calculate Δf if $f_s = 1.5 \times 10^9$ Hz, $c = 3 \times 10^8$ m/s, and $u = 50$ m/s (*Answer* $\Delta f = 500$ Hz)

The Doppler Shift and Relativity We see from Example 15-10 (and Equations 15-33, 15-34, and 15-35) that the magnitude of the Doppler shift in frequency depends on whether it is the source or the receiver that is moving relative to the medium. For sound, these two situations are physically different. For example, if you move relative to still air, you feel air rushing past you. In your reference frame, there is a wind. For sound waves in air, therefore, we can tell whether the source or receiver is moving by noting if there is a wind in the reference frame of the source or the receiver. However, light and other electromagnetic waves propagate through empty space in which there is no medium. There is no "wind" to tell us whether the source or receiver is moving. According to Einstein's theory of relativity, absolute motion cannot be detected, and all observers measure the same speed c for light, independent of their motion relative to the source. Thus Equation 15-35 cannot be correct for the Doppler shift for light. Two modifications must be made in calculating the relativistic Doppler effect for light. First, the speed of waves passing a receiver is c independent of the motion of the receiver. Second, the time interval between the emission of successive wave crests, which is $T_s = 1/f_s$ in the reference frame of the source, is different in the reference frame of the receiver when the two reference frames are in relative motion because of relativistic time dilation and length contraction (Equations R-9 and R-3). (We will discuss the relativistic Doppler effect in Chapter 39.) The result is that the frequency received depends only on the relative speed of approach (or recession) u, and is related to the frequency emitted by

$$f_r = \sqrt{\frac{c \pm u}{c \mp u}}\, f_s \qquad \text{15-37}$$

Choose the signs that give an up-shift in frequency when the source and receiver are approaching, and vice versa. (The upper signs are used if the source and receiver are approaching, and the lower signs are used if they are separating.) Again, when $u << c$, $\Delta f/f_s \approx \pm u/c$, as given by Equation 15-36.

Shock Waves

In our derivations of the Doppler-shift expressions, we assumed that the speed u of the source was less than the wave speed v. If a source moves with speed greater than the wave speed, then there will be no waves in front of the source. Instead, the waves pile up behind the source to form a shock wave. In the case of sound waves, this shock wave is heard as a sonic boom when it arrives at the receiver.

(a)

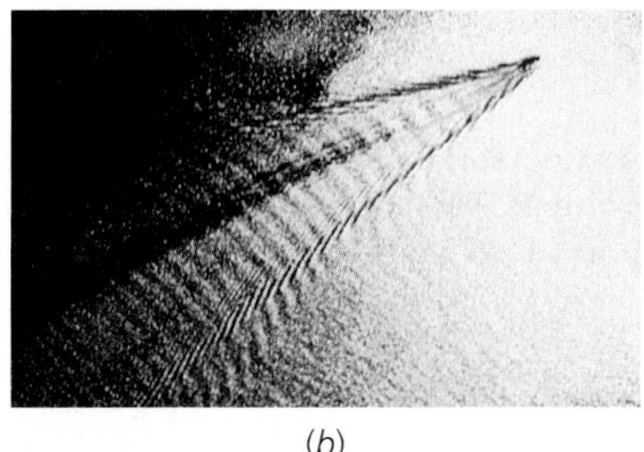

(b)

(c)

(a) Shock waves from a supersonic airplane. (b) Bow waves from a boat. (c) Shock waves produced by a bullet traversing a helium balloon.

Figure 15-26 shows a source originally at point P_1 moving to the right with speed u. After some time t, the wave emitted from point P_1 has traveled a distance vt. The source has traveled a distance ut and will be at point P_2. The line from this new position of the source to the wavefront emitted when the source was at P_1 makes an angle θ with the path of the source, given by

$$\sin\theta = \frac{vt}{ut} = \frac{v}{u} \qquad 15\text{-}38$$

Thus the shock wave is confined to a cone that narrows as u increases. The ratio of the source speed u to the wave speed v is called the Mach number:

$$\text{Mach number} = \frac{u}{v} \qquad 15\text{-}39$$

Equation 15-38 also applies to the electromagnetic radiation called Cerenkov radiation, which is given off when a charged particle moves in a medium with speed u that is greater than the speed of light v in that medium. (According to the special theory of relativity, it is impossible for a particle to move faster than c, the speed of light in vacuum. In a medium such as glass however, electrons and other particles can move faster than the speed of light in that medium.) The blue glow surrounding the fuel elements of a nuclear reactor is an example of Cerenkov radiation.

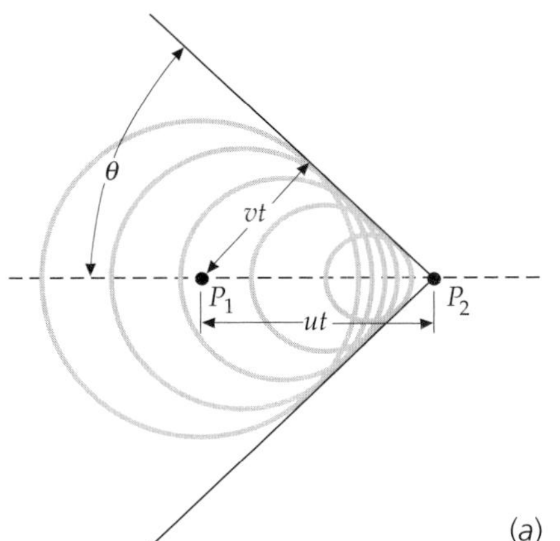

(a)

(b)

FIGURE 15-26 **(a) Source moving with a speed u that is greater than the wave speed v. The envelope of the wavefronts forms a cone with the source at the apex. (b) Waves in a ripple tank produced by a source moving with a speed $u > v$.**

A Sonic Boom — EXAMPLE 15-13 Try It Yourself

A supersonic plane flying due east at an altitude of 15 km passes directly over point *P*. The sonic boom is heard at point *P* when the plane is 22 km east of point *P*. What is the speed of the supersonic plane?

PICTURE THE PROBLEM The speed of the plane is related the sine of the Mach angle (Equation 15-38). Draw a picture so the sine of the Mach angle can be calculated.

Cover the column to the right and try these on your own before looking at the answers.

Steps	Answers
1. Sketch the position of the plane (Figure 15-27) both at the instant the sonic boom is heard at point *P* and at the instant that sound was produced. Label the distance the sound travels $v\Delta t$, and the distance the plane travels $u\Delta t$.	FIGURE 15-27
2. From your sketch and Equation 15-38, calculate u.	$\tan\theta = \dfrac{15\text{ km}}{22\text{ km}}$ so $\theta = 34.3°$ $\sin\theta = \dfrac{v\Delta t}{u\Delta t} = \dfrac{v}{u}$ so $u = \dfrac{v}{\sin\theta} = \boxed{604\text{ m/s}}$

SUMMARY

1. In wave motion, energy and momentum are transported from one point in space to another without the transport of matter.
2. The relation $v = f\lambda$ holds for all harmonic waves.

Topic	Relevant Equations and Remarks	
1. Transverse and Longitudinal Waves	In transverse waves, such as waves on a string, the disturbance is perpendicular to the direction of propagation. In longitudinal waves, such as sound waves, the disturbance is along the direction of propagation.	
2. Speed of Waves	The wave speed v is independent of the motion of the wave source. The speed of a wave relative to the medium depends on the density and elastic properties of the medium.	
Waves on a string	$v = \sqrt{F_T/\mu}$	**15-3**
Sound waves	$v = \sqrt{B/\rho}$	**15-4**

Sound waves in a gas	$v = \sqrt{\gamma RT/M}$	**15-5**
	where T is the absolute temperature, $T = t_C + 273$	**15-6**
	R is the universal gas constant, $R = 8.314\ \mathrm{J/mol\cdot K}$	**15-7**
	M is the molar mass of the gas, which for air is 29×10^{-3} kg/mol, and γ is a constant that depends on the kind of gas. For a diatomic gas such as air, $\gamma = 1.4$. For a monatomic gas such as helium, $\gamma = 1.67$.	
Electromagnetic waves	The speed of electromagnetic waves in vacuum is a universal constant $c = 3 \times 10^8$ m/s	
***3. Wave Equation**	$\frac{\partial^2 y}{\partial x^2} = \frac{1}{v^2}\frac{\partial^2 y}{\partial t^2}$	**15-9*b***
4. Harmonic Waves		
Wave function	$y(x,t) = A \sin(kx \pm \omega t)$ where A is the amplitude, k is the wave number, and ω is the angular frequency. Use − for a wave traveling in the positive x direction, and + for a wave traveling in the negative x direction.	**15-13**
Wave number	$k = \frac{2\pi}{\lambda}$	**15-12**
Angular frequency	$\omega = 2\pi f = \frac{2\pi}{T}$	**15-15**
Speed	$v = f\lambda = \omega/k$	**15-10, 15-14**
Energy	The energy in a harmonic wave is proportional to the square of the amplitude.	
Power for harmonic waves on a string	$P_{av} = \frac{1}{2}\mu v \omega^2 A^2$	**15-18**
5. Harmonic Sound Waves	Sound waves can be considered to be either displacement waves or pressure waves. The human ear is sensitive to sound waves of frequencies from about 20 Hz to 20 kHz. In a harmonic sound wave, the pressure and displacement are 90° out of phase.	
Amplitudes	The pressure and displacement amplitudes are related by $p_0 = \rho\omega v s_0$ where ρ is the density of the medium.	**15-22**
Energy density	$\eta_{av} = \frac{(\Delta E)_{av}}{\Delta V} = \frac{1}{2}\rho\omega^2 s_0^2$	**15-24**
6. Intensity	The intensity of a wave is the average power per unit area. $I = \frac{P_{av}}{A}$	**15-25**

Topic	Relevant Equations and Remarks	
Average energy density η_{av} of a sound wave	$I = \eta_{av}\, v = \frac{1}{2}\rho\omega^2 s_0^2 v = \frac{1}{2}\frac{p_0^2}{\rho v}$	15-28
*Intensity level β in dB	Sound intensity levels are measured on a logarithmic scale. $\beta = 10 \log \frac{I}{I_0}$ where $I_0 = 10^{-12}\ \mathrm{W/m^2}$ is approximately the threshold of hearing.	15-29
7. Reflection and Refraction	When a wave is incident on a boundary surface that separates two regions of differing wave speed, part of the wave is reflected and part is transmitted.	
8. Diffraction	If a wavefront is partially blocked by an obstacle, the unblocked part of the wavefront diffracts (bends) into the region behind the obstacle.	
Ray approximation	If a wavefront is partially blocked by an obstacle, almost all of the diffraction occurs for that part of the wavefront that passes within a few wavelengths of the edge. For those parts of the wavefront that pass farther from the edge than a few wavelengths, diffraction is negligible and the wave propagates in straight lines in the direction of the incident rays.	
9. Doppler Effect	When a sound source and receiver are in relative motion, the received frequency f_r is higher than the frequency of the source f_s if their separation is decreasing, and lower if their separation is increasing.	
Moving source	$\lambda = \frac{v \pm u_s}{f_s}$	15-32†
Moving receiver	$f_r = \frac{v \pm u_r}{\lambda}$	15-34†
Either source or receiver moving	$f_r = \frac{v \pm u_r}{v \pm u_s} f_s$ or $\frac{f_r}{v \pm u_r} = \frac{f_s}{v \pm u_s}$ Choose the signs that give an up-shift in frequency for an approaching source or receiver, and vice versa.	15-35†
Small speeds of source or receiver	$\frac{\Delta f}{f_s} \approx \pm\frac{u}{v} \quad (u << v)$	15-36†
Relativistic Doppler shift	$f_r = \sqrt{\frac{c \pm u}{c \mp u}}\, f_s$ Choose the signs that give an up-shift in frequency for an approaching source or receiver, and vice versa.	15-37
10. Shock Waves	When the source speed is greater than the wave speed, the waves behind the source are confined to a cone of angle θ given by	
Mach angle	$\sin\theta = \frac{v}{u}$	15-38
Mach number	$\text{Mach number} = \frac{u}{v}$	15-39

† Equations 15-32 through 15-36 are valid *only* in the reference frame of the medium. If the medium is moving, the wave speed v is replaced by $v' = v \pm u_m$, where u_m is the speed of the medium.

PROBLEMS

- • Single-concept, single-step, relatively easy
- •• Intermediate-level, may require synthesis of concepts
- ••• Challenging
- SSM Solution is in the *Student Solutions Manual*
- iSOLVE Problems available on iSOLVE online homework service
- iSOLVE ✓ These "Checkpoint" online homework service problems ask students additional questions about their confidence level, and how they arrived at their answer

In a few problems, you are given more data than you actually need; in a few other problems, you are required to supply data from your general knowledge, outside sources, or informed estimates.

Use v = 340 m/s for the speed of sound in air unless otherwise indicated.

Conceptual Problems

1 • SSM A rope hangs vertically from the ceiling. Do waves on the rope move faster, slower, or at the same speed as they move from bottom to top? Explain.

2 • A traveling wave passes a point of observation. At this point, the time between successive crests is 0.2 s. Which of the following is true? (*a*) The wavelength is 5 m. (*b*) The frequency is 5 Hz. (*c*) The velocity of propagation is 5 m/s. (*d*) The wavelength is 0.2 m. (*e*) There is not enough information to justify any of these statements.

3 • True or false: The energy in a wave is proportional to the square of the amplitude of the wave.

4 • A rope hangs vertically. You shake the bottom back and forth, creating a sinusoidal wave train. Is the wavelength near the top the same as, less than, or greater than the wavelength near the bottom?

5 • SSM The crack of a bullwhip is caused by the speed of the tip breaking the sound barrier. Explain how the tapered shape of the whip helps the tip move much faster than the hand holding the whip.

6 • True or false: A 60-dB sound has twice the intensity of a 30-dB sound.

7 • If the source and receiver are at rest relative to each other but the wave medium is moving relative to them, will there be any Doppler shift in frequency?

8 • The frequency of a car horn is f_0. What frequency is observed if both the car and the observer are at rest, but a wind blows toward the observer? (*a*) f_0. (*b*) Greater than f_0. (*c*) Less than f_0. (*d*) It could be either greater or less than f_0. (*e*) It could be f_0 or greater than f_0, depending on how wind speed compares to speed of sound.

9 •• SSM Stars often occur in pairs revolving around their common center of mass. If one of the stars is a black hole, it is invisible. Explain how the existence of such a black hole might be inferred from the light observed from the other, visible star.

10 • When a guitar string is plucked, is the wavelength of the wave it produces in air the same as the wavelength of the wave on the string?

11 • True or false:
(*a*) Wave pulses on strings are transverse waves.
(*b*) Sound waves in air are transverse waves of compression and rarefaction.
(*c*) The speed of sound in air at 20°C is twice that at 5°C.

12 • Sound travels at 340 m/s in air and 1500 m/s in water. A sound of 256 Hz is made under water. In the air, the frequency will be (*a*) the same, but the wavelength will be shorter, (*b*) higher, but the wavelength will stay the same, (*c*) lower, but the wavelength will be longer, (*d*) lower, and the wavelength will be shorter, (*e*) the same, and the wavelength too will stay the same.

13 • SSM While out on patrol, the battleship *Rodger Young* hits a mine and begins to burn, ultimately exploding. Sailor Abel jumps into the water and begins swimming away from the doomed ship, while Sailor Baker gets into a life raft. Comparing their experiences later, Abel tells Baker, "I was swimming underwater, and heard a big explosion from the ship. When I surfaced, I heard a second explosion. What do you think it could be?" Baker says, "I think it was your imagination—I only heard one explosion." Explain why Baker only heard one explosion, while Abel heard two.

14 •• Figure 15-28 shows a wave pulse at time $t = 0$ moving to the right. At this particular time, which segments of the string are moving up? Which are moving down? Is there any segment of the string at the pulse that is instantaneously at rest? Answer these questions by sketching the pulse at a slightly later time and a slightly earlier time to see how the segments of the string are moving.

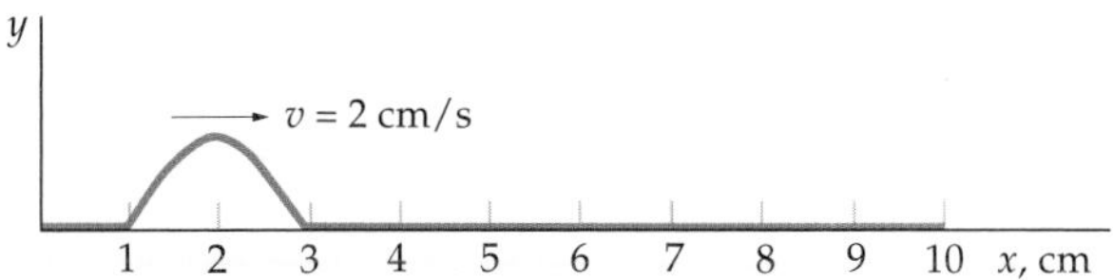

FIGURE 15-28 Problems 14, 15

15 •• Make a sketch of the velocity of each string segment versus position for the pulse shown in Figure 15-28.

16 •• In a classic physics experiment, a bell is placed in a sealed jar and rung while the air is slowly removed. After a while, the bell becomes inaudible. This is commonly cited as proof that sound waves can't travel through a vacuum, but in fact the sound becomes inaudible well before the jar is completely evacuated. Can you give another reason why the sound from the bell can't be heard?

17 •• SSM The explosion of a depth charge beneath the surface of the water is recorded by a helicopter hovering above its surface, as shown in Figure 15-29. Along which path, A, B, or C, will the sound wave take the least time to reach the helicopter?

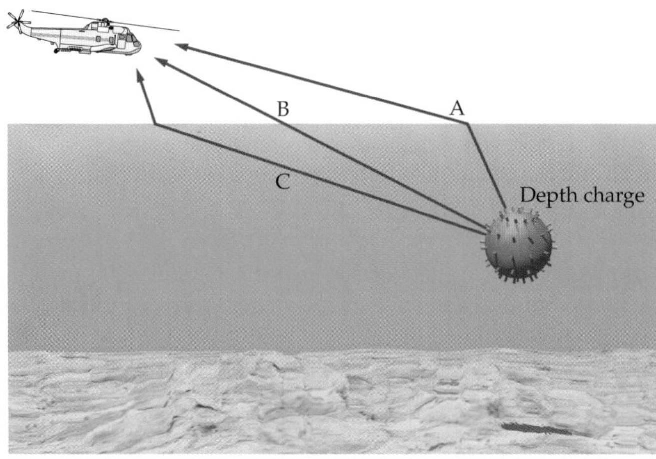

FIGURE 15-29 Problem 17

Estimation and Approximation

18 •• Normal human speech has a sound intensity level of about 65 dB at a distance of 1 m. Estimate the power of human speech.

19 •• A man drops a stone from a high bridge and hears it strike the water below exactly 4 s later. (*a*) Estimate the distance to the water based on the assumption that the travel time for the sound to reach the man is negligible. (*b*) Improve your estimate by using your result from Part (*a*) for the distance to the water to estimate the time it takes for sound to travel this distance and then calculate the distance the rock falls in 4 s minus this time. (*c*) Calculate the exact distance and compare your result with your previous estimates.

20 •• SSM Estimate the speed of the bullet as it passes through the helium balloon in Figure 15-30 from the angle of its shock cone.

FIGURE 15-30 Problem 20

21 •• The new student townhouses at a local college are in the form of a semicircle half-enclosing the track field. To estimate the speed of sound in air, an ambitious physics student stood at the center of the semicircle and clapped his hands rhythmically at a frequency at which he couldn't hear the echo of the clap, because it reached him at the same time as his next clap. This frequency was about 2.5 claps/s. Once he established this frequency, he paced off the distance to the townhouses, which was 30 double paces. Assuming that the distance of a double-pace stride is the same as his height (5 ft 11 in), estimate the speed of sound in air using this data. How far off is this from the commonly accepted value?

Speed of Waves

22 • ISOLVE✓ (*a*) The bulk modulus for water is 2.0×10^9 N/m^2. Use it to find the speed of sound in water. (*b*) The speed of sound in mercury is 1410 m/s. What is the bulk modulus for mercury ($\rho = 13.6 \times 10^3$ kg/m^3)?

23 • SSM ISOLVE✓ Calculate the speed of sound waves in hydrogen gas at $T = 300$ K. (Take $M = 2$ g/mol and $\gamma = 1.4$.)

24 • A steel wire 7 m long has a mass of 100 g. It is under a tension of 900 N. What is the speed of a transverse wave pulse on this wire?

25 • Transverse waves travel at 150 m/s on a wire of length 80 cm that is under a tension of 550 N. What is the mass of the wire?

26 • SSM A wave pulse propagates along a wire in the positive x direction at 20 m/s. What will be the pulse velocity if we (*a*) double the length of the wire but keep the tension and mass per unit length constant? (*b*) double the tension while holding the length and mass per unit length constant? (*c*) double the mass per unit length while holding the other variables constant?

27 • ISOLVE A steel piano wire is 0.7 m long and has a mass of 5 g. It is stretched with a tension of 500 N. (*a*) What is the speed of transverse waves on the wire? (*b*) To reduce the wave speed by a factor of 2 without changing the tension, what mass of copper wire would have to be wrapped around the steel wire?

28 •• A common method for estimating the distance to a lightning bolt is to begin counting when the flash is observed and continue until the thunder clap is heard. The number of seconds counted is then divided by 3 to get the distance in kilometers. (*a*) What is the velocity of sound in kilometers per second? (*b*) How accurate is this procedure? (*c*) Is a correction for the time it takes for the light to reach you important? (The speed of light is 3×10^8 m/s.)

29 •• SSM (*a*) Compute the derivative of the speed of a wave on a string with respect to the tension dv/dF, and show that the differentials dv and dF obey $dv/v = \frac{1}{2}dF/F$. (*b*) A wave moves with a speed of 300 m/s on a wire that is under a tension of 500 N. Using dF to approximate a change in tension, determine how much the tension must be changed to increase the speed to 312 m/s.

30 •• (*a*) Compute the derivative of the velocity of sound in air with respect to the absolute temperature, and show that the differentials dv and dT obey $dv/v = \frac{1}{2}dT/T$. (*b*) Use this result to compute the percentage change in the velocity of sound when the temperature changes from 0 to 27°C. (*c*) If the speed of sound is 331 m/s at 0°C, what is it (approximately) at 27°C? How does this approximation compare with the result of an exact calculation?

31 ••• In this problem, you will derive a convenient formula for the speed of sound in air at temperature t in Celsius degrees. Begin by writing the temperature as $T = T_0 + \Delta T$, where $T_0 = 273$ K corresponds to 0°C and $\Delta T = t$, the Celsius temperature. The speed of sound is a function of T, $v(T)$. To a first-order approximation, you can write $v(T) \approx v(T_0) + (dv/dT)_{T_0}\,\Delta T$, where $(dv/dT)_{T_0}$ is the derivative evaluated at $T = T_0$. Compute this derivative, and show that the result leads to $v = (331 \text{ m/s})\left(1 + \dfrac{t}{2T_0}\right) = (331 + 0.606t)$ m/s.

32 •• Various stories of psychic phenomena can often be explained by considering physical phenomena. For example, there is the often repeated story of a man being woken from a deep sleep for no reason, getting out of bed and walking to the window just in time to hear the sound of the explosion of a munitions plant across town. The story is often cited to give credence to the idea of clairvoyance, but can be explained instead by assuming that the man was woken by the tremor of the sound wave traveling through the earth, and then walked to the window in time to hear the sound wave traveling through the air. If he took 3 s to move from his bed to the window, and the average speed of sound through solid rock is 3000 m/s, how far was his house from the munitions plant?

33 ••• While studying physics in her dorm room, a student is listening to a live radio broadcast of a baseball game. She is 1.6 km due south of the baseball field. Over her radio, the student hears a noise generated by the electromagnetic pulse of a lightning bolt. Two seconds later, she hears over the radio the thunder picked up by the microphone at the baseball field. Four seconds after she hears the noise of the electromagnetic pulse over the radio, thunder rattles her windows. Where, relative to the ballpark, did the lightning bolt occur?

34 ••• SSM Weather station Beta is located 0.75 mi due east of weather station Alpha. Observers at the two stations see a lightning strike to the north of the stations; observers at station Alpha hear the thunder 3.4 s after seeing the strike, while observers at Beta hear it 2.5 s after seeing the strike. Locate the coordinates of the lightning strike relative to the position of station Alpha.

35 ••• A coiled spring, such as a Slinky, is stretched to a length L. It has a force constant k and a mass m. (*a*) Show that the velocity of longitudinal compression waves along the spring is given by $v = L\sqrt{k/m}$. (*b*) Show that this is also the velocity of transverse waves along the spring if the natural length of the spring is much less than L.

The Wave Equation

36 • Show explicitly that the following functions satisfy the wave equation: (*a*) $y(x,t) = k(x + vt)^3$; (*b*) $y(x, t) = Ae^{ik(x - vt)}$ where A and k are constants and $i = \sqrt{-1}$; and (*c*) $y(x,t) = \ln k(x - vt)$.

37 • SSM Show that the function $y = A \sin kx \cos \omega t$ satisfies the wave equation.

Harmonic Waves on a String

38 • ISOLVE One end of a string 6 m long is moved up and down with simple harmonic motion at a frequency of 60 Hz. The waves reach the other end of the string in 0.5 s. Find the wavelength of the waves on the string.

39 • Equation 15-13 expresses the displacement of a harmonic wave as a function of x and t in terms of the wave parameters k and ω. Write the equivalent expressions that contain the following pairs of parameters instead of k and ω: (*a*) k and v, (*b*) λ and f, (*c*) λ and T, (*d*) λ and v, and (*e*) f and v.

40 • SSM Equation 15-10 applies to all types of periodic waves, including electromagnetic waves such as light waves and microwaves, which travel at 3×10^8 m/s in a vacuum. (*a*) The range of wavelengths of light to which the eye is sensitive is about 4×10^{-7} to 7×10^{-7} m. What are the frequencies that correspond to these wavelengths? (*b*) Find the frequency of a microwave that has a wavelength of 3 cm.

41 • ISOLVE A harmonic wave on a string with a mass per unit length of 0.05 kg/m and a tension of 80 N has an amplitude of 5 cm. Each section of the string moves with simple harmonic motion at a frequency of 10 Hz. Find the power propagated along the string.

42 • ISOLVE✓ A rope 2 m long has a mass of 0.1 kg. The tension is 60 N. A power source at one end sends a harmonic wave with an amplitude of 1 cm down the rope. The wave is extracted at the other end without any reflection. What is the frequency of the power source if the power transmitted is 100 W?

43 •• The wave function for a harmonic wave on a string is $y(x, t) = (0.001 \text{ m}) \sin(62.8 \text{ m}^{-1} x + 314 \text{ s}^{-1} t)$. (*a*) In what direction does this wave travel, and what is its speed? (*b*) Find the wavelength, frequency, and period of this wave. (*c*) What is the maximum speed of any string segment?

44 •• A harmonic wave with a frequency of 80 Hz and an amplitude of 0.025 m travels along a string to the right with a speed of 12 m/s. (*a*) Write a suitable wave function for this wave. (*b*) Find the maximum speed of a point on the string. (*c*) Find the maximum acceleration of a point on the string.

45 •• ISOLVE Waves of frequency 200 Hz and amplitude 1.2 cm move along a 20-m string that has a mass of 0.06 kg and a tension of 50 N. (*a*) What is the average total energy of the waves on the string? (*b*) Find the power transmitted past a given point on the string.

46 •• SSM In a real string, a wave loses some energy as it travels down the string. Such a situation can be described by a wave function whose amplitude $A(x)$ depends on x: $y = A(x) \sin (kx - \omega t) = (A_0e^{-bx}) \sin (kx - \omega t)$ (*a*) What is the power transported by the wave at the origin? (*b*) What is the power transported by the wave at point x, where $x > 0$?

47 •• Power is to be transmitted along a stretched wire by means of transverse harmonic waves. The wave speed is 10 m/s and the linear mass density of the wire is 0.01 kg/m. The power source oscillates with an amplitude of 0.50 mm. (*a*) What average power is transmitted along the wire if the frequency is 400 Hz? (*b*) The power transmitted can be increased by increasing the tension in the wire, the frequency of the source, or the amplitude of the waves. By how much would each of these quantities have to increase to cause an increase in power by a factor of 100 if it is the only quantity changed? (*c*) Which of the quantities would probably be the easiest to vary?

48 ••• SSM Two very long strings are tied together at the point $x = 0$. In the region $x < 0$, the wave speed is v_1, while in the region $x > 0$, the speed is v_2. A sinusoidal wave is incident from the left ($x < 0$); part of the wave is reflected and part is transmitted. For $x < 0$, the displacement of the wave is describable by $y(x,t) = A \sin(k_1x - \omega t) + B \sin(k_1x + \omega t)$, while for $x > 0$, $y(x,t) = C \sin(k_2x - \omega t)$, where $\omega/k_1 = v_1$ and $\omega/k_2 = v_2$. (*a*) If we assume that both the wave function y and its first spatial derivative $\partial y/\partial x$ must be continuous at $x = 0$, show that $C/A = 2/(1 + v_1/v_2)$, and that $B/A = (1 - v_1/v_2)/(1 + v_1/v_2)$. (*b*) Show that $B^2 + (v_1/v_2)C^2 = A^2$.

Harmonic Sound Waves

49 • SSM A sound wave in air produces a pressure variation given by

$$p(x,t) = 0.75 \cos \frac{\pi}{2}(x - 340t)$$

where p is in pascals, x is in meters, and t is in seconds. Find (*a*) the pressure amplitude of the sound wave, (*b*) the wavelength, (*c*) the frequency, and (*d*) the speed.

50 • ISOLVE (*a*) Middle C on the musical scale has a frequency of 262 Hz. What is the wavelength of this note in air? (*b*) The frequency of the C an octave above middle C is twice that of middle C. What is the wavelength of this note in air?

51 • (*a*) What is the displacement amplitude for a sound wave having a frequency of 100 Hz and a pressure amplitude of 10^{-4} atm? (*b*) The displacement amplitude of a sound wave of frequency 300 Hz is 10^{-7} m. What is the pressure amplitude of this wave?

52 • ISOLVE✓ (*a*) Find the displacement amplitude of a sound wave of frequency 500 Hz at the pain-threshold pressure amplitude of 29 Pa. (*b*) Find the displacement amplitude of a sound wave with the same pressure amplitude but a frequency of 1 kHz.

53 • A typical loud sound wave with a frequency of 1 kHz has a pressure amplitude of about 10^{-4} atm. (*a*) At $t = 0$, the pressure is a maximum at some point x_1. What is the displacement at that point at $t = 0$? (*b*) What is the maximum value of the displacement at any time and place? (Take the density of air to be 1.29 kg/m³.)

54 • SSM An octave represents a change in frequency by a factor of two. Over how many octaves can a typical person hear?

Waves in Three Dimensions: Intensity

55 • A piston at one end of a long tube filled with air at room temperature and normal pressure oscillates with a frequency of 500 Hz and an amplitude of 0.1 mm. The area of the piston is 100 cm². (*a*) What is the pressure amplitude of the sound waves generated in the tube? (*b*) What is the intensity of the waves? (*c*) What average power is required to keep the piston oscillating (neglecting friction)?

56 • A spherical source radiates sound uniformly in all directions. At a distance of 10 m, the sound intensity level is 10^{-4} W/m². (*a*) At what distance from the source is the intensity 10^{-6} W/m²? (*b*) What power is radiated by this source?

57 • SSM ISOLVE A loudspeaker at a rock concert generates 10^{-2} W/m² at 20 m at a frequency of 1 kHz. Assume that the speaker spreads its energy uniformly in three dimensions. (*a*) What is the total acoustic power output of the speaker? (*b*) At what distance will the intensity be at the pain threshold of 1 W/m²? (*c*) What is the intensity at 30 m?

58 •• When a pin of mass 0.1 g is dropped from a height of 1 m, 0.05 percent of its energy is converted into a sound pulse with a duration of 0.1 s. (*a*) Estimate the range at which the dropped pin can be heard if the minimum audible intensity is 10^{-11} W/m². (*b*) Your result in (*a*) is much too large in practice because of background noise. If you assume that the intensity must be at least 10^{-8} W/m² for the sound to be heard, estimate the range at which the dropped pin can be heard. (In both parts, assume that the intensity is $P/4\pi r^2$.)

*Intensity Level

59 • What is the intensity level in decibels of a sound wave of intensity (*a*) 10^{-10} W/m² and (*b*) 10^{-2} W/m²?

60 • ISOLVE Find the intensity of a sound wave if (*a*) $\beta = 10$ dB and (*b*) $\beta = 3$ dB. (*c*) Find the pressure amplitudes of sound waves in air for each of these intensities.

61 • SSM The sound level of a dog's bark is 50 dB. The intensity of a rock concert is 10,000 times that of the dog's bark. What is the sound level of the rock concert?

62 • ISOLVE Two sounds differ by 30 dB. The intensity of the louder sound is I_L and that of the softer sound is I_s. The value of the ratio I_L/I_s is (*a*) 1000, (*b*) 30, (*c*) 9, (*d*) 100, (*e*) 300.

63 • Show that if the intensity is doubled, the intensity level increases by 3.0 dB.

64 • SSM What fraction of the acoustic power of a noise would have to be eliminated to lower its sound intensity level from 90 to 70 dB?

65 •• ISOLVE A spherical source radiates sound uniformly in all directions. At a distance of 10 m, the sound intensity level is 80 dB. (*a*) At what distance from the source is the intensity level 60 dB? (*b*) What power is radiated by this source?

66 •• A spherical source of intensity I_0 radiates sound uniformly in all directions. Its intensity level is β_1 at a distance r_1 and β_2 at a distance r_2. Find β_2/β_1.

67 •• ISOLVE✓ A loudspeaker at a rock concert generates 10^{-2} W/m^2 at 20 m at a frequency of 1 kHz. Assume that the speaker spreads its energy uniformly in all directions. (*a*) What is the intensity level at 20 m? (*b*) What is the total acoustic power output of the speaker? (*c*) At what distance will the intensity level be at the pain threshold of 120 dB? (*d*) What is the intensity level at 30 m?

68 •• An article on noise pollution claims that sound intensity levels in large cities have been increasing by about 1 dB annually. (*a*) To what percentage increase in intensity does this correspond? Does this increase seem plausible? (*b*) In about how many years will the intensity of sound double if it increases at 1 dB annually?

69 •• Three noise sources produce intensity levels of 70, 73, and 80 dB when acting separately. When the sources act together, their intensities add. (*a*) Find the sound intensity level in decibels when the three sources act at the same time. (*b*) Discuss the effectiveness of eliminating the two least intense sources in reducing the intensity level of the noise.

70 •• SSM If you double the distance between a source of sound and a receiver, the intensity at the receiver drops by approximately (*a*) 2 dB, (*b*) 3 dB, (*c*) 6 dB, (*d*) Amount cannot be determined from the information given.

71 ••• ISOLVE Everyone at a party is talking equally loudly. If only one person were talking, the sound level would be 72 dB. Find the sound level when all 38 people are talking.

72 ••• SSM When a violinist pulls the bow across a string, the force with which the bow is pulled is fairly small, about 0.6 N. Suppose the bow travels across the A string, which vibrates at 440 Hz, at 0.5 m/s. A listener 35 m from the performer hears a sound of 60 dB intensity. With what efficiency is the mechanical energy of bowing converted to sound energy? (Assume that the sound radiates uniformly in all directions.)

73 ••• The noise level in an empty examination hall is 40 dB. When 100 students are writing an exam, the sounds of heavy breathing and pens traveling rapidly over paper cause the noise level to rise to 60 dB (not counting the occasional groans). Assuming that each student contributes an equal amount of noise power, find the noise level to the nearest decibel when 50 students have left.

The Doppler Effect

In Problems 74 through 79, a source emits sounds of frequency 200 Hz that travel through still air at 340 m/s.

74 • The sound source described moves with a speed of 80 m/s relative to still air toward a stationary listener. (*a*) Find the wavelength of the sound in the region between the source and the listener. (*b*) Find the frequency heard by the listener.

75 • Consider the situation in Problem 74 from the reference frame in which the source is at rest. In this frame, the listener moves toward the source with a speed of 80 m/s, and there is a wind blowing at 80 m/s from the listener to the source. (*a*) What is the speed at which the sound travels from the source to the listener in this frame? (*b*) Find the wavelength of the sound in the region between the source and the listener. (*c*) Find the frequency heard by the listener.

76 • The source moves away from the stationary listener at 80 m/s. (*a*) Find the wavelength of the sound waves in the region between the source and the listener. (*b*) Find the frequency heard by the listener.

77 • ISOLVE✓ The listener moves at 80 m/s relative to still air toward the stationary source. (*a*) What is the wavelength of the sound in the region between the source and the listener? (*b*) What is the frequency heard by the listener?

78 • Consider the situation in Problem 77 in a reference frame in which the listener is at rest. (*a*) What is the wind velocity in this frame? (*b*) What is the speed of the sound as it travels from the source to the listener in this frame, that is, relative to the listener? (*c*) Find the wavelength of the sound in the region between the source and the listener in this frame. (*d*) Find the frequency heard by the listener.

79 • The listener moves at 80 m/s relative to the still air away from the stationary source. Find the frequency heard by the listener.

80 • A jet is traveling at Mach 2.5 at an altitude of 5000 m. (*a*) What is the angle that the shock wave makes with the track of the jet? (Assume that the speed of sound at this altitude is still 340 m/s.) (*b*) Where is the jet when a person on the ground hears the shock wave?

81 • If you are running at top speed toward a source of sound at 1000 Hz, estimate the frequency of the sound that you hear. Suppose that you can recognize a change in frequency of 3 percent. Can you use your sense of pitch to estimate your running speed?

82 •• ISOLVE A radar device emits microwaves with a frequency of 2.00 GHz. When the waves are reflected from a car moving directly away from the emitter, a frequency difference of 293 Hz is detected. Find the speed of the car.

83 •• SSM The Doppler effect is routinely used to measure the speed of winds in storm systems. A weather station uses a Doppler radar system of frequency $f = 625$ MHz to bounce a radar pulse off of the raindrops in a swirling thunderstorm system 50 km away; the reflected radar pulse is found to be up-shifted in frequency by 325 Hz. Assuming the wind is headed directly toward the radar antenna, how fast are the winds in the storm system moving? (The radar system can only measure the radial component of the velocity.)

84 •• ISOLVE A stationary destroyer is equipped with sonar that sends out pulses of sound at 40 MHz. Reflected pulses are received from a submarine directly below with a time delay of 80 ms at a frequency of 39.958 MHz. If the speed of sound in seawater is 1.54 km/s, find (*a*) the depth of the submarine and (*b*) its vertical speed.

85 •• A police radar unit transmits microwaves of frequency 3×10^{10} Hz. The speed of these waves in air is 3.0×10^8 m/s. Suppose a car is receding from the stationary police car at a speed of 140 km/h. What is the frequency difference between the transmitted signal and the signal received from the receding car?

86 •• Suppose the police car of Problem 85 is moving in the same direction as the other vehicle at a speed of 60 km/h. What then is the difference in frequency between the emitted and the reflected signals?

87 •• ISOLVE At time $t = 0$, a supersonic plane is directly over point P, flying due west at an altitude of 12 km and a speed of Mach 1.6. Where is the plane when the sonic boom is heard?

88 •• A small radio of mass 0.10 kg is attached to one end of an air track by a spring. The radio emits a sound of 800 Hz. A listener at the other end of the air track hears a sound whose frequency varies between 797 and 803 Hz. (*a*) Determine the energy of the vibrating mass–spring system. (*b*) If the spring constant is 200 N/m, what is the amplitude of vibration of the mass and what is the period of the oscillating system?

89 •• A sound source of frequency f_0 moves with speed u_s relative to still air toward a receiver who is moving with speed u_r relative to still air away from the source. (*a*) Write an expression for the received frequency f'. (*b*) Use the result that $(1 - x)^{-1} \approx 1 + x$ to show that if both u_s and u_r are small compared to v, then the received frequency is approximately

$$f' \approx \left(1 + \frac{u_s - u_r}{v}\right) f_0 = \left(1 + \frac{u_{rel}}{v}\right) f_0$$

where $u_{rel} = u_s - u_r$ is the relative velocity of approach of the source and receiver.

90 •• Two students with vibrating 440-Hz tuning forks walk away from each other with equal speeds. How fast must they walk so that they each hear a frequency of 438 Hz from the other fork?

91 •• A physics student walks down a long hall carrying a vibrating 512-Hz tuning fork. The end of the hall is closed so that sound reflects from it. The student hears a sound of 516 Hz from the wall. How fast is the student walking?

92 •• SSM A small speaker radiating sound at 1000 Hz is tied to one end of an 0.8-m-long rod that is free to rotate about its other end. The rod rotates in the horizontal plane at 4.0 rad/s. Derive an expression for the frequency heard by a stationary observer far from the rotating speaker.

93 •• A balloon carried along by a 36-km/h wind emits a sound of 800 Hz as it approaches a tall building. (*a*) What is the frequency of the sound heard by an observer at the window of this building? (*b*) What is the frequency of the reflected sound heard by a person riding in the balloon?

94 •• ISOLVE✓ A car is approaching a reflecting wall. A stationary observer behind the car hears a sound of frequency 745 Hz from the car horn and a sound of frequency 863 Hz from the wall. (*a*) How fast is the car traveling? (*b*) What is the frequency of the car horn? (*c*) What frequency does the car driver hear reflected from the wall?

95 •• The driver of a car traveling at 100 km/h toward a vertical cliff briefly sounds the horn. Exactly 1 s later she hears the echo and notes that its frequency is 840 Hz. How far from the cliff was the car when the driver sounded the horn and what is the frequency of the horn?

96 •• You are on a transatlantic flight traveling due west at 800 km/h. A Concorde flying at Mach 1.6 and 3 km to the north of your plane is also on an east-to-west course. What is the distance between the two planes when you hear the sonic boom from the Concorde?

97 •• SSM The Hubble space telescope has been used to determine the existence of planets orbiting around distant stars. The planet orbiting the star will cause the star to "wobble" with the same period as the planet's orbit; because of this, light from the star will be Doppler-shifted up and down periodically. Estimate the maximum and minimum wavelengths of light of nominal wavelength 500nm emitted by the sun that is Doppler-shifted by the motion of the sun due to the planet Jupiter.

98 ••• ISOLVE✓ A physics student drops a vibrating 440-Hz tuning fork down the elevator shaft of a tall building. When the student hears a frequency of 400 Hz, how far has the tuning fork fallen?

99 •• The SuperKamiokande neutrino detector in Japan is a water tank the size of a 14-story building. It detects neutrinos, the "ghost particles" of physics, by the shock wave produced when a neutrino imparts most of its energy to an electron, which then goes flying off at near light-speed through the water. If the maximum angle of the Cerenkov shock-wave cone is 48.75°, what is the speed of light in water?

General Problems

100 • At time $t = 0$, the shape of a wave pulse on a string is given by the function

$$y(x,0) = \frac{0.12\ \text{m}^3}{(2.00\ \text{m})^2 + x^2}$$

where x is in meters. (*a*) Sketch $y(x, 0)$ versus x. Give the wave function $y(x,t)$ at a general time t if (*b*) the pulse is moving in the positive x direction with a speed of 10 m/s and (*c*) the pulse is moving in the negative x direction with a speed of 10 m/s.

101 • ISOLVE A wave with a frequency of 1200 Hz propagates along a wire that is under a tension of 800 N. Its wavelength is 24 cm. What will be the wavelength if the tension is decreased to 600 N and the frequency is kept constant?

102 • ISOLVE✓ In a common lecture demonstration of wave pulses, a length of rubber tubing is tied at one end to a fixed post and is passed over a pulley to a weight hanging at the other end. Suppose that the distance from the fixed support to the pulley is 10 m, the mass of this length of tubing is 0.7 kg, and the suspended weight is 110 N. If the tubing is given a transverse blow at one end, how long will it take the resulting pulse to reach the other end?

103 • A boat traveling at 10 m/s on a still lake makes a bow wave at an angle of 20° with its direction of motion. What is the speed of the bow wave?

104 • If a wavelength is much larger than the diameter of a loudspeaker, the speaker radiates in all directions, much like a point source. On the other hand, if the wavelength is much smaller than the loudspeaker diameter, the sound travels in a beam in the forward direction—and does not spread out. Find the frequency of a sound wave that has a wavelength (*a*) 10 times the diameter of a 30-cm speaker and (*b*) one-tenth the diameter of a 30-cm speaker. (*c*) Repeat this problem for a 6-cm speaker.

105 • A whistle of frequency 500 Hz moves in a circle of radius 1 m at 3 rev/s. What are the maximum and minimum frequencies heard by a stationary listener in the plane of the circle and 5 m away from its center?

106 • Ocean waves move toward the beach with a speed of 8.9 m/s and a crest-to-crest separation of 15.0 m. You are in a small boat anchored off shore. (*a*) What is the frequency of the ocean waves? (*b*) You now lift anchor and head out to sea at a speed of 15 m/s. What frequency of the waves do you observe?

107 •• A 12.0-m wire of mass 85 g is stretched under a tension of 180 N. A pulse is generated at the left end of the wire, and 25 ms later a second pulse is generated at the right end of the wire. Where do the pulses first meet?

108 •• SSM ISOLVE ✓ Find the speed of a car the tone of whose horn will drop by 10 percent as it passes you.

109 •• A loudspeaker driver 20 cm in diameter is vibrating at 800 Hz with an amplitude of 0.025 mm. Assuming that the air molecules in the vicinity have the same amplitude of vibration, find (*a*) the pressure amplitude immediately in front of the driver, (*b*) the sound intensity, and (*c*) the acoustic power being radiated.

110 •• ISOLVE A plane, harmonic, acoustical wave that oscillates in air with an amplitude of 1 μm has an intensity of 10 mW/m^2. What is the frequency of the sound wave?

111 •• Water flows at 7 m/s in a pipe of radius 5 cm. A plate having an area equal to the cross-sectional area of the pipe is suddenly inserted to stop the flow. Find the force exerted on the plate. Take the speed of sound in water to be 1.4 km/s. *Hint: When the plate is inserted, a pressure wave propagates through the water at the speed of sound* v_s. *The mass of water brought to a stop in time* Δt *is the water in a length of tube equal to* $v_s\Delta t$.

112 •• A high-speed flash photography setup to capture a picture of a bullet exploding a soap bubble is shown in Figure 15-31. The shock wave from the bullet is to be detected by a microphone that will trigger the flash. The microphone is placed on a track that is parallel to the path of the bullet, at the same height as the bullet. The track is used to adjust the position of the microphone. If the bullet is traveling at 1.25 times the speed of sound, and the distance between the lab bench and the track is 0.35 m, how far back from the soap bubble must the microphone be set to trigger the flash? (Assume that the flash itself is instantaneous once the microphone is triggered.)

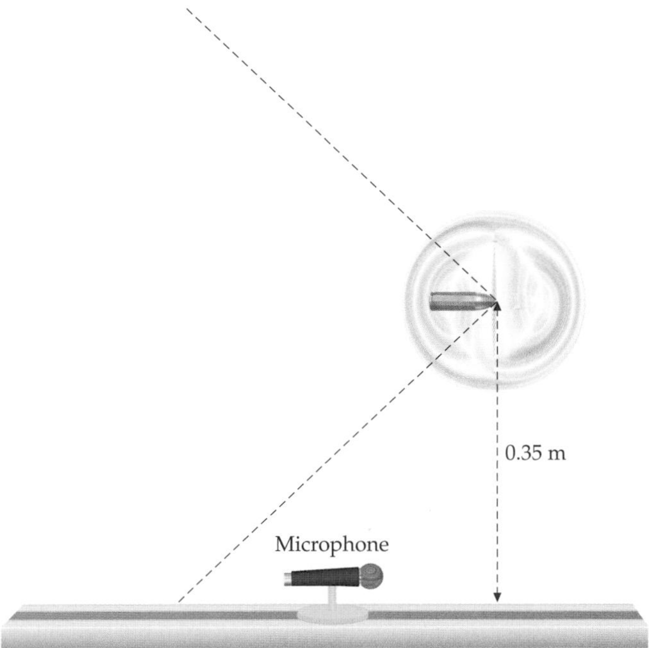

FIGURE 15-31 Problem 112

113 •• A column of precision marchers keeps in step by listening to the band positioned at the head of the column. The beat of the music is for 100 paces/min. A television camera shows that only the marchers at the front and the rear of the column are actually in step. The marchers in the middle section are striding forward with the left foot when those at the front and rear are striding forward with the right foot. The marchers are so well trained, however, that they are all certain that they are in proper step with the music. Explain the source of the problem, and calculate the length of the column.

114 •• A bat flying toward an obstacle at 12 m/s emits brief, high-frequency sound pulses at a repetition frequency of 80 Hz. What is the interval between the arrival times of the echo pulses heard by the bat?

115 •• SSM Laser ranging to the moon is done routinely to accurately determine the earth–moon distance. However, to determine the distance accurately, corrections must be made for the speed of light in the earth's atmosphere, which is 99.997 percent of the speed of light in vacuum. Assuming that the earth's atmosphere is 8 km high, estimate the length of the correction.

116 •• A tuning fork attached to a stretched wire generates transverse waves. The vibration of the fork is perpendicular to the wire. Its frequency is 400 Hz and the amplitude of its oscillation is 0.50 mm. The wire has a linear mass density of 0.01 kg/m and is under a tension of 1 kN. Assume that there are no reflected waves. (*a*) Find the period and frequency of waves on the wire. (*b*) What is the speed of the waves? (*c*) What are the wavelength and wave number? (*d*) Write a suitable wave function for the waves on the wire. (*e*) Calculate the maximum speed and acceleration of a point on the wire. (*f*) At what average rate must energy be supplied to the fork to keep it oscillating at a steady amplitude?

117 ••• If a loop of chain is spun at high speed, it will roll like a hoop without collapsing. Consider a chain of linear mass density μ that is rolling without slipping at a high speed v_0. (*a*) Show that the tension in the chain is $F = \mu v_0^2$. (*b*) If the chain rolls over a small bump, a transverse wave pulse will be generated in the chain. At what speed will it travel along the chain? (*c*) How far around the loop (in degrees) will a transverse wave pulse travel in the time the hoop rolls through one complete revolution?

118 ••• A long rope with a mass per unit length of 0.1 kg/m is under a constant tension of 10 N. A motor at the point $x = 0$ drives one end of the rope with harmonic motion at 5 oscillations per second and an amplitude of 4 cm. (*a*) What is the wave speed? (*b*) What is the wavelength? (*c*) What is the maximum transverse linear momentum of a 1-mm segment of the rope? (*d*) What is the maximum net force on a 1-mm segment of the rope?

119 ••• SSM A heavy rope 3 m long is attached to the ceiling and is allowed to hang freely. (*a*) Show that the speed of transverse waves on the rope is independent of its mass and length but does depend on the distance y from the bottom according to the formula $v = \sqrt{gy}$. (*b*) If the bottom end of the rope is given a sudden sideways displacement, how long does it take the resulting wave pulse to go to the ceiling, reflect, and return to the bottom of the rope?

120 ••• In this problem you will derive an expression for the potential energy of a segment of a string carrying a traveling wave (Figure 15-32). The potential energy of a segment equals the work done by the tension in stretching the string, which is $\Delta U = F(\Delta \ell - \Delta x)$, where F is the tension, $\Delta \ell$ is the length of the stretched segment, and Δx is its original length. From the figure we see that

$$\Delta \ell = \sqrt{(\Delta x)^2 + (\Delta y)^2} = \Delta x[1 + (\Delta y/\Delta x)^2]^{1/2}$$

(*a*) Use the binomial expansion to show that $\Delta \ell - \Delta x \approx \frac{1}{2}(\Delta y/\Delta x)^2\,\Delta x$, and therefore $\Delta U \approx \frac{1}{2}F(\Delta y/\Delta x)^2\,\Delta x$. (*b*) Compute dy/dx from the wave function in Equation 15-13 and show that $\Delta U \approx \frac{1}{2}Fk^2\cos^2(kx - \omega t)\,\Delta x$.

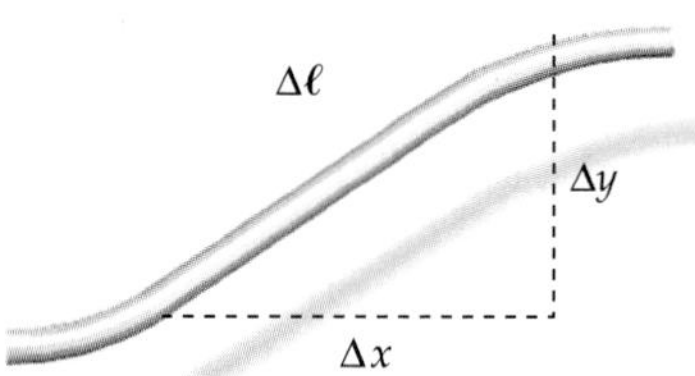

FIGURE 15-32
Problem 120

CHAPTER

16

Superposition and Standing Waves

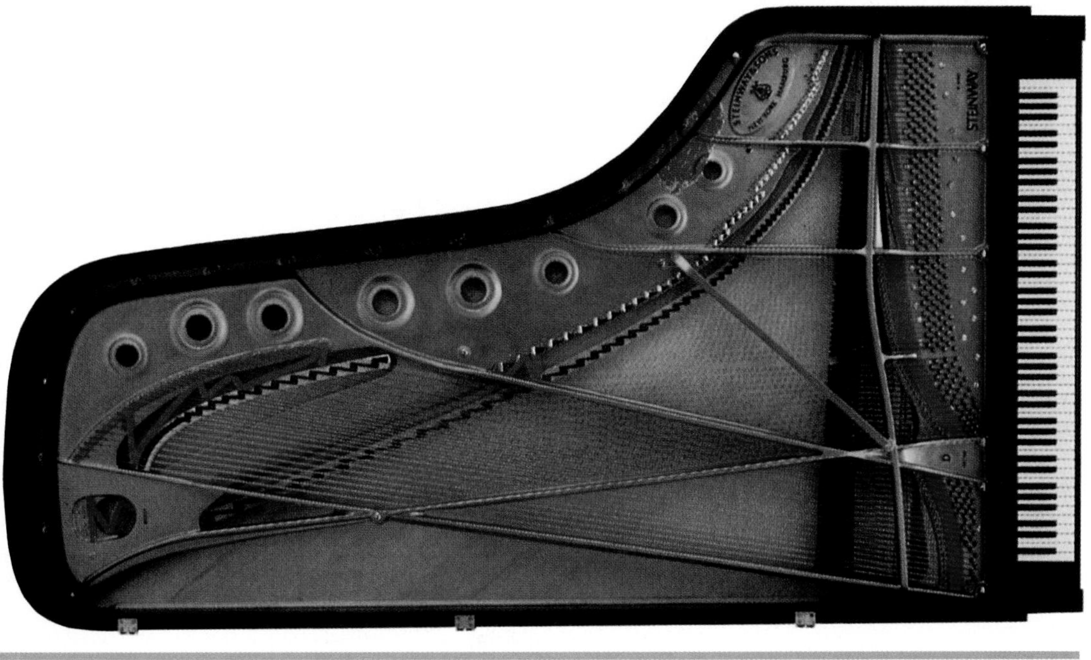

THE STRINGS IN THIS STEINWAY GRAND PIANO VIBRATE WHEN STRUCK BY THE HAMMERS, WHICH ARE CONTROLLED BY THE KEYS. THE LONGER STRINGS VIBRATE AT LOWER FREQUENCIES THAN THE SHORTER STRINGS.

? **What other factors come into play when tuning a piano? (See Example 16-6.)**

When two or more waves overlap in space, their individual disturbances (represented mathematically by their wave functions) superimpose and add algebraically, creating a resultant wave. This property of waves is called the principle of superposition. Under certain circumstances the superposition of harmonic waves of the same frequency produces sustained wave patterns in space. This phenomenon is called interference. Interference and diffraction are what distinguish wave motion from particle motion. Thomas Young's observation in 1801 of interference of light led to the understanding that light propagates via wave motion, not via particle motion as had been proposed by Newton. The observation of interference of electron waves by C. J. Davisson and L. H. Germer in 1927 led to our understanding of the wave nature of electrons and other material objects. These ideas are integral to understanding quantum physics, which is presented in Chapter 34.

➢ **In this chapter, we begin with the superposition of wave pulses on a string and then consider the superposition and interference of harmonic waves. We examine the phenomenon of beats, which result from the interference of two waves of slightly different frequencies, and study standing waves, which occur when harmonic waves are confined in space. We also consider the analysis of complex musical tones in terms of their component harmonic waves, and**

the inverse problem of the synthesis of harmonic waves to produce complex tones. We conclude with a qualitative discussion of the extension of harmonic analysis to aperiodic waves such as wave pulses.

16-1 Superposition of Waves

Figure 16-1*a* shows small wave pulses moving in opposite directions on a string. The shape of the string when they meet can be found by adding the displacements produced by each pulse separately. The **principle of superposition** is a property of wave motion which states:

> When two or more waves overlap, the resultant wave is the algebraic sum of the individual waves.

PRINCIPLE OF SUPERPOSITION

Mathematically, when there are two pulses on the string, the total wave function is the algebraic sum of the individual wave functions.

In the special case of two pulses that are identical except that one is inverted relative to the other, as in Figure 16-1*b*, there is an instant when the pulses exactly overlap and add to zero. At this instant the string is horizontal, but it is not stationary. At the right edge of the overlap region the string is moving upward and at the left edge it is moving downward. A short time later the individual pulses emerge, each continuing in its original direction.

Superposition is a characteristic and unique property of wave motion. There is no analogous situation in Newtonian particle motion; that is, two Newtonian particles never overlap or add together in this way.

*Superposition and the Wave Equation

The principle of superposition follows from the fact that the wave equation (Equation 15-9) is linear for small transverse displacements. That is, the function $y(x,t)$ and its derivatives occur only to the first power. The defining property of a linear equation is that if y_1 and y_2 are two solutions of the equation, then the linear combination

$$y_3 = C_1y_1 + C_2y_2 \qquad 16\text{-}1$$

is also a solution, where C_1 and C_2 are any constants. The linearity of the wave equation can be shown by the direct substitution of y_3 into the equation. The result is the mathematical statement of the principle of superposition. If any two waves satisfy a wave equation, then their algebraic sum also satisfies the same wave equation.

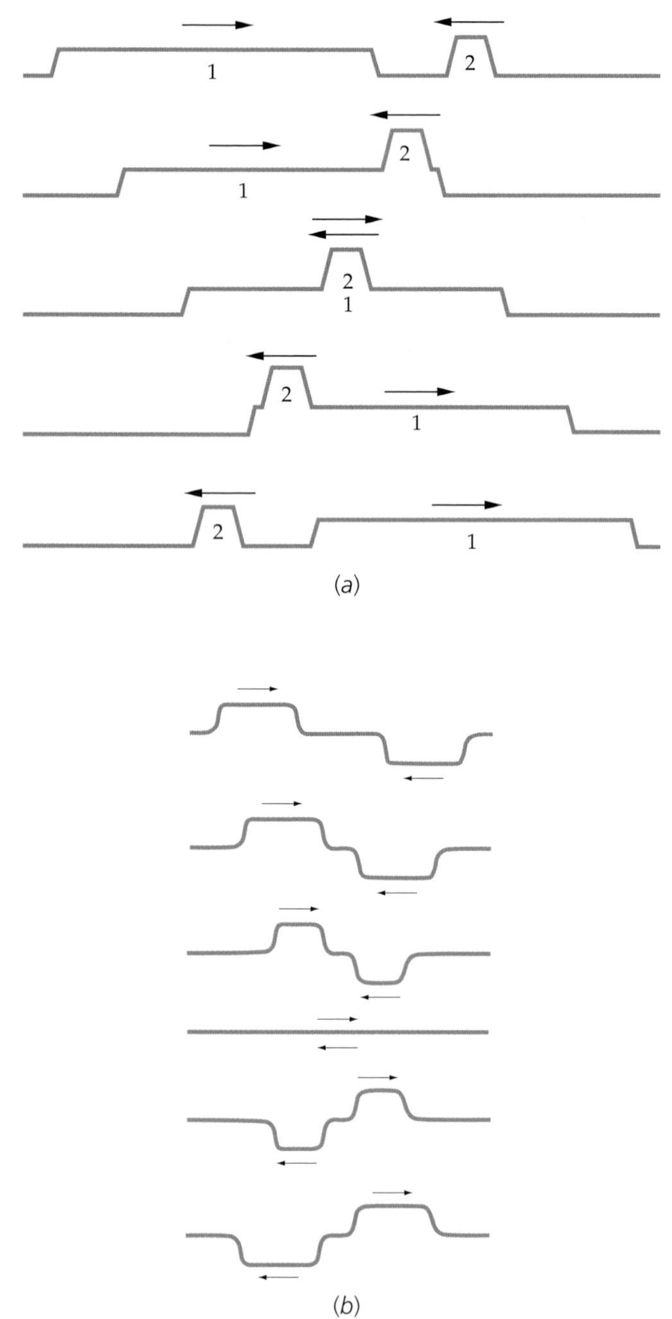

FIGURE 16-1 **Wave pulses moving in opposite directions on a string. The shape of the string when the pulses meet is found by adding the displacements of each separate pulse. (*a*) Superposition of pulses having displacements in the same direction. (*b*) Superposition of pulses having equal but opposite displacements. Here the algebraic addition of the displacement amounts to the subtraction of the magnitudes.**

SUPERPOSITION AND THE WAVE EQUATION **EXAMPLE 16-1**

Show that if functions y_1 and y_2 both satisfy wave equation

$$\frac{\partial^2 y}{\partial x^2} = \frac{1}{v^2}\frac{\partial^2 y}{\partial t^2} \qquad \text{(Equation 15-9}b\text{)}$$

then the function y_3 given by Equation 16-1 also satisfies it.

PICTURE THE PROBLEM Substitute y_3 into the wave equation, assume that y_1 and y_2 each satisfy the wave equation, and show that, as a consequence, the linear combination $C_1y_1 + C_2y_2$ satisfies the wave equation.

1. Substitute the expression for y_3 in Equation 16-1 into the left side of the wave equation, then break it into separate terms for y_1 and y_2:

$$\frac{\partial^2 y_3}{\partial x^2} = \frac{\partial^2}{\partial x^2}(C_1y_1 + C_2y_2) = C_1\frac{\partial^2 y_1}{\partial x^2} + C_2\frac{\partial^2 y_2}{\partial x^2}$$

2. Both y_1 and y_2 satisfy the wave function. Write the wave equation for both y_1 and y_2:

$$\frac{\partial^2 y_1}{\partial x^2} = \frac{1}{v^2}\frac{\partial^2 y_1}{\partial t^2} \quad \text{and} \quad \frac{\partial^2 y_2}{\partial x^2} = \frac{1}{v^2}\frac{\partial^2 y_2}{\partial t^2}$$

3. Substitute the step 2 results into the step 1 result and factor out any common terms:

$$\frac{\partial^2 y_3}{\partial x^2} = C_1\frac{1}{v^2}\frac{\partial^2 y_1}{\partial t^2} + C_2\frac{1}{v^2}\frac{\partial^2 y_2}{\partial t^2} = \frac{1}{v^2}\left(C_1\frac{\partial^2 y_1}{\partial t^2} + C_2\frac{\partial^2 y_2}{\partial t^2}\right)$$

4. Move the constants inside the arguments of the derivatives and express the sum of the derivatives as the derivative of the sum:

$$\frac{\partial^2 y_3}{\partial x^2} = \frac{1}{v^2}\left(\frac{\partial^2 C_1y_1}{\partial t^2} + \frac{\partial^2 C_2y_2}{\partial t^2}\right) = \frac{1}{v^2}\frac{\partial^2}{\partial t^2}(C_1y_1 + C_2y_2)$$

5. The argument of the derivative in step 4 is y_3:

$$\therefore \quad \boxed{\frac{\partial^2 y_3}{\partial x^2} = \frac{1}{v^2}\frac{\partial^2 y_3}{\partial t^2}}$$

Interference of Harmonic Waves

The result of the superposition of two harmonic waves of the same frequency depends on the phase difference δ between the waves. Let $y_1(x,t)$ be the wave function for a harmonic wave traveling to the right with amplitude A, angular frequency ω, and wave number k:

$$y_1 = A\sin(kx - \omega t) \qquad \text{16-2}$$

For this wave function, we have chosen the phase constant to be zero.† If we have another harmonic wave also traveling to the right with the same amplitude, frequency, and wave number, then the general equation for its wave function can be written

$$y_2 = A\sin(kx - \omega t + \delta) \qquad \text{16-3}$$

where δ is the phase constant. The two waves described by Equations 16-2 and 16-3 differ in phase by δ. Figure 16-2 shows a plot of the two wave functions versus position for a fixed time. The resultant wave is the sum

$$y_1 + y_2 = A\sin(kx - \omega t) + A\sin(kx - \omega t + \delta) \qquad \text{16-4}$$

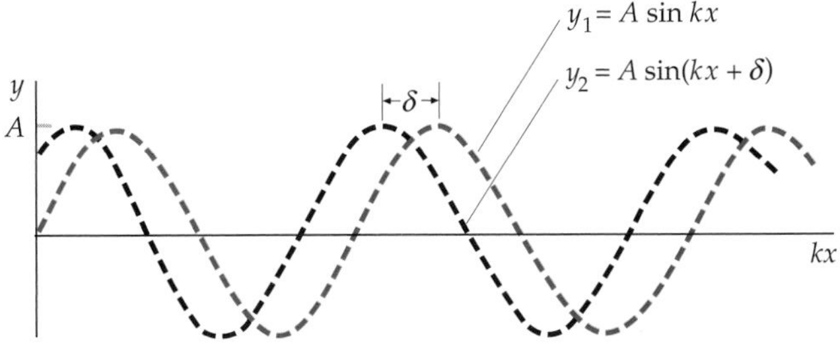

FIGURE 16-2 Displacement versus position for two harmonic waves having the same amplitude, frequency, and wavelength, but differing in phase by δ.

† This choice is convenient but not mandatory. If, for example, we chose $t = 0$ when the displacement was maximum at $x = 0$, we would write $y_1 = A\cos(kx - \omega t) = A\sin(kx - \omega t - \pi/2)$.

We can simplify Equation 16-4 by using the trigonometric identity

$$\sin \theta_1 + \sin \theta_2 = 2 \cos \tfrac{1}{2}(\theta_1 - \theta_2) \sin \tfrac{1}{2}(\theta_1 + \theta_2) \qquad \text{16-5}$$

For this case, $\theta_1 = kx - \omega t$ and $\theta_2 = kx - \omega t + \delta$, so that

$$\tfrac{1}{2}(\theta_1 - \theta_2) = -\tfrac{1}{2}\delta$$

and

$$\tfrac{1}{2}(\theta_1 + \theta_2) = kx - \omega t + \tfrac{1}{2}\delta$$

Thus, Equation 16-4 becomes

$$y_1 + y_2 = \left[2A \cos \tfrac{1}{2}\delta\right] \sin(kx - \omega t + \tfrac{1}{2}\delta) \qquad \text{16-6}$$

SUPERPOSITION OF TWO WAVES OF THE SAME AMPLITUDE AND FREQUENCY

where we have used $\cos(-\tfrac{1}{2}\delta) = \cos \tfrac{1}{2}\delta$. We see that the result of the superposition of two harmonic waves of equal wave number and frequency is a harmonic wave having the same wave number and frequency. The resultant wave has amplitude $2A \cos \tfrac{1}{2}\delta$ and a phase equal to half the difference between the phases of the original waves. The phenomenon of two or more waves of the same, or almost the same, frequency superposing to produce an observable pattern in the intensity is called **interference.** In this example, the intensity, which is proportional to the square of the amplitude, is uniform. If the two waves are in phase, then $\delta = 0$, $\cos 0 = 1$, and the amplitude of the resultant wave is $2A$. The interference of two waves in phase is called **constructive interference** (Figure 16-3). If the two waves are 180° out of phase, then $\delta = \pi$, $\cos(\pi/2) = 0$, and the amplitude of the resultant wave is zero. The interference of two waves 180° out of phase is called **destructive interference** (Figure 16-4).

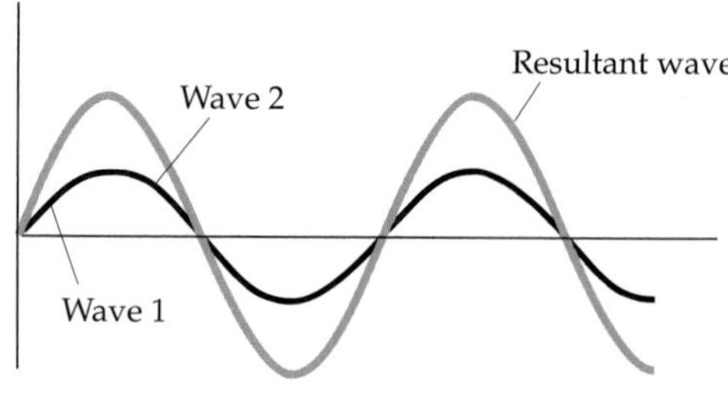

FIGURE 16-3 Constructive interference. When two waves are in phase, the amplitude of the resultant wave is the sum of the amplitudes of the individual waves.

EXERCISE Two waves with the same frequency, wavelength, and amplitude are traveling in the same direction. (*a*) If they differ in phase by $\pi/2$ and each has an amplitude of 4.0 cm, what is the amplitude of the resultant wave? (*b*) For what phase difference δ will the resultant amplitude be equal to 4.0 cm? (*Answer* (*a*) 5.66 cm (*b*) 120° or 240°)

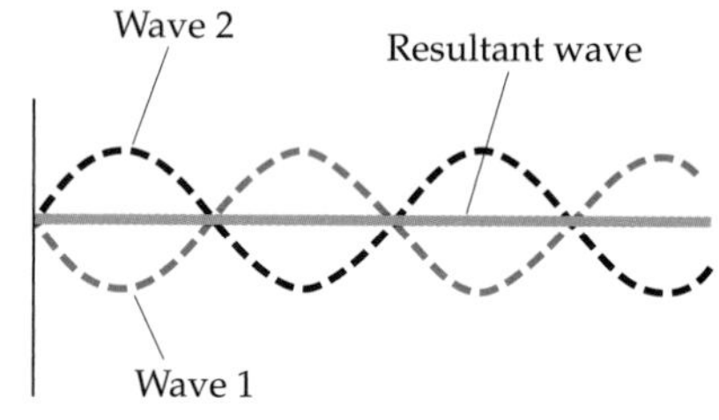

FIGURE 16-4 Destructive interference. When two waves have a phase difference of π, the amplitude of the resultant wave is the difference between the amplitudes of the individual waves. If the original waves have equal amplitudes, they cancel completely.

Beats The interference of two sound waves with slightly different frequencies produces the interesting phenomenon known as **beats.** Consider two sound waves that have angular frequencies of ω_1 and ω_2 and the same pressure amplitude p_0. What do we hear? At a fixed point, the spatial dependence of the wave merely contributes a phase constant, so we can neglect it. The pressure at the ear due to either wave acting alone will be a simple harmonic function of the type

$$p_1 = p_0 \sin \omega_1 t$$

and

$$p_2 = p_0 \sin \omega_2 t$$

where we have chosen sine functions for convenience and have assumed that the waves are in phase at time $t = 0$. Using the trigonometry identity

$$\sin \theta_1 + \sin \theta_2 = 2 \cos \tfrac{1}{2}(\theta_1 - \theta_2) \sin \tfrac{1}{2}(\theta_1 + \theta_2)$$

for the sum of two sine functions, we obtain for the resultant wave

$$p = p_0 \sin \omega_1 t + p_0 \sin \omega_2 t = 2p_0 \cos \tfrac{1}{2}(\omega_1 - \omega_2)t \sin \tfrac{1}{2}(\omega_1 + \omega_2)t$$

If we write $\omega_{av} = \frac{1}{2}(\omega_1 + \omega_2)$ for the average angular frequency and $\Delta\omega = \omega_1 - \omega_2$ for the difference in angular frequencies, the resultant wave function is

$$p = 2p_0 \cos(\tfrac{1}{2}\Delta\omega\, t)\sin \omega_{av} t = 2p_0 \cos(2\pi\tfrac{1}{2}\Delta f\, t)\sin 2\pi f_{av} t \qquad 16\text{-}7$$

where $\Delta f = \Delta\omega/(2\pi)$ and $f_{av} = \omega_{av}/(2\pi)$.

Figure 16-5 shows a plot of pressure variations as a function of time. The waves are originally in phase and add constructively at time $t = 0$. Because their frequencies differ, the waves gradually become out of phase, and at time t_1 they are 180° out of phase and interfere destructively.† An equal time interval later (time t_2 in the figure), the two waves are again in phase and interfere constructively. The greater the difference in the frequencies of the two waves, the more rapidly they oscillate in and out of phase.

The tone we hear has a frequency of $f_{av} = \frac{1}{2}(f_1 + f_2)$ and amplitude $2p_0 \cos(2\pi\frac{1}{2}\Delta f\, t)$. (For some values of t the amplitude is negative. Since $-\cos\theta = \cos(\theta + \pi)$, a change in the sign of the amplitude is equivalent to a 180° phase change.) The amplitude oscillates with the frequency $\frac{1}{2}\Delta f$. Since the sound intensity is proportional to the square of the amplitude, the sound is loud whenever the amplitude function is either a maximum or a minimum. Thus the frequency of this variation in intensity, called the **beat frequency,** is twice $\frac{1}{2}\Delta f$:

$$f_{beat} = \Delta f \qquad 16\text{-}8$$

BEAT FREQUENCY

FIGURE 16-5 Beats. (*a*) Two waves of different but nearly equal frequencies that are in phase at $t_0 = 0$ are 180° out of phase at some time later t_1. At a still later time, t_2, they are back in phase. (*b*) The resultant of the two waves shown in (*a*). The frequency of the resultant wave is about the same as those of the original waves, but the amplitude is modulated as indicated by the dashed envelope. The amplitude is maximum at times t_0 and t_2 and zero at times t_1 and t_3.

The beat frequency equals the difference in the individual frequencies of the two waves. If we simultaneously strike two tuning forks having the frequencies 241 Hz and 243 Hz, we will hear a pulsating tone at the average frequency of 242 Hz that has a maximum intensity 2 times per second; that is, the beat frequency is 2 Hz. The ear can detect up to about 15 to 20 beats per second. Above this frequency, the fluctuations in loudness are too rapid to be distinguished.

The phenomenon of beats is often used to compare an unknown frequency with a known frequency, as when a tuning fork is used to tune a piano string. Pianos are tuned by simultaneously ringing the tuning fork and striking a key, while at the same time adjusting the tension of the piano string until the beats are far apart, indicating that the difference in frequency of the two sound generators is very small.

TUNING A GUITAR — EXAMPLE 16-2

When a 440-Hz (concert A) tuning fork is struck simultaneously with the playing of the A string of a slightly out-of-tune guitar, 3 beats per second are heard. The guitar string is tightened a little to increase its frequency. As the guitar string is slowly tightened, you hear the beat frequency slowly increase. What is the frequency of the guitar string after it is tightened?

PICTURE THE PROBLEM Because 3 beats per second were heard initially, the original frequency of the guitar string was either 443 Hz or 437 Hz. Had it been 437 Hz, slowly increasing the string's frequency by tightening it would decrease the beat frequency.

Because the beat frequency increases as the tension increases, from 3 to 6 beats per second, the original frequency must have been 443 Hz.

$$f = f_A + f_{beat} = 440\text{ Hz} + 3\text{ Hz} = \boxed{443\text{ Hz}}$$

† Complete cancellation occurs only when the pressure amplitudes of the two waves are equal.

Phase Difference due to Path Difference A common cause of a phase difference between two waves is different path lengths between the sources of the waves and the point of interference. Suppose that two sources oscillate in phase (for example, positive crests leave the sources at the same time) and emit harmonic waves of the same frequency and wavelength. Now consider a point in space for which the path lengths to the two sources differ. If the path difference is one wavelength, as is the case in Figure 16-6*a*, or an integral number of wavelengths, the interference is constructive. If the path difference is one half of a wavelength or an odd number of half wavelengths, as in Figure 16-6*b*, the maximum of one wave falls at the minimum of the other and the interference is destructive.

The wave functions for waves from two sources oscillating in phase can be written

$$p_1 = p_0 \sin(kx_1 - \omega t)$$

and

$$p_2 = p_0 \sin(kx_2 - \omega t)$$

The phase difference for these two wave functions is

$$\delta = (kx_2 - \omega t) - (kx_1 - \omega t) = k(x_2 - x_1) = k\Delta x$$

Using $k = 2\pi/\lambda$, we have

$$\delta = k\Delta x = 2\pi\frac{\Delta x}{\lambda} \qquad \text{16-9}$$

PHASE DIFFERENCE DUE TO PATH DIFFERENCE

S_1 P_1 S_2 (*a*)

S_1 P_2 S_2 (*b*)

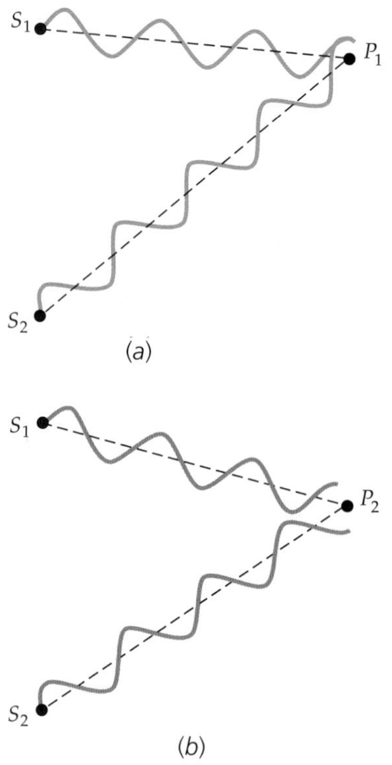

FIGURE 16-6 **Waves from two sources S_1 and S_2 that are in phase when they meet at a point P_1. (*a*) When the path difference is one wavelength λ, the waves are in phase at P_1 and interfere constructively. (*b*) When the path difference is $\frac{1}{2}\lambda$, the waves at P_2 are out of phase by 180° and therefore interfere destructively. If the waves are of equal amplitude at P_2, they will cancel completely at this point.**

A Resultant Sound Wave **EXAMPLE 16-3**

Two sound sources oscillate in phase. At a point 5.00 m from one source and 5.17 m from the other, the amplitude of the sound from each source separately is p_0. Find the amplitude of the resultant wave if the frequency of the sound waves is (*a*) 1000 Hz, (*b*) 2000 Hz, and (*c*) 500 Hz. (Use 340 m/s for the speed of sound.)

PICTURE THE PROBLEM The amplitude of the resultant wave due to superposition of two waves differing in phase by δ is given by $A = 2p_0 \cos\frac{1}{2}\delta$ (Equation 16-6), where p_0 is the amplitude of either wave and $\delta = 2\pi\,\Delta x/\lambda$ is the phase difference. We are given the path difference, $\Delta x = 5.17\text{ m} - 5\text{ m} = 0.17\text{ m}$, so all that is needed is the wavelength λ.

(*a*) 1. The wavelength equals the speed divided by the frequency. Calculate λ for $f = 1000$ Hz:

$$\lambda = \frac{v}{f} = \frac{340\text{ m/s}}{1000\text{ Hz}} = 0.34\text{ m}$$

2. For $\lambda = 0.34$ m, the given path difference ($\Delta x = 0.17$ m) is $\frac{1}{2}\lambda$, so we expect destructive interference. Use this value of λ to calculate the phase difference δ and use δ to calculate the amplitude A:

$$\delta = 2\pi\frac{\Delta x}{\lambda} = 2\pi\frac{0.17\text{ m}}{0.34\text{ m}} = \pi$$

so

$$A = 2p_0 \cos\frac{1}{2}\delta = 2p_0 \cos\frac{\pi}{2} = \boxed{0}$$

(*b*) 1. Calculate λ for $f = 2000$ Hz:

$$\lambda = \frac{v}{f} = \frac{340\text{ m/s}}{2000\text{ Hz}} = 0.17\text{ m}$$

2. For $\lambda = 0.17$ m, the path difference equals λ, so we expect constructive interference. Calculate the phase difference and amplitude:

$$\delta = 2\pi\frac{\Delta x}{\lambda} = 2\pi\frac{0.17\text{ m}}{0.17\text{ m}} = 2\pi$$

so

$$A = 2p_0 \cos \tfrac{1}{2}\delta = 2p_0 \cos \pi = \boxed{-2p_0}$$

(*c*) 1. Calculate λ for $f = 500$ Hz:

$$\lambda = \frac{v}{f} = \frac{340\text{ m/s}}{500\text{ Hz}} = 0.68\text{ m}$$

2. Calculate the phase difference and amplitude:

$$\delta = 2\pi\frac{\Delta x}{\lambda} = 2\pi\frac{0.17\text{ m}}{0.68\text{ m}} = \frac{\pi}{2}$$

so

$$A = 2p_0 \cos \frac{1}{2}\delta = 2p_0 \cos \frac{\pi}{4} = \boxed{\sqrt{2}p_0}$$

REMARKS In part (*b*), A is found to be negative. Equation 16-6 can be written $y_1 + y_2 = A' \sin (kx - \omega t + \frac{\delta}{2})$, which can also be written $y_1 + y_2 = -A' \sin (kx - \omega t + \frac{\delta}{2} + \pi)$. A phase shift of $\pi = 180°$ is equivalent to multiplying by -1.

SOUND INTENSITY OF TWO LOUDSPEAKERS — EXAMPLE 16-4

Two identical loudspeakers face each other at a distance of 180 cm and are driven by a common audio oscillator at 680 Hz. Locate the points between the speakers along a line joining them for which the sound intensity is (*a*) maximum and (*b*) minimum. (Neglect the variation in intensity from either speaker with distance, and use 340 m/s for the speed of sound.)

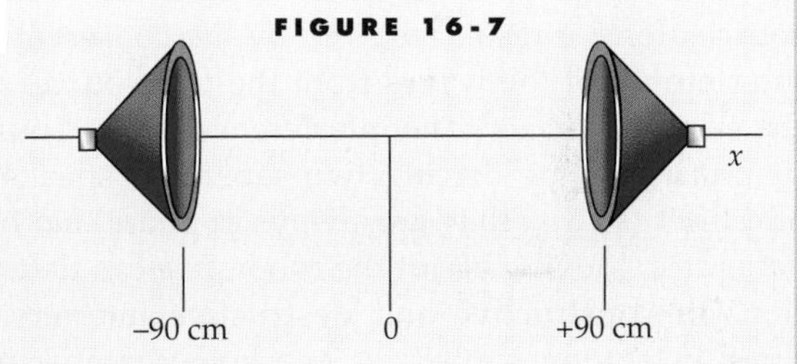

FIGURE 16-7

PICTURE THE PROBLEM We choose the origin to be at the midpoint between the speakers (Figure 16-7). Since this point is equidistant from the speakers, it is a point of maximum intensity. When we move a distance x toward one of the speakers, the path difference is $2x$. The intensity will be maximum when $2x = 0, \lambda, 2\lambda, 3\lambda, \ldots$, and minimum when $2x = (1/2)\lambda, (3/2)\lambda, (5/2)\lambda, \ldots$.

(*a*) 1. The intensity will be maximum when $2x$ equals an integral number of wavelengths:

$$2x = 0, \pm\lambda, \pm 2\lambda, \pm 3\lambda, \ldots$$

2. Calculate the wavelength:

$$\lambda = \frac{v}{f} = \frac{340\text{ m/s}}{680\text{ Hz}} = 0.5\text{ m} = 50\text{ cm}$$

3. Solve for x using the calculated wavelength:

$$x = 0, \pm\tfrac{1}{2}\lambda, \pm\lambda, \pm\tfrac{3}{2}\lambda, \ldots$$

$$= \boxed{0, \pm 25\text{ cm}, \pm 50\text{ cm}, \pm 75\text{ cm}}$$

(*b*) 1. The intensity will be minimum when $2x$ equals an odd number of half wavelengths:

$$2x = \pm\tfrac{1}{2}\lambda, \pm\tfrac{3}{2}\lambda, \pm\tfrac{5}{2}\lambda, \ldots$$

2. Solve for x using the calculated wavelength:

$$x = \pm\tfrac{1}{4}\lambda, \pm\tfrac{3}{4}\lambda, \pm\tfrac{5}{4}\lambda, \ldots$$

$$= \boxed{\pm 12.5\text{ cm}, \pm 37.5\text{ cm}, \pm 62.5\text{ cm}, \pm 87.5\text{ cm}}$$

REMARKS The maxima and minima will be relative maxima and relative minima because the amplitude from the near speaker will be slightly greater than that from the far speaker. Only seven terms were used for the maxima and only eight terms for the minima because any additional terms would be at a distance beyond one speaker.

Figure 16-8*a* shows the wave pattern produced by two point sources in a ripple tank that are oscillating in phase. Each source produces waves with circular wavefronts. The circular wavefronts shown all have the same phase and are separated by one wavelength. We can construct a similar pattern with a compass by drawing circular arcs representing the wave crests from each source at some particular time (Figure 16-8*b*). Where the crests from each source overlap, the waves interfere constructively. At these points, the path lengths from the two sources are either equal or they differ by an integral number of wavelengths. The dashed lines indicate the points that are equidistant from the sources or whose path differences are one wavelength, two wavelengths, or three wavelengths. At each point along any of these lines the interference is constructive, so these are lines of interference maxima. Between the lines of interference maxima are lines of interference minima. On a line of interference minima, the path length from any point on the line to each of the two sources differs by an odd number of half wavelengths. Throughout the region where the two waves are superposed, the amplitude of the resultant wave is given by $A = 2p_0 \cos \frac{1}{2}\delta$, where p_0 is the amplitude of each wave separately and δ is related to the path difference Δr by $\delta = 2\pi \Delta r/\lambda$ (Equation 16-9).

Figure 16-9 shows the intensity I of the resultant wave from two sources as a function of path difference Δx. At points where the interference is constructive, the amplitude of the resultant wave is twice that of either wave alone, and since the intensity is proportional to the square of the amplitude, the intensity is $4I_0$, where I_0 is the intensity due to either source alone. At points of destructive interference, the intensity is zero. The average intensity, shown by the dashed line in the figure, is twice the intensity due to either source alone, a result required by the conservation of energy. The interference of the waves from the two sources thus redistributes the energy in space. The interference of two sound sources can be demonstrated by driving two separated speakers with the same amplifier (so that they are always in phase) fed by an audio-signal generator. Moving about the room, one can detect by ear the positions of constructive and destructive interference.† This is best done in a room called an anechoic chamber, where reflections (echoes) off the walls of the room are minimized.

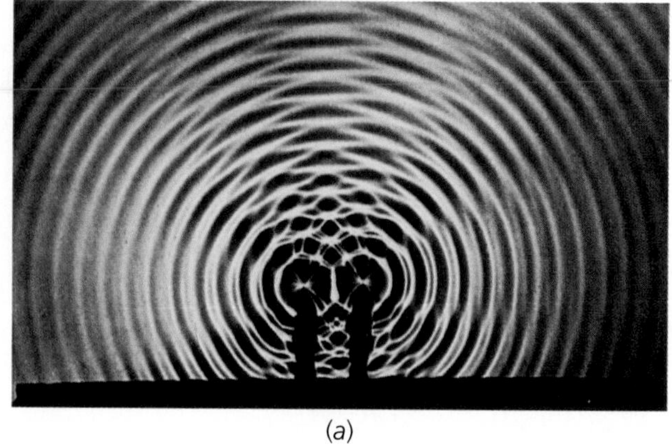

(a)

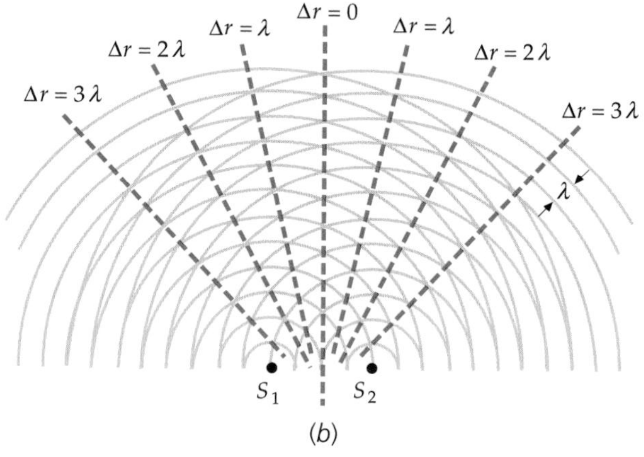

(b)

FIGURE 16-8 (*a*) Water waves in a ripple tank produced by two nearby sources oscillating in phase. (*b*) Drawing of wave crests for the sources in (*a*). The dashed lines indicate points for which the path difference is an integral number of wavelengths.

Coherence Two sources need not be in phase to produce an interference pattern. Consider two sources that are 180° out of phase. (Two speakers that are in phase can be made to be out of phase by 180° merely by switching the leads to one of the speakers.) The intensity pattern is the same as that in Figure 16-9 except that the maxima and minima are interchanged. At points for which the distance differs by an integral number of wavelengths, the interference is destructive because the waves are 180° out of phase. At points where the path difference is an odd number of half wavelengths, the waves are now in phase because the 180° phase difference of the sources is offset by the 180° phase difference due to the path difference.

Similar interference patterns will be produced by any two sources whose phase difference remains constant. Two

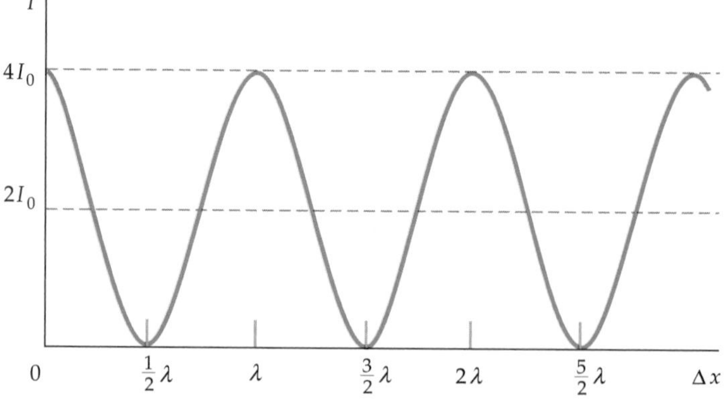

FIGURE 16-9 Intensity versus path difference for two sources that are in phase. I_0 is the intensity due to each source individually.

† In this demonstration, the sound intensity will be not quite zero at the points of destructive interference because of sound reflections from the walls or objects in the room.

sources that remain in phase or maintain a constant phase difference are said to be **coherent.** Coherent sources of water waves in a ripple tank are easy to produce by driving both sources with the same motor. Coherent sound sources are obtained by driving two speakers with the same signal source and amplifier.

Wave sources whose difference in phase is not constant but varies randomly are said to be **incoherent sources.** There are many examples of incoherent sources, such as two speakers driven by different amplifiers or two violins played by different violinists. For incoherent sources, the interference at a particular point varies rapidly back and forth from constructive to destructive, and no interference pattern is maintained long enough to be observed. The resultant intensity of waves from two or more incoherent sources is simply the sum of the intensities due to the individual sources.

16-2 Standing Waves

If waves are confined in space, like the waves on a piano string, sound waves in an organ pipe, or light waves in a laser, reflections at both ends cause the waves to travel in both directions. These superposing waves interfere in accordance with the principle of superposition. For a given string or pipe, there are certain frequencies for which superposition results in a stationary vibration pattern called a **standing wave.** Standing waves have important applications in musical instruments and in quantum theory.

Standing Waves on Strings

String Fixed at Both Ends If we fix both ends of a string and move a portion of the string up and down with simple harmonic motion of small amplitude, we find that at certain frequencies, standing-wave patterns such as those shown in Figure 16-10 are produced. The frequencies that produce these patterns are called the **resonance frequencies** of the string system. Each such frequency, with its accompanying wave function, is called a **mode of vibration.** The lowest resonance frequency is called the **fundamental** frequency f_1. It produces the standing-wave pattern shown in Figure 16-10*a*, which is called the **fundamental mode** of vibration or the **first harmonic.** The second lowest frequency f_2 produces the pattern shown in Figure 16-10*b*. This mode of vibration has a frequency twice that of the fundamental frequency and is called the second harmonic. The third lowest frequency f_3 is three times the fundamental frequency, and it produces the third harmonic pattern shown in Figure 16-10*c*. The set of all resonant frequencies is called the **resonant frequency spectrum** of the string.

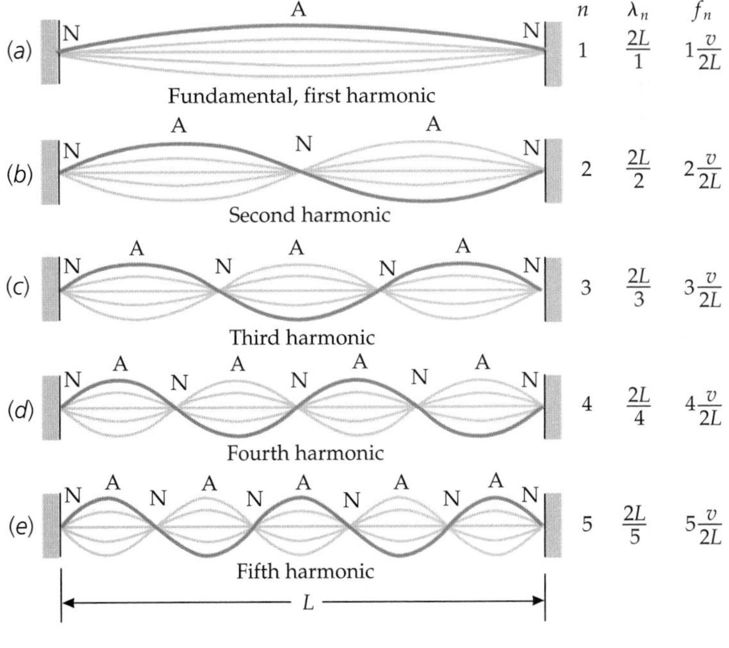

FIGURE 16-10 Standing waves on a string that is fixed at both ends. Points labeled A are antinodes and those labeled N are nodes. In general, the *n*th harmonic has *n* antinodes.

Not all resonant frequencies are called harmonics. Only if each frequency of a resonant frequency spectrum is an integral multiple of the fundamental (lowest) frequency are the frequencies referred to as harmonics. Many systems that support standing waves have resonant frequency spectra in which the resonant frequencies are not integral multiples of the lowest frequency. In all resonant frequency spectra the lowest resonant frequency is called the fundamental frequency (or just the fundamental), the next lowest resonant frequency is called the first **overtone,** the next lowest the second overtone, and so forth. This terminology has its roots in music. Only if each resonant frequency is an integral multiple of the fundamental frequency are they referred to as harmonics.

We note from Figure 16-10 that for each harmonic there are certain points on the string (the midpoint in Figure 16-10*b*, for example) that do not move. Such points are called **nodes.** Midway between each pair of nodes is a point of maximum amplitude of vibration called an **antinode.** Both fixed ends of the string are, of course, nodes. (If one end is attached to a tuning fork or other vibrator rather than being fixed, it will still be approximately a node because the amplitude of the vibration at that end is so much smaller than the amplitude at the antinodes.) We note that the first harmonic has one antinode, the second harmonic has two antinodes, and so on.

We can relate the resonance frequencies to the wave speed in the string and the length of the string. The distance from a node to the nearest antinode is one-fourth of the wavelength. Therefore, the length of the string L equals one-half the wavelength in the fundamental mode of vibration (Figure 16-11) and, as Figure 16-10 reveals, L equals two half-wavelengths for the second harmonic, three half-wavelengths for the third harmonic, and so forth. In general, if λ_n is the wavelength of the nth harmonic, we have

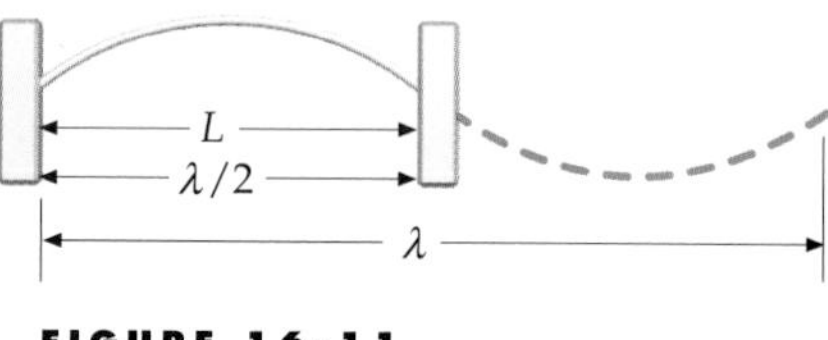

FIGURE 16-11

$$L = n\frac{\lambda_n}{2}, \qquad n = 1, 2, 3, \ldots \qquad 16\text{-}10$$

STANDING-WAVE CONDITION, BOTH ENDS FIXED

This result is known as the **standing-wave condition.** We can find the frequency of the nth harmonic from the fact that the wave speed v equals the frequency f_n times the wavelength. Thus,

$$f_n = \frac{v}{\lambda_n} = \frac{v}{2L/n} \qquad n = 1, 2, 3, \ldots$$

or

$$f_n = n\frac{v}{2L} = nf_1 \qquad n = 1, 2, 3, \ldots \qquad 16\text{-}11$$

RESONANCE FREQUENCIES, BOTH ENDS FIXED

where $f_1 = v/(2L)$ is the fundamental frequency.

You shouldn't bother to memorize Equation 16-11. Just sketch Figure 16-10 to remind yourself of the standing-wave condition, $\lambda_n = 2L/n$, and then use $v = f_n\lambda_n$.

PROBLEM-SOLVING GUIDELINE

We can understand standing waves in terms of resonance. Consider a string of length L that is attached at one end to a vibrator (Figure 16-12) and is fixed at the other end. The first wave crest sent out by the vibrator travels down the string a distance L to the fixed end, where it is reflected and inverted. It then travels back a distance L and is again reflected and inverted at the vibrator. The total time for the round trip is $2L/v$. If this time equals the period

FIGURE 16-12

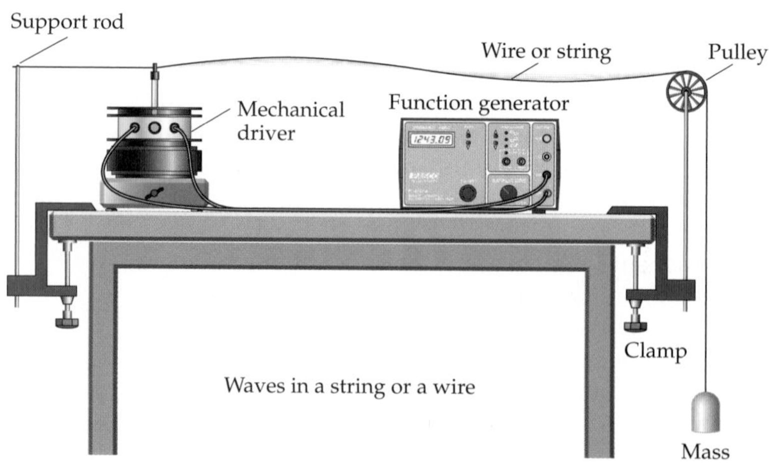

of the vibrator, the twice-reflected wave crest exactly overlaps the second wave crest produced by the vibrator, and the two crests interfere constructively, producing a crest with twice the original amplitude. The combined wave crest travels down the string and back and is added to by the third crest produced by the vibrator, increasing the amplitude three-fold, and so on. Thus, the vibrator is in resonance with the string. The wavelength is equal to $2L$ and the frequency is equal to $v/(2L)$.

Resonance also occurs at other vibrator frequencies. The vibrator is in resonance with the string if the time it takes for the first wave crest to travel the distance $2L$ is any integer n times the period T_n of the vibrator. That is, if $2L/v = nT_n$, where $2L/v$ is the round trip time for a wave crest. Thus,

$$f_n = \frac{1}{T_n} = n\frac{v}{2L}, \qquad n = 1, 2, 3, \ldots$$

is the condition for resonance. This is the same result we found by fitting an integral number of half-wavelengths into the distance L. Various damping effects, such as the loss of energy during reflection and the imperfect flexibility of the string, put a limit on the maximum amplitude that can be reached.

The resonance frequencies given by Equation 16-11 are also called the **natural frequencies** of the string. When the frequency of the vibrator is not one of the natural frequencies of the vibrating string, standing waves are not produced. After the first wave travels the distance $2L$ and is reflected from the fork, it differs in phase from the wave being generated at the vibrator (Figure 16-13). When this resultant wave has traveled the distance $2L$ and is again reflected at the vibrator, it will differ in phase from the next wave generated. In some cases, the new resultant wave will have an amplitude greater than that of the previous wave, in other cases the new amplitude will be less. On the average, the amplitude will not increase but will remain on the order of the amplitude of the first wave generated, which is the amplitude of the vibrator. This amplitude is very small compared with the amplitudes attained at resonance frequencies.

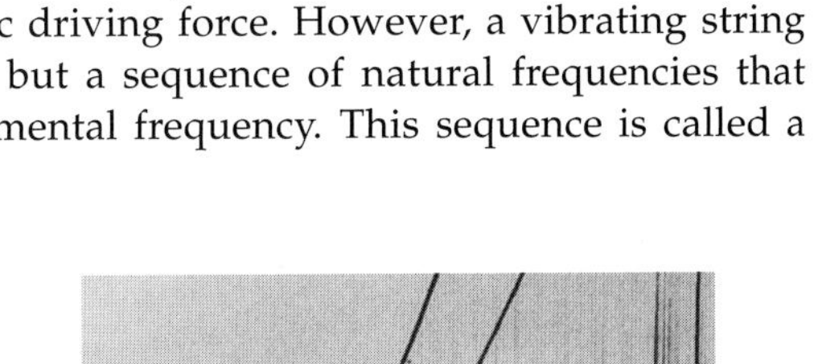

FIGURE 16-13 Waves on a string produced by a mechanical wave driver whose frequency is not in resonance with the natural frequencies of the string. The wave leaving the wave driver for the first time (dashed red line) is not in phase with the waves that have been reflected two or more times (gray lines), and these waves are not in phase with each other, so there is no buildup in amplitude. The resultant wave (black line) has about the same amplitude as the individual waves, which is about the amplitude of the driver.

The resonance of standing waves is analogous to the resonance of a simple harmonic oscillator with a harmonic driving force. However, a vibrating string has not just one natural frequency but a sequence of natural frequencies that are integral multiples of the fundamental frequency. This sequence is called a **harmonic series.**

Turbulent winds set up standing waves in the Tacoma Narrows suspension bridge, leading to its collapse on November 7, 1940, just four months after it had been opened for traffic.

Give Me an A — EXAMPLE 16-5

A string is stretched between two fixed supports 0.7 m apart and the tension is adjusted until the fundamental frequency of the string is concert A, 440 Hz. What is the speed of transverse waves on the string?

PICTURE THE PROBLEM The wave speed equals the frequency times the wavelength. For a string fixed at both ends, in the fundamental mode there is a single antinode in the middle of the string. Thus the length of the string equals one half-wavelength.

1. The wave speed is related to the frequency and wavelength: $v = f_1\lambda_1$

2. For the fundamental, the length of the string is one half-wavelength: $L = \lambda_1/2$

3. Use this wavelength and the given frequency to find the speed: $v = f_1\lambda_1 = f_1 2L = (440\text{ Hz}) \times 2(0.7\text{ m})$

$$= \boxed{616\text{ m/s}}$$

EXERCISE The speed of transverse waves on a stretched string is 200 m/s. If the string is 5 m long, find the frequencies of the fundamental and the second and third harmonics. (*Answer* $f_1 = 20$ Hz, $f_2 = 40$ Hz, $f_3 = 60$ Hz)

Testing Piano Wire — EXAMPLE 16-6

You have a summer job at a music shop, helping the owner build instruments. He asks you to test a new wire for possible use in pianos. He tells you that the 3-m-long wire has a linear mass density of 0.0025 kg/m, and he has found two adjacent resonant frequencies at 252 Hz and at 336 Hz. He wants you to determine the fundamental frequency of the wire and determine whether or not the wire is a good choice for piano strings. You know that safety issues start to arise if the tension in the wire gets above 700 N.

PICTURE THE PROBLEM The tension F_T is found from $v = \sqrt{F_T/\mu}$, where the speed v can be found from $v = f\lambda$ using any harmonic. The wavelength for the fundamental is twice the length of the wire. To find the fundamental frequency let 252 Hz be the nth harmonic. Then $f_n = nf_1$ and $f_{n+1} = (n+1)f_1$, where $f_{n+1} =$ 336 Hz. We can solve these two equations for f_1.

1. The tension is related to the wave speed: $v = \sqrt{F_T/\mu}$ or $F_T = \mu v^2$

2. Use the fundamental f_1, with $\lambda_1 = 2L$, to obtain the speed: $v = f_1\lambda_1 = f_1(2L)$

3. Combine the two previous results to find the tension: $F_T = \mu v^2 = \mu f_1^2(2L)^2$

4. The consecutive harmonics f_n and f_{n+1} are related to the fundamental frequency f_1: $nf_1 = 252\text{ Hz}$

$$(n+1)f_1 = 336\text{ Hz}$$

5. Dividing these equations eliminates f_1 and allows us to determine n: $\dfrac{n}{n+1} = \dfrac{252\text{ Hz}}{336\text{ Hz}} = 0.75 = \dfrac{3}{4}$

$$4n = 3n + 3, \quad \text{so} \quad n = 3$$

6. Solve for f_1:	$f_n = nf_1$ so $f_1 = \frac{f_n}{n} = \frac{f_3}{3} = \frac{252\text{ Hz}}{3} = 84\text{ Hz}$
7. Using the step 3 result, solve for F_T:	$F_T = \mu f_1^2(2L)^2 = (0.0025\text{ kg/m})(84\text{ Hz})^2(6\text{ m})^2$ $= 635\text{ N}$
8. Is the tension safe?	The tension is less than 700 N. It seems it is safe to use as long as the tension is not increased significantly

String Fixed at One End, Free at the Other Figure 16-14 shows a string that has one end fixed and one end attached to a massless ring that is free to slide up and down on a friction-free pole. The vertical motion of the end of the string attached to the ring is unconstrained, so it is said to be a free end. The ring is massless, that is, a finite vertical force on it by the string would give the ring an infinite vertical acceleration. This acceleration will remain finite if the slope of the string at its free end remains horizontal. This means that the free end of the string is an antinode. In the fundamental mode of vibration for a string fixed at one end only, there is a node at the fixed end and an antinode at the free end, so $L = \lambda_1/4$ (Figure 16-15). (The distance from a node to an adjacent antinode is equal to one-quarter wavelength.)

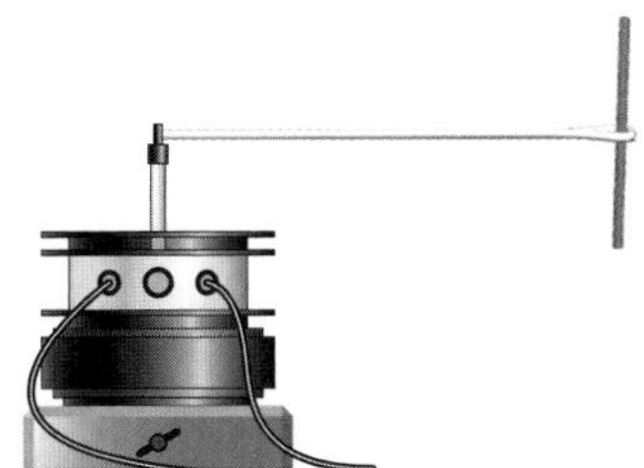

FIGURE 16-14 An approximation of a string fixed at one end and free at the other end can be produced by connecting the "free" end of the string to a ring that is free to move on a post. The end attached to the mechanical wave driver is approximately fixed because the amplitude of the driver is very small.

In each mode of vibration shown in Figure 16-16 there are an odd number of quarter-wavelengths in the length L. That is, $L = n\lambda_n/4$, where $n = 1, 3, 5, \ldots$. The standing-wave condition can thus be written

$$L = n\frac{\lambda_n}{4}, \qquad n = 1, 3, 5, \ldots \qquad \text{16-12}$$

STANDING-WAVE CONDITION, ONE END FREE

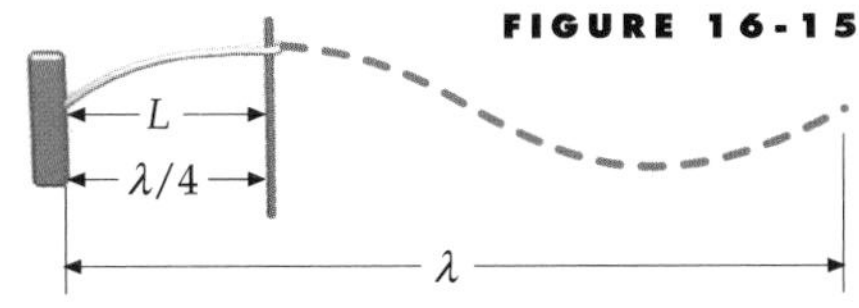

FIGURE 16-15

so $\lambda_n = 4L/n$. The resonance frequencies are therefore given by

$$f_n = \frac{v}{\lambda_n} = n\frac{v}{4L} = nf_1, \qquad n = 1, 3, 5, \ldots \qquad \text{16-13}$$

RESONANCE FREQUENCIES, ONE END FREE

where

$$f_1 = \frac{v}{4L} \qquad \text{16-14}$$

is the fundamental frequency. The natural frequencies of this system occur in the ratios 1:3:5:7:..., which means that the even harmonics are missing.

Again, an easy way to remember the resonance frequencies given by Equation 16-13 is to sketch Figure 16-16 to remind yourself of the standing-wave condition and use $v = f_n\lambda_n$

PROBLEM-SOLVING GUIDELINE

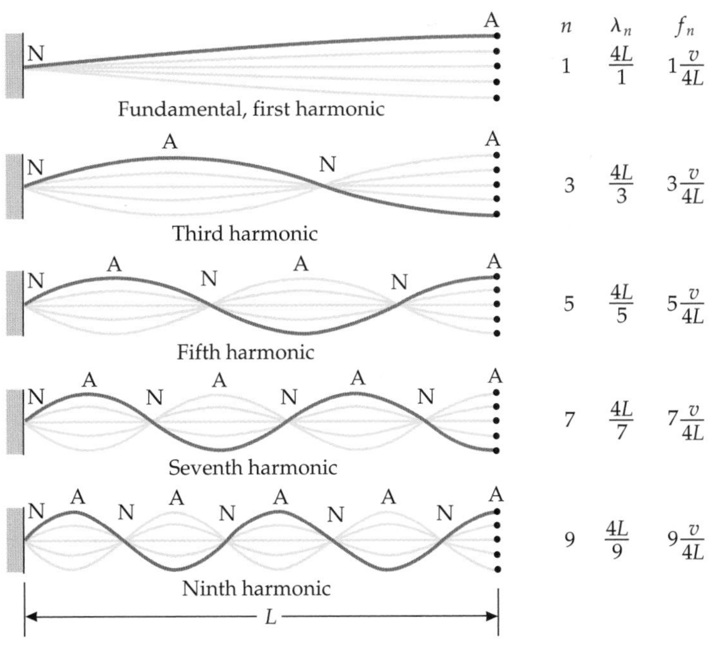

FIGURE 16-16 Standing waves on a string fixed at only one end. The free end is an antinode.

Wave Functions for Standing Waves If a string vibrates in its nth mode, each point on the string moves with simple harmonic motion. Its displacement $y_n(x,t)$ is given by

$$y_n(x,t) = A_n(x)\cos(\omega_n t + \delta_n)$$

where ω_n is the angular frequency, δ_n is the phase constant, which depends on the initial conditions, and $A_n(x)$ is the amplitude, which depends on the position x of the segment. The function $A_n(x)$ is the shape of the string when $\cos(\omega_n t + \delta_n) = 1$ (the instant that the vibration has its maximum displacement). The amplitude of a string vibrating in its nth mode is described by

$$A_n(x) = A_n \sin k_n x \qquad \text{16-15}$$

where $k_n = 2\pi/\lambda_n$ is the wave number. The wave function for a standing wave in the nth harmonic can thus be written

$$y_n(x,t) = A_n \sin(k_n x)\cos(\omega_n t + \delta_n) \qquad \text{16-16}$$

It is useful to remember the two conditions necessary for standing-wave motion, which are as follows:

1. Each point on the string either remains at rest or oscillates in simple harmonic motion. (Those points remaining at rest are at nodes.)
2. The motions of any two points on the string not at nodes oscillate either in phase or 180° out of phase.

NECESSARY CONDITIONS FOR A STANDING WAVE MOTION ON A LENGTH OF STRING

STANDING WAVES — **EXAMPLE 16-7** Try It Yourself

(*a*) The wave functions for two waves that have equal amplitude, frequency, and wavelength, but that travel in opposite directions, are given by $y_1 = y_0 \sin(kx - \omega t)$ and $y_2 = y_0 \sin(kx + \omega t)$. Show that the superposition of these two waves is a standing wave. (*b*) A standing wave on a string that is fixed at both ends is given by $y(x,t) = (0.024\ \text{m}) \sin(52.3\ \text{m}^{-1}\, x) \cos(480\ \text{s}^{-1}\, t)$. Find the speed of waves on the string and find the distance between adjacent nodes for the standing waves.

PICTURE THE PROBLEM To show that the superposition of the two given waves is a standing wave is to show that the algebraic sum of y_1 and y_2 can be written in the form of Equation 16-16. To find the wave speed and the wavelength we compare the given wave function with Equation 16-16 and identify the wave number and angular frequency. Knowing these, we can determine the wavelength and wave speed.

Cover the column to the right and try these on your own before looking at the answers.

Steps	Answers
(*a*) 1. Write Equation 16-16. If the sum of y_1 and y_2 can be written in this form, then the superposition of the two traveling waves is a standing wave.	$y(x,t) = A \sin kx \cos \omega t$
2. Add the two wave functions and use the trigonometric identity $\sin\theta_1 + \sin\theta_2 = 2\sin\frac{1}{2}(\theta_1 + \theta_2)\cos\frac{1}{2}(\theta_1 - \theta_2)$.	$y = y_0 \sin(kx - \omega t) + y_0 \sin(kx + \omega t)$ $= 2y_0 \sin kx \cos \omega t$

Note: This is of the form given by Equation 16-16 with $A = 2y_0$, so the superposition is a standing wave.

(*b*) 1. Identify the wave number and the angular frequency. $k = \boxed{52.3\ \text{m}^{-1}}$, $\omega = \boxed{480\ \text{s}^{-1}}$

2. Calculate the speed from $v = \omega/k$. $v = \boxed{9.18\ \text{m/s}}$

3. Find the wavelength $\lambda = 2\pi/k$, and use it to find the distance between nodes. $\frac{\lambda}{2} = \boxed{6.01\ \text{cm}}$

Standing Sound Waves

An organ pipe is a familiar example of the use of standing waves in air columns. In the flue-type organ pipe, a stream of air is directed against the sharp edge of an opening (point A in Figure 16-17). The complicated swirling motion of the air near the edge sets up vibrations in the air column. The resonance frequencies of the pipe depend on the length of the pipe and on whether the top is stopped (closed) or open.

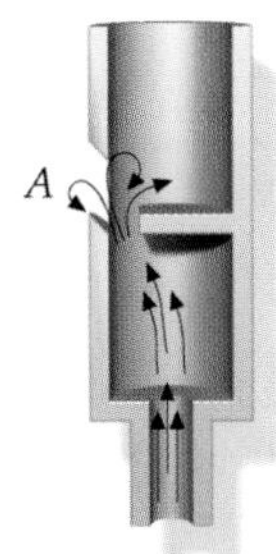

FIGURE 16-17 Flue-type organ pipe. Air is blown against the edge, causing a swirling motion of the air near point A that excites standing waves in the pipe. There is a pressure node near point A, which is open to the atmosphere.

In an open organ pipe, the pressure remains at one atmosphere near each open end. Since the pressure just beyond the ends does not vary, we say that there is a pressure node near each end. If the sound wave in the tube is a one-dimensional wave, which is largely correct if the tube diameter is much smaller than the wavelength, then the pressure node is at the opening of the tube. In practice, however, the pressure nodes lie slightly beyond the ends of the tube. The effective length of the pipe is $L_{\text{eff}} = L + \Delta L$ where ΔL is the end correction, which is of the order of the tube diameter. The standing-wave condition for this system is the same as that for a string fixed at both ends, where L is replaced by L_{eff} (the effective length of the tube), and all the same equations apply.

In a stopped organ pipe (open at one end, closed at the other), there is a pressure node near the opening (point A in Figure 16-17 and a pressure antinode at the closed end. The standing-wave condition for this system is the same as that for a string with one end fixed and one end free. The effective length of the tube is equal to an odd integer times $\lambda/4$. That is, the wavelength of the fundamental mode is 4 times the effective length of the tube, and only the odd harmonics are present.

As we saw in Chapter 15, a sound wave can be thought of as either a pressure wave or a displacement wave. The pressure and displacement variations in a sound wave are 90° out of phase. Thus, in a standing sound wave, the pressure nodes are displacement antinodes and vice versa. Near the open end of an organ pipe there is a pressure node and a displacement antinode, whereas at a stopped end there is a pressure antinode and a displacement node.

STANDING SOUND WAVES IN AN AIR COLUMN I **EXAMPLE 16-8** **Try It Yourself**

If the speed of sound is 340 m/s, what are the allowed frequencies and wavelengths for standing sound waves in an unstopped (open at both ends) organ pipe whose effective length is 1 m?

PICTURE THE PROBLEM There is a displacement antinode (and a pressure node) at each end. Therefore, the effective length of the pipe is equal to an integral number of half-wavelengths.

Cover the column to the right and try these on your own before looking at the answers.

Steps	Answers
1. Calculate the fundamental wavelength from $\lambda_1 = 2L_{eff}$.	$\lambda_1 = 2L_{eff} = 2\text{ m}$
2. Use your value of λ_1 to calculate the fundamental frequency f_1.	$f_1 = \frac{v}{\lambda_1} = 170\text{ Hz}$
3. Write expressions for the frequencies f_n and wavelengths λ_n of the other harmonics in terms of n.	$\boxed{f_n = nf_1 = n(170\text{ Hz}), \quad n = 1, 2, 3, \ldots}$ $\boxed{\lambda_n = \frac{2L}{n} = \frac{2\text{ m}}{n}, \quad n = 1, 2, 3, \ldots}$

Standing Sound Waves in an Air Column II **EXAMPLE 16-9**

When a tuning fork of frequency 500 Hz is held above a tube that is partly filled with water as in Figure 16-18, resonances are found when the water level is at distances L = 16.0, 50.5, 85.0, and 119.5 cm from the top of the tube. (*a*) What is the speed of sound in air? (*b*) How far from the open end of the tube is the displacement antinode?

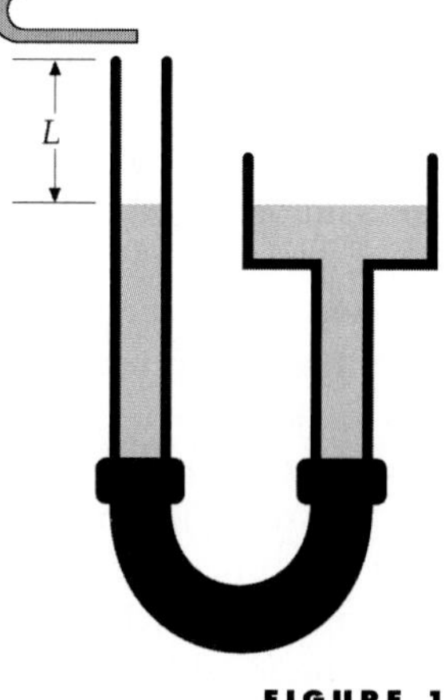

FIGURE 16-18

PICTURE THE PROBLEM Sound waves of frequency 500 Hz are excited in the air column whose length L can be adjusted (by adjusting the water level). The air column is stopped at one end, open at the other. Thus at resonance, the number of quarter-wavelengths in the effective length L_{eff} of the tube is equal to an odd integer (Figure 16-19). There is a displacement node at the surface of the water and a displacement antinode a short distance above the open end of the tube. Since the frequency is fixed, so is the wavelength. The speed is then found from $v = f\lambda$, where f is 500 Hz.

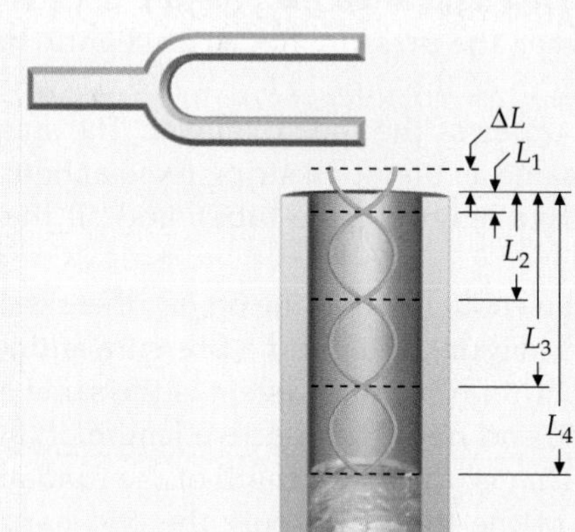

FIGURE 16-19

(*a*) 1. The speed of sound in air is related to the frequency and wavelength:

$$v = f\lambda$$

2. The wavelength is twice the distance between successive water levels at which resonance occurs:

$$\lambda = 2\,(L_{n+1} - L_n),\ n = 1, 2, 3, 4$$

3. The distance between successive levels is found from the data given in the problem:

$$L_{n+1} - L_n = L_4 - L_3 = 119.5\text{ cm} - 85\text{ cm} = 34.5\text{ cm}$$

so

$$\lambda = 2(34.5\text{ cm}) = 69\text{ cm} = 0.69\text{ m}$$

4. Substitute the values of f and λ to determine v:

$$v = f\lambda = (500\text{ Hz})(0.69\text{ m}) = \boxed{345\text{ m/s}}$$

(*b*) There will be a displacement antinode one quarter-wavelength above the displacement node at the surface of the water. Thus, the distance from the highest water level supporting resonance and the displacement antinode above the opening of the tube is one-quarter wavelength. $\frac{1}{4}\lambda = L_1 + \Delta L$:

$$\Delta L = \tfrac{1}{4}\lambda - L_1 = \tfrac{1}{4}(69.0\text{ cm}) - (16.0\text{ cm})$$

$$= \boxed{1.25\text{ cm}}$$

Most musical wind instruments are much more complicated than simple cylindrical tubes. The conical tube, which is the basis for the oboe, bassoon, English horn, and saxophone, has a complete harmonic series with its fundamental wavelength equal to twice the length of the cone. Brass instruments are combinations of cones and cylinders. The analysis of these instruments is extremely complex. The fact that they have nearly harmonic series is a triumph of educated trial and error rather than mathematical calculation.

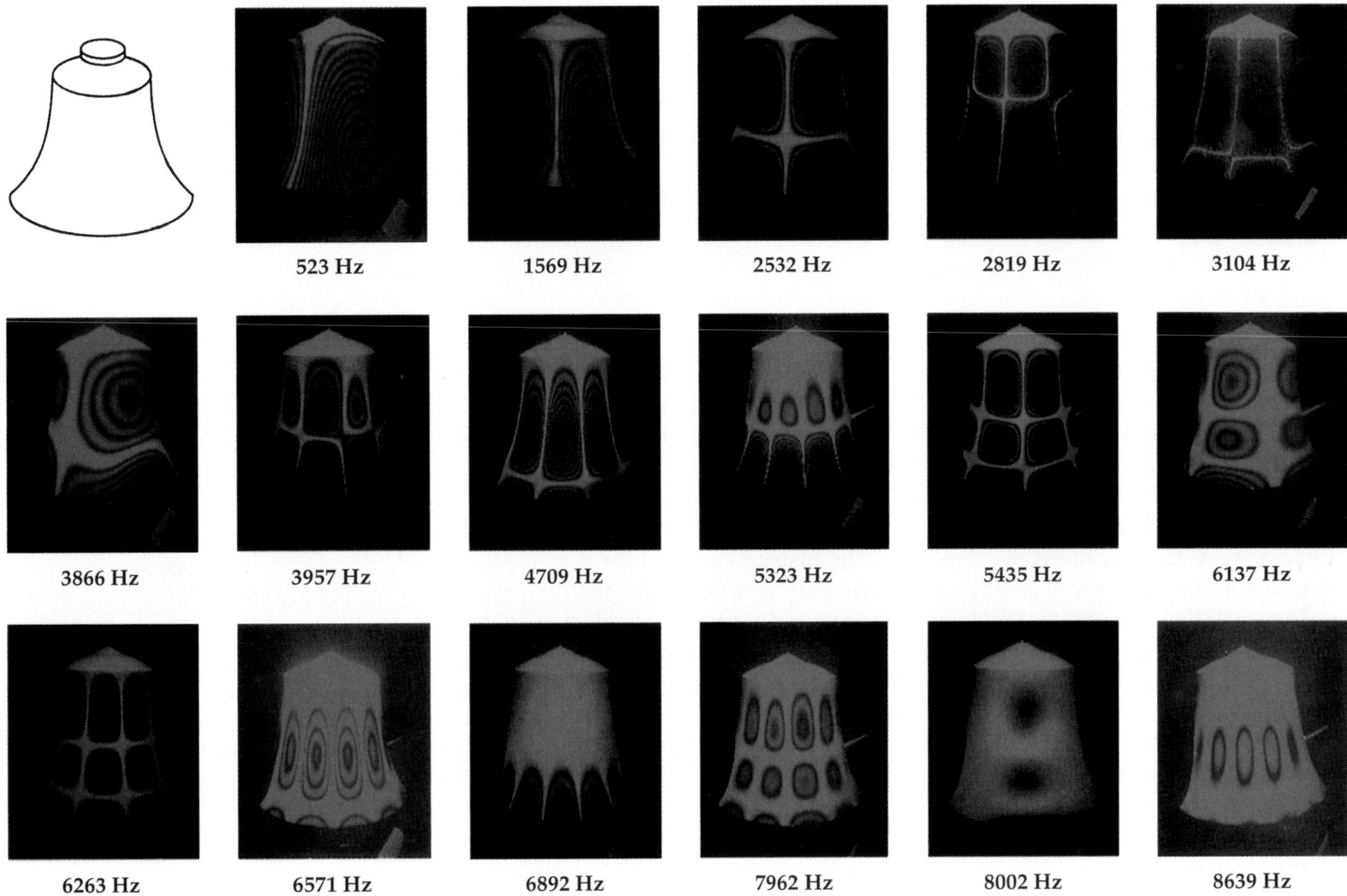

Holographic interferograms showing standing waves in a handbell. The "bull's eyes" locate the antinodes.

*16-3 The Superposition of Standing Waves

As we saw in the preceding section, there is a set of natural resonance frequencies that produce standing waves for sound waves in air columns or vibrating strings that are fixed at one or both ends. For example, for a string fixed at both ends, the frequency of the fundamental mode of vibration is $f_1 = v/(2L)$, where L is the length of the string and v is the wave speed and the wave function is Equation 16-16:

$$y_1(x,t) = A_1 \sin k_1 x \cos(\omega_1 t + \delta_1)$$

In general, a vibrating system does not vibrate in a single harmonic mode. Instead, the motion consists of a mixture of the allowed harmonics. The wave function is a linear combination of the harmonic wave functions:

$$y(x,t) = \sum_n A_n \sin(k_n x) \cos(\omega_n t + \delta_n) \qquad \text{16-17}$$

where $k_n = 2\pi/\lambda_n$, $\omega_n = 2\pi f_n$, and A_n and δ_n are constants. The constants A_n and δ_n depend on the initial position and velocity of the string. If a harp string, for example, is plucked at the center and released, as in Figure 16-20, the initial shape of the string is symmetric about the point $x = \frac{1}{2}L$ and the initial velocity is zero throughout the length of the string. The motion of the string after it has been released will remain symmetric about $x = \frac{1}{2}L$. Only the odd harmonics, which are also symmetric about $x = \frac{1}{2}L$, will be excited. The even harmonics, which are antisymmetric about $x = \frac{1}{2}L$, are not excited; that is, the constant A_n is zero for all even n. The shapes of the first four harmonics are shown in Figure 16-21. Most of the energy of the plucked string is associated with the fundamental, but small amounts of energy are associated with the third, fifth, and other odd harmonic modes. Figure 16-22 shows an approximation to the initial shape of the string using the superposition of only the first three odd harmonics.

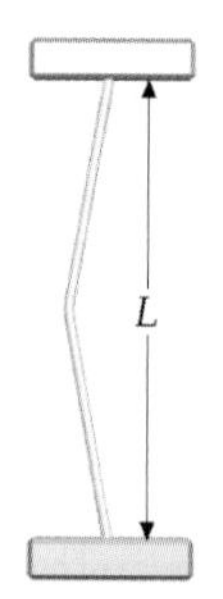

FIGURE 16-20 A string plucked at the center. When it is released, its vibration is a linear combination of standing waves.

*16-4 Harmonic Analysis and Synthesis

When a clarinet and an oboe play the same note, say, concert A, they sound quite different. Both notes have the same **pitch,** a physiological sensation of the highness or lowness of the note that is strongly correlated with frequency. However, the notes differ in what is called **tone quality.** The principal reason for the difference in tone quality is that, although both the clarinet and oboe are producing vibrations at the same fundamental frequency, each instrument is also producing harmonics whose relative intensities depend on the instrument and how it is played. If the sound produced by each instrument were entirely at the fundamental frequency of the instrument, they would sound identical.

Figure 16-23 shows plots of the pressure variations versus time for the sound from a tuning fork, a clarinet, and an oboe, each playing the same note. These patterns are called **waveforms.** The waveform for the sound from the tuning fork is nearly a pure sine wave, but those from the clarinet and the oboe are clearly more complex.

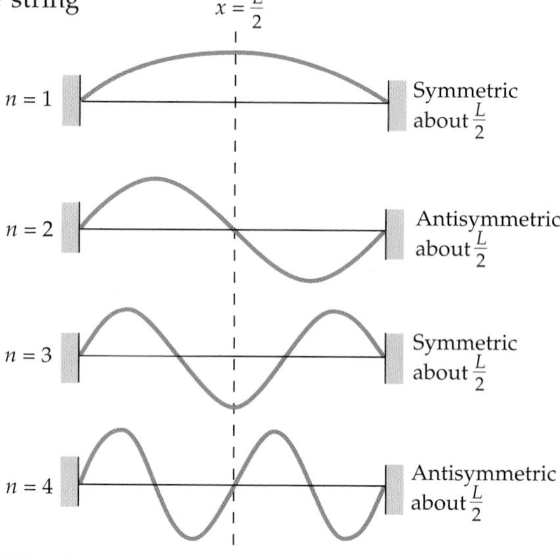

FIGURE 16-21 The first four harmonics for a string fixed at both ends. The odd harmonics are symmetrical about the center of the string, whereas the even harmonics are not. When a string is plucked at the center, it vibrates only in its odd harmonics.

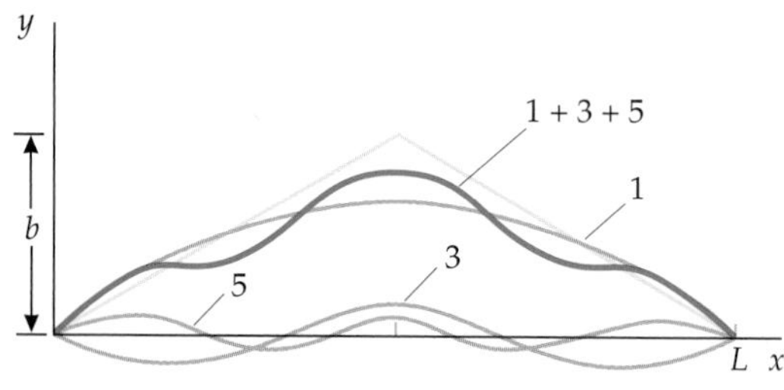

FIGURE 16-22 Approximating the shape of a string plucked at the center, as in Figure 16-20, using harmonics. The red line is an approximation of the original shape of the string based on the first three odd harmonics. The height of the string is exaggerated in this drawing to show the relative amplitudes of the harmonics. Most of the energy is associated with the fundamental, but there is some energy in the third, fifth, and other odd harmonics.

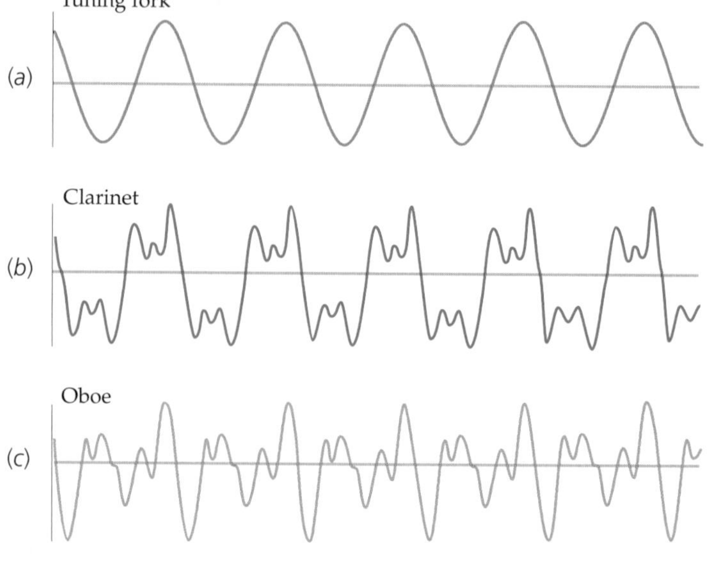

FIGURE 16-23 Waveforms of (*a*) a tuning fork, (*b*) a clarinet, and (*c*) an oboe, each at a fundamental frequency of 440 Hz and at approximately the same intensity.

Waveforms can be analyzed in terms of the harmonics that comprise them by means of **harmonic analysis.** (Harmonic analysis is also called **Fourier analysis** after the French mathematician J.B.J. Fourier, who developed the techniques for analyzing periodic functions.) Figure 16-24 shows a plot of the relative intensities of the harmonics of the waveforms in Figure 16-23. The waveform of the sound from the tuning fork contains only the fundamental frequency. That for the sound from the clarinet contains the fundamental, large amounts of the third, fifth, and seventh harmonics, and lesser amounts of the second, fourth, and sixth harmonics. For the sound from the oboe, there is more intensity in the second and third harmonics than in the fundamental.

The inverse of harmonic analysis is **harmonic synthesis,** which is the construction of a periodic wave from harmonic components. Figure 16-25*a* shows the first three odd harmonics used to synthesize a square wave and Figure 16-25*b* shows the square wave that results from the sum of the three harmonics. The more harmonics used in a synthesis, the closer the approximation will be to the actual waveform (the gray line in the Figure). The relative amplitudes of the harmonics needed to synthesize the square wave are shown in Figure 16-26.

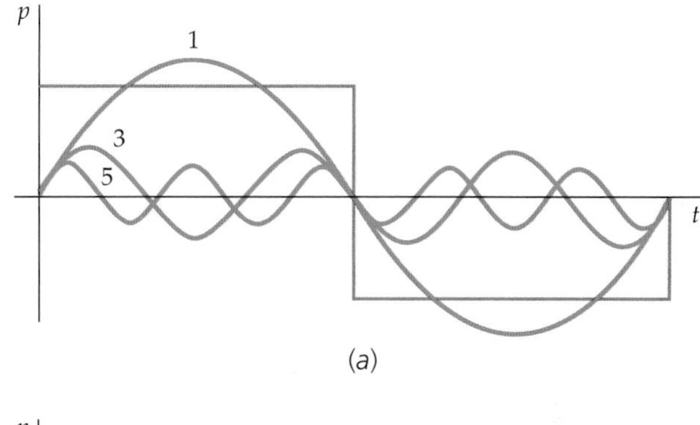

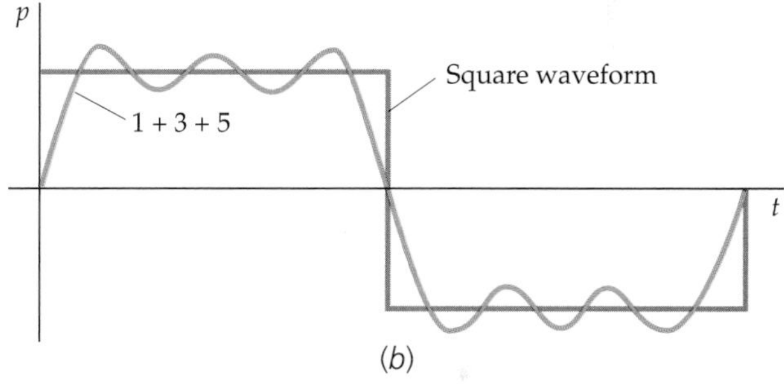

FIGURE 16-25 (*a*) **The first three odd harmonics of a single sine wave, used to synthesize a square wave.** (*b*) **The approximation of a square wave that results from summing the first three odd harmonics in** (*a*).

*16-5 Wave Packets and Dispersion

The waveforms discussed in Section 16.4 are periodic in time. Pulses, which are aperiodic, can also be represented by a group of harmonic waves of different frequencies. However, the synthesis of a pulse requires a continuous

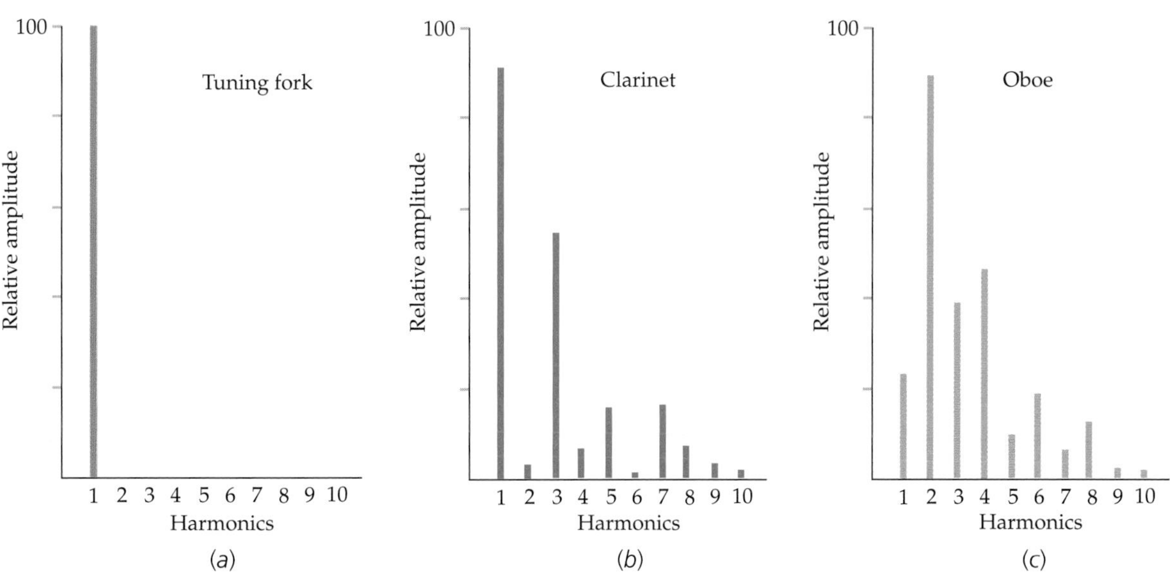

FIGURE 16-24 Relative intensities of the harmonics in the waveforms shown in Figure 16-23 for (*a*) **the tuning fork,** (*b*) **the clarinet, and** (*c*) **the oboe.**

distribution of frequencies rather than a discrete set of harmonics as in Figure 16-26. Such a group is called a **wave packet.** The characteristic feature of a wave pulse is that it has a beginning and an end, whereas a harmonic wave repeats over and over. If the duration Δt of the pulse is very short, the range of frequencies $\Delta\omega$ needed to describe the pulse is very large. The general relation between Δt and $\Delta\omega$ is

$$\Delta\omega\, \Delta t \sim 1 \qquad \text{16-18}$$

where the tilde $\sim$ means "of the order of."

The exact value of this product depends on just how the quantities $\Delta\omega$ and Δt are defined. For any reasonable definitions, $\Delta\omega$ and $1/\Delta t$ have the same order of magnitude. A wave pulse produced by a source of short duration Δt, like the crack of a bat on a ball, has a narrow width in space $\Delta x = v\,\Delta t$, where v is the wave speed. Each harmonic wave of frequency ω has a wave number $k = \omega/v$. A range of frequencies $\Delta\omega$ implies a range of wave numbers $\Delta k = \Delta\omega/v$. Substituting $v\,\Delta k$ for $\Delta\omega$ in Equation 16-18 gives $v\,\Delta k\,\Delta t \sim 1$ or

$$\Delta k\, \Delta x \sim 1 \qquad \text{16-19}$$

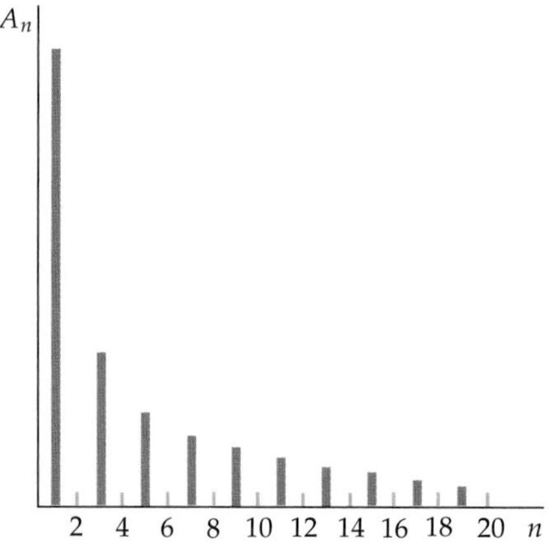

FIGURE 16-26 Relative amplitudes A_n of the first ten harmonics needed to synthesize a square wave. The more harmonics that are used, the closer the approximation is to the square wave.

If a wave packet is to maintain its shape as it travels, all of the component harmonic waves that make up the packet must travel with the same speed. This occurs if the speed of the component waves in a given medium is independent of frequency or wavelength. Such a medium is called a **nondispersive medium.** Air is, to an excellent approximation, a nondispersive medium for sound waves, but solids and liquids generally are not. (Probably the most familiar example of dispersion is the formation of a rainbow, due to the fact that the velocity of light waves in water depends slightly on the frequency of the light, so the different colors, corresponding to different frequencies, have slightly different angles of refraction.)

When the wave speed in a dispersive medium depends only slightly on the frequency (or wavelength), a wave packet changes shape very slowly as it travels, and it covers a considerable distance as a recognizable entity. But the speed of the packet, called the **group velocity,** is not the same as the (average) speed of the individual component harmonic waves, called the **phase velocity.** (By the speed of an individual harmonic wave we mean the speed of its wavefronts. Because wavefronts are lines or surfaces of constant phase, their speed is called the phase velocity of the wave.)

SUMMARY

1. The principle of superposition, which holds for all electromagnetic waves in empty space, for waves on a flexible taut string in the small angle approximation, and for sound waves of small amplitude, follows from the linearity of the corresponding wave equations.
2. Interference is an important wave phenomenon that applies to all coherent superposing waves. It follows from the principle of superposition. Diffraction and interference distinguish wave motion from particle motion.
3. The standing-wave conditions can be recalled by sketching a string or tube and drawing waves that have nodes at a fixed or stopped end, and antinodes at a free or open end.

Topic	Relevant Equations and Remarks
1. Superposition and Interference	The superposition of two harmonic waves of equal amplitude, wave number, and frequency but phase difference δ results in a harmonic wave of the same wave number and frequency, but differing in phase and amplitude from each of the two waves $y = y_1 + y_2 = y_0 \sin(kx - \omega t) + y_0 \sin(kx - \omega t + \delta) = [2y_0 \cos \frac{1}{2}\delta] \sin(kx - \omega t + \frac{1}{2}\delta)$ **16-6**
Constructive interference	If waves are in phase or differ in phase by an integer times 2π, then the amplitudes of the waves add and the interference is constructive.
Destructive interference	If waves differ in phase by π or by an odd integer times π, then the amplitudes subtract and the interference is destructive.
Beats	Beats are the result of the interference of two waves of slightly different frequencies. The beat frequency equals the difference in the frequencies of the two waves: $f_{\text{beat}} = \Delta f$ **16-8**
Phase difference δ due to path difference Δx	$\delta = k\,\Delta x = 2\pi \frac{\Delta x}{\lambda}$ **16-9**
2. Standing Waves	Standing waves occur for certain frequencies and wavelengths when waves are confined in space. They occur only if each point of the system oscillates in simple harmonic motion and any two moving points move either in phase or 180° out of phase.
Wavelength	The distance between a node and an adjacent antinode is a quarter-wavelength.
String fixed at both ends	For a string fixed at both ends, there is a node at each end so that an integral number of half-wavelengths must fit into the length of the string. The standing-wave condition in this case is $L = n\frac{\lambda_n}{2}, \quad n = 1, 2, 3, \ldots$ **16-10**
Standing wave function for a string fixed at both ends	The allowed waves form a harmonic series, with the frequencies given by $f_n = \frac{v}{\lambda_n} = n\frac{v}{2L} = nf_1, \quad n = 1, 2, 3, \ldots$ **16-11** where $f_1 = v/2L$ is the lowest frequency, called the fundamental.
Organ pipe open at both ends	Standing sound waves in the air in a pipe that is open at both ends have a pressure node (and a displacement antinode) near each end so that the standing wave condition is the same as for a string fixed at both ends.
String fixed at one end and free at the other	For a string with one end fixed and one end free, there is a node at the fixed end and an antinode at the free end, so that an integral number of quarter-wavelengths must fit into the length of the string. The standing-wave condition in this case is $L = n\frac{\lambda_n}{4}, \quad n = 1, 3, 5, \ldots$ **16-12** Only the odd harmonics are present. Their frequencies are given by $f_n = \frac{v}{\lambda_n} = n\frac{v}{4L} = nf_1, \quad n = 1, 3, 5, \ldots$ **16-13** where $f_1 = v/4L$.
Organ pipe open at one end and stopped at the other	Standing sound waves in a pipe that is open at one end and stopped at the other end have a displacement antinode at the open end and a displacement node at the stopped end. The standing wave condition is the same as for a string fixed at one end.

Wave Functions for Standing Waves	$y_n(x,t) = A_n \sin(k_n x) \cos(\omega_n t + \delta_n)$	**16-16**
	where $k_n = 2\pi/\lambda_n$ and $\omega_n = 2\pi f_n$.	
	The necessary conditions for standing waves on a string are 1. Each point on the string either remains at rest or oscillates with simple harmonic motion. (Those points remaining at rest are nodes.) 2. The motions of any two points on the string that are not nodes oscillate either in phase or 180° out of phase.	
***3. Superposition of Standing Waves**	A vibrating system typically does not vibrate in a single harmonic mode but in a superposition of the allowed harmonic modes.	
***4. Harmonic Analysis and Synthesis**	Sounds of different tone quality contain different mixtures of harmonics. The analysis of a particular tone in terms of its harmonic content is called harmonic analysis. Harmonic synthesis is the construction of a tone by the addition of harmonics.	
***5. Wave Packets**	A wave pulse can be represented by a continuous distribution of harmonic waves. The range of frequencies $\Delta\omega$ is related to the width in time Δt, and the range of wave numbers Δk is related to the width in space Δx.	
Frequency and time ranges	$\Delta\omega\, \Delta t \sim 1$	**16-18**
Wave number and space ranges	$\Delta k\, \Delta x \sim 1$	**16-19**
***6. Dispersion**	In a nondispersive medium, the phase velocity is independent of frequency, and a pulse (wave packet) travels without change in shape. In a dispersive medium, the phase velocity varies with frequency, and the pulse changes shape as it moves. The pulse moves with a velocity called the group velocity of the packet.	

PROBLEMS

- • Single-concept, single-step, relatively easy
- •• Intermediate-level, may require synthesis of concepts
- ••• Challenging
- SSM Solution is in the *Student Solutions Manual*
- iSOLVE Problems available on iSOLVE online homework service
- iSOLVE✓ These "Checkpoint" online homework service problems ask students additional questions about their confidence level, and how they arrived at their answer

In a few problems, you are given more data than you actually need; in a few other problems, you are required to supply data from your general knowledge, outside sources, or informed estimates.

Conceptual Problems

1 •• SSM Two rectangular wave pulses are traveling in opposite directions along a string. At $t = 0$, the two pulses are as shown in Figure 16-27. Sketch the wave functions for $t = 1$, 2, and 3 s.

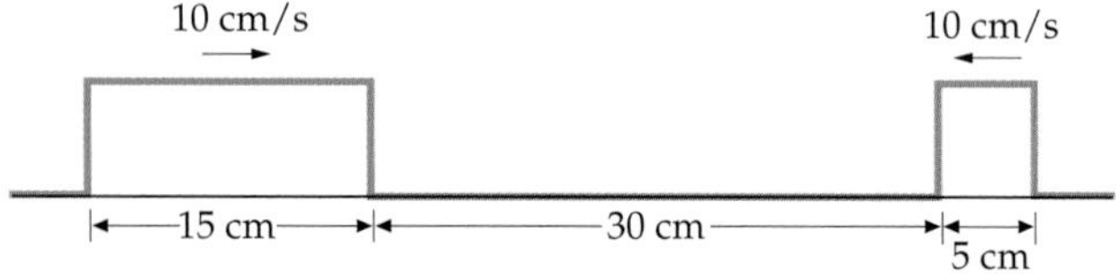

FIGURE 16-27 Problems 1, 2

2 •• Repeat Problem 1 for the case in which the pulse on the right is inverted.

3 • Beats are produced by the superposition of two harmonic waves only if (*a*) their amplitudes and frequencies are equal, (*b*) their amplitudes are the same but their frequencies differ slightly, (*c*) their frequencies differ slightly even if their amplitudes are not equal, (*d*) their frequencies are equal but their amplitudes differ slightly.

4 • True or false:

(*a*) The frequency of the third harmonic is three times that of the first harmonic.

(*b*) The frequency of the fifth harmonic is five times that of the fundamental.

(*c*) In a pipe that is open at one end and closed at the other, the even harmonics are not excited.

5 •• Standing waves result from the superposition of two waves of (*a*) the same amplitude, frequency, and direction of propagation, (*b*) the same amplitude and frequency and opposite directions of propagation, (*c*) the same amplitude, slightly different frequency, and the same direction of propagation, (*d*) the same amplitude, slightly different frequency, and opposite directions of propagation.

6 • SSM The resonant frequencies of a violin string are all integer multiples of the fundamental frequency, while the resonant frequencies of a circular drumhead are irregularly spaced. Given this information, explain the difference in the sounds of a violin and a drum.

7 • An organ pipe open at both ends has a fundamental frequency of 400 Hz. If one end of this pipe is now closed, the fundamental frequency will be (*a*) 200 Hz, (*b*) 400 Hz, (*c*) 546 Hz, (*d*) 800 Hz.

8 •• A string fixed at both ends resonates at a fundamental frequency of 180 Hz. Which of the following will reduce the fundamental frequency to 90 Hz? (*a*) Double the tension and double the length. (*b*) Halve the tension and keep the length fixed. (*c*) Keep the tension fixed and double the length. (*d*) Keep the tension fixed and halve the length.

9 •• How do the resonance frequencies of an organ pipe change when the air temperature increases?

10 • SSM When two waves moving in opposite directions superimpose as in Figure 16-1, does either impede the progress of the other?

11 • When a guitar string is plucked, is the wavelength of the wave it produces in air the same as the wavelength of the wave on the string?

12 • When two waves interfere constructively or destructively, is there any gain or loss in energy? Explain.

13 • A musical instrument consists of drinking glasses partially filled with water that are struck with a small mallet. Explain how this works.

14 •• During an organ recital, the air compressor that drives the organ pipes suddenly fails. An enterprising physics student in the audience comes to the rescue by connecting a tank of pure nitrogen gas under high pressure to the output of the compressor. What effect, if any, will this change have on the operation of the organ? What if the tank contained helium?

15 •• SSM When the tension on a piano wire is increased, which of the following occurs? (*a*) Its wavelength decreases. (*b*) Its wavelength remains the same while its frequency increases. (*c*) Its wavelength and frequency increase. (*d*) None of the above occur.

16 •• The following instructions are given for connecting stereo speakers to an amplifier so that they are in phase: "After both speakers are connected, play a monophonic record or program with the bass control turned up and the treble control turned down. While listening to the speakers, turn the balance control so that first one speaker is heard separately, then the two together, and then the other separately. If the bass is stronger when both speakers play together, they are connected properly. If the bass is weaker when both play together than when each plays separately, reverse the connections on one speaker." Explain why this method works. In particular, explain why a stereo source is not used and why only the bass is compared.

17 •• The constant γ for helium (and all monatomic gases) is 1.67. If a man inhales helium and then speaks, he sounds like Alvin of the Chipmunks. Why?

18 •• SSM Figure 16-28 is a photograph of two pieces of very finely woven silk placed one on top of the other. Where the pieces overlap, a series of light and dark lines are seen. This moiré pattern can also be seen when a scanner is used to copy photos from a book or newspaper. What causes the moiré pattern, and how is it similar to the phenomenon of interference?

FIGURE 16-28 Problem 18

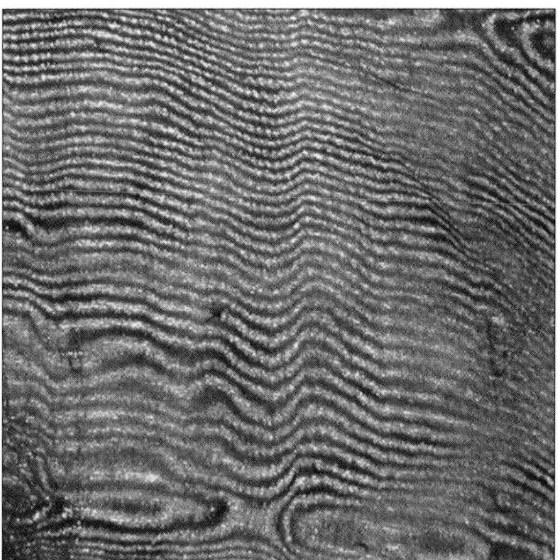

Estimation and Approximation

19 •• About how accurately can you tune a piano string to a tuning-fork frequency?

20 • SSM The shortest pipes used in organs are about 7.5 cm long. (*a*) What is the fundamental frequency of a pipe this long that is open at both ends? (*b*) For such a pipe, what is the highest harmonic that is within the audible range? (The normal range of hearing is about 20 to 20,000 Hz.)

21 •• On a windy day, a drain pipe will sometimes resonate. Estimate the resonance frequency of a drain pipe on a single-story house. How much might this frequency change from winter to summer in your region?

Superposition and Interference

22 • Two waves traveling on a string in the same direction both have a frequency of 100 Hz, a wavelength of 2 cm, and an amplitude of 0.02 m. What is the amplitude of the resultant wave if the original waves differ in phase by (*a*) $\pi/6$ and (*b*) $\pi/3$?

23 • Two waves having the same frequency, wavelength, and amplitude are traveling in the same direction. If they differ in phase by $\pi/2$ and each has an amplitude of 0.05 m, what is the amplitude of the resultant wave?

24 • SSM ISOLVE Two sound sources oscillate in phase with the same amplitude A. They are separated in space by $\lambda/3$. What is the amplitude of the resultant wave formed from the two sources at a point that is on the line that passes through the sources but is not between the sources?

25 • Two sound sources oscillate in phase with a frequency of 100 Hz. At a point 5.00 m from one source and 5.85 m from the other, the amplitude of the sound from each source separately is A. (*a*) What is the phase difference in the sound waves from the two sources at that point? (*b*) What is the amplitude of the resultant wave at that point?

26 • SSM With a compass, draw circular arcs representing wave crests originating from each of two point sources a distance $d = 6$ cm apart for $\lambda = 1$ cm. Connect the intersections corresponding to points of constant path difference and label the path difference for each line. (See Figure 16-8.)

27 • Two speakers separated by some distance emit sound waves of the same frequency. At some point P, the intensity due to each speaker separately is I_0. The path distance from P to one of the speakers is $\frac{1}{2}\lambda$ greater than that from P to the other speaker. What is the intensity at P if (*a*) the speakers are coherent and in phase, (*b*) the speakers are incoherent, and (*c*) the speakers are coherent but have a phase difference of π rad?

28 • Answer the questions of Problem 27 for a point P' for which the distance to the far speaker is 1λ greater than the distance to the near speaker. Assume that the intensity at point P' due to each speaker separately is again I_0.

29 • Two speakers separated by some distance emit sound waves of the same frequency, but the speakers are out of phase by 90°. Let r_1 be the distance from some point to speaker 1 and r_2 be the distance from that point to speaker 2. Find the smallest value of $r_2 - r_1$ at which the sound at that point will be (*a*) maximum and (*b*) minimum. (Express your answers in terms of the wavelength.)

30 •• SSM Show that, if the separation between two sound sources radiating coherently in phase is less than half a wavelength, complete destructive interference will not be observed in any direction.

31 •• ISOLVE A transverse wave of frequency 40 Hz propagates down a string. Two points 5 cm apart are out of phase by $\pi/6$. (*a*) What is the wavelength of the wave? (*b*) At a given point, what is the phase difference between two displacements for times 5 ms apart? (*c*) What is the wave velocity?

32 •• It is thought that the brain determines the direction of the source of a sound by sensing the phase difference between the sound waves striking the eardrums. A distant source emits sound of frequency 680 Hz. When you are directly facing a sound source there should be no phase difference. Estimate the phase difference between the sounds received by each ear as you turn from facing directly toward the source through 90°.

33 •• ISOLVE✓ Sound source A is located at $x = 0, y = 0$, and sound source B is placed at $x = 0$, $y = 2.4$ m. The two sources radiate coherently in phase. An observer at $x = 40$ m, $y = 0$ notes that as she takes a few steps in either the positive or negative y direction away from $y = 0$, the sound intensity diminishes. What are the lowest and the next higher frequencies of the sources that can account for that observation?

34 •• Suppose that the observer in Problem 33 finds herself at a point of minimum intensity at $x = 40$ m, $y = 0$. What are then the lowest and next higher frequencies of the sources consistent with this observation?

35 •• SSM Two harmonic water waves of equal amplitudes but different frequencies, wave vectors, *and velocities* are superposed on each other. The total displacement of the wave can be written as $y(x,t) = A[\cos(k_1x - \omega_1t) + \cos(k_2x - \omega_2t)]$, where $\omega_1k_1 = v_1$ (the speed of the first wave) and $\omega_2/k_2 = v_2$ (the speed of the second wave). (*a*) Show that $y(x,t)$ can be written in the form $y(x,t) = 2A\cos[(\Delta k/2)x - (\Delta\omega/2)t]\cos(k_{av}x - \omega_{av}t)$ where $\omega_{av} = (\omega_1 + \omega_2)/2$. $k_{av} = (k_1 + k_2)/2$, $\Delta\omega = \omega_1 - \omega_2$, and $\Delta k = k_1 - k_2$. The factor $2A\cos[(\Delta k/2)x - (\Delta\omega/2)t]$ is called the *envelope* of the wave. (*b*) Using a spreadsheet program or graphing calculator, make a graph of $y(x,t)$ for $A = 1$, $\omega_1 = 1$ rad/s, $k_1 = 1$ m^{-1}, $\omega_2 = 0.9$ rad/s, and $k_2 = 0.8$ m^{-1} at $t = 0$ s, 0.5 s, and 1 s for x between 0 m and 50 m. (*c*) What is the speed at which the envelope moves?

36 •• Two point sources that are in phase are separated by a distance d. An interference pattern is detected along a line parallel to the line through the sources and a large distance D from the sources, as shown in Figure 16-29. (*a*) Show that the path difference from the two sources to some point on the line at a small angle θ is given approximately by $\Delta s = d\sin\theta$. *Hint: Assume that* D >> d, *so the lines from the sources to* P *are approximately parallel.* (*b*) Show that the distance y_m from the central maximum point to the mth interference maximum is given approximately by $y_m = m(D\lambda/d)$.

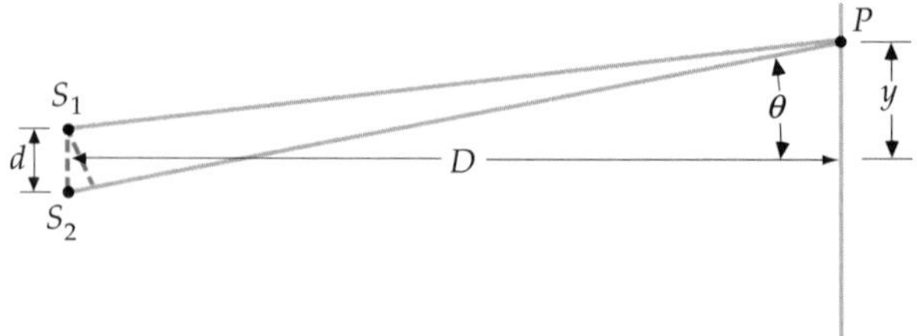

FIGURE 16-29 Problems 36–40

37 •• ISOLVE✓ Two sound sources radiating in phase at a frequency of 480 Hz interfere such that maxima are heard at angles of 0° and 23° from a line perpendicular to that joining the two sources. Find the separation between the two sources and any other angles at which a maximum intensity will be heard. (Use the result of Problem 36.)

38 ••• SSM Two loudspeakers are driven in phase by an audio amplifier at a frequency of 600 Hz. The speakers are on the y axis, one at $y = +1.00$ m and the other at $y = -1.00$ m. A listener begins at $y = 0$ a very large distance D away and walks along a line parallel to the y axis. (See Problem 36.) (*a*) At what angle θ will she first hear a minimum in the sound intensity? (*b*) At what angle will she first hear a maximum (after $\theta = 0$)? (*c*) How many maxima can she possibly hear if she keeps walking in the same direction?

39 ••• ISOLVE Two sound sources driven in phase by the same amplifier are 2 m apart on the y axis. At a point a very large distance from the y axis, constructive interference is first heard at an angle $\theta_1 = 0.140$ rad with the x axis and is next heard at $\theta_2 = 0.283$ rad (see Figure 16-29). (*a*) What is the wavelength of the sound waves from the sources? (*b*) What is the frequency of the sources? (*c*) At what other angles is constructive interference heard? (*d*) What is the smallest angle for which the sound waves cancel?

40 ••• The two sound sources from Problem 39 are now driven 90° out-of-phase, but at the same frequency as in Problem 39. At what angles are constructive and destructive interference heard?

41 ••• A radio telescope consists of two antennas separated by a distance of 200 m. Both antennas are tuned to a particular frequency, such as 20 MHz. The signals from each antenna are fed into a common amplifier, but one signal first passes through a phase adjuster that delays its phase by a chosen amount so that the telescope can "look" in different directions. When the phase delay is zero, plane radio waves that are incident vertically on the antennas produce signals that add constructively at the amplifier. What should be the phase delay so that signals coming from an angle $\theta = 10°$ with the vertical (in the plane formed by the vertical and the line joining the antennas) will add constructively at the amplifier?

Beats

42 • ISOLVE When two tuning forks are struck simultaneously, 4 beats per second are heard. The frequency of one fork is 500 Hz. (*a*) What are the possible values for the frequency of the other fork? (*b*) A piece of wax is placed on the 500-Hz fork to lower its frequency slightly. Explain how the measurement of the new beat frequency can be used to determine which of your answers to Part (*a*) is the correct frequency of the second fork.

43 •• SSM Two ambulances move toward each other on a straight road. Each travels at a speed of 50 mph. The siren on each ambulance produces a sound wave of frequency 500 Hz. (*a*) The driver of each ambulance hears the other's siren and hears a beat note between the frequency of his own siren and the siren of the other ambulance. What is the frequency of the beat note? (*b*) A passerby stands midway between the two ambulances. What is the frequency of the beat note between the two sirens that he hears?

Standing Waves

44 • SSM ISOLVE✓ A string fixed at both ends is 3 m long. It resonates in its second harmonic at a frequency of 60 Hz. What is the speed of transverse waves on the string?

45 • ISOLVE A string 3 m long and fixed at both ends is vibrating in its third harmonic. The maximum displacement of any point on the string is 4 mm. The speed of transverse waves on this string is 50 m/s. (*a*) What are the wavelength and frequency of this wave? (*b*) Write the wave function for this wave.

46 • ISOLVE✓ Calculate the fundamental frequency for a 10-m organ pipe that is (*a*) open at both ends and (*b*) closed at one end.

47 • A steel wire having a mass of 5 g and a length of 1.4 m is fixed at both ends and has a tension of 968 N. (*a*) Find the speed of transverse waves on the wire. (*b*) Find the wavelength and frequency of the fundamental. (*c*) Find the frequencies of the second and third harmonics.

48 • A rope 4 m long is fixed at one end; the other end is attached to a light string so that it is free to move. The speed of waves on the rope is 20 m/s. Find the frequency of (*a*) the fundamental, (*b*) the second harmonic, and (*c*) the third harmonic.

49 • A piano wire without windings has a fundamental frequency of 200 Hz. When it is wound with wire, its linear mass density is doubled. What is its new fundamental frequency, assuming that the tension is unchanged?

50 • SSM What is the greatest length that an organ pipe can have in order to have its fundamental note in the audible range (20 to 20,000 Hz) if (*a*) the pipe is closed at one end and (*b*) it is open at both ends?

51 •• ISOLVE✓ The wave function $y(x,t)$ for a certain standing wave on a string fixed at both ends is given by $y(x,t) = 4.2 \sin 0.20x \cos 300t$, where y and x are in centimeters and t is in seconds. (*a*) What are the wavelength and frequency of this wave? (*b*) What is the speed of transverse waves on this string? (*c*) If the string is vibrating in its fourth harmonic, how long is it?

52 •• The wave function $y(x,t)$ for a certain standing wave on a string fixed at both ends is given by $y(x,t) = (0.05\text{ m}) \sin 2.5\text{ m}^{-1}\, x \cos 500\text{ s}^{-1}\, t$. (*a*) What are the speed and amplitude of the two traveling waves that result in this standing wave? (*b*) What is the distance between successive nodes on the string? (*c*) What is the shortest possible length of the string?

53 •• A 2.51-m-long string has the wave function given in Problem 52. (*a*) Sketch the position of the string at the times $t = 0$, $t = T/4$, $t = T/2$, and $t = 3T/4$, where $T = 1/f$ is the period of the vibration. (*b*) Find T in seconds. (*c*) At a time t when the string is horizontal, that is, $y(x) = 0$ for all x, what has become of the energy in the wave?

54 •• SSM ISOLVE Three successive resonance frequencies for a certain string are 75, 125, and 175 Hz. (*a*) Find the ratios of each pair of successive resonance frequencies. (*b*) How can you tell that these frequencies are for a string fixed at one end only rather than for a string fixed at both ends? (*c*) What is the fundamental frequency? (*d*) Which harmonics are these resonance frequencies? (*e*) If the speed of transverse waves on this string is 400 m/s, find the length of the string.

55 •• The space above the water in a tube like that shown in Example 16-9 is 120 cm long. Near the open end there is a loudspeaker that is driven by an audio oscillator whose frequency can be varied from 10 to 5000 Hz. (*a*) What is the lowest frequency of the oscillator that will produce resonance within the tube? (*b*) What is the highest frequency that will produce resonance? (*c*) How many different frequencies of the oscillator will produce resonance? (Neglect the end correction.)

56 •• A 460-Hz tuning fork causes resonance in the tube in Example 16-9 when the top of the tube is 18.3 and 55.8 cm above the water surface. (*a*) Find the speed of sound in air. (*b*) What is the end correction to adjust for the fact that the antinode does not occur exactly at the end of the open tube?

57 •• SSM ISOLVE✓ At 16°C, the fundamental frequency of an organ pipe is 440.0 Hz. What will be the fundamental frequency of the pipe if the temperature increases to 32°C? Would it be better to construct the pipe with a material that expands substantially as the temperature increases or should the pipe be made of material that maintains the same length at all normal temperatures?

58 •• The end correction for a circular pipe is approximately $\Delta L = 0.3186D$, where D is the pipe diameter. Find the length of a pipe open at both ends that will produce a middle C (256 Hz) as its fundamental mode for pipes of diameter $D =$ 1 cm, 10 cm, and 30 cm.

59 •• ISOLVE A violin string of length 40 cm and mass 1.2 g has a frequency of 500 Hz when it is vibrating in its fundamental mode. (*a*) What is the wavelength of the standing wave on the string? (*b*) What is the tension in the string? (*c*) Where should you place your finger to increase the frequency to 650 Hz?

60 •• The G string on a violin is 30 cm long. When played without fingering, it vibrates at a frequency of 196 Hz. The next higher notes on the C-major scale are A (220 Hz), B (247 Hz), C (262 Hz), and D (294 Hz). How far from the end of the string must a finger be placed to play each of these notes?

61 •• A string with a mass density of 4×10^{-3} kg/m is under a tension of 360 N and is fixed at both ends. One of its resonance frequencies is 375 Hz. The next higher resonance frequency is 450 Hz. (*a*) What is the fundamental frequency of this string? (*b*) Which harmonics are the ones given? (*c*) What is the length of the string?

62 •• ISOLVE✓ A string fastened at both ends has successive resonances with wavelengths of 0.54 m for the nth harmonic and 0.48 m for the $(n + 1)$th harmonic. (*a*) Which harmonics are these? (*b*) What is the length of the string?

63 •• The strings of a violin are tuned to the tones G, D, A, and E, which are separated by a fifth from one another. That is, $f(D) = 1.5f(G)$, $f(A) = 1.5f(D) = 440$ Hz, and $f(E) = 1.5f(A)$. The distance between the two fixed points, the bridges at the scroll and over the body of the instrument, is 30 cm. The tension on the E string is 90 N. (*a*) What is the mass per meter of the E string? (*b*) To prevent distortion of the instrument over time, it is important that the tension on all strings be the same. Find the masses per meter of the other strings.

64 •• An ambulance is driving at 50 mph towards the brick wall of the hospital, which reflects the sound of the siren back toward the ambulance. When the ambulance is stationary, the siren's frequency is 500 Hz. (*a*) What is the spatial period of the standing wave caused by the sound of the siren and its reflection? (*b*) A doctor standing between the ambulance and the wall will hear the siren grow alternately louder and softer as the ambulance drives toward her. Why is this?

65 •• To tune a violin, the violinist first tunes the A string to the correct pitch of 440 Hz and then bows two adjoining strings simultaneously and listens for a beat pattern. While bowing the A and E strings, the violinist hears a beat frequency of 3 Hz and notes that the beat frequency increases as the tension on the E string is increased. (The E string is to be tuned to 660 Hz.) (*a*) Why is a beat produced by these two strings bowed simultaneously? (*b*) What is the frequency of the E string vibration when the beat frequency is 3 Hz? (*c*) If the tension on the E string is 80.0 N when the beat frequency is 3 Hz, what tension corresponds to perfect tuning of that string?

66 •• Suppose that you carry a small oscillator and speaker as you walk very slowly down a long hall. The speaker emits a sound of frequency 680 Hz, which is reflected from the walls at each end of the hall. As you walk along, you note that the sound intensity that you hear passes through successive maxima and minima. What distance must you walk to pass from one maximum to the next?

67 •• SSM Show that the standing wave function $A' \sin kx \cos(\omega t + \delta)$ can be written as the sum of two harmonic wave functions—one for a wave traveling in the positive x direction and the other for a wave of the same amplitude traveling in the negative x direction. The traveling waves each have the same wave number and angular frequency as does the standing wave.

68 •• A 2-m string is fixed at one end and is vibrating in its third harmonic with amplitude 3 cm and frequency 100 Hz. (*a*) Write the wave function for this vibration. (*b*) Write an expression for the kinetic energy of a segment of the string of length dx at a point x at some time t. At what time is this kinetic energy maximum? What is the shape of the string at this time? (*c*) Find the maximum kinetic energy of the string by integrating your expression for Part (*b*) over the total length of the string.

69 •• SSM A commonly used physics experiment that examines resonances of transverse waves on a string is shown in Figure 16-30. A weight is attached to the end of a string draped over a pulley; the other end of the string is attached to a mechanical oscillator that moves the string up and down at a set frequency f. The length L between the oscillator and the pulley is fixed. For certain values of the weight the string resonates. If $L = 1$ m, $f = 80$ Hz, and the mass density of the string is $\mu = 0.75$ g/m, what weights are needed for each of the first three modes (standing waves) of the string?

FIGURE 16-30 Problem 69

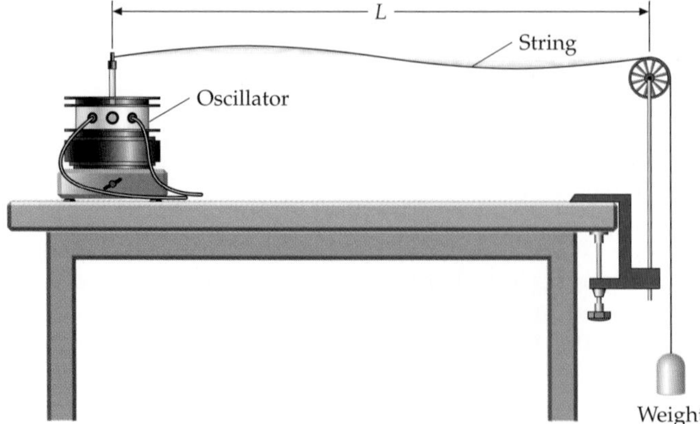

*Wave Packets

70 • ISOLVE Information used by computers is transmitted along a cable in the form of short electric pulses at the rate of 10^7 pulses per second. (*a*) What is the maximum duration of each pulse if no two pulses overlap? (*b*) What is the range of frequencies to which the receiving equipment must respond?

71 • SSM A tuning fork of frequency f_0 begins vibrating at time $t = 0$ and is stopped after a time interval Δt. The waveform of the sound at some later time is shown as a function of x. Let N be the (approximate) number of cycles in this waveform. (*a*) How are N, f_0, and Δt related? (*b*) If Δx is the length in space of this wave packet, what is the wavelength in terms of Δx and N? (*c*) What is the wave number k in terms of N and Δx? (*d*) The number N is uncertain by about ± 1 cycle. Use Figure 16-31 to explain why. (*e*) Show that the uncertainty in the wave number due to the uncertainty in N is $2\pi/\Delta x$.

FIGURE 16-31 Problem 71

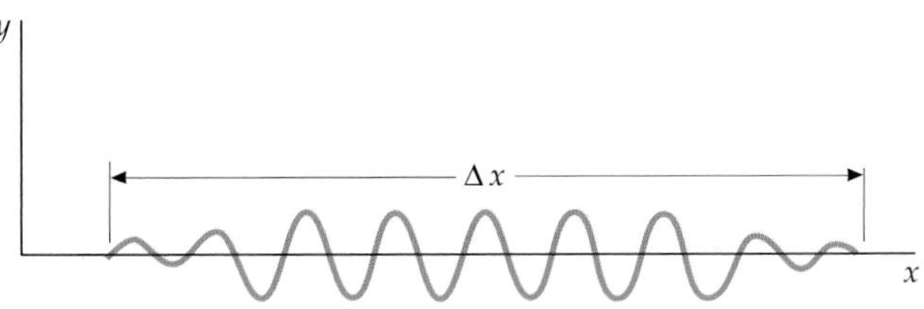

General Problems

72 • Middle C on the equal-temperament scale used by modern instrument makers has a frequency of 261.63 Hz. If a 7-g piano wire that is 80 cm long is to be tuned so that 261.63 is its fundamental frequency, what should be the tension in the wire?

73 • ISOLVE The ear canal, which is about 2.5 cm long, roughly approximates a pipe that is open at one end and closed at the other. (*a*) What are the resonance frequencies of the ear canal? (*b*) Describe the possible effect of the resonance modes of the ear canal on the threshold of hearing.

74 • A 4-m-long, 160-g rope is fixed at one end and is tied to a light string at the other end. Its tension is 400 N. (*a*) What are the wavelengths of the fundamental and the next two harmonics? (*b*) What are the frequencies of these standing waves?

75 •• Two waves from two coherent sources have the same wavelength λ, frequency ω, and amplitude A. What is the path difference if the resultant wave at some point has amplitude A?

76 •• ISOLVE A 35-m string has a linear mass density of 0.0085 kg/m and is under a tension of 18 N. Find the frequencies of the lowest four harmonics if (*a*) the string is fixed at both ends and (*b*) the string is fixed at one end and attached to a long, thin, massless thread at the other end.

77 •• ISOLVE✓ You find an abandoned mine shaft and decide to measure its depth. Using an audio oscillator of variable frequency, you note that you can produce successive resonances at frequencies of 63.58 and 89.25 Hz. What is the depth of the shaft?

78 •• A string 5 m long that is fixed at one end only is vibrating in its fifth harmonic with a frequency of 400 Hz. The maximum displacement of any segment of the string is 3 cm. (*a*) What is the wavelength of this wave? (*b*) What is the wave number k? (*c*) What is the angular frequency? (*d*) Write the wave function for this standing wave.

79 •• The wave function for a standing wave on a string is described by $y(x,t) = 0.02 \sin 4\pi x \cos 60\pi t$, where y and x are in meters and t is in seconds. Determine the maximum displacement and maximum speed of a point on the string at (*a*) $x = 0.10$ m, (*b*) $x = 0.25$ m, (*c*) $x = 0.30$ m, and (*d*) $x = 0.50$ m.

80 •• A 2.5-m-long wire with a mass of 0.10 kg is fixed at both ends and is under tension of 30 N. When the nth harmonic is excited, there is a node 0.50 m from one end. (*a*) What is n? (*b*) What are the frequencies of the first three allowed modes of vibration?

81 •• SSM In an early method used to determine the speed of sound in gases, powder was spread along the bottom of a horizontal, cylinderical glass tube. One end of the tube was closed by a piston that oscillated at a known frequency f. The other end was closed by a movable piston whose position was adjusted until resonance occurred. At resonance, the powder collected in equally spaced piles along the bottom of the tube. (*a*) Explain why the powder collects in this way. (*b*) Derive a formula that gives the speed of sound in the gas in terms of f and the distance between the piles of powder. (*c*) Give suitable values for the frequency f and the distance between the piles of powder. (*d*) Give suitable values for the frequency f and the length L of the tube for which the speed of sound could be measured in either air or helium.

82 •• In a lecture demonstration of standing waves, a string is attached to a tuning fork that vibrates at 60 Hz and sets up transverse waves of that frequency on the string. The other end of the string passes over a pulley, and the tension is varied by attaching weights to that end. The string has approximate nodes at the tuning fork and at the pulley. (*a*) If the string has a linear mass density of 8 g/m and is 2.5 m long (from the tuning fork to the pulley), what must be the tension for the string to vibrate in its fundamental mode? (*b*) Find the tension necessary for the string to vibrate in its second, third, and fourth harmonic.

83 •• ISOLVE✓ Three successive resonance frequencies in an organ pipe are 1310, 1834, and 2358 Hz. (*a*) Is the pipe closed at one end or open at both ends? (*b*) What is the fundamental frequency? (*c*) What is the length of the pipe?

84 •• ISOLVE✓ A wire of mass 1 g and length 50 cm is stretched with a tension of 440 N. It is then placed near the open end of the tube in Example 16-9 and stroked with a violin bow so that it oscillates at its fundamental frequency. The water level in the tube is then lowered until a resonance is obtained, which occurs at 18 cm below the top of the tube. Use the data given to determine the speed of sound in air. Why is this method not very accurate?

85 •• A standing wave on a rope is represented by the wave function $y(x,t) = 0.02 \sin \frac{1}{2}\pi x \cos 40\pi t$, where x and y are in meters and t is in seconds. (*a*) Write wave functions for two traveling waves that, when superimposed, will produce the resultant standing-wave pattern. (*b*) What is the distance between the nodes of the standing wave? (*c*) What is the velocity of a segment of the rope at $x = 1$ m? (*d*) What is the acceleration of a segment of the rope at $x = 1$ m?

86 •• ISOLVE Two identical speakers emit sound waves of frequency 680 Hz uniformly in all directions. The total audio output of each speaker is 1 mW. A point P is 2.00 m from one speaker and 3.00 m from the other. (*a*) Find the intensities I_1 and I_2 from each speaker separately at point P. (*b*) If the speakers are driven coherently and are in phase, what is the intensity at point P? (*c*) If they are driven coherently but are 180° out of phase, what is the intensity at point P? (*d*) If the speakers are incoherent, what is the intensity at point P?

87 •• Three waves with the same frequency, wavelength, and amplitude are traveling in the same direction. The three waves are given by $y_1(x,t) = 0.05 \sin\left(kx - \omega t - \frac{\pi}{3}\right)$, $y_2(x,t) = 0.05 \sin(kx - \omega t)$, and $y_3(x,t) = 0.05 \sin\left(kx - \omega t + \frac{\pi}{3}\right)$. Find the resultant wave.

88 •• A plane wave has the form $f(x, y, t) = A \cos(k_x x + k_y y - \omega t)$. Show that the direction in which the wave is traveling makes an angle $\theta = \tan^{-1}(k_y/k_x)$ with the positive x direction and that the wave speed is $v = \omega/\sqrt{k_x^2 + k_y^2}$.

89 •• SSM The speed of sound is proportional to the square root of the absolute temperature T (Equation 15-5). (*a*) Show that if the temperature changes by a small amount ΔT, the fundamental frequency of an organ pipe changes by approximately Δf, where $\Delta f/f = \frac{1}{2}\Delta T/T$. (*b*) Suppose that an organ pipe that is closed at one end has a fundamental frequency of 200 Hz when the temperature is 20°C. What will be its fundamental frequency when the temperature is 30°C? (Ignore any change in the length of the pipe due to thermal expansion.)

90 •• Two traveling wave pulses on a string are represented by the wave functions

$$y_1(x,t) = \frac{0.02}{2 + (x - 2t)^2}$$

and

$$y_2(x,t) = \frac{-0.02}{2 + (x + 2t)^2}$$

where x is in meters and t is in seconds. (*a*) Using a spreadsheet program or graphing calculator, make a graph of each wave function separately as a function of x at $t = 0$ and describe the behavior of each as time increases. (*b*) Find the resultant wave function at $t = 0$. (*c*) Find the resultant wave function at $t = 1$ s. (*d*) Graph the resultant wave function at $t = 1$ s.

91 •• If you put your ear and your hand near the end of a long, open-ended tube and snap your fingers, you will hear a sound similar to that of a guitar string being plucked. (Tubes of about 1-m length are best.) (*a*) Explain what causes this sound. (*b*) What effective tube length do you need to make a sound like that of a guitar string with a pitch of A above middle C (440 Hz)?

92 •• The kinetic energy of a segment of length Δx and mass Δm of a vibrating string is given by $\Delta K = \frac{1}{2}\Delta m(\partial y/\partial t)^2 = \frac{1}{2}\mu(\partial y/\partial t)^2\,\Delta x$, where $\mu = \Delta m/\Delta x$. (*a*) Find the total kinetic energy of the nth mode of vibration of a string of length L fixed at both ends. (*b*) Give the maximum kinetic energy of the string. (*c*) What is the wave function when the kinetic energy has its maximum value? (*d*) Show that the maximum kinetic energy in the nth mode is proportional to $n^2A_n^2$.

93 •• (*a*) Show that when the tension in a string fixed at both ends is changed by a small amount dF, the frequency of the fundamental is changed by approximately df, where $df/f = \frac{1}{2}dF/F$. Does this result apply to all harmonics? (*b*) Use this result to find the percentage change in the tension needed to increase the frequency of the fundamental of a piano wire from 260 to 262 Hz.

94 •• SSM Two sources of harmonic waves on the x axis have a phase difference that is proportional to time: $\delta_s = Ct$, where C is a constant. The amplitude of the wave from each source at some point P on the x axis is A_0. (*a*) Write the wave functions for each of the two waves at point P, assuming this point to be a distance x_1 from one source and $x_1 + \Delta x$ from the other. (*b*) Find the resultant wave function and show that its amplitude is $2A_0 \cos\left[\frac{1}{2}(\delta + \delta_0)\right]$, where δ is the phase difference at P due to the path difference. (*c*) Using a spreadsheet program or graphing calculator, graph the intensity at point P versus time for a zero path difference. (Let I_0 be the intensity due to each wave separately.) What is the time average of the intensity? (*d*) Make the same graph for the intensity at a point for which the path difference is $\lambda/2$.

95 ••• The wave functions of two standing waves on a string of length L are $y_1(x,t) = A_1 \cos \omega_1 t \sin k_1 x$ and $y_2(x,t) = A_2 \cos \omega_2 t \sin k_2 x$, where $k_n = n\pi/L$ and $\omega_n = n\omega_1$. The wave function of the resultant wave is $y_r(x,t) = y_1(x,t) + y_2(x,t)$. (*a*) Find the velocity of a segment dx of the string. (*b*) Find the kinetic energy of this segment. (*c*) By integration, find the total kinetic energy of the resultant wave. Notice the disappearance of the cross terms so that the total kinetic energy is proportional to $(n_1A_1)^2 + (n_2A_2)^2$.

96 ••• A 2-m wire fixed at both ends is vibrating in its fundamental mode. The tension in the wire is 40 N and the mass of the wire is 0.1 kg. At the midpoint of the wire, the amplitude is 2 cm. (*a*) Find the maximum kinetic energy of the wire. (*b*) At the instant that the transverse displacement is given by (0.02 m) $\sin(\pi x/2)$, what is the kinetic energy of the wire? (*c*) At what position on the wire does the kinetic energy per unit length have its largest value? (*d*) Where does the potential energy per unit length have its maximum value?

97 ••• In principle, a wave with almost any arbitrary shape can be expressed as a sum of harmonic waves of different frequencies. (*a*) Consider the function defined by

$$f(x) = \frac{4}{\pi}\left(\frac{\cos x}{1} + \frac{\cos 3x}{3} + \frac{\cos 5x}{5} + \cdots\right)$$
$$= \frac{4}{\pi}\sum_{n=0}^{\infty}(-1)^n \frac{\cos[(2n + 1)x]}{2n + 1}$$

Write a spreadsheet program to calculate this series using a finite number of terms, and make three graphs of the function in the range $x = 0$ to $x = 4\pi$. For the first graph approximate the sum from $n = 0$ to $n = \infty$ with the first term of the sum. For the second and third graphs use only the first five term and the first ten terms, respectively. This function is sometimes called the *square wave* (or *θ function*). (*b*) What is the relation between this function and Liebnitz' series for π,

$$\frac{\pi}{4} = 1 - \frac{1}{3} + \frac{1}{5} - \frac{1}{7} + \ldots ?$$

98 ••• Write a spreadsheet program to calculate and graph the function

$$f(x) = \frac{4}{\pi}\left(\sin x - \frac{\sin 3x}{9} + \frac{\sin 5x}{25} - \cdots\right)$$
$$= \frac{4}{\pi}\sum_n \frac{(-1)^n \sin(2n+1)x}{(2n+1)^2}$$

What kind of wave is this?

99 ••• If you clap your hands at the end of a long, cylindrical tube, the echo you hear back will not sound like the handclap; instead, you will hear what sounds like a whistle, initially at a very high frequency, but descending rapidly down to almost nothing. This "culvert whistler" can be explained by thinking of the sound from the clap as a single compression radiating outward from the hands. The echoes of the handclap arriving at your ear have traveled along different paths through the tube, as shown in Figure 16-32. The first echo to arrive travels straight down and straight back along the tube, while the second echo reflects once off of the center of the tube going out, and again going back, the third echo reflects twice at points 1/4 and 3/4 of the distance, etc. The tone of the sound you hear reflects the frequency at which these echoes reach your ears. (*a*) Show that the time delay between the n_{th} echo and the $n{+}1_{\text{th}}$ is

$$\Delta t_n = \frac{2}{v}\left(\sqrt{(2n)^2 r^2 + L^2} - \sqrt{[2(n-1)]^2 r^2 + L^2}\right),$$

where v is the speed of sound, L is the length of the tube and r is its radius. (*b*) Using a spreadsheet program or graphing calculator, graph Δt_n versus n for $L = 90$ m, $r = 1$ m. (These are the approximate length and diameter of the long tube in the San Francisco Exploratorium.) Go to at least $n = 100$. (*c*) From your graph, explain why the frequency decreases over time. What are the highest and lowest frequencies you will hear in the whistler?

FIGURE 16-32 Problem 99

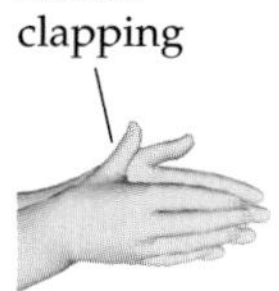

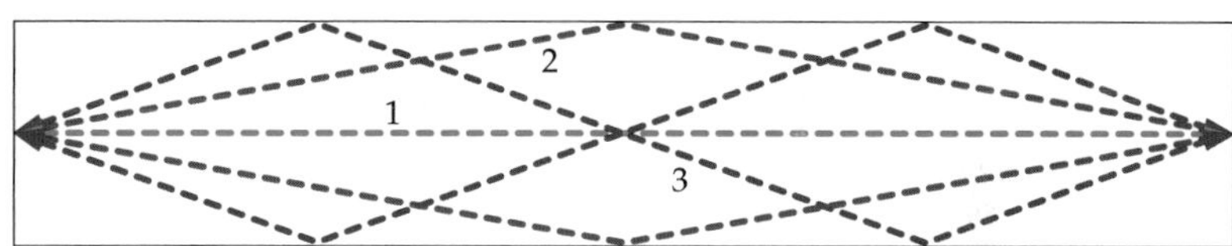

PART III THERMODYNAMICS

CHAPTER 17

Temperature and the Kinetic Theory of Gases

THESE HELIUM BALLOONS ARE WELL INFLATED ON A WARM SUMMER DAY.

? What might happen to them if they are taken indoors to an air-conditioned room? (See Example 17-6.)

Temperature is familiar to us as the measure of the hotness or coldness of objects or of our surroundings.

➢ **In this chapter, we will show that a consistent temperature scale can be defined in terms of the properties of gases at low densities, and that temperature is a measure of the average internal molecular kinetic energy of an object.**

17-1 Thermal Equilibrium and Temperature

Our sense of touch can usually tell us if an object is hot or cold. Early in childhood we learn that to make a cold object warmer, we place it in contact with a hot object. To make a hot object cooler, we place it in contact with a cold object.

When an object is heated or cooled, some of its physical properties change. Most solids and liquids expand when they are heated. A gas, if its pressure is kept constant, will also expand when it is heated, or, if its volume is kept constant, its pressure will rise. If an electrical conductor is heated, its electrical resistance changes. (This is discussed in Chapter 25.) A physical property that changes with temperature is called a **thermometric property.** A change in a thermometric property indicates a change in the temperature of the object.

Suppose that we place a warm copper bar in close contact with a cold iron bar so that the copper bar cools and the iron bar warms. We say that the two bars are in **thermal contact.** The copper bar contracts slightly as it cools, and the iron bar expands slightly as it warms. Eventually this process stops and the lengths of the bars remain constant. The two bars are then in **thermal equilibrium** with each other.

Suppose instead that we place the warm copper bar in a cool running stream. The bar cools until it stops contracting, at the point at which the bar and the water are in thermal equilibrium. Next we place a cold iron bar in the stream on the side opposite the copper bar. The iron bar will warm until it and the water are also in thermal equilibrium. If we remove the bars and place them in thermal contact with each other, we find that their lengths do not change. They are in thermal equilibrium with each other. Though it is common sense, there is no logical way to deduce this fact, which is called the **zeroth law of thermodynamics** (Figure 17-1):

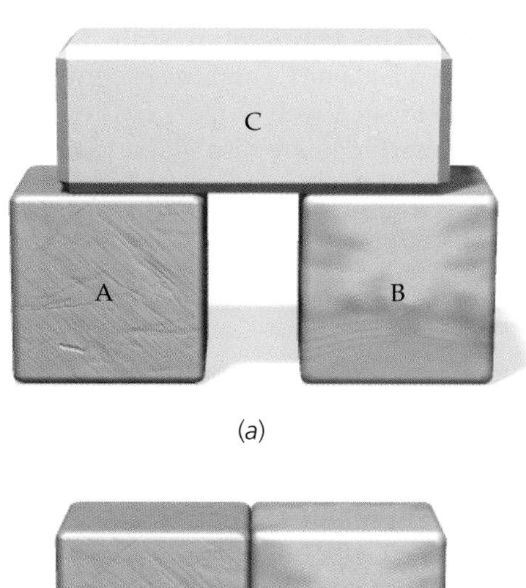

(*a*)

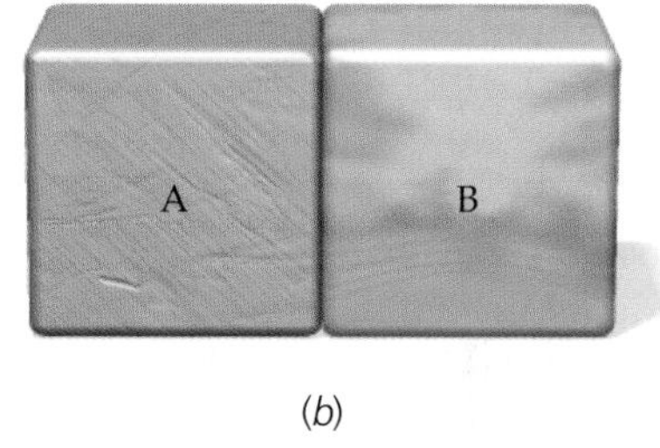

(*b*)

FIGURE 17-1 The zeroth law of thermodynamics. (*a*) Systems A and B are in thermal contact with system C but not with each other. When A and B are each in thermal equilibrium with C, they are in thermal equilibrium with each other, which can be checked by placing them in contact with each other as in (*b*).

> If two objects are in thermal equilibrium with a third, then they are in thermal equilibrium with each other.

ZEROTH LAW OF THERMODYNAMICS

Two objects are defined to have the same *temperature* if they are in thermal equilibrium with each other. The zeroth law, as we will see, enables us to define a temperature scale.

17-2 The Celsius and Fahrenheit Temperature Scales

Any thermometric property can be used to establish a temperature scale. The common mercury thermometer consists of a glass bulb and tube containing a fixed amount of mercury.† When this thermometer is put in contact with a warmer object, the mercury expands, increasing the length of the mercury column (the glass expands too, but by a negligible amount). We can create a scale along the glass tube as follows. First the thermometer is placed in ice and water in equilibrium‡ at a pressure of 1 atm. When the thermometer is in thermal equilibrium with the ice water, the position of the mercury column is marked on the glass tube. This is the **ice-point temperature** (also called the **normal freezing point** of water). Next, the thermometer is placed in boiling water at a pressure of 1 atm. When the thermometer is in thermal equilibrium with the boiling water, the new position of the mercury column is marked. This is the **steam-point temperature** (also called the **normal boiling point** of water).

† Because mercury is highly toxic, mercury thermometers are no longer sold in the United States. Today, alcohol is commonly used in thermometers.

‡ Water and ice in equilibrium provide a constant-temperature bath. When ice is placed in warm water, the water cools as some of the ice melts. Eventually, thermal equilibrium is reached and no more ice melts. If the system is heated slightly, some more of the ice melts, but the temperature does not change as long as some ice remains.

The **Celsius temperature scale** defines the ice-point temperature as zero degrees Celsius (0°C) and the steam-point temperature as 100°C. The space between the 0° and 100° marks is divided into 100 equal intervals (degrees). Degree markings are also extended below and above these points. If L_t is the length of the mercury column, the Celsius temperature t_C is given by

$$t_C = \frac{L_t - L_0}{L_{100} - L_0} \times 100° \qquad 17\text{-}1$$

where L_0 is the length of the mercury column when the thermometer is in an ice bath and L_{100} is its length when the thermometer is in a steam bath. The normal temperature of the human body measured on the Celsius scale is about 37°C.

The **Fahrenheit temperature scale** (which is used in the United States) defines the ice-point temperature as 32°F and the steam-point temperature as 212°F.† To convert temperatures between Fahrenheit and Celsius scales, we note there are 100 Celsius degrees and 180 Fahrenheit degrees between the ice and steam points. A temperature change of one Celsius degree therefore equals a change of 1.8 = 9/5 Fahrenheit degrees. To convert a temperature from one scale to the other, we must also take into account the fact that the zero temperatures of the two scales are not the same. The general relation between a Fahrenheit temperature t_F and Celsius temperature t_C is

$$t_C = \tfrac{5}{9}(t_F - 32°) \qquad 17\text{-}2$$

FAHRENHEIT–CELSIUS CONVERSION

FIGURE 17-2 A bimetallic strip. When heated or cooled, the two metals expand or contract by different amounts, causing the strip to bend.

CONVERTING FAHRENHEIT AND CELSIUS TEMPERATURES **EXAMPLE 17-1**

(*a*) Find the temperature on the Celsius scale equivalent to 41°F. (*b*) Find the temperature on the Fahrenheit scale equivalent to 37.0°C.

(*a*) Apply Equation 17-2 with $t_F = 41°\text{F}$: $t_C = \frac{5}{9}(t_F - 32°) = \frac{5}{9}(41° - 32°) = \frac{5}{9}(9°) = \boxed{5°\text{C}}$

(*b*) 1. Solve Equation 17-2 for t_F in terms of t_C: $t_F = \frac{9}{5}t_C + 32°$

2. Substitute $t_C = 37°\text{C}$: $t_F = \frac{9}{5}(37.0°) + 32° = 66.6° + 32° = \boxed{98.6°\text{F}}$

EXERCISE (*a*) Find the Celsius temperature equivalent to 68°F. (*b*) Find the Fahrenheit temperature equivalent to −40°C. (*Answer* (*a*) 20°C (*b*) −40°F)

Other thermometric properties can be used to set up thermometers and construct temperature scales. Figure 17-2 shows a bimetallic strip consisting of two different metals bonded together. When the strip is heated or cooled, it bends to accommodate the difference in the thermal expansion of the two metals. Figure 17-3 shows a thermometer consisting of a bimetallic coil with a pointer attached to indicate the temperature. When the thermometer is heated, the coil bends and the pointer moves. Like mercury thermometers, it is calibrated by dividing the interval between the ice point and the steam point into 100 Celsius degrees (or 180 Fahrenheit degrees).

† When the German physicist Daniel Fahrenheit devised his temperature scale, he wanted all measurable temperatures to be positive. Originally, he chose 0°F for the coldest temperature he could obtain with a mixture of ice and salt water and 96°F (a convenient number with many factors for subdivision) for the temperature of the human body. He then modified his scale slightly to make the ice-point and steam-point temperatures whole numbers. This resulted in the average temperature of the human body being between 98° and 99°F.

FIGURE 17-3 (*a*) A thermometer using a bimetallic strip in the form of a coil. (The red pointer is attached to one end of the coil.) When the temperature of the coil increases, the needle rotates clockwise because the outer metal expands more than the inner metal. (*b*) A home thermostat controls the central air conditioner. When the air gets warmer, the coil expands, the glass bulb mounted on it tilts, and mercury in the tube slides to close an electrical switch, turning on the air conditioning. A slide lever (at the lower right), used to rotate the coil mount, is used to set the desired temperature. The circuit will be broken when the cooler air causes the bimetallic coil to contract.

(*a*)

Glass Bulb
Mercury switch

Bimetalic
strip coil

Slide
lever

(*b*)

17-3 Gas Thermometers and the Absolute Temperature Scale

When different types of thermometers are calibrated in ice water and steam, they agree (by definition) at 0°C and 100°C, but they give slightly different readings at points in between. Discrepancies increase markedly above the steam point and below the ice point. However, in one group of thermometers, gas thermometers, measured temperatures agree closely with each other even far from the calibration points. In a **constant-volume gas thermometer,** the gas volume is kept constant, and change in gas pressure is used to indicate a change in temperature (Figure 17-4). An ice-point pressure P_0 and steam-point pressure P_{100} are determined by placing the thermometer in ice–water and water–steam baths, and the interval between is divided into 100 equal degrees (for the Celsius scale). If the pressure is P_t in a bath whose temperature is to be determined, that temperature in degrees Celsius is defined to be

$$t_C = \frac{P_t - P_0}{P_{100} - P_0} \times 100° \qquad 17\text{-}3$$

Suppose we measure a specific temperature, say the boiling point of sulfur at 1 atm pressure, using four constant-volume gas thermometers, each containing one of four gasses—air, hydrogen, nitrogen, and oxygen. The thermometers are calibrated, meaning values for P_{100} and P_0 are determined for each. Each thermometer is then immersed in boiling sulfur, and when it is in thermal equilibrium with the sulfur, the pressure in the thermometer is measured. Next, the

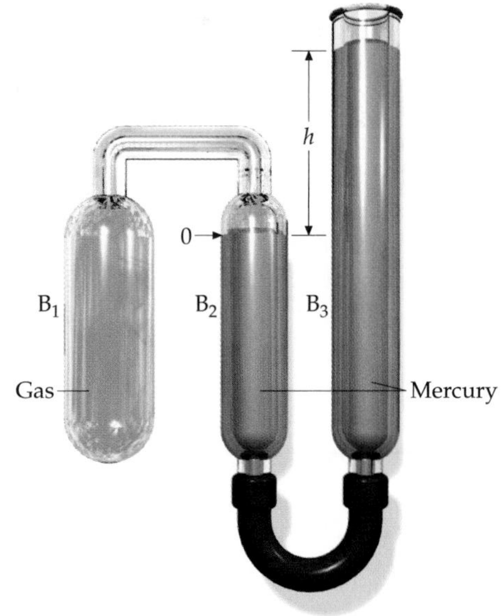

FIGURE 17-4 A constant-volume gas thermometer. The volume is kept constant by raising or lowering tube B_3 so that the mercury in tube B_2 remains at the zero mark. The temperature is chosen to be proportional to the pressure of the gas in tube B_1, which is indicated by the height h of the mercury column in tube B_3.

temperature is calculated using Equation 17-3. Will this process give the same result for each of the four thermometers? Surprisingly perhaps, the answer is yes. All four thermometers measure the same temperature so long as the density of the gas in each is sufficiently low.

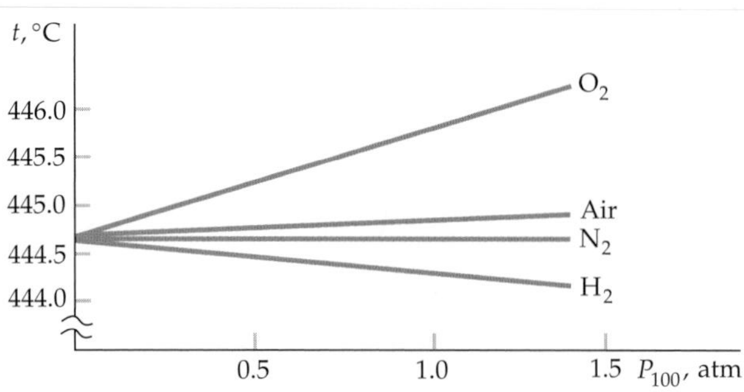

FIGURE 17-5 Temperature of the boiling point of sulfur measured with constant-volume gas thermometers filled with various gases. Increasing or decreasing the amount of gas in the thermometer varies the pressure P_{100} at the steam point of water. As the amount of gas is reduced, the temperatures measured by all the thermometers approach the value 444.60°C.

One measure of the density of the gas in the thermometer is its pressure at the steam point, P_{100}. If we vary the amount of gas in a constant-volume gas thermometer, by either adding or removing gas, we change both P_{100} and P_0. As a result, each time the amount of gas is varied, the thermometer must be recalibrated. Figure 17-5 shows the results of measurements of the boiling point of sulfur using four constant-volume gas thermometers, each filled with air, hydrogen, nitrogen, or oxygen. For each thermometer the measured temperature is plotted as a function of the steam-point pressure P_{100} of the thermometer. As the amount of a gas is reduced, its density and the steam-point pressure both decrease. We see that agreement among the thermometers is very close at low gas densities (low P_{100}). In the limit as gas density goes to zero, all gas thermometers give the same value for the temperature of boiling sulfur. This low-density temperature measurement is independent of the properties of any particular gas. Of course, there is nothing special about the boiling point of sulfur. Constant-volume gas thermometers at low densities are in agreement at any temperature. Thus, low-density gas thermometers can be used to define temperature.

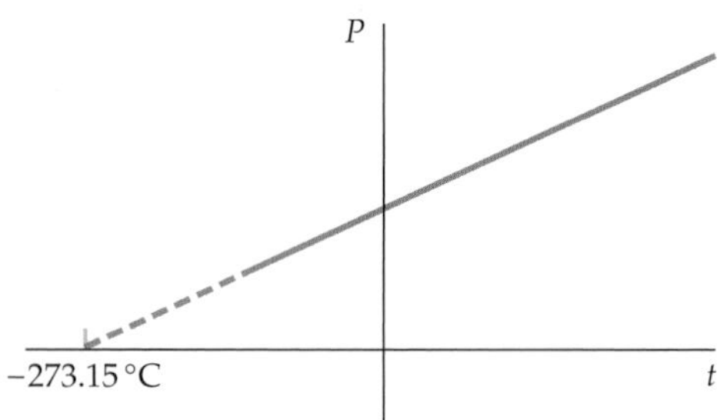

FIGURE 17-6 Plot of pressure versus temperature as measured by a constant-volume gas thermometer. When extrapolated to zero pressure, the plot intersects the temperature axis at the value −273.15°C.

Now consider a series of temperature measurements with a constant-volume gas thermometer that has a very small but fixed amount of gas. According to Equation 17-3, the pressure in the thermometer P_t varies linearly with the measured temperature t_C. Figure 17-6 shows a plot of pressure versus measured temperature in a constant-volume gas thermometer. When we extrapolate this straight line to zero pressure, the temperature approaches −273.15°C. This limit is the same no matter what kind of gas is used.

A reference state that is much more precisely reproducible than either the ice or steam points is the **triple point of water**—the unique temperature and pressure at which water, water vapor, and ice coexist in equilibrium (see Figure 17-7). This equilibrium state occurs at 4.58 mmHg and 0.01°C. The **ideal-gas temperature scale** is defined so that the temperature of the triple point is 273.16 kelvins (K). (The kelvin is a degree unit that is the same size as the Celsius degree.) The temperature T of any other state is defined to be proportional to the pressure in a constant-volume gas thermometer:

$$T = \frac{273.16\ \text{K}}{P_3} P \qquad \text{17-4}$$

IDEAL-GAS TEMPERATURE SCALE

where P is the observed pressure of the gas in the thermometer and P_3 is the pressure when the thermometer is immersed in a water–ice–vapor bath at its triple point. The value of P_3 depends

FIGURE 17-7 H_2O at its triple point. The spherical flask contains water, ice, and water vapor in equilibrium.

on the amount of gas in the thermometer. The ideal-gas temperature scale, defined by Equation 17-4, has the advantage that any measured temperature does not depend on the properties of the particular gas that is used, but depends only on the general properties of gases.

The lowest temperature that can be measured with a gas thermometer is about 1 K, and requires helium for the gas. Below this temperature helium liquefies; all other gases liquefy at higher temperatures (see Table 17-1). In Chapter 19 we will see that the second law of thermodynamics can be used to define the **absolute temperature scale** independent of the properties of any substance, and with no limitations on the range of temperatures that can be measured. Temperatures as low as a millionth of a kelvin have been measured. The absolute scale so defined is identical to that defined by Equation 17-4 for the range of temperatures for which gas thermometers can be used. The symbol T is used when referring to absolute temperature.

Because the Celsius degree and the kelvin are the same size, temperature *differences* are the same on both the Celsius and the absolute temperature scales (also called the **Kelvin scale**). That is, a temperature *change* of 1 K is identical to a temperature *change* of 1C°.† The two scales differ only in the choice of zero temperature. To convert from degrees Celsius to kelvins, we merely add 273.15:‡

$$T = t_C + 273.15 \text{ K} \qquad 17\text{-}5$$

CELSIUS–ABSOLUTE CONVERSION

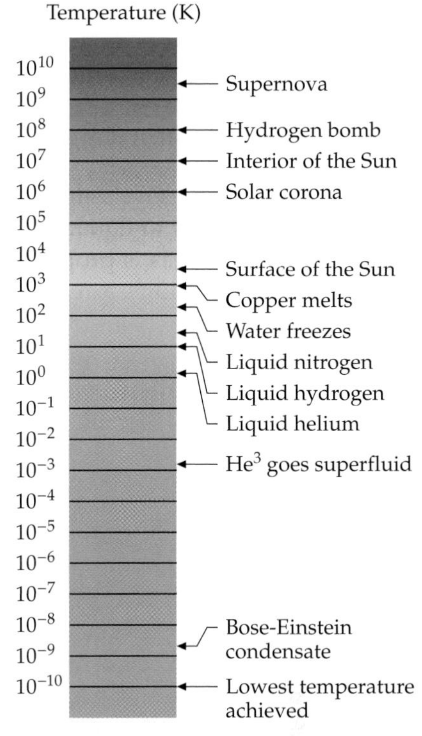

TABLE 17-1 The temperatures of various places and phenomena.

Although the Celsius and Fahrenheit scales are convenient for everyday use, the absolute scale is much more convenient for scientific purposes, partly because many formulas are more simply expressed in it, and partly because the absolute temperature can be given a more fundamental interpretation.

CONVERTING FROM KELVIN TO FAHRENHEIT **EXAMPLE 17-2**

What is the Kelvin temperature corresponding to 70°F?

PICTURE THE PROBLEM First convert to degrees Celsius, then to kelvins.

1. Convert to degrees Celsius: $t_C = \frac{5}{9}(70° - 32°) = 21.1°\text{C}$

2. To find the Kelvin temperature we add 273: $T = t_C + 273 = 21.1 + 273 = \boxed{294 \text{ K}}$

EXERCISE The "high-temperature" superconductor $YBa_2Cu_3O_7$ becomes superconducting when the temperature is lowered to 92 K. Find the superconducting threshold temperature in degrees Fahrenheit. (*Answer:* −294°F)

17-4 The Ideal-Gas Law

The properties of gases at low densities allow the definition of the ideal-gas temperature scale. If we compress such a gas while keeping its temperature constant, the pressure increases. Similarly, if a gas expands at constant temperature, its pressure decreases. To a good approximation, the product of the pressure and

† We write 1C° to indicate a *temperature change* of one Celsius degree, in contrast to 1°C, which means a temperature of one degree Celsius.

‡ For most purposes, we can round off the temperature of absolute zero to −273°C.

volume of a low-density gas is constant at a constant temperature. This result was discovered experimentally by Robert Boyle (1627–1691), and is known as **Boyle's law:**

$$PV = \text{constant (constant temperature)}$$

A more general law exists that reproduces Boyle's law as a special case. According to Equation 17-4, the absolute temperature of a low-density gas is proportional to its pressure at constant volume. In addition—a result discovered experimentally by Jacques Charles (1746–1823) and Joseph Gay-Lussac (1778–1850)—the absolute temperature of a low-density gas is proportional to its volume at constant pressure. We can combine these two results by stating

$$PV = CT \qquad 17\text{-}6$$

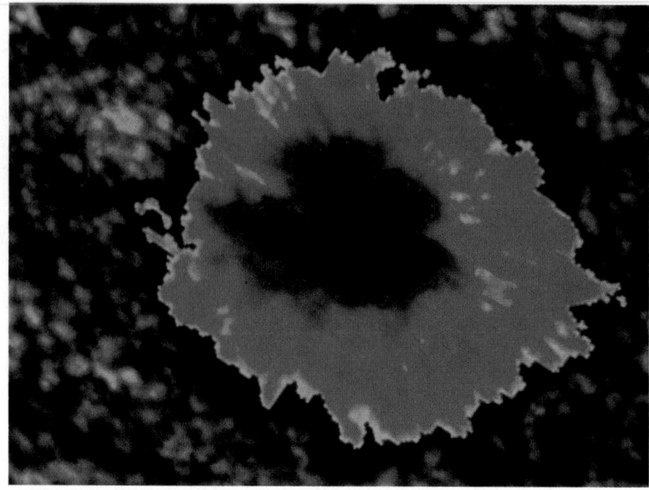

The atmosphere of Venus is almost entirely CO_2. However, measurements by the Pioneer Venus Orbiter show an atomic hydrogen cloud surrounding Venus. The crescent-shaped image shows atomic oxygen, and the bars show hydrogen extending far above the atmosphere. Since the escape speed on Venus is 10.3 km/s, slightly smaller than the escape speed on earth, and since the Venusian atmosphere is considerably warmer than earth's atmosphere, all of the hydrogen in the atmosphere at the time of the formation of Venus should have escaped by now.

where C is a constant of proportionality. We can see that this constant is proportional to the amount of gas by considering the following. Suppose that we have two containers with identical volumes, each holding the same amount of the same kind of gas at the same temperature and pressure. If we consider the two containers as one system, we have twice the amount of gas at twice the volume, but at the same temperature and pressure. We have thus doubled the quantity $PV/T = C$ by doubling the amount of gas. We can therefore write C as a constant k times the number of molecules in the gas N:

$$C = kN$$

Equation 17-6 then becomes

$$PV = NkT \qquad 17\text{-}7$$

The constant k is called **Boltzmann's constant.** It is found experimentally to have the same value for any kind of gas:

$$k = 1.381 \times 10^{-23}\ \text{J/K} = 8.617 \times 10^{-5}\ \text{eV/K} \qquad 17\text{-}8$$

An amount of gas is often expressed in moles. A **mole** (mol) of any substance is the amount of that substance that contains Avogadro's number N_A of atoms or molecules, defined as the number of carbon atoms in 12 g of ^{12}C:

$$N_A = 6.022 \times 10^{23} \qquad 17\text{-}9$$

AVOGADRO'S NUMBER

If we have n moles of a substance, then the number of molecules is

$$N = nN_A \qquad 17\text{-}10$$

Equation 17-7 is then

$$PV = nN_AkT = nRT \qquad 17\text{-}11$$

where $R = N_Ak$ is called the **universal gas constant.** Its value, which is the same for all gases, is

$$R = N_Ak = 8.314\ \text{J/(mol}\cdot\text{K)} = 0.08206\ \text{L}\cdot\text{atm/(mol}\cdot\text{K)} \qquad 17\text{-}12$$

Figure 17-8 shows plots of $PV/(nT)$ versus the pressure P for several gases. For all gases, $PV/(nT)$ is nearly constant over a large range of pressures. Even oxygen, which varies the most in this graph, changes by only about 1 percent between 0 and 5 atm. An **ideal gas** is defined as one for which $PV/(nT)$ is constant for all pressures. The pressure, volume, and temperature of an ideal gas are related by

$$PV = nRT \qquad \text{17-13}$$

IDEAL-GAS LAW

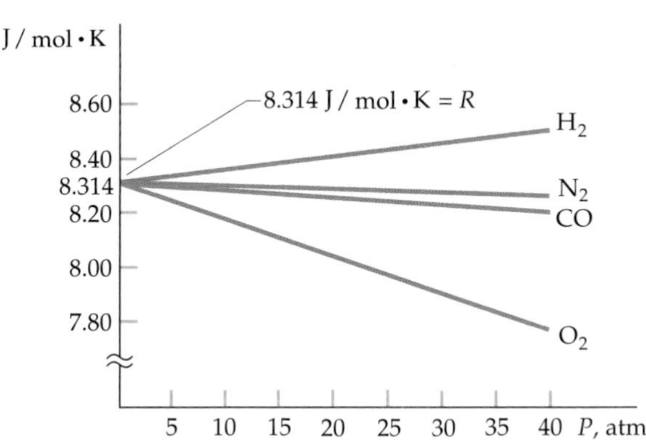

FIGURE 17-8 Plot of PV/nT versus P for real gases. In these plots, varying the amount of gas varies the pressure. The ratio PV/nT approaches the same value, 8.314 J/(mol·K), for all gases as we reduce the density, and thereby the pressure, of the gas. This value is the universal gas constant R.

Equation 17-13, which relates the variables P, V, and T, is known as the ideal-gas law, and is an example of an **equation of state.** It describes the properties of real gases with low densities (and therefore low pressures). At higher densities, corrections must be made to this equation. In Chapter 20 we discuss another equation of state, the van der Waals equation, which includes such corrections. For any gas at any density, there is an equation of state relating P, V, and T for a given amount of gas. Thus the state of a given amount of gas is determined by any two of the three **state variables** P, V, and T.

VOLUME OF AN IDEAL GAS — **EXAMPLE 17-3**

What volume is occupied by 1 mol of an ideal gas at a temperature of 0°C and a pressure of 1 atm?

We can find the volume using the ideal-gas law, with $T = 273$ K:

$$V = \frac{nRT}{P} = \frac{(1\ \text{mol})(0.0821\ \text{L·atm/[mol·K]})(273\ \text{K})}{1\ \text{atm}} = \boxed{22.4\ \text{L}}$$

REMARKS Note that by writing R in L·atm/(mol·K), we could write P in atmospheres to get V in liters.

EXERCISE Find (*a*) the number of moles n and (*b*) the number of molecules N in 1 cm³ of a gas at 0°C and 1 atm. (*Answer* (*a*) $n = 4.46 \times 10^{-5}$ mol (*b*) $N = 2.68 \times 10^{19}$ molecules)

The temperature of 0°C = 273 K and the pressure of 1 atm are often referred to as **standard conditions.** We see from Example 17-3 that under standard conditions, 1 mol of an ideal gas occupies a volume of 22.4 L.

Figure 17-9 shows plots of P versus V at several constant temperatures T. These curves are called **isotherms.** The isotherms for an ideal gas are hyperbolas. For a fixed amount of gas, we can see from Equation 17-13 that the quantity PV/T is constant. Using the subscripts 1 for the initial values and 2 for the final values, we have

$$\frac{P_2V_2}{T_2} = \frac{P_1V_1}{T_1} \qquad \text{17-14}$$

IDEAL-GAS LAW FOR FIXED AMOUNT OF GAS

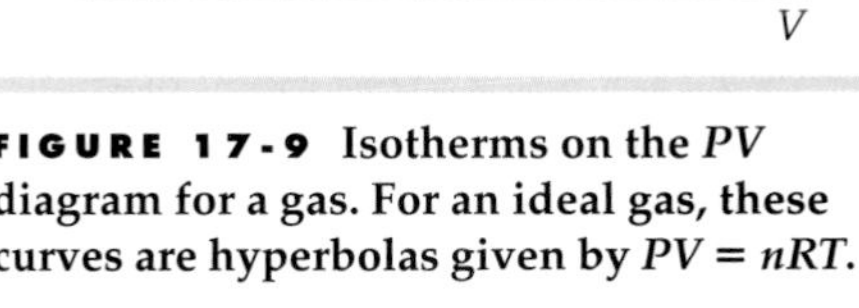

FIGURE 17-9 Isotherms on the PV diagram for a gas. For an ideal gas, these curves are hyperbolas given by $PV = nRT$.

HEATING AND COMPRESSING A GAS — EXAMPLE 17-4

A gas has a volume of 2 L, a temperature of 30°C, and a pressure of 1 atm. When the gas is heated to 60°C and compressed to a volume of 1.5 L, what is its new pressure?

PICTURE THE PROBLEM Since the amount of gas is fixed, the pressure can be found using Equation 17-14. Let subscripts 1 and 2 refer to the initial and final states, respectively.

1. Express the pressure P_2 in terms of P_1 and the initial and final volumes and temperatures:

$$\frac{P_1V_1}{T_1} = \frac{P_2V_2}{T_2}, \qquad P_2 = \frac{T_2V_1}{T_1V_2}P_1$$

2. Calculate the initial and final absolute temperatures:

$$T_1 = 273 + 30 = 303\text{ K}$$

$$T_2 = 273 + 60 = 333\text{ K}$$

3. Substitute numerical values in step 1 to find P_2:

$$P_2 = \frac{(333\text{ K})(2\text{ L})}{(303\text{ K})(1.5\text{ L})}(1\text{ atm}) = \boxed{1.47\text{ atm}}$$

EXERCISE How many moles of gas are in the system described in this example? (*Answer* $n = 0.0804$ mol)

The mass per mole of a substance is called its **molar mass** M. (The terms *molecular weight or molecular mass* are also sometimes used.) The molar mass of ^{12}C is, by definition, 12 g/mol or 12×10^{-3} kg/mol. Molar masses of the elements are given in the periodic table in Appendix E. The molar mass of a molecule such as CO_2 is the sum of the molar masses of the elements in the molecule. Because the molar mass of oxygen is 16 g/mol (actually 15.999 g/mol), the molar mass of O_2 is 32 g/mol and that of CO_2 is 12 + 32 = 44 g/mol.

The mass of n moles of a gas is given by

$$m = nM$$

and the density ρ of an ideal gas is

$$\rho = \frac{m}{V} = \frac{nM}{V}$$

Using $n/V = P/RT$ from Equation 17-13, we have

$$\rho = \frac{M}{RT}P \qquad 17\text{-}15$$

At a given temperature, the density of an ideal gas is proportional to its pressure.

THE MASS OF A HYDROGEN ATOM — EXAMPLE 17-5

The molar mass of hydrogen is 1.008 g/mol. What is the mass of one hydrogen atom?

PICTURE THE PROBLEM Let m be the mass of a hydrogen atom. Since there are N_A atoms in a mole, the molar mass M is given by $M = mN_A$. We can use this to solve for m.

The mass of a hydrogen atom is the molar mass divided by Avogadro's number:

$$m = \frac{M}{N_A} = \frac{1.008\ \text{g/mol}}{6.022 \times 10^{23}\ \text{atoms/mol}}$$

$$= \boxed{1.67 \times 10^{-24}\ \text{g/atom}}$$

REMARKS Note that Avogadro's number is essentially the reciprocal of the mass of the hydrogen atom measured in grams.

EXPANDING A GAS AT CONSTANT TEMPERATURE **EXAMPLE 17-6** **Try It Yourself**

A 100-g sample of CO_2 occupies a volume of 55 L at a pressure of 1 atm. (*a*) What is the temperature? (*b*) If the volume is increased to 80 L and the temperature is kept constant, what is the new pressure?

PICTURE THE PROBLEM Both questions can be answered using the ideal-gas law (Equation 17-13) if we first find the number of moles, n.

Cover the column to the right and try these on your own before looking at the answers.

Steps	Answers
(*a*) 1. The number of moles n is calculated from the mass of the sample m and the molar mass M of CO_2: The molar mass, from information in Appendix C, is 44 g/mol.	$n = \frac{m}{M} = 2.27\ \text{mol}$
2. Find the temperature T from the ideal-gas law.	$T = \frac{PV}{nR} = \boxed{295\ \text{K}}$
(*b*) Use PV = constant to find the new pressure for $V = 80$ L.	$P_2 = \boxed{0.688\ \text{atm}}$

EXERCISE If the temperature is decreased at constant pressure, what happens to the volume? (*Answer* It decreases)

17-5 The Kinetic Theory of Gases

The description of the behavior of a gas in terms of the macroscopic state variables P, V, and T can be related to simple averages of microscopic quantities such as the mass and speed of the molecules in the gas. The resulting theory is called **the kinetic theory of gases.**

From the point of view of kinetic theory, a gas consists of a large number of molecules making elastic collisions with each other and with the walls of a container. In the absence of external forces (we may neglect gravity), there is no preferred position for a molecule in the container,† and no preferred direction for its velocity vector. The molecules are separated, on the average, by distances that are large compared with their diameters, and they exert no forces on each other except when they collide. (This final assumption is equivalent to assuming a very low gas density, which, as we saw in the last section, is the same as assuming that the gas is an ideal gas. Because momentum is conserved, the collisions the

† Because of gravity, the density of molecules at the bottom of the container is slightly greater than at the top. As discussed in Chapter 13, the density of air decreases by half at a height of about 5.5 km, so the variation over a normal sized container is negligible.

molecules make with each other have no effect on the total momentum in any direction—thus such collisions can be neglected.)

Calculating the Pressure Exerted by a Gas

The pressure that a gas exerts on its container is due to collisions between gas molecules and the container walls. This pressure is a force per unit area and, by Newton's second law, this force is the rate of change of momentum of the gas molecules colliding with the wall.

Consider a rectangular container of volume V containing N gas molecules, each of mass m moving with a speed v. Let us calculate the force exerted by these molecules on the right-hand wall, which is perpendicular to the x axis and has area A. The molecules hitting this wall in a time interval Δt are those that are within distance $v_x\Delta t$ of the wall (Figure 17-10) and are moving to the right. Thus, the number of molecules hitting the wall during time Δt is the number per unit volume N/V times the volume $v_x\Delta t\ A$ times $\frac{1}{2}$ because, on average, only half the molecules are moving to the right. That is,

$$\text{Molecules that hit the wall} = \frac{1}{2}\frac{N}{V}v_x\Delta t\ A$$

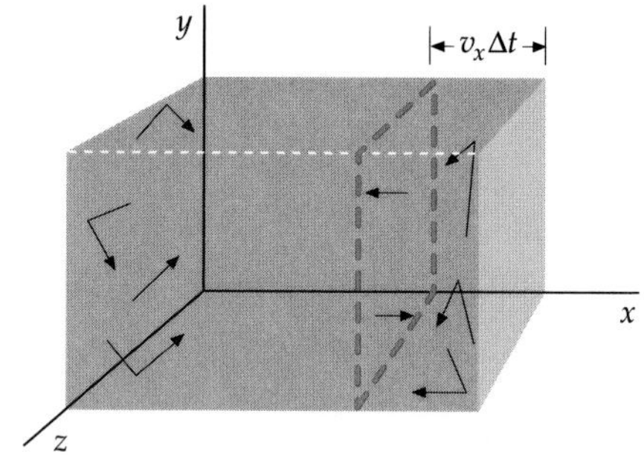

FIGURE 17-10 Gas molecules in a rectangular container. In a time interval Δt, the molecules closer to the right wall than the distance $v_x\Delta t$ will hit the right wall if they are moving to the right.

The x component of momentum of a molecule is $+mv_x$ before it hits the wall, and $-mv_x$ after an elastic collision with the wall. The change in momentum has the magnitude $2mv_x$. The magnitude of the total change in momentum $|\Delta\vec{p}|$ of all molecules during a time interval Δt is $2mv_x$ times the number of molecules that hit the wall during this interval:

$$|\Delta\vec{p}| = (2mv_x) \times \left(\frac{1}{2}\frac{N}{V}v_x\Delta t\ A\right) = \frac{N}{V}mv_x^2 A\ \Delta t \qquad \text{17-16}$$

The magnitude of the force exerted by the wall on the molecules and by the molecules on the wall is $|\Delta\vec{p}|/\Delta t$. The pressure is the magnitude of this force divided by the area A:

$$P = \frac{F}{A} = \frac{1}{A}\frac{|\Delta\vec{p}|}{\Delta t} = \frac{N}{V}mv_x^2$$

or

$$PV = Nmv_x^2 \qquad \text{17-17}$$

To allow for the fact that all the molecules in a container do not have the same speed, we merely replace v_x^2 with the average value $(v_x^2)_{\text{av}}$. Then, writing Equation 17-17 in terms of the kinetic energy $\frac{1}{2}mv_x^2$ associated with motion along the x axis, we have

$$PV = 2N(\tfrac{1}{2}mv_x^2)_{\text{av}} \qquad \text{17-18}$$

The Molecular Interpretation of Temperature

Comparing Equation 17-18 with Equation 17-7, which was obtained experimentally for any gas at very low densities, we can see that

$$PV = NkT = 2N(\tfrac{1}{2}mv_x^2)_{\text{av}}$$

or

$$(\tfrac{1}{2}mv_x^2)_{av} = \tfrac{1}{2}kT \tag{17-19}$$

THE AVERAGE ENERGY ASSOCIATED WITH MOTION IN THE x DIRECTION

Thus, the average kinetic energy associated with motion along the x axis is $\frac{1}{2}kT$. But there is nothing special about the x direction. On the average,

$$(v_x^2)_{av} = (v_y^2)_{av} = (v_z^2)_{av} \tag{17-20}$$

and

$$(v^2)_{av} = (v_x^2)_{av} + (v_y^2)_{av} + (v_z^2)_{av} = 3(v_x^2)_{av}$$

Writing $(v_x^2)_{av} = \frac{1}{3}(v^2)_{av}$ and K_{av} for the average translational kinetic energy of the molecules,† Equation 17-19 becomes

$$K_{av} = (\tfrac{1}{2}mv^2)_{av} = \tfrac{3}{2}kT \tag{17-21}$$

AVERAGE KINETIC ENERGY OF A MOLECULE

The absolute temperature is thus a measure of the average translational kinetic energy of the molecules. The total translational kinetic energy of n moles of a gas containing N molecules is

$$K = N(\tfrac{1}{2}mv^2)_{av} = \tfrac{3}{2}NkT = \tfrac{3}{2}nRT \tag{17-22}$$

KINETIC ENERGY OF TRANSLATION FOR n MOLES OF A GAS

where we've used $Nk = nN_Ak = nR$. Thus, the translational kinetic energy is $\frac{3}{2}kT$ per molecule and $\frac{3}{2}RT$ per mole.

We can use these results to estimate the order of magnitude of the speeds of the molecules in a gas. The average value of v^2 is, by Equation 17-21,

$$(v^2)_{av} = \frac{3kT}{m} = \frac{3N_AkT}{N_Am} = \frac{3RT}{M}$$

where $M = N_Am$ is the molar mass. The square root of $(v^2)_{av}$ is referred to as the **root mean square** (rms) speed:

$$v_{rms} = \sqrt{(v^2)_{av}} = \sqrt{\frac{3kT}{m}} = \sqrt{\frac{3RT}{M}} \tag{17-23}$$

Note that Equation 17-23 is similar to Equation 15-5 for the speed of sound in a gas:

$$v_{sound} = \sqrt{\frac{\gamma RT}{M}} \tag{17-24}$$

where $\gamma = 1.4$ for air. This is not surprising since a sound wave in air is a pressure disturbance propagated by collisions between air molecules.

---

† We include the word *translational* because the molecules may also have rotational or vibrational kinetic energy. Only the translational kinetic energy is relevant to the calculation of the pressure exerted by a gas on the walls of its container.

THE RMS SPEED OF GAS MOLECULES **EXAMPLE 17-7**

Oxygen gas (O_2) has a molar mass of about 32 g/mol and hydrogen gas (H_2) has a molar mass of about 2 g/mol. Calculate (*a*) the rms speed of an oxygen molecule when the temperature is 300 K and (*b*) the rms speed of a hydrogen molecule at the same temperature.

PICTURE THE PROBLEM (*a*) We find v_{rms} using Equation 17-23. For the units to work out right, we use $R = 8.31\ \text{J}/(\text{mol}\cdot\text{K})$, and we express the molecular mass of O_2 in kg/mol. (*b*) Since v_{rms} is proportional to $1/\sqrt{M}$, and the molar mass of hydrogen is one-sixteenth that of oxygen, the rms speed of hydrogen is 4 times that of oxygen.

(*a*) Substitute the given values into Equation 17-23:

$$v_{\text{rms}} = \sqrt{\frac{3RT}{M}} = \sqrt{\frac{3(8.31\ \text{J}/[\text{mol}\cdot\text{K}])(300\ \text{K})}{32 \times 10^{-3}\ \text{kg/mol}}}$$

$$= \boxed{483\ \text{m/s}}$$

(*b*) Use $v_{\text{rms}} \propto 1/\sqrt{M}$ to calculate v_{rms} for hydrogen:

$$\frac{v_{\text{rms}}(H_2)}{v_{\text{rms}}(O_2)} = \frac{\sqrt{M_{O_2}}}{\sqrt{M_{H_2}}}$$

so

$$v_{\text{rms}}(H_2) = \sqrt{\frac{M_{O_2}}{M_{H_2}}}\, v_{\text{rms}}(O_2) = \sqrt{\frac{32\ \text{g/mol}}{2\ \text{g/mol}}}\,(483\ \text{m/s})$$

$$= \boxed{1930\ \text{m/s}}$$

REMARKS The rms speed of oxygen molecules is 483 m/s = 1080 mi/h, about 1.4 times the speed of sound in air, which at 300 K is about 347 m/s.

EXERCISE Find the rms speed of a nitrogen molecule ($M = 28$ g/mol) at 300 K. (*Answer* 516 m/s)

The Equipartition Theorem

We have seen that the average kinetic energy associated with translational motion in any direction is $\frac{1}{2}kT$ per molecule (Equation 17-21) (or, equivalently, $\frac{1}{2}RT$ per mole), where k is Boltzmann's constant. If the energy of a molecule associated with its motion in one direction is momentarily increased, say, by a collision between the molecule and a moving piston during a compression, collisions between that molecule and other molecules will quickly redistribute the added energy. When the gas is again in equilibrium, energy will be equally partitioned among the translational kinetic energies associated with motion in the x, y, and z directions. This sharing of the energy equally among the three terms in the translational kinetic energy is a special case of the **equipartition theorem,** a result that follows from classical statistical mechanics. Each component of position and momentum (including angular position and angular momentum) that appears as a squared term in the expression for the energy of the system is called a **degree of freedom.** Typical degrees of freedom are associated with the kinetic energy of translation, rotation, and vibration, and with the potential energy of vibration. The equipartition theorem states that:

> When a substance is in equilibrium, there is an average energy of $\frac{1}{2}kT$ per molecule or $\frac{1}{2}RT$ per mole associated with each degree of freedom.

EQUIPARTITION THEOREM

In Chapter 18 we will use the equipartition theorem to relate the measured heat capacities of gases to their molecular structure.

Mean Free Path

The average speed of molecules in a gas at normal pressures is several hundred meters per second, yet if somebody across the room from you opens a perfume bottle, you don't detect the odor for several minutes. The reason for the time delay is that the perfume molecules do not travel directly toward you, but instead travel a zigzag path due to collisions with the air molecules. The average distance λ traveled by a molecule between collisions is called its **mean free path.** (The reason you smell the perfume at all is due to air currents (convection). The time for a perfume molecule to diffuse across a room is of the order of weeks.)

The mean free path of a gas molecule is related to its size, to the size of the surrounding gas molecules, and to the density of the gas. Consider one gas molecule of radius r_1 moving with speed v through a region of stationary molecules (Figure 17-11). The moving molecule will collide with another molecule of radius r_2 if the centers of the two molecules come within a distance $d = r_1 + r_2$ from each other. (If all the molecules are the same type, then d is the molecular diameter.) As the molecule moves, it will collide with any molecule whose center is in a circle of radius d (Figure 17-12). In some time t, the molecule moves a distance vt and collides with every molecule in the cylindrical volume $\pi d^2 vt$. The number of molecules in this volume is $n_V \pi d^2 vt$, where $n_V = N/V$ is the number of molecules per unit volume. (After each collision, the direction of the molecule changes, so the path actually zigs and zags.) The total path length divided by the number of collisions is the mean free path:

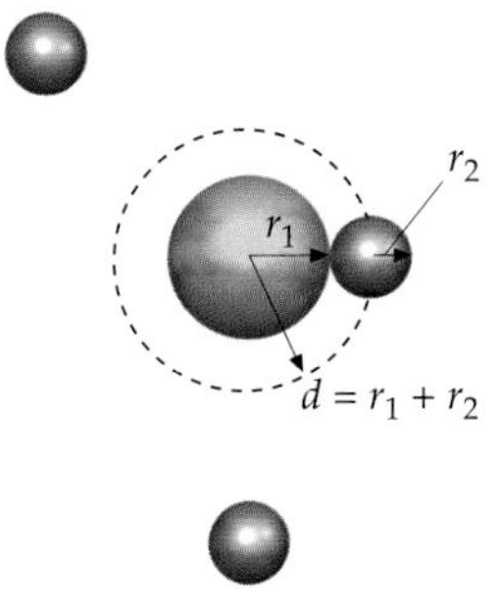

FIGURE 17-11 **Model of a molecule (center sphere) moving in a gas. The molecule of radius r_1 will collide with any molecule of radius r_2 if their centers are a distance $d = r_1 + r_2$ apart, which is any molecule whose center is on a sphere of radius $d = r_1 + r_2$ centered about the molecule.**

$$\lambda = \frac{vt}{n_V \pi d^2 vt} = \frac{1}{n_V \pi d^2}$$

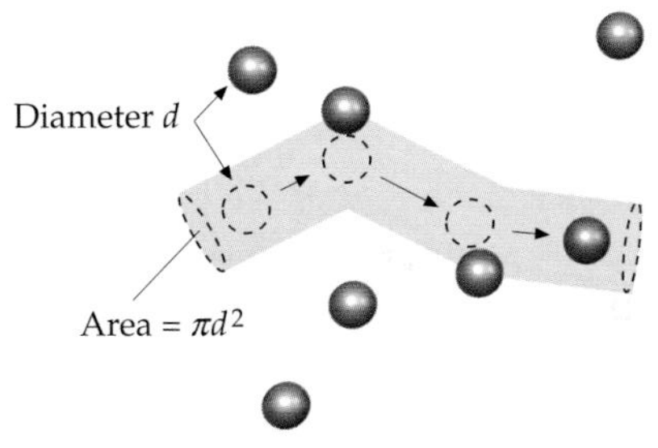

FIGURE 17-12 **Model of a molecule moving with speed v in a gas of similar molecules. The motion is shown during time t. The molecule of diameter d will collide with any similar molecule whose center is in a cylinder of volume $\pi d^2 vt$. In this picture, all collisions are assumed to be elastic and all the molecules but one are assumed to be at rest.**

This calculation of the mean free path assumes that all but one of the gas molecules are stationary, which is not a realistic situation. When the motion of all the molecules is taken into account, the correct expression for the mean free path is given by

$$\lambda = \frac{1}{\sqrt{2}\, n_V \pi d^2} \qquad \text{17-25}$$

The average time between collisions is called the collision time τ. The reciprocal of the collision time, $1/\tau$, is equal to the average number of collisions per second, or the collision frequency. If v_{av} is the average speed, then the average distance traveled between collisions is

$$\lambda = v_{av}\tau \qquad \text{17-26}$$

MEAN FREE PATH OF A CO MOLECULE IN AIR **EXAMPLE 17-8** **Put It in Context**

The local poison control center wants to know more about carbon monoxide and how it spreads through a room. You are asked *(a)* to calculate the mean free path of a carbon monoxide molecule and *(b)* to estimate the mean time between collisions. The molecular mass of carbon monoxide is 28 g/mol. Assume that the CO molecule is traveling in air at 300K and 1 atm, and that the diameters of a CO molecule and air molecules are approximately 3.75×10^{-10} m.

PICTURE THE PROBLEM (*a*) Since d is given, we can find λ from $\lambda = 1/(\sqrt{2}\, n_V \pi d^2)$ using the ideal gas law to find $n_V = N/V$ (*b*) We can estimate the collision time by using v_{rms} for the average speed.

(*a*) 1. Write λ in terms of the number density n_V and the molecular diameter d:

$$\lambda = \frac{1}{\sqrt{2}\, n_V \pi d^2}$$

2. Use the equation $PV = NkT$ to calculate $n_V = N/V$:

$$n_V = \frac{N}{V} = \frac{P}{kT} = \frac{101.3 \times 10^3\ \text{Pa}}{(1.38 \times 10^{-23}\ \text{J/K})(300\ \text{K})}$$

$$= 2.45 \times 10^{25}\ \text{molecules/m}^3$$

3. Substitute this value of n_V and the given value of d to calculate λ:

$$\lambda = \frac{1}{\sqrt{2}\, n_V \pi d^2}$$

$$= \frac{1}{\sqrt{2}\,(2.45 \times 10^{25}/\text{m}^3)\pi(3.75 \times 10^{-10}\ \text{m}^2)^2}$$

$$= \boxed{6.53 \times 10^{-8}\ \text{m}}$$

(*b*) 1. Write τ in terms of the mean free path λ:

$$\tau = \frac{\lambda}{v_{av}}$$

2. Estimate v_{av} by calculating v_{rms}:

$$v_{rms} = \sqrt{\frac{3RT}{M}} = \sqrt{\frac{3(8.31\ \text{J/[mol}\cdot\text{K]})(300\ \text{K})}{28 \times 10^{-3}\ \text{kg/mol}}}$$

$$= 517\ \text{m/s}$$

3. Use $v_{av} \approx v_{rms}$ to estimate τ:

$$\tau = \frac{\lambda}{v_{av}} = \frac{6.53 \times 10^{-8}\ \text{m}}{517\ \text{m/s}} = \boxed{1.26 \times 10^{-10}\ \text{s}}$$

REMARKS Note that we put atmospheric pressure in pascals to get the proper units for λ. The mean free path is about 200 times the diameter of the molecule, and the collision frequency is about $1/\tau \approx 8 \times 10^9$ collisions per second.

*The Distribution of Molecular Speeds

We would not expect all of the molecules in a gas to have the same velocity. The calculation of the pressure of a gas allows us to calculate the square of the average speed and therefore the average energy of molecules in a gas, but it does not yield any details about the *distribution* of molecular velocities. Before we consider this problem, we will discuss the idea of distribution functions in general with some elementary examples from common experience.

Distribution Functions Suppose that a teacher gave a 25-point quiz to a large number N of students. To describe the results, the teacher might give the average score, but this would not be a complete description. If all the students received a score of 12.5, for example, that would be quite different from half the students receiving 25 and the other half zero, but the average score would be the same in both cases. A complete description of the results would be to give the number of students n_i who received a score s_i for all the scores received. Alternatively, one could give the fraction of the students $f_i = n_i/N$ who received the score s_i. Both n_i and f_i, which are functions of the variable s, are called **distribution functions.** The fractional distribution is somewhat more convenient to use. The probability that one of the N students selected at random received the score s_i equals the total number of students who received that score n_i divided by N, that is, the probability equals f_i. Note that

$$\sum_i f_i = \sum_i \frac{n_i}{N} = \frac{1}{N} \sum_i n_i$$

and since $\Sigma n_i = N$,

$$\sum_i f_i = 1 \qquad \text{17-27}$$

Equation 17-27 is called the **normalization condition** for fractional distributions.

To find the average score, we add all the scores and divide by N. Since each score s_i was obtained by $n_i = Nf_i$ students, this is equivalent to

$$s_{av} = \frac{1}{N}\sum_i n_i s_i = \sum_i s_i f_i \qquad \text{17-28}$$

Similarly, the average of any function $g(s)$ is defined by

$$g(s)_{av} = \frac{1}{N}\sum_i g(s_i)n_i = \sum_i g(s_i)f_i \qquad \text{17-29}$$

In particular, the square of the average score of the square of the scores is

$$(s^2)_{av} = \frac{1}{N}\sum_i s_i^2 n_i = \sum_i s_i^2 f_i \qquad \text{17-30}$$

The square root of $(s^2)_{av}$ is called the **root mean square score** or rms score. A possible distribution function is shown in Figure 17-13. For this distribution, the most probable score (that obtained by the most students) is 16, the average score is 14.2, and the rms score is 14.9.

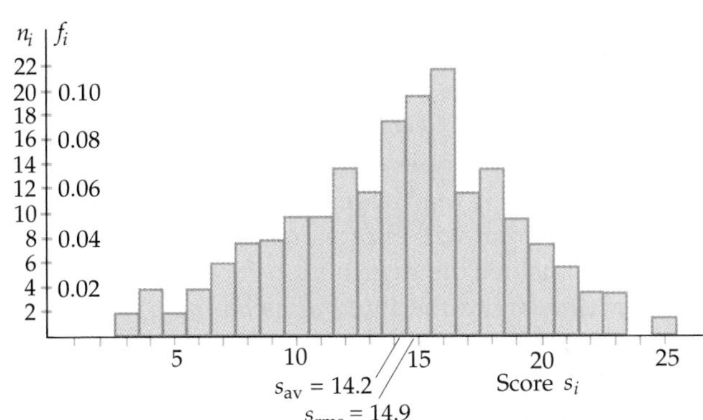

FIGURE 17-13 Grade distribution for a 25-point quiz given to 200 students. n_i is the number of students receiving grade s_i and $f_i = n_i/N$ is the fraction of students receiving grade s_i.

MAKING THE GRADE **EXAMPLE 17-9**

Fifteen students took a 25-point quiz. Their scores were 25, 22, 22, 20, 20, 20, 18, 18, 18, 18, 18, 15, 15, 15, and 10. Find the average score and the rms score.

PICTURE THE PROBLEM The distribution function for this problem is $n_{25} = 1$, $n_{22} = 2$, $n_{20} = 3$, $n_{18} = 5$, $n_{15} = 3$, and $n_{10} = 1$. To find the average score, we use Equation 17-28. To find the rms score, we use Equation 17-30 and then take the square root.

1. By definition, s_{av} is:

$$s_{av} = \frac{1}{N}\sum_i n_i s_i$$
$$= \frac{1}{15}[1(25) + 2(22) + 3(20) + 5(18) + 3(15) + 1(10)]$$
$$= \frac{1}{15}(274) = 18.3$$

2. To calculate s_{rms}, first find the average of s^2:

$$(s^2)_{av} = \frac{1}{N}\sum_i n_i s_i^2$$
$$= \frac{1}{15}[1(25)^2 + 2(22)^2 + 3(20)^2 + 5(18)^2 + 3(15)^2 + 1(10)^2]$$
$$= \frac{1}{15}(5188) = 346$$

3. Take the square root of $(s^2)_{av}$:

$$s_{rms} = \sqrt{(s^2)_{av}} = \boxed{18.6}$$

Now consider the case of a continuous distribution, for example, the distribution of heights in a population. For any finite number N, the number of people who are *exactly* 2 m tall is zero. If we assume that height can be determined to any desired accuracy, there are an infinite number of possible heights, so the probability is zero that anybody has any one particular (exact) height. Therefore, we divide the heights into intervals Δh (for example, Δh might be 1 cm or 0.5 cm) and ask what fraction of people has heights that fall in any particular interval. For very large N, this number is proportional to the size of the interval, provided the interval is sufficiently small. We define the distribution function $f(h)$ as the fraction of the number of people with heights in the interval between h and $h + \Delta h$. Then for N people, $Nf(h)\ \Delta h$ is the number of people whose height is between h and $h + \Delta h$. Figure 17-14 shows a possible height distribution.

The fraction of people with heights in a given interval Δh is the area $f(h)\ \Delta h$. If N is very large, we can choose Δh to be very small, and the histogram will approximate a continuous curve. We can therefore consider the distribution function $f(h)$ to be a continuous function, write the interval as dh, and replace the sums in Equations 17-27 through 17-30 by integrals:

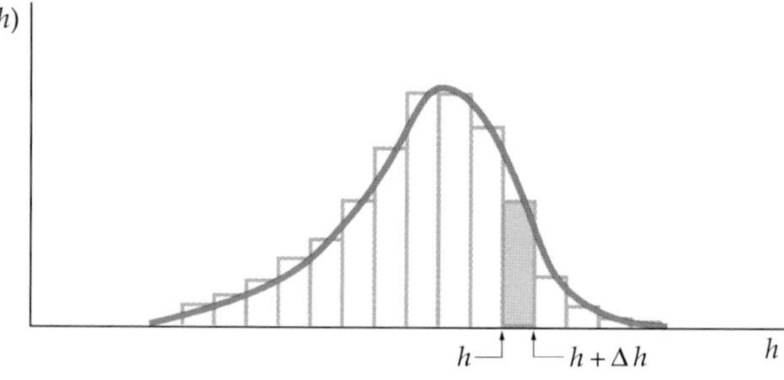

FIGURE 17-14 A possible height distribution function. The fraction of the number of heights between h and $h + \Delta h$ equals the shaded area $f(h)\ \Delta h$. The histogram can be approximated by a continuous curve as shown.

$$\int f(h)\, dh = 1$$

$$h_{av} = \int h f(h)\, dh \qquad 17\text{-}32$$

$$[g(h)]_{av} = \int g(h)\, f(h)\, dh \qquad 17\text{-}33$$

where $g(h)$ is an arbitrary function of h. Thus,

$$(h^2)_{av} = \int h^2 f(h)\, dh \qquad 17\text{-}34$$

The probability of a person selected at random having a height between h and $h + dh$ is $f(h)dh$. A useful quantity characterizing a distribution is the **standard deviation** σ defined by

$$\sigma^2 = [(x - x_{av})^2]_{av} \qquad 17\text{-}35a$$

Expanding the square on the right, we obtain

$$\sigma^2 = [x^2 - 2xx_{av} + x_{av}^2]_{av} = (x^2)_{av} - 2x_{av}x_{av} + x_{av}^2$$

or

$$\sigma^2 = (x^2)_{av} - x_{av}^2 \qquad 17\text{-}35b$$

The standard deviation measures the spread of the values about the average value. For most distributions there will be few values that differ from x_{av} by more than a few multiples of σ. For the familiar bell-shaped distribution (called a normal distribution), about 68 percent of the values are expected to fall within $x_{av} \pm \sigma$.

In Example 17-7, we found that the rms value was greater than the average value. This is a general feature for any distribution (unless all the values are identical, in which case $x_{rms} = x_{av}$). We can see this from Equation 17-35b by noting that $x_{rms}^2 = (x^2)_{av}$. Then $\sigma^2 = (x^2)_{av} - x_{av}^2 = x_{rms}^2 - x_{av}^2$. Since σ^2 and x_{rms} are always positive, x_{rms} must always be greater than $|x_{av}|$.

For the familiar bell-shaped distribution (called a normal distribution), 68 percent of the values fall within $x_{av} \pm \sigma$, 95 percent fall within $x_{av} \pm 2\sigma$, and 99.7 percent fall within $x_{av} \pm 3\sigma$. This is known as the 68, 95, 99.7 rule.

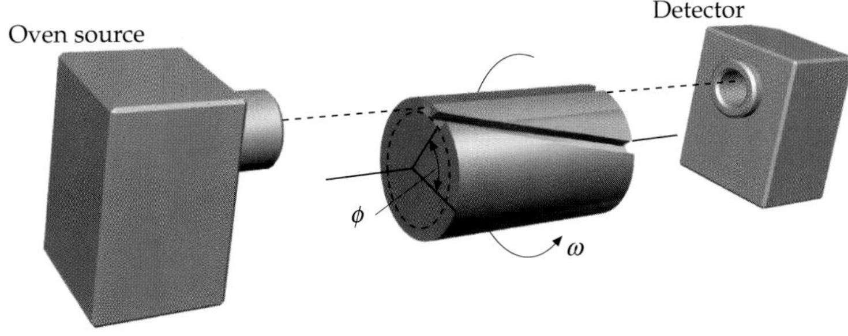

FIGURE 17-15 Schematic diagram of the apparatus for determining the speed distribution of the molecules of a gas. A substance is vaporized in an oven and the vapor molecules are allowed to escape through a hole in the oven wall into a vacuum chamber. The molecules are collimated into a narrow beam by a series of slits (not shown). The beam is aimed at a detector that counts the number of molecules that are incident on it in a given period of time. A rotating cylinder stops most of the beam. Small slits in the cylinder (only one of which is depicted here) allow the passage of molecules that have a narrow range of speeds that is determined by the angular velocity of rotation of the cylinder. Varying the angular velocity of the cylinder and counting the number of molecules that reach the detector for each angular velocity give a measure of the number of molecules in each range of speeds.

The Maxwell–Boltzmann Distribution The distribution of the molecular speeds of a gas can be measured directly using the apparatus illustrated in Figure 17-15. In Figure 17-16, these speeds are shown for two different temperatures. The quantity $f(v)$ in Figure 17-16 is called the **Maxwell–Boltzmann speed distribution function.** In a gas of N molecules, the number with speeds in the range between v and $v + dv$ is dN, given by

$$dN = N f(v)\, dv \qquad 17\text{-}36$$

The fraction $dN/N = f(v)\, dv$ in a particular range dv is illustrated by the shaded region in the figure. The Maxwell–Boltzmann speed distribution function can be derived using statistical mechanics. The result is

$$f(v) = \frac{4}{\sqrt{\pi}}\left(\frac{m}{2kT}\right)^{3/2} v^2 e^{-mv^2/(2kT)} \qquad 17\text{-}37$$

MAXWELL–BOLTZMANN SPEED DISTRIBUTION FUNCTION

The most probable speed v_{max} is that speed for which $f(v)$ is maximum. It is left as a problem to show that

$$v_{max} = \sqrt{\frac{2kT}{m}} = \sqrt{\frac{2RT}{M}} \qquad 17\text{-}38$$

Comparing Equation 17-38 with Equation 17-23, we see that the most probable speed is slightly less than the rms speed.

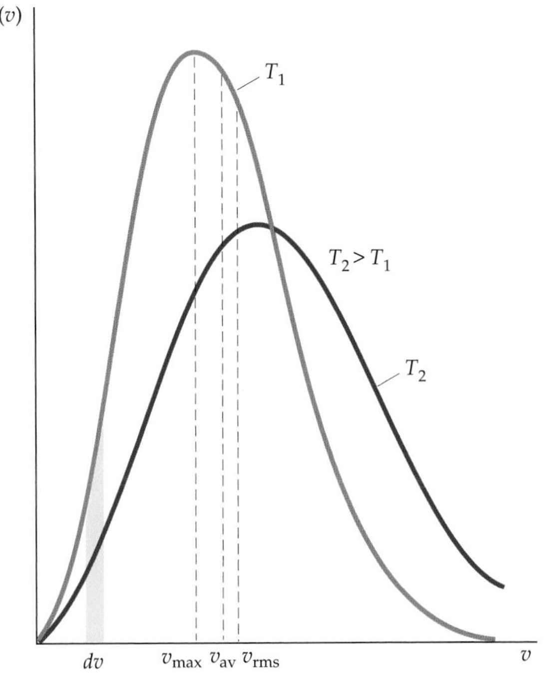

FIGURE 17-16 Distributions of molecular speeds in a gas at two temperatures, T_1 and $T_2 > T_1$. The shaded area $f(v)\, dv$ equals the fraction of the number of molecules having a particular speed in a narrow range of speeds dv. The mean speed v_{av} and the rms speed v_{rms} are both slightly greater than the most probable speed v_{max}.

USING THE MAXWELL–BOLTZMANN DISTRIBUTION **EXAMPLE 17-10**

Calculate the average value of v^2 for the molecules in a gas using the Maxwell–Boltzmann distribution function.

PICTURE THE PROBLEM The average value of v^2 is calculated from Equation 17-34 with v replacing h and $f(v)$ given by Equation 17-37.

1. By definition, $(v^2)_{av}$ is: $$(v^2)_{av} = \int_0^\infty v^2 f(v)\, dv$$

2. Use Equation 17-37 for $f(v)$:

$$(v^2)_{av} = \int_0^\infty v^2 \frac{4}{\sqrt{\pi}}\left(\frac{m}{2kT}\right)^{3/2} v^2 e^{-mv^2/(2kT)}\, dv$$

$$= \frac{4}{\sqrt{\pi}}\left(\frac{m}{2kT}\right)^{3/2} \int_0^\infty v^4\, e^{-mv^2/(2kT)}\, dv$$

3. The integral in step 2 can be found in standard integral tables:

$$\int_0^\infty v^4 e^{-mv^2/(2\,kT)}\, dv = \frac{3}{8}\sqrt{\pi}\left(\frac{2kT}{m}\right)^{5/2}$$

4. Use this result to calculate $(v^2)_{av}$:

$$(v^2)_{av} = \frac{4}{\sqrt{\pi}}\left(\frac{m}{2kT}\right)^{3/2} \frac{3}{8}\sqrt{\pi}\left(\frac{2kT}{m}\right)^{5/2} = \boxed{\frac{3kT}{m}}$$

REMARKS Note that our result agrees with $v_{rms} = \sqrt{3kT/m}$ from Equation 17-23.

In Example 17-6 we found that the rms speed of hydrogen molecules is about 1.93 km/s. This is about one-sixth of the escape speed at the surface of the earth, which we found to be 11.2 km/s in Section 11-3. So why is there no free hydrogen in the earth's atmosphere? As we can see from Figure 17-16, a considerable fraction of the molecules of a gas in equilibrium have speeds greater than the rms speed. When the rms speed of the molecules of a particular gas is as great as 15 to 20 percent of the escape speed for a planet, enough of the molecules have speeds greater than the escape speed so that the gas does not remain in the atmosphere of that planet very long before escaping. Thus, there is virtually no hydrogen gas in the earth's atmosphere. The rms speed of oxygen molecules, on the other hand, is about one-fourth that of hydrogen molecules, which makes it only about 4 percent of the escape speed at the surface of the earth. Therefore, only a negligible fraction of the oxygen molecules have speeds greater than the escape speed, and oxygen remains in the earth's atmosphere.

The Energy Distribution The Maxwell–Boltzmann speed distribution as given by Equation 17-37 can also be written as an energy distribution. We write the number of molecules with energy E in the range between E and $E + dE$ as

$$dN = NF(E)\, dE$$

where $F(E)$ is the energy distribution function. This will be the same number as given by Equation 17-37 if the energy E is related to the speed v by $E = \frac{1}{2}mv^2$. Then

$$dE = mv\, dv$$

and

$$Nf(v)\, dv = NF(E)\, dE$$

We can write

$$f(v)\, dv = Cv^2 e^{-mv^2/(2kT)}\, dv = Cve^{-E/(kT)}\, v\, dv = C\left(\frac{2E}{m}\right)^{1/2} e^{-E/(kT)} \frac{dE}{m}$$

where $C = (4/\sqrt{\pi})[m/(2kT)]^{3/2}$ (from Equation 17-37). The energy distribution function $F(E)$ is thus given by

$$F(E) = \frac{4}{\sqrt{\pi}}\left(\frac{m}{2kT}\right)^{3/2}\left(\frac{2}{m}\right)^{1/2}\frac{1}{m}E^{1/2}e^{-E/(kT)}$$

Simplifying, we obtain the **Maxwell–Boltzmann energy distribution function:**

$$F(E) = \frac{2}{\sqrt{\pi}}\left(\frac{1}{kT}\right)^{3/2} E^{1/2}e^{-E/(kT)} \qquad \text{17-39}$$

MAXWELL–BOLTZMANN ENERGY DISTRIBUTION FUNCTION

In the language of statistical mechanics, the energy distribution is considered to be the product of two factors: one, called the **density of states,** is proportional to $E^{1/2}$; the other is the probability of a state being occupied, which is $e^{-E/(kT)}$ and is called the **Boltzmann factor.**

SUMMARY

Topic	Relevant Equations and Remarks	
1. Celsius and Fahrenheit Scales	On the Celsius scale, the ice point is defined to be 0°C and the steam point is 100°C. On the Fahrenheit scale, the ice point is 32°F and the steam point is 212°F. Temperatures on the Fahrenheit and Celsius scales are related by	
	$t_C = \frac{5}{9}(t_F - 32°)$	**17-2**
2. Gas Thermometers	Gas thermometers have the property that they all agree with each other in the measurement of any temperature as long as the density of the gas is very low. The ideal-gas temperature T is defined by	
	$T = \frac{273.16\text{ K}}{P_3}P$	**17-4**
	where P is the observed pressure of the gas in the thermometer and P_3 is the pressure when the thermometer is immersed in a water–ice–vapor bath at its triple point.	
3. Kelvin Temperature Scale	The absolute temperature or temperature in kelvins is related to the Celsius temperature by	
	$T = t_C + 273.15\text{ K}$	**17-5**
4. Ideal Gas	At low densities, all gases obey the ideal-gas law.	
Equation of state	$PV = nRT$	**17-13**
Universal gas constant	$R = N_Ak = 8.314\text{ J/(mol·K)}$ $= 0.08206\text{ L·atm/(mol·K)}$	**17-12**
Boltzmann's constant	$k = 1.381 \times 10^{-23}\text{ J/K} = 8.617 \times 10^{-5}\text{ eV/K}$	**17-8**
Avogadro's number	$N_A = 6.022 \times 10^{23}$	**17-9**
Equation for a fixed amount of gas	A form of the ideal-gas law that is useful for solving problems involving a fixed amount of gas is	
	$\frac{P_2V_2}{T_2} = \frac{P_1V_1}{T_1}$	**17-14**

5. Kinetic Theory of Gases

Molecular interpretation of temperature	The absolute temperature T is a measure of the average molecular translational kinetic energy.
Equipartition theorem	When a system is in equilibrium, there is an average energy of $\frac{1}{2}kT$ per molecule (or $\frac{1}{2}RT$ per mole) associated with each degree of freedom.
Average kinetic energy	For an ideal gas, the average translational kinetic energy of the molecules is $K_{av} = (\frac{1}{2}mv^2)_{av} = \frac{3}{2}kT$ **17-21**
Total kinetic energy	The total translational kinetic energy of n moles of a gas containing N molecules is given by $K = N(\frac{1}{2}mv^2)_{av} = \frac{3}{2}NkT = \frac{3}{2}nRT$ **17-22**
rms speed of molecules	The rms speed of a molecule of a gas is related to the absolute temperature by $v_{rms} = \sqrt{(v^2)_{av}} = \sqrt{\frac{3kT}{m}} = \sqrt{\frac{3RT}{M}}$ **17-23** where m is the mass of the molecule and M is the molar mass.
Mean free path	The mean free path of a molecule is related to its diameter d and the number of molecules per unit volume n_V by $\lambda = \frac{1}{\sqrt{2}n_V \pi d^2}$ **17-25**

***6. Maxwell–Boltzmann Distribution**	$f(v) = \frac{4}{\sqrt{\pi}}\left(\frac{m}{2kT}\right)^{3/2} v^2 e^{-mv^2/(2kT)}$	**17-37**
Energy distribution	$F(E) = \frac{2}{\sqrt{\pi}}\left(\frac{1}{kT}\right)^{3/2} E^{1/2} e^{-E/(kT)}$	**17-39**

PROBLEMS

In a few problems, you are given more data than you actually need; in a few other problems, you are required to supply data from your general knowledge, outside sources, or informed estimates.

- • Single-concept, single-step, relatively easy
- •• Intermediate-level, may require synthesis of concepts
- ••• Challenging
- SSM Solution is in the *Student Solutions Manual*
- iSOLVE Problems available on iSOLVE online homework service
- iSOLVE ✓ These "Checkpoint" online homework service problems ask students additional questions about their confidence level, and how they arrived at their answer

Conceptual Problems

1 • SSM True or false:

(*a*) Two objects in thermal equilibrium with each other must be in thermal equilibrium with a third object.

(*b*) The Fahrenheit and Celsius temperature scales differ only in the choice of the zero temperature.

(*c*) The kelvin is the same size as the Celsius degree.

(*d*) All thermometers give the same result when measuring the temperature of a particular system.

2 • How can you determine if two bodies are in thermal equilibrium with each other if it is impossible to put them into thermal contact with each other?

3 • "One day I woke up and it was 20°F in my bedroom," said Mert to his old friend Mort. "That's nothing," replied Mort. "My room was once −5°C." Which room was colder?

4 •• Two identical vessels contain different ideal gases at the same pressure and temperature. It follows that (*a*) the number of gas molecules is the same in both vessels, (*b*) the total mass of gas is the same in both vessels, (*c*) the average speed of the gas molecules is the same in both vessels, (*d*) none of these answers is correct.

5 •• Figure 17-17 shows a plot of volume versus temperature for a process that takes an ideal gas from point A to point B. What happens to the pressure of the gas?

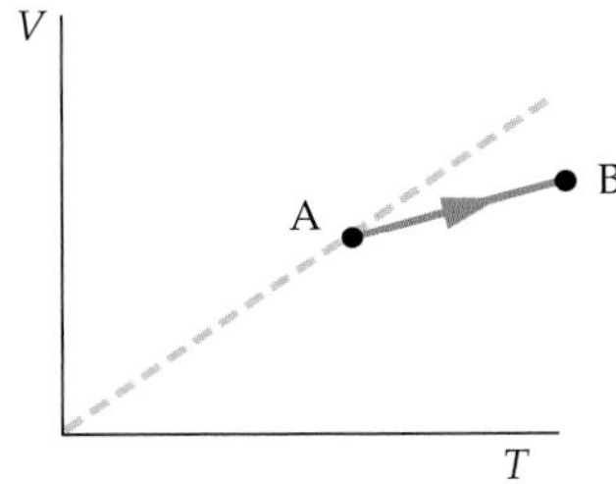

FIGURE 17-17
Problem 5

6 •• SSM Figure 17-18 shows a plot of pressure versus temperature for a process that takes an ideal gas from point A to point B. What happens to the volume of the gas?

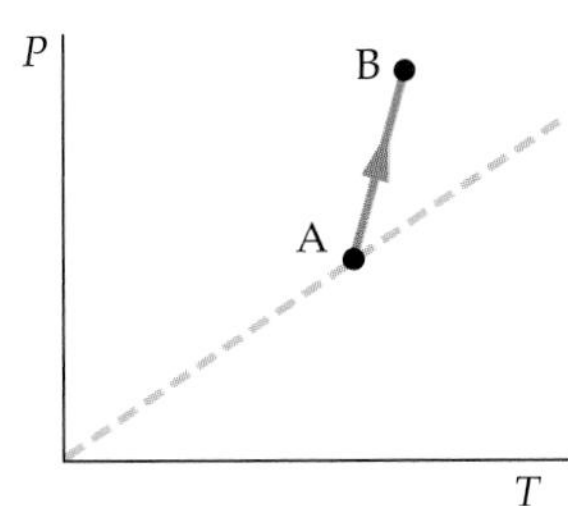

FIGURE 17-18
Problem 6

7 • True or false: The absolute temperature of a gas is a measure of the average translational kinetic energy of the gas molecules.

8 • By what factor must the absolute temperature of a gas be increased to double the rms speed of its molecules?

9 • How does the average translational kinetic energy of a molecule of a gas change if the pressure is doubled while the volume is kept constant? If the volume is doubled while the pressure is kept constant?

10 •• A vessel holds an equal number of moles of helium and methane, CH_4. The ratio of the rms speeds of the helium atoms to the CH_4 molecules is (*a*) 1, (*b*) 2, (*c*) 4, (*d*) 16.

11 • True or false: If the pressure of a gas increases, the temperature must increase.

12 • Why might the Celsius and Fahrenheit scales be more convenient than the absolute scale for ordinary, nonscientific purposes?

13 • SSM The temperature of the interior of the sun is said to be about 10^7 degrees. Do you think that this is degrees Celsius or kelvins, or does it matter?

14 • ISOLVE✓ If the temperature of an ideal gas is doubled while maintaining constant pressure, the average speed of the molecules (*a*) remains constant, (*b*) increases by a factor of 4, (*c*) increases by a factor of 2, (*d*) increases by a factor of $\sqrt{2}$.

15 • ISOLVE✓ If both temperature and volume of an ideal gas are halved, the pressure (*a*) diminishes by a factor of 2, (*b*) remains constant, (*c*) increases by a factor of 2, (*d*) diminishes by a factor of $\sqrt{2}$.

16 • The average translational kinetic energy of the molecules of an ideal gas depends on (*a*) the number of moles of the gas and its temperature, (*b*) the pressure of the gas and its temperature, (*c*) the pressure of the gas only, (*d*) the temperature of the gas only.

17 • If a vessel contains equal amounts, by weight, of helium and argon, which of the following are true?

(*a*) The pressure exerted by the two gases on the walls of the container is the same.
(*b*) The average speed of a helium atom is the same as that of an argon atom.
(*c*) The number of helium atoms and argon atoms in the vessel are equal.
(*d*) None of these statements is correct.

18 •• Two rooms, A and B, have equal volumes and are connected by an open door. Room A, which is air-conditioned, is 5C° cooler than room B. Which room has more air in it?

19 • Two different gases are at the same temperature. What can you say about the rms speeds of the gas molecules? What can you say about the average kinetic energies of the molecules?

20 •• Explain in terms of molecular motion why the pressure on the walls of a container increases when a gas is heated at constant volume.

21 •• SSM Explain in terms of molecular motion why the pressure on the walls of a container increases when the volume of a gas is reduced at constant temperature.

22 •• Oxygen has a molar mass of 32 g/mol and nitrogen has a molar mass of 28 g/mol. The oxygen and nitrogen molecules in a room have

(*a*) equal average kinetic energies, but the oxygen molecules are faster.
(*b*) equal average kinetic energies, but the oxygen molecules are slower.
(*c*) equal average kinetic energies and speeds.
(*d*) equal average speeds, but the oxygen molecules have a higher average kinetic energy.
(*e*) equal average speeds, but the oxygen molecules have a lower average kinetic energy.
(*f*) None of these answers is correct.

23 •• As any low-temperature physicist knows, liquid nitrogen is relatively cheap, while liquid helium is very expensive. One reason for this is that while nitrogen is the most common constituent of the atmosphere, helium is only found sealed in underground pockets of rock. Use ideas from this chapter to explain why this is true.

Estimation and Approximation

24 •• SSM A stoppered test tube that has a volume of 10 ml has 1 ml of water at its bottom and is at a temperature of 100°C and initially at a pressure of 1 atm (1.01×10^5 N/m²). Estimate the pressure inside the test tube when the water is completely boiled away.

25 ••• In Chapter 11, we found that the escape speed at the surface of a planet of radius R is $v_e = \sqrt{2gR}$, where g is the acceleration due to gravity at the surface of the planet. If the rms speed of a gas is greater than about 15 to 20 percent of the escape velocity of a planet, virtually all of the molecules of that gas will escape the atmosphere of the planet.

(*a*) At what temperature is v_{rms} for O_2 equal to 15 percent of the escape speed for the earth?
(*b*) At what temperature is v_{rms} for H_2 equal to 15 percent of the escape speed for the earth?
(*c*) Temperatures in the upper atmosphere reach 1000 K. How does this account for the low abundance of hydrogen in the earth's atmosphere?
(*d*) Compute the temperatures for which the rms speeds of O_2 and H_2 are equal to 15 percent of the escape velocity at the surface of the moon, where g is about one-sixth of its value on earth and R = 1738 km. How does this account for the absence of an atmosphere on the moon?

26 •• The escape velocity on Mars is 5.0 km/s and the surface temperature is typically 0°C. Calculate the rms speeds for (*a*) H_2, (*b*) O_2, and (*c*) CO_2 at this temperature. (*d*) Based on the criterion given in the chapter, are H_2, O_2, and CO_2 likely to be found in the atmosphere of Mars?

27 •• SSM Repeat Problem 26 for Jupiter, whose escape velocity is 60 km/s and whose temperature is typically −150°C.

Jupiter as seen from about twelve million miles. Because the escape speed at the surface of Jupiter is about 600 km/s, Jupiter easily retains hydrogen in its atmosphere.

Temperature Scales

28 • A certain ski wax is rated for use between −12 and −7°C. What is this temperature range on the Fahrenheit scale?

29 • ISOLVE The melting point of gold (Au) is 1945.4°F. Express this temperature in degrees Celsius.

30 • SSM ISOLVE What is the Celsius temperature corresponding to the normal temperature of the human body, 98.6°F?

31 • ISOLVE✓ The length of the column of mercury in a thermometer is 4.0 cm when the thermometer is immersed in ice water and 24.0 cm when the thermometer is immersed in boiling water. (*a*) What should be the length at room temperature, 22.0°C? (*b*) If the mercury column is 25.4 cm long when the thermometer is immersed in a chemical solution, what is the temperature of the solution?

32 • The temperature of the interior of the sun is about 10^7 K. What is this temperature on (*a*) the Celsius scale and (*b*) the Fahrenheit scale?

33 • The boiling point of nitrogen N_2 is 77.35 K. Express this temperature in degrees Fahrenheit.

34 • ISOLVE The pressure of a constant-volume gas thermometer is 0.400 atm at the ice point and 0.546 atm at the steam point. (*a*) When the pressure is 0.100 atm, what is the temperature? (*b*) What is the pressure at 444.6°C, the boiling point of sulfur?

35 • SSM ISOLVE A constant-volume gas thermometer reads 50 torr at the triple point of water. (*a*) What will be the pressure when the thermometer measures a temperature of 300 K? (*b*) What ideal-gas temperature corresponds to a pressure of 678 torr?

36 • A constant-volume gas thermometer has a pressure of 30 torr when it reads a temperature of 373 K. (*a*) What is its triple-point pressure P_3? (*b*) What temperature corresponds to a pressure of 0.175 torr?

37 • At what temperature do the Fahrenheit and Celsius temperature scales give the same reading?

38 • ISOLVE Sodium melts at 371 K. What is the melting point of sodium on the Celsius and Fahrenheit temperature scales?

39 • The boiling point of oxygen at 1 atm is 90.2 K. What is the boiling point of oxygen on the Celsius and Fahrenheit scales?

40 •• On the Réaumur temperature scale, the melting point of ice is 0°R and the boiling point of water is 80°R. Derive expressions for converting temperatures on the Réaumur scale to the Celsius and Fahrenheit scales.

41 ••• SSM A thermistor is a solid-state device whose resistance varies greatly with temperature. Its temperature dependence is given approximately by $R = R_0e^{B/T}$, where R is in ohms (Ω), T is in kelvins, and R_0 and B are constants that can be determined by measuring R at calibration points such as the ice point and the steam point. (*a*) If $R = 7360\ \Omega$ at the ice point and 153 Ω at the steam point, find R_0 and B. (*b*) What is the resistance of the thermistor at $t = 98.6°F$? (*c*) What is the rate of change of the resistance with temperature (dR/dT) at the ice point and the steam point? (*d*) At which temperature is the thermistor most sensitive?

The Ideal-Gas Law

42 • ISOLVE✓ A gas is kept at constant pressure. If its temperature is changed from 50 to 100°C, by what factor does the volume change?

43 • ISOLVE✓ A 10-L vessel contains gas at a temperature of 0°C and a pressure of 4 atm. How many moles of gas are in the vessel? How many molecules?

44 •• ISOLVE✓ A pressure as low as 1×10^{-8} torr can be achieved using an oil diffusion pump. How many molecules are there in 1 cm^3 of a gas at this pressure if its temperature is 300 K?

45 •• SSM You copy the following paragraph from a Martian physics textbook: "1 *snorf* of an ideal gas occupies a volume of 1.35 *zaks*. At a temperature of 22 *glips*, the gas has a pressure of 12.5 *klads*. At a temperature of −10 *glips*, the same gas now has a pressure of 8.7 *klads*." Determine the temperature of absolute zero in *glips*.

46 •• ISOLVE A motorist inflates the tires of her car to a gauge pressure of 180 kPa on a day when the temperature is −8.0°C. When she arrives at her destination, the tire pressure has increased to 245 kPa. What is the temperature of the tires if we assume that (*a*) the tires do not expand or (*b*) that the tires expand by 7 percent?

47 •• A room is 6 m by 5 m by 3 m. (*a*) If the air pressure in the room is 1 atm and the temperature is 300 K, find the number of moles of air in the room. (*b*) If the temperature rises by 5 K and the pressure remains constant, how many moles of air leave the room?

48 •• SSM ISOLVE The boiling point of helium at 1 atm is 4.2 K. What is the volume occupied by helium gas due to evaporation of 10 g of liquid helium at 1 atm pressure and a temperature of (*a*) 4.2 K and (*b*) 293 K?

49 •• A container with a volume of 6.0 L holds 10 g of liquid helium. As the container warms to room temperature, what is the pressure exerted by the gas on its walls?

50 •• SSM ISOLVE An automobile tire is filled to a gauge pressure of 200 kPa when its temperature is 20°C. (Gauge pressure is the difference between the actual pressure and atmospheric pressure.) After the car has been driven at high speeds, the tire temperature increases to 50°C. (*a*) Assuming that the volume of the tire does not change and that air behaves as an ideal gas, find the gauge pressure of the air in the tire. (*b*) Calculate the gauge pressure if the volume of the tire expands by 10 percent.

51 •• Calculate the mass density of air at a temperature of 24°C and a pressure of 1 atm (1.01×10^5 N/m^2) using the ideal-gas law. Air is roughly 74 percent N_2 and 26 percent O_2.

52 •• ISOLVE✓ A scuba diver is 40 m below the surface of a lake, where the temperature is 5°C. He releases an air bubble with a volume of 15 cm^3. The bubble rises to the surface, where the temperature is 25°C. What is the volume of the bubble right before it breaks the surface? *Hint: Remember that the pressure also changes.*

53 •• A hot-air balloon has a volume of 1.5 m^3 and is open at the bottom. If the air inside the balloon is at a temperature of 75°C, while the temperature of the air outside the balloon is 24°C, at a pressure of about 1 atm, what is the net force on the balloon and its contents? (Neglect the weight of the balloon itself.)

54 ••• A helium balloon is used to lift a load of 110 N. The weight of the balloon's skin is 50 N and the volume of the balloon when fully inflated is 32 m^3. The temperature of the air is 0°C and the atmospheric pressure is 1 atm. The balloon is inflated with sufficient helium gas that the net force on the balloon and its load is 30 N. Neglect changes of temperature with altitude. (*a*) How many moles of helium gas are contained in the balloon? (*b*) At what altitude will the balloon be fully inflated? (*c*) Does the balloon ever reach the altitude at which it is fully inflated? (*d*) If the answer to (*c*) is affirmative, what is the maximum altitude attained by the balloon?

Kinetic Theory of Gases

55 • SSM (*a*) Find v_{rms} for an argon atom if 1 mol of the gas is confined to a 1-L container at a pressure of 10 atm. (For argon, $M = 40 \times 10^{-3}$ kg/mol.) (*b*) Compare this with v_{rms} for a helium atom under the same conditions. (For helium, $M = 4 \times 10^{-3}$ kg/mol.)

56 • ISOLVE✓ Find the total translational kinetic energy of 1 L of oxygen gas held at a temperature of 0°C and a pressure of 1 atm.

57 • ISOLVE Find the rms speed and the average kinetic energy of a hydrogen atom at a temperature of 10^7 K. (At this temperature, which is of the order of the temperature in the interior of a star, the hydrogen is ionized and consists of a single proton.)

58 • SSM In one model of a solid, the material is assumed to consist of a regular array of atoms in which each atom has a fixed equilibrium position and is connected by springs to its neighbors. Each atom can vibrate in the x, y, and z directions. The total energy of an atom in this model is

$$E = \tfrac{1}{2}mv_x^2 + \tfrac{1}{2}mv_y^2 + \tfrac{1}{2}mv_z^2 + \tfrac{1}{2}kx^2 + \tfrac{1}{2}ky^2 + kz^2$$

What is the average energy of an atom in the solid when the temperature is T? What is the total energy of 1 mol of such a solid?

59 • Show that the mean free path for a molecule in an ideal gas at temperature T and pressure P is given by

$$\lambda = \frac{kT}{\sqrt{2}P\pi d^2}$$

60 •• ISOLVE✓ A pressure as low as $P = 7 \times 10^{-11}$ Pa has been obtained. Suppose that a chamber contains helium at this pressure and at room temperature (300 K). Estimate the mean free path λ and the collision time τ for helium in the chamber. Take the diameter of a helium molecule to be 10^{-10} m.

61 •• SSM Oxygen (O_2) is confined to a cubic container 15 cm on a side at a temperature of 300 K. Compare the average kinetic energy of a molecule of the gas to the change in its gravitational potential energy if it falls from the top of the container to the bottom.

*The Distribution of Molecular Speeds

62 •• Show that $f(v)$ given by Equation 17-37 is maximum when $v = \sqrt{2kT/m}$. *Hint: Set $df/dv = 0$ and solve for v.*

63 •• SSM $f(v)$ is defined in Equation 17-37. Because $f(v)\,dv$ gives the fraction of molecules that have speeds in the range dv, the integral of $f(v)\,dv$ over all the possible ranges of speeds must equal 1. Given the integral

$$\int_0^\infty v^2 e^{-av^2}\,dv = \frac{\sqrt{\pi}}{4}a^{-3/2}$$

show that $\int_0^\infty f(v)\,dv = 1$, where $f(v)$ is given by Equation 17-37.

64 •• Given the integral

$$\int_0^\infty v^3 e^{-av^2}\,dv = \frac{a^{-2}}{2}$$

calculate the average speed v_{av} of molecules in a gas using the Maxwell–Boltzmann distribution function.

65 •• SSM Current experiments in atomic trapping and cooling can create low-density gases of rubidium and other atoms with temperatures in the nanokelvin (10^{-9} K) range. These atoms are trapped and cooled using magnetic fields and lasers in ultrahigh vacuum chambers. One method that is used to measure the temperature of a trapped gas is to turn the trap off and measure the time it takes for molecules of the gas to fall a given distance! Consider a gas of rubidium atoms at a temperature of 120 nK. Calculate how long it would take an atom traveling at the rms speed of the gas to fall a distance of 10 cm if (*a*) it were initially moving directly downward and (*b*) if it were initially moving directly upward. Assume that the atom doesn't collide with any others along its trajectory.

General Problems

66 • At what temperature will the rms speed of an H_2 molecule equal 331 m/s?

67 •• (*a*) If 1 mol of a gas in a container occupies a volume of 10 L at a pressure of 1 atm, what is the temperature of the gas in kelvins? (*b*) The container is fitted with a piston so that the volume can change. When the gas is heated at constant pressure, it expands to a volume of 20 L. What is the temperature of the gas in kelvins? (*c*) The volume is fixed at 20 L, and the gas is heated at constant volume until its temperature is 350 K. What is the pressure of the gas?

68 •• ISOLVE✓ A cubic metal box with sides of 20 cm contains air at a pressure of 1 atm and a temperature of 300 K. The box is sealed so that the volume is constant and it is heated to a temperature of 400 K. Find the net force on each wall of the box.

69 •• SSM Water, H_2O, can be converted into H_2 and O_2 gases by electrolysis. How many moles of these gases result from the electrolysis of 2 L of water?

70 •• A massless cylinder 40 cm long rests on a horizontal frictionless table. The cylinder is divided into two equal sections by a membrane. One section contains nitrogen and the other contains oxygen. The pressure of the nitrogen is twice that of the oxygen. How far will the cylinder move if the membrane is removed?

71 •• A cylinder contains a mixture of nitrogen gas (N_2) and hydrogen gas (H_2). At a temperature T_1 the nitrogen is completely dissociated but the hydrogen does not dissociate at all, and the pressure is P_1. If the temperature is doubled to $T_2 = 2T_1$, the pressure is tripled due to complete dissociation of hydrogen. If the mass of hydrogen is m_H, find the mass of nitrogen m_N.

72 •• SSM Three insulated vessels of equal volume V are connected by thin tubes that can transfer gas but do not transfer heat. Initially all vessels are filled with the same type of gas at a temperature T_0 and pressure P_0. Then the temperature in the first vessel is doubled and the temperature in the second vessel is tripled. The temperature in the third vessel remains unchanged. Find the final pressure P' in the system in terms of the initial pressure P_0.

73 •• A constant-volume gas thermometer with a triple-point pressure $P_3 = 500$ torr is used to measure the boiling point of some substance. When the thermometer is placed in thermal contact with the boiling substance, its pressure is 734 torr. Some of the gas in the thermometer is then allowed to escape so that its triple-point pressure is 200 torr. When it is again placed in thermal contact with the boiling substance, its pressure is 293.4 torr. Again, some of the gas is removed from the thermometer so that its triple-point pressure is 100 torr. When the thermometer is placed in thermal contact with the boiling substance once again, its pressure is 146.65 torr. Find the ideal-gas temperature of the boiling substance.

74 •• SSM The mean free path for O_2 molecules at a temperature of 300 K at 1 atm pressure ($p = 1.01 \times 10^5$ Pa) is $\lambda = 7.1 \times 10^{-8}$ m. Use this data to estimate the size of an O_2 molecule.

75 •• An experimental balloon contains hydrogen gas (H_2) at a temperature of 300 K and a pressure of 1 atm (1.01×10^5 N/m^2). (*a*) Calculate the mean-free path of a hydrogen molecule. Assume that a H_2 molecule is effectively spherical, with a mean diameter of 1.6×10^{-10} m. (*b*) Calculate the available volume per molecule (V/N), and find the average distance between each molecule and its nearest neighboring molecule (approximately the cube root of the available volume). Which is larger, the mean free path or the average nearest-neighbor distance between molecules?

76 ••• A cylinder 2.4 m tall is filled with 0.1 mol of an ideal gas at standard temperature and pressure (Figure 17-19). The top of the cylinder is then sealed with a piston whose mass is 1.4 kg and the piston is allowed to drop until it is in equilibrium. (*a*) Find the height of the piston, assuming that the temperature of the gas does not change as it is compressed. (*b*) Suppose that the piston is pushed down below its equilibrium position by a small amount and then released. Assuming that the temperature of the gas remains constant, find the frequency of vibration of the piston.

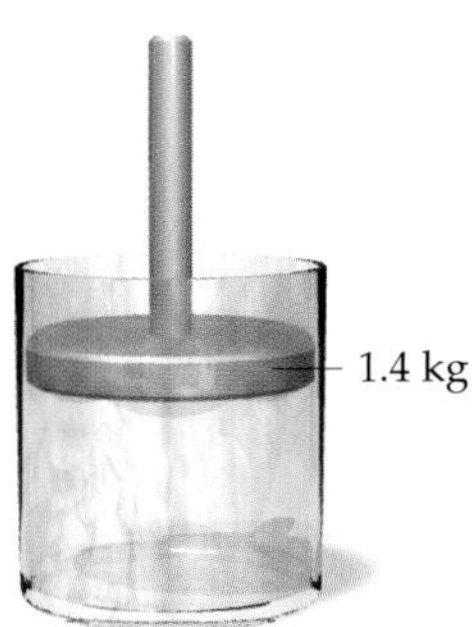

FIGURE 17-19
Problem 76

77 ••• SSM The table below gives values of

$$\frac{4}{\sqrt{\pi}}\int_0^x z^2 e^{-z^2}\,dz$$

for different values of x. Use the table to answer the following questions: (*a*) For O_2 gas at 273 K, what fraction of molecules have speeds less than 400 m/s? (*b*) For the same gas, what percentage of molecules have speeds between 190 m/s and 565 m/s?

x	$\frac{4}{\sqrt{\pi}}\int_0^x z^2 e^{-z^2}\,dz$
0.1	7.48×10^{-4}
0.2	5.88×10^{-3}
0.3	0.019
0.4	0.044
0.5	0.081
0.6	0.132

x	$\frac{4}{\sqrt{\pi}}\int_0^x z^2 e^{-z^2}\,dz$
0.7	0.194
0.8	0.266
0.9	0.345
1.0	0.438
1.5	0.788
2.0	0.954

CHAPTER

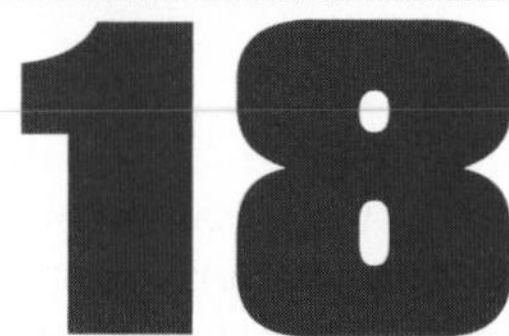

Heat and the First Law of Thermodynamics

THE WARM LEMONADE IN THIS PITCHER IS COOLED BY ADDING ICE. HEAT IS TRANSFERRED FROM THE LEMONADE TO THE ICE BECAUSE OF A DIFFERENCE IN TEMPERATURE.

? How much ice should you add to a cup of lemonade to reduce the temperature of the lemonade from 20°C to 0°C? (See Example 18-4.)

Heat is energy that is being transferred from one system to another because of a difference in temperature. In the seventeenth century, Galileo, Newton, and other scientists generally supported the theory of the ancient Greek atomists who considered thermal energy to be a manifestation of molecular motion. In the next century, methods were developed for making quantitative measurements of the amount of heat that leaves or enters an object, and it was found that if objects are in thermal contact, the amount of heat that leaves one object equals the amount that enters the other. This discovery led to the caloric theory of heat as a conserved material substance. In this theory, an invisible fluid called "caloric" flowed out of one object and into another and this "caloric" could be neither created nor destroyed.

The caloric theory reigned until the nineteenth century, when it was found that friction between objects could generate an unlimited amount of thermal energy, deposing of the idea that caloric was a substance present in a fixed amount.

The modern theory of heat did not emerge until the 1840s, when James Joule (1818–1889) demonstrated that the increase or decrease of a given amount of thermal energy was always accompanied by the decrease or increase of an equivalent quantity of mechanical energy. Thermal energy, therefore, is not itself conserved. Instead, thermal energy is a form of internal energy, and it is energy that is conserved. ➢ **In this chapter, we define heat capacity, and examine how heating a system can cause either a change in its temperature or a change in its phase. We then examine the relationship between heat conduction, work, and internal energy of a system and express the law of conservation of energy for the thermal systems as the first law of thermodynamics. Finally, we shall see how the heat capacity of a system is related to its molecular structure.**

18-1 Heat Capacity and Specific Heat

When energy is transferred to a substance by heating it, the temperature of the substance usually rises.[†] The amount of heat energy Q needed to raise the temperature of a substance is proportional to the temperature change and to the mass of the substance:

$$Q = C\Delta T = mc\Delta T \qquad \text{18-1}$$

where C is the **heat capacity,** which is defined as the amount of energy transferred via heating necessary to raise the temperature of a substance by one degree. The **specific heat** c is the heat capacity per unit mass:

$$c = \frac{C}{m} \qquad \text{18-2}$$

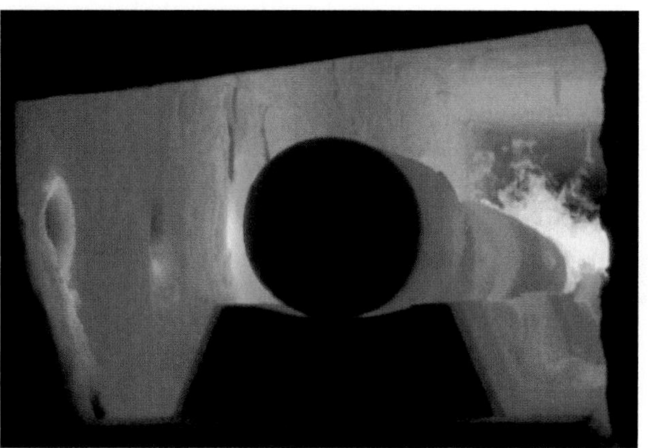

Steel ingots in a twin-tube tunnel furnace. The three 53-cm diameter carbon steel ingots seen here have been heated for about 7 hours to approximately 1340°C. Each 3200-kg ingot sits on a furnace car that transports it through the 81-m furnace, which is divided into twelve separate heating zones so that the temperature of the ingot is increased gradually to prevent cracking. The ingots, glowing a yellow-whitish color, exit the furnace to be milled into large, heavy-walled pipes.

The historical unit of heat energy, the **calorie,** was originally defined to be the amount of heat needed to raise the temperature of one gram of water one Celsius degree.[‡] Because we now recognize that heat is a measure of energy transfer, we can define the calorie in terms of the SI unit of energy, the joule:

$$1 \text{ cal} = 4.184 \text{ J} \qquad \text{18-3}$$

The U.S. customary unit of heat is the **Btu** (for British thermal unit), which was originally defined to be the amount of energy needed to raise the temperature of 1 pound of water by 1°F. The Btu is related to the calorie and to the joule by

$$1 \text{ Btu} = 252 \text{ cal} = 1.054 \text{ kJ} \qquad \text{18-4}$$

The original definition of the calorie implies that the specific heat of water is[§]

$$\begin{aligned} c_{\text{water}} &= 1 \text{ cal/(g}\cdot\text{C}^\circ) = 1 \text{ kcal/(kg}\cdot\text{C}^\circ) \\ &= 1 \text{ kcal/(kg}\cdot\text{K}) = 4.184 \text{ kJ/(kg}\cdot\text{K}) \end{aligned} \qquad \text{18-5}a$$

Similarly, from the definition of the Btu, the specific heat of water in U.S. customary units is

$$c_{\text{water}} = 1 \text{ Btu/(lb}\cdot\text{F}^\circ) \qquad \text{18-5}b$$

† An exception occurs during a change in phase, as when water freezes or evaporates. Changes of phase are discussed in Section 18-2.

‡ The kilocalorie is then the amount of heat energy needed to raise the temperature of 1 kg of water by 1°C. The "calorie" used in measuring the energy equivalent of foods is actually the kilocalorie.

§ Careful measurement shows that the specific heat of water varies by about 1 percent over the temperature range from 0 to 100°C. We will usually neglect this small variation.

The heat capacity per mole is called the **molar specific heat** c',

$$c' = \frac{C}{n}$$

where n is the number of moles. Since $C = mc$, the molar specific heat c' and specific heat c are related by

$$c' = \frac{C}{n} = \frac{mc}{n} = Mc \qquad 18\text{-}6$$

where $M = m/n$ is the molar mass. Table 18-1 lists the specific heats and molar specific heats of some solids and liquids. Note that the molar heats of all the metals are about the same. We will discuss the significance of this in Section 18-7.

RAISING THE TEMPERATURE **EXAMPLE 18-1**

How much heat is needed to raise the temperature of 3 kg of copper by 20 C°?

PICTURE THE PROBLEM The amount of heat needed to raise the temperature of the substance (copper) is proportional to the temperature change (20 C°) and to the mass (3 kg) of the substance.

The required heat is given by Equation 18-1 with $c = 0.386$ kJ/kg·K from Table 18-1:

$$Q = mc\,\Delta T = (3\text{ kg})(0.386\text{ kJ/kg·K})(20\text{ K}) = \boxed{23.2\text{ kJ}}$$

REMARKS Note that we use $\Delta T = 20$ C° = 20 K. Alternatively, we could express the specific heat as 0.386 kJ/kg·C° and write the temperature change as 20 C°.

EXERCISE A 2-kg aluminum block is originally at 10°C. If 36 kJ of heat energy are added to the block, what is its final temperature? (*Answer* 30°C)

TABLE 18-1

Specific Heats and Molar Specific Heats of Some Solids and Liquids

Substance	c, kJ/kg·K	c, kcal/kg·K or Btu/lb·F°	c′, J/mol·K
Aluminum	0.900	0.215	24.3
Bismuth	0.123	0.0294	25.7
Copper	0.386	0.0923	24.5
Glass	0.840	0.20	—
Gold	0.126	0.0301	25.6
Ice (−10°C)	2.05	0.49	36.9
Lead	0.128	0.0305	26.4
Silver	0.233	0.0558	24.9
Tungsten	0.134	0.0321	24.8
Zinc	0.387	0.0925	25.2
Alcohol (ethyl)	2.4	0.58	111
Mercury	0.140	0.033	28.3
Water	4.18	1.00	75.2

Large bodies of water, such as lakes or oceans, tend to moderate fluctuations of the air temperature nearby because the bodies of water can absorb or release large quantities of thermal energy while undergoing only very small changes in temperature.

We see from Table 18-1 that the specific heat of water is considerably larger than that of other substances. Thus, water is an excellent material for storing thermal energy, as in a solar heating system. It is also an excellent coolant, as in a car engine.

Calorimetry

To measure the specific heat of an object we can first heat it to some known temperature, say the boiling point of water, then transfer it to a water bath of known mass and initial temperature, and, finally, measure the final equilibrium temperature of the object (and the bath). If the system is isolated from its surroundings (by insulating the container, for example), then the heat leaving the object will equal the heat entering the water and its container. This procedure is called **calorimetry,** and the insulated water container is called a **calorimeter.**

Let m be the mass of the object, let c be its specific heat, and let T_{io} be its initial temperature. If T_f is the final temperature of the object in its water bath, the heat leaving the object is

$$Q_{out} = mc(T_{io} - T_f)$$

Similarly, if T_{iw} is the initial temperature of the water and container, and T_f is their final equilibrium temperature, then the heat absorbed by the water and container is

$$Q_{in} = m_w c_w (T_f - T_{iw}) + m_c c_c (T_f - T_{iw})$$

where m_w and $c_w = 4.18\ \mathrm{kJ/kg{\cdot}K}$ are the mass and specific heat of the water, and m_c and c_c are the mass and specific heat of the container. (Note that we have chosen the temperature differences so that the heat in and heat out are both positive quantities.) Setting these amounts of heat equal yields the specific heat c of the object:

$$Q_{out} = Q_{in} \qquad \text{18-7}$$

$$mc\,(T_{io} - T_f) = m_w c_w (T_f - T_{iw}) + m_c c_c (T_f - T_{iw})$$

Because only temperature differences occur in Equation 18-7, and because the kelvin and Celsius degree are the same size, it doesn't matter whether we use kelvins or Celsius degrees.

MEASURING SPECIFIC HEAT **EXAMPLE 18-2**

To measure the specific heat of lead, you heat 600 g of lead shot to 100°C and place it in an aluminum calorimeter of mass 200 g that contains 500 g of water initially at 17.3°C. If the final temperature of the mixture is 20.0°C, what is the specific heat of lead? [The specific heat of the aluminum container is 0.900 kJ/kg·K.]

PICTURE THE PROBLEM We set the heat leaving the lead equal to the heat entering the water and container and solve for the specific heat of lead c_{Pb}.

1. Write the heat leaving the lead in terms of its specific heat: $Q_{Pb} = m_{Pb}c_{Pb}|\Delta T_{Pb}|$

2. Find the heat absorbed by the water: $Q_w = m_w c_w \Delta T_w$

3. Find the heat absorbed by the container: $Q_c = m_c c_c \Delta T_c$

4. Set the heat out equal to the heat in:

$$Q_{Pb} = Q_w + Q_c$$

$$m_{Pb}c_{Pb}|\Delta T_{Pb}| = m_w c_w \Delta T_w + m_c c_c \Delta T_c$$

where

$\Delta T_c = \Delta T_w = 2.7\text{ K}$ and $|\Delta T_{Pb}| = 80\text{ K}$

5. Solve for c_{Pb}:

$$c_{Pb} = \frac{(m_w c_w + m_c c_c)\Delta T_w}{m_{Pb}|\Delta T_{Pb}|}$$

$$= \frac{[(0.5\text{ kg})(4.18\text{ kJ/kg·K}) + (0.2\text{ kg})(0.9\text{ kJ/kg·K})](2.7\text{ K})}{(0.6\text{ kg})(80\text{ K})}$$

$$= \boxed{0.128\text{ kJ/kg·K}}$$

REMARKS Note that the specific heat of lead is considerably less than that of water.

18-2 Change of Phase and Latent Heat

If heat is added to ice at 0°C, the temperature of the ice does not change. Instead, the ice melts. Melting is an example of a **phase change.** Common types of phase changes include fusion (liquid to solid), melting (solid to liquid), vaporization (liquid to vapor or gas), condensation (gas or vapor to liquid), and sublimation (solid directly to vapor, as when solid carbon dioxide [dry ice] changes to vapor). There are other types of phase changes as well, such as the change of a solid from one crystalline form to another. For example, carbon under intense pressure becomes a diamond.

Although melting indicates that the ice has experienced a change in phase, the temperature of the ice does not change.

Molecular theory can help us to understand why temperature remains constant during a phase change. The molecules in a liquid are close together and exert attractive forces on each other, whereas molecules in a gas are far apart. Because of this molecular attraction, it takes energy to remove molecules from a liquid to form a gas. Consider a pot of water sitting over a flame on the stove. At first, as the water is heated, the motion of its molecules increases and the temperature rises. When the temperature reaches the boiling point, the molecules can no longer increase their kinetic energy and remain in the liquid. As the liquid water vaporizes, the added heat energy is used to overcome the attractive forces between the water molecules as they spread farther apart in the gas phase. The added energy thus increases the potential energy of the molecules rather than their kinetic energy. Because temperature is a measure of the average translational *kinetic* energy of molecules, the temperature doesn't change.

For a pure substance, a change in phase at a given pressure occurs only at a particular temperature. For example, pure water at a pressure of 1 atm changes from solid to liquid at 0°C (the normal melting point of water) and from liquid to gas at 100°C (the normal boiling point of water).

The heat energy required to melt a substance of mass m with no change in its temperature is proportional to the mass of the substance:

$$Q_f = mL_f \qquad 18\text{-}8$$

where L_f is called the **latent heat of fusion** of the substance. At a pressure of 1 atm, the latent heat of fusion for water is 333.5 kJ/kg = 79.7 kcal/kg. If the phase change is from liquid to gas, the heat required is

$$Q_v = mL_v \qquad 18\text{-}9$$

where L_v is the **latent heat of vaporization.** For water at a pressure of 1 atm, the latent heat of vaporization is 2.26 MJ/kg = 540 kcal/kg. Table 18-2 gives the normal melting and boiling points, and the latent heats of fusion and vaporization at 1 atm, for various substances.

TABLE 18-2

Normal Melting Point (MP), Latent Heat of Fusion (L_f), Normal Boiling Point (BP), and Latent Heat of Vaporization (L_v) for Various Substances at 1 atm

Substance	MP, K	L_f, kJ/kg	BP, K	L_v, kJ/kg
Alcohol, ethyl	159	109	351	879
Bromine	266	67.4	332	369
Carbon dioxide	—	—	194.6†	573†
Copper	1356	205	2839	4726
Gold	1336	62.8	3081	1701
Helium	—	—	4.2	21
Lead	600	24.7	2023	858
Mercury	234	11.3	630	296
Nitrogen	63	25.7	77.35	199
Oxygen	54.4	13.8	90.2	213
Silver	1234	105	2436	2323
Sulfur	388	38.5	717.75	287
Water	273.15	333.5	373.15	2257
Zinc	692	102	1184	1768

† These values are for sublimation. Carbon dioxide does not have a liquid state at 1 atm.

CHANGING ICE INTO STEAM — EXAMPLE 18-3 Try It Yourself

How much heat is needed to change 1.5 kg of ice at −20 C° and 1 atm into steam?

PICTURE THE PROBLEM The heat required to change the ice into steam consists of four parts: Q_1, the heat needed to warm the ice from −20°C to 0°C; Q_2, the heat needed to melt the ice; Q_3, the heat needed to warm the water from 0°C to 100°C; and Q_4, the heat needed to vaporize the water. In calculating Q_1 and Q_3, we will assume that the specific heats are constant, with the values 2.05 kJ/kg·K for ice and 4.18 kJ/kg·K for water.

Cover the column to the right and try these on your own before looking at the answers.

Steps	Answers
1. Use $Q_1 = mc\,\Delta T$ to find the heat needed to warm the ice to 0°C.	$Q_1 = 61.5\text{ kJ} = 0.0615\text{ MJ}$

2. Use L_f from Table 18-2 to find the heat Q_2 needed to melt the ice. $Q_2 = 500\text{ kJ} = 0.500\text{ MJ}$

3. Find the heat Q_3 needed to warm the water from 0°C to 100°C. $Q_3 = 627\text{ kJ} = 0.627\text{ MJ}$

4. Use L_v from Table 18-2 to find the heat Q_4 needed to vaporize the water. $Q_4 = 3.39\text{ MJ}$

5. Sum your results to find the total heat Q. $Q = Q_1 + Q_2 + Q_3 + Q_4 = \boxed{4.58\text{ MJ}}$

REMARKS Notice that most of the heat was needed to vaporize the water, and that the amount needed to melt the ice was almost as much as that needed to raise the temperature of the water by 100 C°. A graph of temperature versus time for the case in which the heat is added at a constant rate of 1 kJ/s is shown in Figure 18-1. Note that it takes considerably longer to vaporize the water than it does to melt the ice or to raise the temperature of the water. When all of the water has vaporized, the temperature again rises as heat is added.

EXERCISE An 830-g piece of lead is heated to its melting point of 600 K. How much additional heat energy must be added to melt the lead? (*Answer* 20.5 kJ)

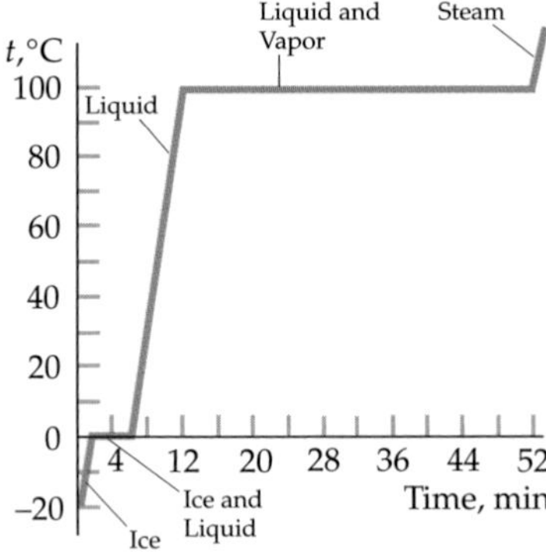

FIGURE 18-1

A COOL DRINK **EXAMPLE 18-4** **Put It in Context**

A 2-liter pitcher of lemonade has been sitting on the picnic table in the sun all day at 33°C. You pour 0.24 kg into a Styrofoam cup and add 2 ice cubes (each 0.025 kg at 0°C). (*a*) Assuming no heat lost to the surroundings, what is the final temperature of the lemonade? (*b*) What is the final temperature if you add 6 ice cubes?

PICTURE THE PROBLEM We set the heat lost by the lemonade equal to the heat gained by the ice cubes. Let T_f be the final temperature of the lemonade and water. We assume that lemonade has the same specific heat as water.

(*a*) 1. Write the heat lost by the lemonade in terms of the final temperature T_f: $Q_{out} = m_L c|\Delta T| = m_L c(T_{Li} - T_f)$

2. Write the heat gained by the ice cubes and resulting water in terms of the final temperature: $Q_{in} = m_{ice}L_f + m_{ice}c\Delta T_w = m_{ice}L_f + m_{ice}cT_f$

3. Set the heat lost equal to the heat gained and solve for T_f:

$$Q_{out} = Q_{in}$$

$$m_L c(T_{Li} - T_f) = m_{ice}L_f + m_{ice}cT_f$$

so

$$T_f = \frac{m_L c T_{Li} - m_{ice}L_f}{(m_L + m_{ice})c}$$

$$= \frac{(0.24\text{ kg})(4.18\text{ kJ/(kg}\cdot\text{C°}))(33\text{°C}) - (0.05\text{ kg})(333.5\text{ kJ/kg})}{(0.29\text{ kg})(4.18\text{ kJ/(kg}\cdot\text{C°}))}$$

$$= \boxed{13.6\text{°C}}$$

(*b*) 1. For 6 ice cubes, $m_{\text{ice}} = 0.15$ kg. Find the final temperature as in step 3 of Part (*a*):

$$T_f = \frac{m_L c T_{Li} - m_{\text{ice}} L_f}{(m_L + m_{\text{ice}})c}$$

$$= \frac{(0.24\ \text{kg})(4.18\ \text{kJ/kg}\cdot\text{C}^\circ)(33^\circ\text{C}) - (0.15\ \text{kg})(333.5\ \text{kJ/kg})}{(0.39\ \text{kg})(4.18\ \text{kJ/kg}\cdot\text{C}^\circ)}$$

$$= -10.4^\circ\text{C}$$

2. This cannot be correct! No amount of ice at 0°C can lower the temperature of warm lemonade to below 0°C. Our calculation is wrong because our assumption in step 2 of Part (*a*) that all of the ice melts was wrong. Instead, the heat given off by the lemonade as it cools from 32°C to 0°C is not enough to melt all of the ice. The final temperature is thus:

$$T_f = \boxed{0^\circ\text{C}}$$

PLAUSIBILITY CHECK Let's calculate how much ice is melted. For the lemonade to cool from 33°C to 0°C, it must give off heat in the amount $Q_{\text{out}} = (0.24\ \text{kg})(4.18\ \text{kJ/kg}\cdot\text{C}^\circ)(33^\circ\text{C}) = 33.1$ kJ. The mass of ice that this amount of heat will melt is $m_{\text{ice}} = Q_{\text{in}}/L_f = 33.1\ \text{kJ}/(333.5\ \text{kJ/kg}) = 0.10$ kg. This is the mass of only 4 ice cubes. Adding more than 4 ice cubes does not lower the temperature below 0°C. It merely increases the amount of ice in the ice-lemonade mixture. In problems like this one, we should first find out how much ice must be melted to reduce the temperature of the liquid to 0°C. If less than that amount is added, we can proceed as in Part (*a*). If more ice is added, the final temperature is 0°C.

18-3 Joule's Experiment and the First Law of Thermodynamics

We can raise the temperature of a system by adding heat, but we can also raise its temperature by doing work on it.

Figure 18-2 is a diagram of the apparatus Joule used in a famous experiment in which he determined the amount of work needed to raise the temperature of 1 g of water by 1 C°. Here the system is a thermally insulated container of 1 g of water. Joule's apparatus converts the potential energy of falling weights into work done on the water by an attached paddle, as shown in the figure. Joule found that he could raise the temperature of his water sample by 1 F° by dropping 772 pounds of attached weights a distance of one foot. Converting to modern units, Joule found that it takes about 4.184 J (the energy units adopted by the scientific community in 1948) to raise the temperature of 1 g of water by 1 C°. The result that 4.184 J of mechanical energy is equivalent to 1 cal of heat energy is known as the **mechanical equivalence of heat.**

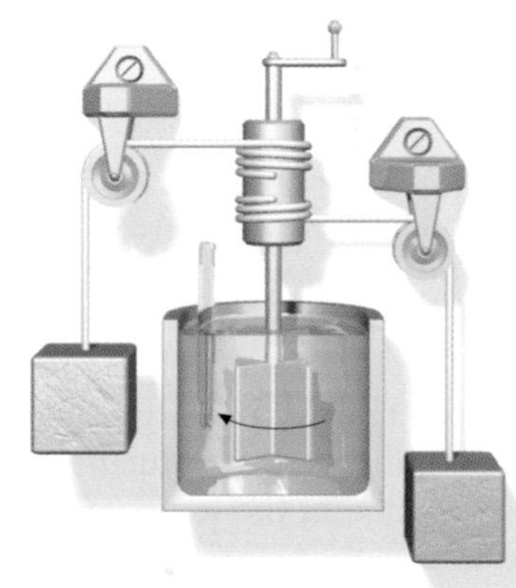

FIGURE 18-2 Schematic diagram for Joule's experiment. Insulating walls to prevent heat transfer enclose water. As the weights fall at constant speed, they turn a paddle wheel, which does work on the water. If friction is negligible, the work done by the paddle wheel on the water equals the loss of mechanical energy of the weights, which is determined by calculating the loss in the potential energy of the weights.

There are other ways of doing work on this system. For example, we could drop the insulated container of water from some height *h*, letting the system make an inelastic collision with the ground, or we could do mechanical work to generate electricity and then use the electricity to heat the water (Figure 18-3). In all such experiments, the same amount of work is required to produce a given temperature change. By the conservation of energy, the work done equals the increase in the internal energy of the system.

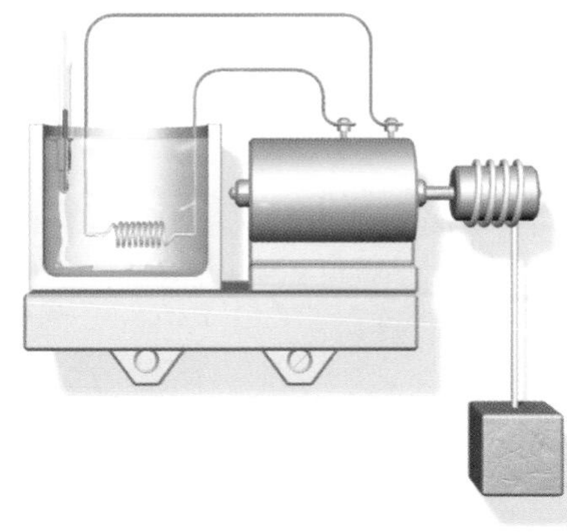

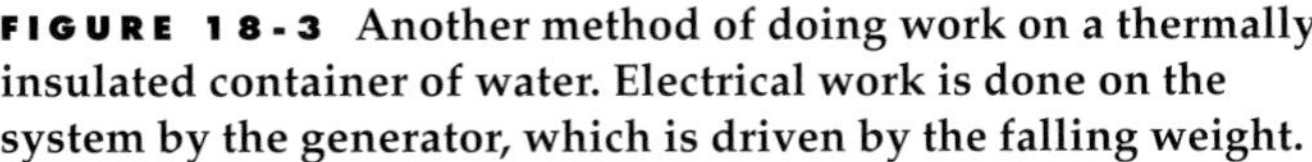

FIGURE 18-3 Another method of doing work on a thermally insulated container of water. Electrical work is done on the system by the generator, which is driven by the falling weight.

HEATING WATER BY DROPPING IT — EXAMPLE 18-5

You drop a thermally insulated container of water from a height h to the ground. If the collision is perfectly inelastic and all of the mechanical energy lost goes into the internal energy of the water, what must h be for the temperature of the water to increase by 1 C°?

PICTURE THE PROBLEM The kinetic energy of the water just before it hits the ground equals its original potential energy mgh. During the collision, this energy is converted into thermal energy Q, which in turn causes a rise in temperature given by $Q = mc\Delta T$.

1. Set the potential energy equal to the thermal energy: $mgh = mc\Delta T$

2. Solve for the height h: $h = \frac{c\Delta T}{g}$

3. Substitute $c = 4.18$ kJ/kg·K and $\Delta T = 1$ C° $= 1$ K: $h = \frac{(4.18\text{ kJ/kg·K})(1\text{ K})}{9.81\text{ N/kg}} = 0.426\text{ km}$

$= \boxed{426\text{ m}}$

REMARKS Note that h is independent of the mass of the water. It is also rather large, which illustrates one of the difficulties with Joule's experiment—a large amount of work must be done to produce a measurable change in the temperature of the water.

Suppose that we perform Joule's experiment but replace the insulating walls of the container with conducting walls. We find that the work needed to produce a given change in the temperature of the system depends on how much heat is added to or subtracted from the system by conduction through the walls. However, if we sum the work done on the system and the net heat added to the system, the result is always the same for a given temperature change. That is, the sum of the heat transfer *into* the system and the work done *on* the system equals the change in the internal energy of the system. This is the **first law of thermodynamics,** which is simply a statement of the conservation of energy.

Let W_{on} stand for the work done by the surroundings *on* the system, and let W_{by} stand for the work done *by* the system on its surroundings. For example, suppose our system is a gas confined to a cylinder by a piston. If the piston compresses the gas, the surroundings do positive work on the gas, and W_{on} is positive. (However, if the gas expands against the piston, the surroundings do negative work on the gas, and W_{on} is negative.) Also, let Q_{in} stand for the heat transferred into the system. If heat is transferred into the system then Q_{in} is positive; if heat is transferred out of the system then Q_{in} is negative (Figure 18-4). Using these conventions, and denoting the internal energy by ΔE_{int}, the first law of thermodynamics is written

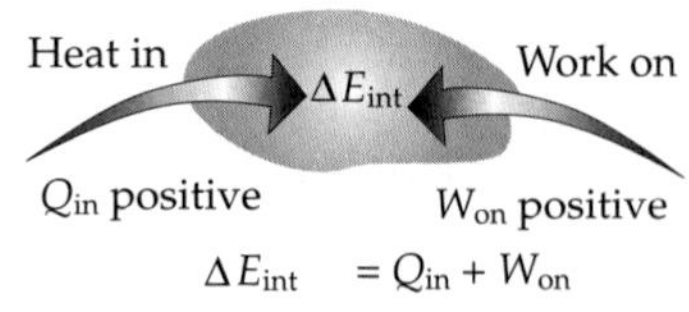

FIGURE 18-4 Sign convention for the first law of thermodynamics.

$$\Delta E_{int} = Q_{in} + W_{on} \qquad 18\text{-}10$$

The change in the internal energy of the system equals the heat transferred into the system plus the work done on the system.

FIRST LAW OF THERMODYNAMICS

Equation 18-10 is the same as the work energy theorem $W_{ext} = \Delta E_{sys}$ of Chapter 7 (Equation 7-9), except we have added the heat term Q_{in} and called the energy of the system ΔE_{int}.

STIRRING THE WATER **EXAMPLE 18-6**

You do 25 kJ of work on a system consisting of 3 kg of water by stirring it with a paddle wheel. During this time, 15 kcal of heat leaves the system due to poor insulation. What is the change in the internal energy of the system?

PICTURE THE PROBLEM We express all energies in joules and apply the first law of thermodynamics.

1. ΔE_{int} is found from the first law of thermodynamics: $\Delta E_{\text{int}} = Q_{\text{in}} + W_{\text{on}}$

2. Heat is *removed* from the system, thus the heat *added* is negative: $Q_{\text{in}} = -15\text{ kcal} = -(15\text{ kcal})\left(\frac{4.18\text{ kJ}}{1\text{ kcal}}\right)$

$= -62.7\text{ kJ}$

3. The work done *on* the system is positive, thus: $W_{\text{on}} = 25\text{ kJ}$

4. Substitute these quantities and solve for ΔE_{int}: $\Delta E_{\text{int}} = Q_{\text{in}} + W_{\text{on}} = (-62.7\text{ kJ}) + (25\text{ kJ})$

$= \boxed{-37.7\text{ kJ}}$

REMARKS The internal energy decreases because the heat loss exceeds the work gain.

It is important to understand that the internal energy E_{int} is a function of the state of the system, just as P, V, and T are functions of the state of the system. Consider a gas in some initial state (P_i, V_i). The temperature T_i is determined by the equation of state. For example, if the gas is ideal, the $T_i = P_iV_i/(nR)$. The internal energy $E_{\text{int,i}}$ also depends only on the state of the gas, which is determined by any two state variables such as P and V, P and T, or V and T. If we compress the gas or let it expand, add to or remove heat from it, do work on it or let it do work, the gas will move through a sequence of states; that is, it will have different values of the state functions P, V, T, and E_{int}. If the gas is then returned to its original state (P_i, V_i), the temperature T and the internal energy E_{int} must equal their original values.

On the other hand, the net heat input Q and the work W done by the gas are not functions of the state of the system. There are no functions Q or W associated with any particular state of the gas. We could take the gas through a sequence of states beginning and ending at state (P_i, V_i) during which the gas did positive work and absorbed an equal amount of heat. Or we could take it though a different sequence of states such that work was done on the gas and heat was removed from the gas. **It is correct to say that a system has a large amount of internal energy, but it is not correct to say that a system has a large amount of heat or a large amount of work.** Heat is not something that is contained in a system. Rather, it is a measure of the energy that flows from one system to another because of a difference in temperature. Work is a measure of the energy that flows from one system to another because the point of contact of a force exerted by one system on the other undergoes a displacement not perpendicular to the force.

For very small amounts of heat added, work done, or changes in internal energy, it is customary to write Equation 18-10 as

$$dE_{\text{int}} = dQ_{\text{in}} + dW_{\text{on}} \tag{18-11}$$

In this equation, dE_{int} is the differential of the internal-energy function. However, neither dQ_{in} nor dW_{on} is a differential of any function. Instead, dQ_{in} merely represents a small amount of heat added to the system, and dW_{on} represents a small amount of work done by the system.

18-4 The Internal Energy of an Ideal Gas

The translational kinetic energy K of the molecules in an *ideal* gas is related to the absolute temperature T by Equation 17-22 in Chapter 17:

$$K = \tfrac{3}{2}nRT$$

where n is the number of moles of gas and R is the universal gas constant. If the internal energy of a gas is just this translational kinetic energy, then $E_{int} = K$, and

$$E_{int} = \tfrac{3}{2}nRT \qquad 18\text{-}12$$

Then the internal energy will depend only on the temperature of the gas, and not on its volume or pressure. If the molecules have other types of energy in addition to translational kinetic energy, such as energy of rotation, the internal energy will be greater than that given by Equation 18-12. But according to the equipartition theorem (Chapter 17, Section 5) the average energy associated with any degree of freedom will be $\frac{1}{2}RT$ per mole ($\frac{1}{2}kT$ per molecule), so again, the internal energy will depend only on the temperature and not on the volume or pressure.

We can imagine that the internal energy of a *real* gas might include other kinds of energy, which depend on the pressure and volume of the gas. Suppose, for example, that nearby gas molecules exert attractive forces on each other. Work is then required to increase the separation of the molecules. Then, if the average distance between the molecules is increased, the potential energy associated with the molecular attraction will increase. The internal energy of the gas will then depend on the volume of the gas as well as on its temperature.

Joule, using an apparatus like the one shown in Figure 18-5, performed a simple but interesting experiment to determine whether or not the internal energy of a gas depends on its volume. Initially, the compartment on the left in Figure 18-5 contains a gas and the compartment on the right has been evacuated. A stopcock that is initially closed connects the two compartments. The whole system is thermally insulated from its surroundings by rigid walls so that no heat can go into or out of the system *and* no work can be done. When the stopcock is opened, the gas rushes into the evacuated chamber. This process is called a **free expansion.** Eventually, the gas reaches thermal equilibrium with itself. Since no work has been done on the gas and no heat has been transferred to it, the final internal energy of the gas must equal its initial internal energy. If the gas molecules exert attractive forces on one another, the potential energy associated with these forces will increase as the volume increases. Since energy is conserved, the kinetic energy of translation must therefore decrease, which will result in a decrease in the temperature of the gas.

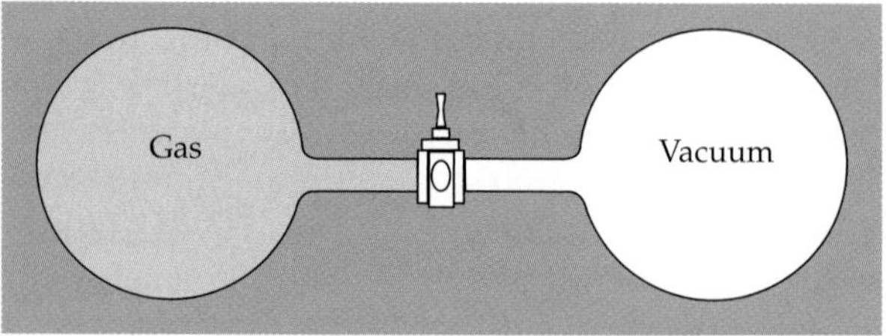

FIGURE 18-5 Free expansion of a gas. When the stopcock on the gas is opened, the gas expands rapidly into the evacuated chamber. Because no work is done on the gas and the whole system is thermally insulated, the initial and final internal energies of the gas are equal.

When Joule did this experiment, he found the final temperature to be equal to the initial temperature. Subsequent experiments verified this result for low gas densities. This implies that for a gas at low density—that is, for an ideal gas—the temperature depends only on the internal energy, or as we usually think of it, the internal energy depends only on the temperature. However, if the experiment is done with a large amount of gas initially in the left compartment so that the density is high, then the temperature after expansion is slightly lower than the temperature before the expansion. This indicates that there is a small attraction between the gas molecules of a real gas.

18-5 Work and the *PV* Diagram for a Gas

In many types of engines, a gas expanding against a moveable piston does work. For example, in a steam engine, water is heated in a boiler to produce steam. The steam then does work as it expands and drives a piston. In an automobile engine, a mixture of gasoline vapor and air is ignited, causing it to explode. The resulting high temperatures and pressures cause the gas to expand rapidly, driving a piston and doing work. In this section, we will see how we can mathematically describe the work done by an expanding gas.

Quasi-Static Processes

Figure 18-6 shows an ideal gas confined in a container with a tightly fitting piston that we assume to be frictionless. If the piston moves, the volume of the gas changes. The temperature or pressure or both must also change since these three variables are related by the equation of state $PV = nRT$. If we suddenly push in the piston to compress the gas, the pressure will initially be greater near the piston than far from it. Eventually the gas will settle down to a new equilibrium pressure and temperature. We cannot determine such macroscopic variables as T, P, or E_{int} for the entire gas system until equilibrium is restored in the gas. However, if we move the piston slowly in small steps and allow equilibrium to be reestablished after each step, we can compress or expand the gas in such a way that the gas is never far from an equilibrium state. In this kind of process, called a **quasi-static process,** the gas moves through a series of equilibrium states. In practice, it is possible to approximate quasi-static processes fairly well.

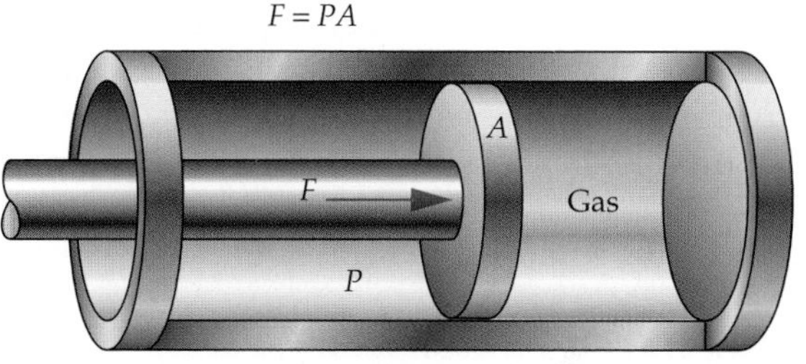

FIGURE 18-6 Gas confined in a thermally insulated cylinder with a movable piston. If the piston moves a distance dx, the volume of the gas changes by $dV = A\,dx$. The work done by the gas is $PA\,dx = P\,dV$, where P is the pressure.

Let us begin with a gas at a high pressure, and let it expand quasi-statically. The magnitude of the force F exerted by the gas on the piston is PA, where A is the area of the piston and P is the gas pressure. As the piston moves a small distance dx, the work done *by* the gas on the piston is

$$dW_{\text{by gas}} = F_x dx = PA dx = P dV \qquad \text{18-13}$$

where $dV = A\,dx$ is the increase in the volume of the gas. During the expansion the piston exerts a force of magnitude PA on the gas, but opposite in direction to the force of the gas on the piston. Thus, work done by the piston *on* the gas is just the negative of the work done *by* the gas

$$dW_{\text{on gas}} = -dW_{\text{by gas}} = -P dV \qquad \text{18-14}$$

Note that for an expansion, dV is positive, the gas does work on the piston, so $dW_{\text{on gas}}$ is negative, and for a compression, dV is negative, work is done on the gas, so $dW_{\text{on gas}}$ is positive.

The work done on the gas during an expansion or compression from a volume of V_i to a volume of V_f is

$$W_{\text{on gas}} = -\int_{V_i}^{V_f} P dV \qquad \text{18-15}$$

WORK DONE ON A GAS

To calculate this work we need to know how the pressure varies during the expansion or compression. The various possibilities can be illustrated most easily using a *PV* diagram.

PV Diagrams

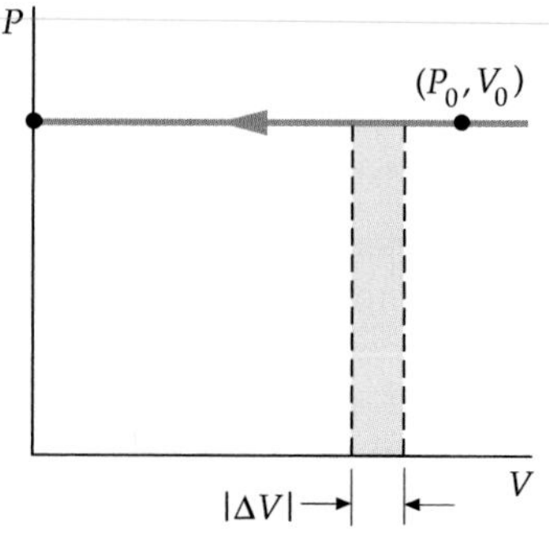

FIGURE 18-7 Each point on a *PV* diagram, such as (P_0, V_0), represents a particular state of the gas. The horizontal line represents states with a constant pressure P_0. The shaded area, $P_0\,|\Delta V|$, represents the work done on the gas as it is compressed an amount $|\Delta V|$.

We can represent the states of a gas on a diagram of P versus V. Because by specifying both P and V we specify the state of the gas, each point on the PV diagram indicates a particular state of the gas. Figure 18-7 shows a PV diagram with a directed horizontal line representing a series of states that all have the same value of P. This line represents a *compression* at constant pressure. Such a process is called an **isobaric compression.** For a volume change of ΔV (ΔV is negative for a compression), we have

$$W_{\text{on}} = -\int_{V_i}^{V_f} P\,dV = -P\int_{V_i}^{V_f} dV = -P\Delta V$$

which is equal to the shaded area under the curve (directed line) in the figure. In general, for a compression the work done on the gas is equal to the area under the P-versus-V curve. (For an expansion the work done on the gas is equal to the negative of the area under the P-versus-V curve.) Because pressures are often given in atmospheres and volumes are often given in liters, it is convenient to have a conversion factor between liter-atmospheres and joules:

$$1\ \text{L}\cdot\text{atm} = (10^{-3}\ \text{m}^3)(101.3 \times 10^3\ \text{N/m}^2) = 101.3\ \text{J} \qquad 18\text{-}16$$

EXERCISE If 5 L of an ideal gas at a pressure of 2 atm is cooled so that it contracts at constant pressure until its volume is 3 L, what is the work done on the gas? (*Answer* 405.2 J)

Figure 18-8 shows three different possible paths on a PV diagram for a gas that is initially in state (P_i, V_i) and is finally in state (P_f, V_f). We assume that the gas is ideal and have chosen the original and final states to have the same temperature so that $P_iV_i = P_fV_f = nRT$. Since the internal energy depends only on the temperature, the initial and final internal energies are the same also.

In Figure 18-8*a*, the gas is heated at constant volume until its pressure is P_f, after which it is cooled at constant pressure until its volume is V_f. The work done on the gas along the constant-volume (vertical) part of path A is zero; along the constant pressure (horizontal) part of the path A, it is $P_f\,|V_f - V_i| = -P_f\,(V_f - V_i)$.

In Figure 18-8*b*, the gas is first cooled at constant pressure until its volume is V_f, after which it is heated at constant volume until its pressure is P_f. The work done on the gas along this path is $P_i|V_f - V_i| = -P_i(V_f - V_i)$, which is much less than that done along the path shown in Figure 18-8*a* as can be seen by comparing the shaded regions in Figure 18-8*a* and Figure 18-8*b*.

In Figure 18-8*c*, path C represents an **isothermal** compression, meaning that the temperature remains constant. (Keeping the temperature constant during the compression requires that heat be transferred out of the gas during the

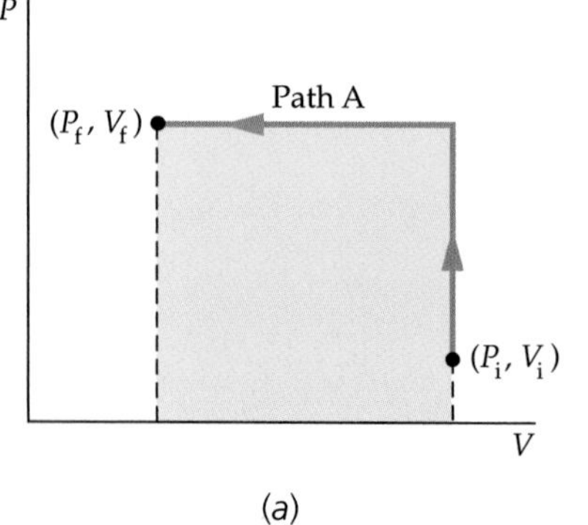

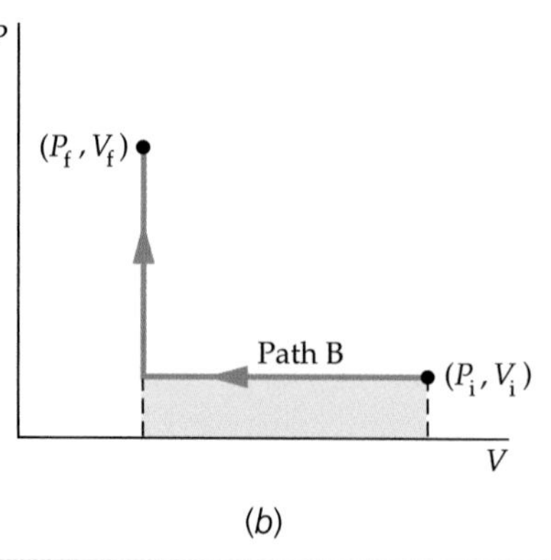

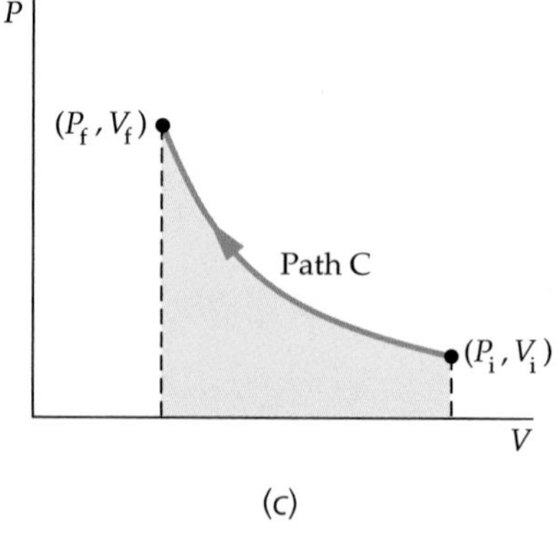

FIGURE 18-8 Three paths on *PV* diagrams connecting an initial state (P_i, V_i) and a final state (P_f, V_f). The corresponding shaded area indicates the work done on the gas along each path.

compression.) We can calculate the work done on the gas along path C by using $P = nRT/V$. Hence, the work done on the gas as it is compressed from V_i to V_f is

$$W_{on} = -\int_{V_i}^{V_f} P\,dV = -\int_{V_i}^{V_f} \frac{nRT}{V}\,dV$$

Since T is constant for an isothermal process, we can remove it from the integral. We then have

$$W_{isothermal} = -nRT\int_{V_i}^{V_f} \frac{dV}{V} = -nRT\ln\frac{V_f}{V_i} = nRT\ln\frac{V_i}{V_f} \qquad 18\text{-}17$$

WORK DONE ON GAS DURING ISOTHERMAL COMPRESSION

We see that the amount of work done on the gas is different for each process illustrated. Since for these states $E_{int,f} = E_{int,i}$, the net amount of heat added must also be different for each of the processes. This discussion illustrates the fact that both the work done and the heat added depend only on how a system moves from one state to another, but the change in the internal energy of the system does not.

WORK DONE ON AN IDEAL GAS — EXAMPLE 18-7

An ideal gas undergoes a cyclic process from point *A* to point *B* to point *C* to point *D* and back to point *A* as shown in Figure 18-9. The gas begins at a volume of 1 L and a pressure of 2 atm and expands at constant pressure until the volume is 2.5 L, after which it is cooled at constant volume until its pressure is 1 atm. It is then compressed at constant pressure until its volume is again 1 L, after which it is heated at constant volume until it is back in its original state. Find the total work done on the gas and the total heat added to it during the cycle.

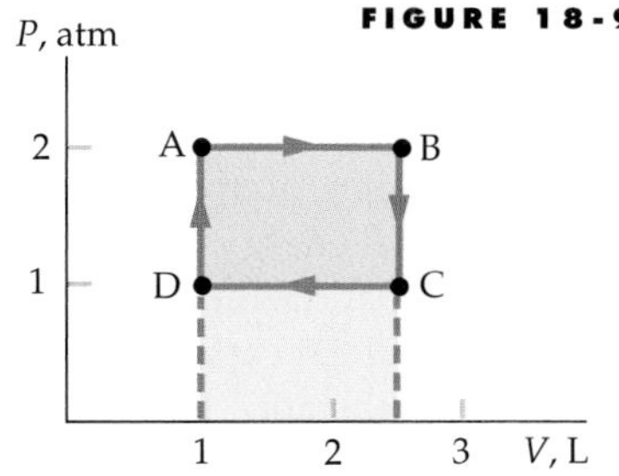

FIGURE 18-9

PICTURE THE PROBLEM We calculate the work done during each step. Since $\Delta E_{int} = 0$ for any complete cycle, the first law of thermodynamics implies that the total heat added to the gas plus the total work done on the gas equals zero.

1. From *A* to *B* the process is an isobaric (constant pressure) expansion, so the work done on the gas is negative. The work done on the gas equals the negative of the shaded area under the *AB* curve, shown in Figure 18-10*a*:

$$W_{AB} = -P\Delta V = -P(V_B - V_A) = -(2\text{ atm})(2.5\text{ L} - 1\text{ L}) = -3\text{ L}\cdot\text{atm}$$

2. Convert the units to joules:

$$W_{AB} = -3\text{ L}\cdot\text{atm} \times \frac{101.3\text{ J}}{1\text{ L}\cdot\text{atm}} = -304\text{ J}$$

3. From *B* to *C* (Figure 18-9) the gas cools at constant volume so the work done is zero:

$$W_{BC} = 0$$

4. As the gas undergoes an isobaric compression from *C* to *D*, the work done on it is positive. This work equals the area under the *CD* curve, shown in Figure 18-10*b*:

$$W_{CD} = -P\Delta V = -P(V_D - V_C) = -(1\text{ atm})(1\text{ L} - 2.5\text{ L}) = 1.5\text{ L}\cdot\text{atm} = \boxed{152\text{ J}}$$

(a)

P, atm
W_{on} = −AREA = −3 L·atm
AREA = (1.5 L)(2 atm) = 3 L·atm

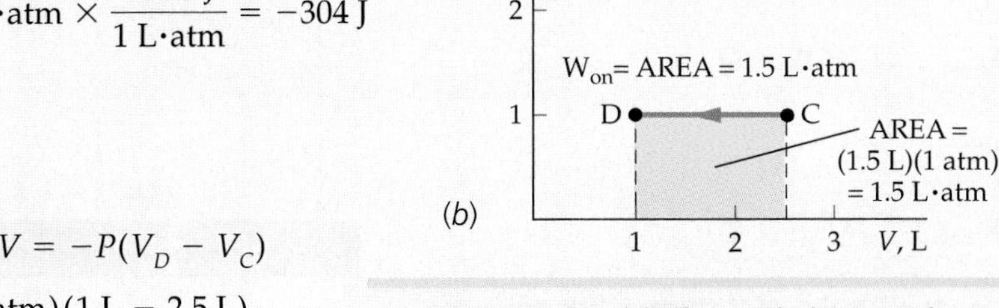

FIGURE 18-10 (*a*) The work done on the gas during the expansion from *A* to *B* is equal to the negative of the area under the curve. (*b*) The work done on the gas during the compression from *C* to *D* is equal to the area under the curve.

5. As the gas is heated back to its original state A, the volume is again constant (Figure 18-9), so no work is done:
   $W_{DA} = 0$

6. The total work done by the gas is the sum of the work done along each step:
   $$W_{\text{total}} = W_{AB} + W_{BC} + W_{CD} + W_{DA}$$
   $$= (-304\text{ J}) + 0 + 152\text{ J} + 0 = \boxed{-152\text{ J}}$$

7. Because the gas is back in its original state, the total change in internal energy is zero:
   $\Delta E_{\text{int}} = 0$

8. The heat added is found from the first law:
   $$\Delta E_{\text{int}} = Q_{\text{in}} + W_{\text{on}}$$
   so
   $$Q_{\text{in}} = \Delta E_{\text{int}} - W_{\text{on}} = 0 - (-152\text{ J}) = \boxed{152\text{ J}}$$

REMARKS The work done by the gas equals the negative of the work done on the gas, so the total work done by the gas during the cycle is +152 J. During the cycle the gas extracts 152 J of heat from its surroundings and does 152 J of work on its surroundings. This leaves the gas in its initial state. The total work done by the gas equals the area enclosed by the cycle in Figure 18-9. Such cyclic processes have important applications for heat engines, as we will see in Chapter 20.

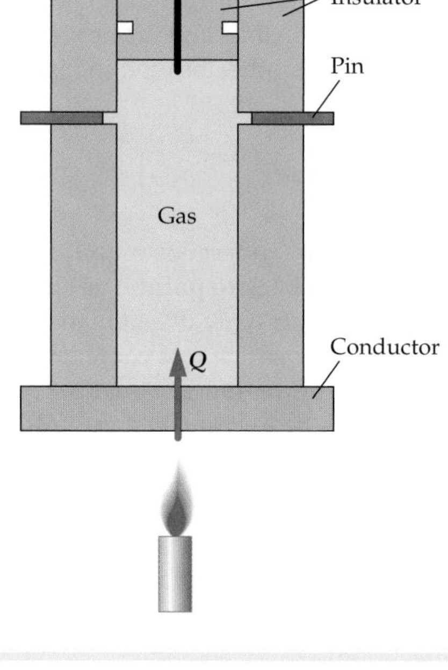

FIGURE 18-11 Heat is added and the pressure remains constant. The gas expands, thus doing positive work on the piston.

18-6 Heat Capacities of Gases

The determination of the heat capacity of a substance provides information about its internal energy, which is related to its molecular structure. For all substances that expand when heated, the heat capacity at constant pressure C_p is greater than the heat capacity at constant volume C_v. If heat is added at constant pressure, the substance expands and does positive work on its surroundings (Figure 18-11). Therefore, it takes more heat to obtain a given temperature change at constant pressure than to obtain the same temperature change when heated at constant volume. The expansion is usually negligible for solids and liquids, so for them $C_p \approx C_v$. But a gas heated at constant pressure readily expands and does a significant amount of work, so $C_p - C_v$ is not negligible.

If heat is added to a gas at constant volume, no work is done (Figure 18-12), so the heat added equals the increase in the internal energy of the gas. Writing Q_v for the heat added at constant volume, we have

$$Q_v = C_v\,\Delta T$$

Since $W = 0$, we have from the first law of thermodynamics

$$\Delta E_{\text{int}} = Q_v + W = Q_v$$

Thus,

$$\Delta E_{\text{int}} = C_v \Delta T$$

Taking the limit as ΔT approaches zero, we obtain

$$dE_{\text{int}} = C_v\,dT \qquad \text{18-18}a$$

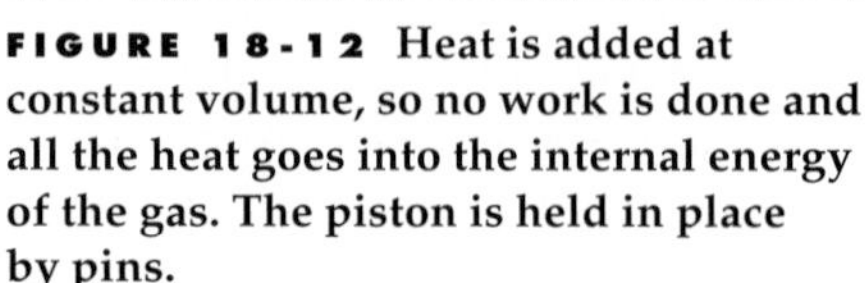

FIGURE 18-12 Heat is added at constant volume, so no work is done and all the heat goes into the internal energy of the gas. The piston is held in place by pins.

and

$$C_v = \frac{dE_{int}}{dT} \qquad \text{18-18}b$$

The heat capacity at constant volume is the rate of change of the internal energy with temperature. Since E_{int} and T are both state functions, Equations 18-18*a* and 18-18*b* hold for any process.

Now let's calculate the difference $C_p - C_v$ for an ideal gas. From the definition of C_p, the heat added at constant pressure is

$$Q_p = C_p\,\Delta T$$

From the first law of thermodynamics,

$$\Delta E_{int} = Q_p + W_{on} = Q_p - P\Delta V$$

Then

$$\Delta E_{int} = C_p\,\Delta T - P\Delta V \quad \text{or} \quad C_p\,\Delta T = \Delta E_{int} + P\Delta V$$

For infinitesimal changes, this becomes

$$C_p\,dT = dE_{int} + P\,dV$$

Using Equation 18-18*a* for dE_{int}, we obtain

$$C_p\,dT = C_v\,dT + P dV \qquad \text{18-19}$$

The pressure, volume, and temperature for an ideal gas are related by

$$PV = nRT$$

Taking the differentials of both sides of the ideal-gas law, we obtain

$$P\,dV + V\,dP = nR\,dT$$

For a constant-pressure process $dP = 0$, so

$$P\,dV = nR\,dT$$

Substituting this into Equation 18-19 gives

$$C_p\,dT = C_v\,dT + nR\,dT = (C_v + nR)\,dT$$

Therefore,

$$C_p = C_v + nR \qquad \text{18-20}$$

which shows that, for an ideal gas, the heat capacity at constant pressure is greater than the heat capacity at constant volume by the amount nR.

Table 18-3 lists measured molar heat capacities c'_p and c'_v for several gases. Note from this table that the ideal gas prediction, $c'_p - c'_v = R$, holds quite well for all gases. The table also shows that c'_v is approximately 1.5R for all monatomic gases, 2.5R for all diatomic gases, and greater than 2.5R for gases consisting of more complex molecules. We can understand these results by considering the molecular model of a gas (Chapter 17.) The total translational kinetic energy of n moles of a gas is $K = \frac{3}{2}nRT$ (Equation 17-22). Thus, if the internal energy of a gas consists of translational kinetic energy only, we have

$$E_{\text{int}} = \tfrac{3}{2}nRT \qquad \text{18-21}$$

The heat capacities are then

$$C_v = \frac{dE_{\text{int}}}{dT} = \frac{3}{2}nR \qquad \text{18-22}$$

C_v FOR AN IDEAL MONATOMIC GAS

and

$$C_p = C_v + nR = \tfrac{5}{2}nR \qquad \text{18-23}$$

C_p FOR AN IDEAL MONATOMIC GAS

The results in Table 18-3 agree well with these predictions for monatomic gases, but for other gases, the heat capacities are greater than those predicted by Equations 18-22 and 18-23. The internal energy for a gas consisting of diatomic or more complicated molecules is evidently greater than $\frac{3}{2}nRT$. The reason is that such molecules can have other types of energy, such as rotational or vibrational energy, in addition to translational kinetic energy.

TABLE 18-3

Molar Heat Capacities in J/mol·K of Various Gases at 25°C

Gas	c'_p	c'_v	c'_v/R	$c'_p - c'_v$	$(c'_p - c'_v)/R$
Monatomic					
He	20.79	12.52	1.51	8.27	0.99
Ne	20.79	12.68	1.52	8.11	0.98
Ar	20.79	12.45	1.50	8.34	1.00
Kr	20.79	12.45	1.50	8.34	1.00
Xe	20.79	12.52	1.51	8.27	0.99
Diatomic					
N_2	29.12	20.80	2.50	8.32	1.00
H_2	28.82	20.44	2.46	8.38	1.01
O_2	29.37	20.98	2.52	8.39	1.01
CO	29.04	20.74	2.49	8.30	1.00
Polyatomic					
CO_2	36.62	28.17	3.39	8.45	1.02
N_2O	36.90	28.39	3.41	8.51	1.02
H_2S	36.12	27.36	3.29	8.76	1.05

HEATING, COOLING, AND COMPRESSING AN IDEAL GAS — EXAMPLE 18-8 Try It Yourself

A system consisting of 0.32 mol of a monatomic ideal gas with $c'_v = \frac{3}{2}RT$ occupies a volume of 2.2 L at a pressure of 2.4 atm, as represented by point A in Figure 18-13. The system is carried through a cycle consisting of three processes:

1. **The gas is heated at constant pressure until its volume is 4.4 L at point B.**
2. **The gas is cooled at constant volume until the pressure decreases to 1.2 atm (point C).**
3. **The gas undergoes an isothermal compression back to point A.**

(a) What is the temperature at points A, B, and C? (b) Find W, Q, and ΔE_{int} for each process and for the entire cycle.

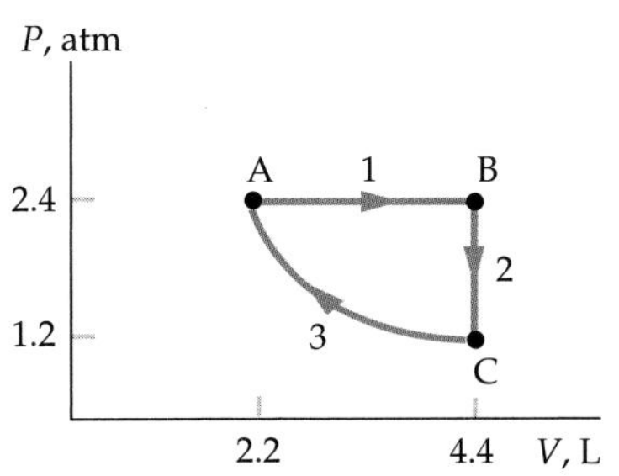

FIGURE 18-13

PICTURE THE PROBLEM You can find the temperatures at all points from the ideal-gas law. You can find the work for each process by finding the area under the curve, and the heat exchanged from the given heat capacity and the initial and final temperatures for each process. In process 3, T is constant, so $\Delta E_{int} = 0$ and the heat input plus the work done on the gas equals zero.

Cover the column to the right and try these on your own before looking at the answers.

Steps	Answers
(a) Find the temperatures at points A, B, and C using the ideal-gas law.	$T_A = T_C = \boxed{201\text{ K}}$, $T_B = \boxed{402\text{ K}}$
(b) 1. For process 1, use $W_1 = -P_c\Delta V$ to calculate the work, and $C_p = \frac{5}{2}nR$ to calculate the heat Q_1. Then use W_1 and Q_1 to calculate $\Delta E_{int,1}$.	$W_1 = -5.28\text{ L·atm} = \boxed{-535\text{ J}}$, $Q_1 = \boxed{1337\text{ J}}$ $\Delta E_{int,1} = Q_1 + W_1 = \boxed{802\text{ J}}$
2. For process 2, use $C_v = \frac{3}{2}nR$ and $T_C - T_B$ from step 1 to find Q_2. Then, since $W_2 = 0$, $\Delta E_{int,2} = Q_2$.	$W_2 = \boxed{0}$, $Q_2 = \boxed{-802\text{ J}}$, $\Delta E_{int,2} = \boxed{-802\text{ J}}$
3. Calculate W_3 from $W = -nRT\ln(V_A/V_C)$ in the isothermal compression. Then, since $\Delta E_{int,3} = 0$, $Q_3 = -W_3$.	$W_3 = \boxed{371\text{ J}}$, $Q_3 = \boxed{-371\text{ J}}$, $\Delta E_{int,3} = \boxed{0}$
4. Find the total work W, the total heat Q, and the total change ΔE_{int} by summing the quantities found in steps 2, 3, and 4.	$W_{total} = W_1 + W_2 + W_3$ $= (-535\text{ J}) + 0 + 371\text{ J} = \boxed{-164\text{ J}}$ $Q_{total} = Q_1 + Q_2 + Q_3$ $= 1337\text{ J} + (-802\text{ J}) + (-371\text{ J})$ $= \boxed{164\text{ J}}$ $\Delta E_{int,total} = \Delta E_{int,1} + \Delta E_{int,2} + \Delta E_{int,3}$ $= 802\text{ J} + (-802\text{ J}) + 0 = \boxed{0}$

REMARKS The total change in internal energy is zero, as it must be for a cyclic process. The total work done on the gas plus the total heat absorbed by the gas equals zero. The total work done on the gas equals the area under the CA curve minus the area under the AB curve, which equals the negative of the area enclosed by the three curves in Figure 18-13.

Heat Capacities and the Equipartition Theorem

According to the equipartition theorem stated in Section 5 of Chapter 17, the internal energy of n moles of a gas should equal $\frac{1}{2}nRT$ for each degree of freedom of the gas molecule. The heat capacity at constant volume of a gas should then be $\frac{1}{2}nR$ times the number of degrees of freedom of the molecule. From Table 18-2, nitrogen, oxygen, hydrogen, and carbon monoxide all have molar heat capacities at constant volume of about $\frac{5}{2}R$. Thus, the molecules in each of these gases have five degrees of freedom. About 1880, Rudolf Clausius speculated that these gases must consist of diatomic molecules that can rotate about two axes, giving them two additional degrees of freedom (Figure 18-14). The two degrees of freedom besides the three for translation are now known to be associated with their rotation about each of the two axes, x' and y', perpendicular to the line joining the atoms. The kinetic energy of a diatomic molecule is therefore

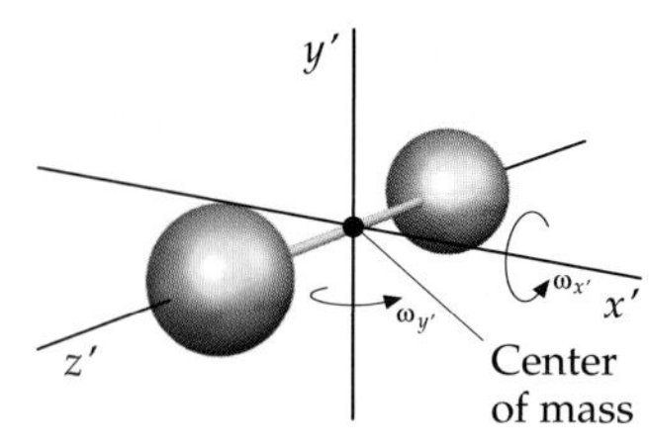

FIGURE 18-14 **Rigid-dumbbell model of a diatomic molecule.**

$$K = \tfrac{1}{2}mv_x^2 + \tfrac{1}{2}mv_y^2 + \tfrac{1}{2}mv_z^2 + \tfrac{1}{2}I_{x'}\omega_{x'}^2 + \tfrac{1}{2}I_{y'}\omega_{y'}^2$$

The total internal energy of n moles of such a gas is then

$$E_{\text{int}} = 5 \times \tfrac{1}{2}nRT = \tfrac{5}{2}nRT \qquad \text{18-24}$$

and the heat capacity at constant volume is

$$C_{\text{v}} = \tfrac{5}{2}nR \qquad \text{18-25}$$

Apparently, diatomic gases do not rotate about the line joining the two atoms—if they did, there would be six degrees of freedom and C_{v} would be $\frac{6}{2}nR = 3nR$, which is contrary to experimental results. Furthermore, monatomic gases apparently do not rotate at all. We will see in Section 18-8 that these puzzling facts are easily explained when we take into account the quantization of energy.

HEATING A DIATOMIC IDEAL GAS **EXAMPLE 18-9**

Two moles of oxygen gas are heated from a temperature of 20°C and a pressure of 1 atm to a temperature of 100°C. Assume that oxygen is an ideal gas. (*a*) How much heat must be supplied if the volume is kept constant during the heating? (*b*) How much heat must be supplied if the pressure is kept constant? (*c*) How much work does the gas do in part (*b*)?

PICTURE THE PROBLEM The heat needed for constant-volume heating is $Q_{\text{v}} = C_{\text{v}}\,\Delta T$, where $C_{\text{v}} = \frac{5}{2}nR$ since oxygen is a diatomic gas. For constant-pressure heating, $Q_{\text{p}} = C_{\text{p}}\,\Delta T$, where $C_{\text{p}} = C_{\text{v}} + nR$. Finally, the amount of work done by the gas equals the negative of the work done on the gas, which can be found from $\Delta E_{\text{int}} = Q + W_{\text{on}}$. Alternatively, $W_{\text{by}} = P\Delta V$.

(*a*) 1. Write the heat needed for constant volume in terms of C_{v} and ΔT:

$$Q_{\text{v}} = C_{\text{v}}\Delta T$$

2. Calculate the heat for $\Delta T = 80\text{ C}° = 80\text{ K}$:

$$Q_V = C_V\,\Delta T = \tfrac{5}{2}nR\Delta T$$

$$= \frac{5}{2}(2\text{ mol})\left(8.314\,\frac{\text{J}}{\text{mol}\cdot\text{K}}\right)(80\text{ K})$$

$$= \boxed{3.33\text{ kJ}}$$

(*b*) 1. Write the heat needed for constant pressure in terms of C_p and ΔT:

$$Q_p = C_p\,\Delta T$$

2. Calculate the heat capacity at constant pressure:

$$C_p = C_v + nR = \tfrac{5}{2}nR + nR = \tfrac{7}{2}nR$$

3. Calculate the heat added at constant pressure for $\Delta T = 80$ K:

$$Q_p = C_p\,\Delta T = \frac{7}{2}(2\ \text{mol})\left(8.314\,\frac{\text{J}}{\text{mol}\cdot\text{K}}\right)(80\ \text{K}) = \boxed{4.66\ \text{kJ}}$$

(*c*) 1. The work W_{on} can be found from the first law of thermodynamics:

$$\Delta E_{int} = Q_{in} + W_{on}, \quad \text{so} \quad W_{on} = \Delta E_{int} - Q_{in}$$

2. The internal energy change equals the heat added at constant volume which was calculated in Part (*a*):

$$\Delta E_{int} = Q_v = C_v\,\Delta T = \tfrac{5}{2}nR\Delta T$$

and

$$Q_p = C_p\,\Delta T = \tfrac{7}{2}nR\Delta T$$

so

$$W_{on} = \Delta E_{int} - Q_p = \tfrac{5}{2}nR\Delta T - \tfrac{7}{2}nR\Delta T = -nR\Delta T$$

$$= -(2\ \text{mol})\left(8.314\,\frac{\text{J}}{\text{mol}\cdot\text{K}}\right)(80\ \text{K})$$

$$= -1.33\ \text{kJ}$$

3. The work done by the gas at constant pressure is then:

$$W_{by} = -W_{on} = \boxed{1.33\ \text{kJ}}$$

REMARKS Note that the change in internal energy is independent of the process. It depends only on the initial and final states.

EXERCISE Find the initial and final volumes of this gas from the ideal-gas law, and use them to calculate the work done by the gas if the heat is added at constant pressure from $W_{by} = P\Delta V$. (*Answer* $V_i = 48.0$ L, $V_f = 61.1$ L, $W = 13.1$ L·atm $= 1.33$ kJ)

18-7 Heat Capacities of Solids

In Section 18-1, we noted that all of the metals listed in Table 18-1 have approximately equal molar specific heats. Experimentally, most solids have molar heat capacities approximately equal to $3R$:

$$c' = 3R = 24.9\ \text{J/mol}\cdot\text{K} \qquad \text{18-26}$$

This result is known as the **Dulong–Petit law.** We can understand this law by applying the equipartition theorem to the simple model for a solid shown in Figure 18-15. According to this model, a solid consists of a regular array of atoms in which each of the atoms has a fixed equilibrium position and is connected by springs to its neighbors. Each atom can vibrate in the x, y, and z directions. The total energy of an atom in a solid is

$$E = \tfrac{1}{2}mv_x^2 + \tfrac{1}{2}mv_y^2 + \tfrac{1}{2}mv_z^2 + \tfrac{1}{2}k_{eff}x^2 + \tfrac{1}{2}k_{eff}y^2 + \tfrac{1}{2}k_{eff}z^2$$

where k_{eff} is the effective force constant of the hypothetical springs. Each atom thus has six degrees of freedom. The equipartition theorem states that a substance

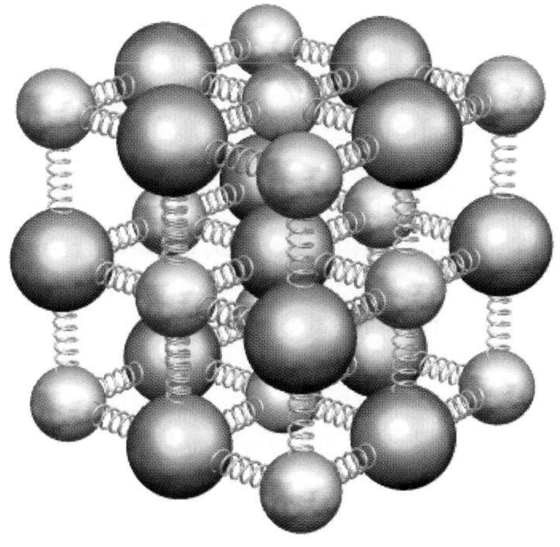

FIGURE 18-15 Model of a solid in which the atoms are connected to each other by springs. The internal energy of the molecule consists of the kinetic and potential energies of vibration.

in equilibrium has an average energy of $\frac{1}{2}RT$ per mole for each degree of freedom. Thus, the internal energy of a mole of a solid is

$$E_{\text{int,m}} = 6 \times \tfrac{1}{2}RT = 3RT \qquad \text{18-27}$$

which means that c' is equal to $3R$.

USING THE DULONG–PETIT LAW **EXAMPLE 18-10**

The molar mass of copper is 63.5 g/mol. Use the Dulong–Petit law to calculate the specific heat of copper.

PICTURE THE PROBLEM The Dulong–Petit law gives the molar specific heat of a solid, c'. The specific heat is then $c = c'/M$ (Equation 18-6), where M is the molar mass.

1. The Dulong–Petit law gives c' in terms of R: $c' = 3R$

2. Using $M = 63.5\text{ g/mol}$ for copper, the specific heat is:

$$c = \frac{c'}{M} = \frac{3R}{M} = \frac{3(8.314\text{ J/mol·K})}{63.5\text{ g/mol}}$$

$$= 0.392\text{ J/(g·K)} = \boxed{0.392\text{ kJ/kg·K}}$$

REMARKS This differs from the measured value of 0.386 kJ/kg·K given in Table 18-1 by less than 2 percent.

EXERCISE The specific heat of a certain metal is measured to be 1.02 kJ/kg·K. (*a*) Calculate the molar mass of this metal, assuming that the metal obeys the Dulong–Petit law. (*b*) What is the metal? (*Answer* (*a*) $M = 24.4\text{ g/mol}$. (*b*) The metal must be magnesium, which has a molar mass of 24.31 g/mol)

18-8 Failure of the Equipartition Theorem

Although the equipartition theorem had spectacular successes in explaining the heat capacities of gases and solids, it had equally spectacular failures. For example, if a diatomic gas molecule like the one in Figure 18-14 rotates about the line joining the atoms, there should be an additional degree of freedom. Similarly, if a diatomic molecule is not rigid, the two atoms should vibrate along the line joining them. We would then have two more degrees of freedom corresponding to kinetic and potential energies of vibration. But according to the measured values of the molar heat capacities in Table 18-3, diatomic gases apparently do not rotate about the line joining them, nor do they vibrate. The equipartition theorem gives no explanation for this, nor for the fact that monatomic atoms apparently do not rotate about any of the three possible perpendicular axes in space. Furthermore, heat capacities are found to depend on temperature, contrary to the predictions of the equipartition theorem. The most spectacular case of the temperature dependence of heat capacity is that of H_2, as shown in Figure 18-16. At temperatures below 70 K, c'_v for H_2 is $\frac{3}{2}R$, which is the same as that for a gas of molecules that translate, but do not rotate or vibrate. At temperatures between

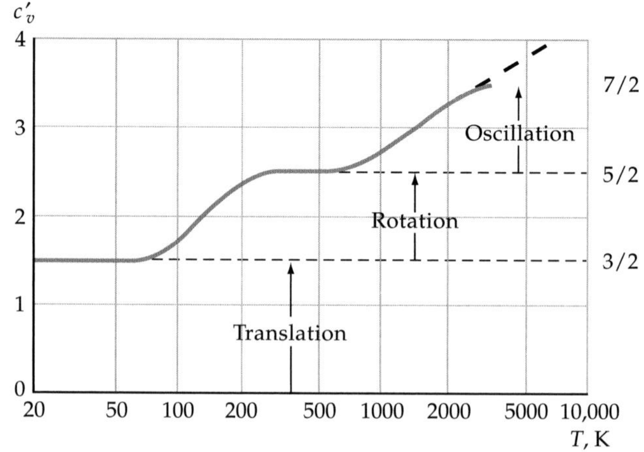

FIGURE 18-16 **Temperature dependence of the molar heat capacity of H_2. (The curve is qualitative in those regions where c'_v is changing.)**

250 K and 700 K, $c_v' = \frac{5}{2}R$, which is that for molecules that translate and rotate but do not vibrate. And at temperatures above 700 K, the H_2 molecules begin to vibrate. However, the molecules dissociate before c_v' reaches $\frac{7}{2}R$. Finally, the equipartition theorem predicts a constant value of $3R$ for the heat capacity of solids. While this result holds for many, although not all, solids at high temperatures, it does not hold at very low temperatures.

The equipartition theorem fails because the energy is **quantized.** That is, a molecule can have only certain values of internal energy, as illustrated schematically by the energy-level diagram in Figure 18-17. The molecule can gain or lose energy only if the gain or loss takes it to another allowed level. For example, the energy that can be transferred between colliding gas molecules is of the order of kT, the typical thermal energy of a molecule. The validity of the equipartition theorem depends on the relative size of kT and the spacing of the allowed energy levels.

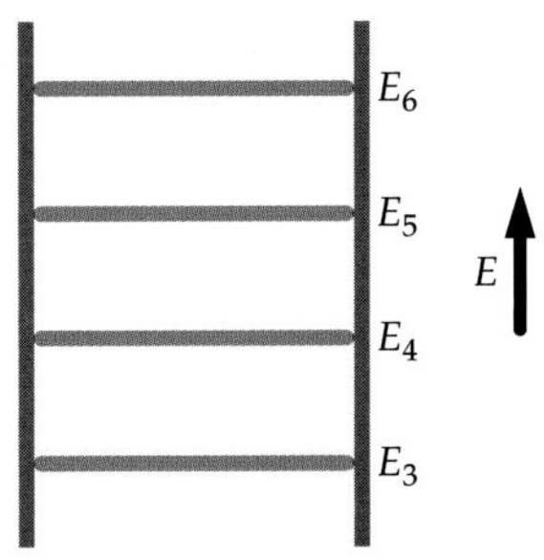

FIGURE 18-17 **Energy-level diagram. A system can have only certain discrete energies.**

If the spacing of the allowed energy levels is large compared with kT, energy cannot be transferred by collisions and the classical equipartition theorem is not valid. If the spacing of the levels is much smaller than kT, energy quantization will not be noticed and the equipartition theorem will hold.

CONDITIONS FOR THE VALIDITY OF THE EQUIPARTITION THEOREM

Consider the rotation of a molecule. The energy of rotation is

$$E = \frac{1}{2}I\omega^2 = \frac{(I\omega)^2}{2I} = \frac{L^2}{2I} \qquad \text{18-28}$$

where I is the moment of inertia of the molecule, ω is its angular velocity, and $L = I\omega$ is its angular momentum. In Section 10-5, we mentioned that angular momentum is quantized, and its magnitude is restricted to

$$L = \sqrt{\ell(\ell + 1)}\hbar \qquad \ell = 0, 1, 2, \ldots \qquad \text{18-29}$$

where $\hbar = h/(2\pi)$, and h is Planck's constant. The energy of a rotating molecule is therefore quantized to the values

$$E = \frac{L^2}{2I} = \frac{\ell(\ell + 1)\hbar^2}{2I} = \ell(\ell + 1)\, E_{0r} \qquad \text{18-30}$$

where

$$E_{0r} = \frac{\hbar^2}{2I} \qquad \text{18-31}$$

is characteristic of the energy gap between levels. If this energy is much less than kT, we expect classical physics and the equipartition theorem to hold. Let us define a critical temperature T_c by

$$kT_c = E_{0r} = \frac{\hbar^2}{2I} \qquad \text{18-32}$$

If T is much greater than this critical temperature, then kT will be much greater than the spacing of the energy levels, which is of the order of kT_c, and we expect classical physics and the equipartition theorem to be valid. If T is less than or of the order of T_c, then kT will not be much greater than the energy-level spacing, and we expect classical physics and the equipartition theorem to break down. Let's estimate T_c for some cases of interest.

1. *Rotation of H_2 about an axis perpendicular to the line joining the H atoms and through the center of mass* (Figure 18-18): The moment of inertia of H_2 about the axis is

$$I_H = 2M_H\left(\frac{r_s}{2}\right)^2 = \frac{1}{2}M_H r_s^2$$

where M_H is the mass of an H atom and r_s is the separation distance. For hydrogen, $M_H = 1.67 \times 10^{-27}$ kg, and $r_s \approx 8 \times 10^{-11}$ m. The critical temperature is then

$$T_c = \frac{\hbar^2}{2kI} = \frac{\hbar^2}{kM_H r_s^2}$$

$$= \frac{(1.05 \times 10^{-34}\,\text{J}\cdot\text{s})^2}{(1.38 \times 10^{-23}\,\text{J/K})(1.67 \times 10^{-27}\,\text{kg})(8 \times 10^{-11}\,\text{m})^2} \approx 75\text{ K}$$

As we see from Figure 18-16, this is approximately the temperature below which the rotational energy does not contribute to the heat capacity.

2. *O_2*: Because the mass of O_2 is about 16 times that of H_2, and the separation is about the same, the critical temperature for O_2 should be about $(75/16) \approx$ 4.6 K. For all temperatures for which O_2 exists as a gas, $T >> T_c$, so kT is much greater than the energy level spacing. Consequently, we expect the equipartition theorem of classical physics to apply.

3. *Rotation of a monatomic gas:* Consider the He atom, which consists of the He nucleus and two electrons. The mass of an electron is about 8000 times smaller than the mass of the He nucleus, but the radius of the nucleus is about 100,000 times smaller than the distance to the electron. Therefore, the moment of inertia of the He atom is almost entirely due to its two electrons. The distance from the He nucleus to its electrons is about half the separation distance of the H atoms in H_2, and the electron mass is about 2000 times smaller than that of the H nucleus. Thus, using $m_e = M_H/2000$ and $r = r_s/2$, we find the moment of inertia of the two electrons in He to be roughly

$$I_{He} = 2m_e r^2 \approx 2\frac{M_H}{2000}\left(\frac{r_s}{2}\right)^2 = \frac{I_H}{2000}$$

The critical temperature for He is thus about 2000 times that of H_2 or about 150,000 K. This is much higher than the dissociation temperature (the temperature at which electrons are stripped from their nuclei) for helium. So, the gap between allowed energy levels is always much greater than kT and the He molecules cannot be induced to rotate by collisions occurring in the gas. Other monatomic gases have slightly greater moments of inertia because they have more electrons, but their critical temperatures are still tens of thousands of kelvins. Therefore, their molecules also cannot be induced to rotate by collisions occurring in the gas.

4. *Rotation of a diatomic gas about the axis joining the atoms:* We see from our discussion of monatomic gases that the moment of inertia for a diatomic gas molecule about this axis will also be due mainly to the electrons and will be of the same order of magnitude as for a monatomic gas. Again, the critical temperature, T_c, calculated in order for this rotation to occur due to collisions between molecules in the gas, exceeds the gas's dissociation temperature, making rotation under those circumstances impossible.

It is interesting to note that the successes of the equipartition theorem in explaining the measured heat capacities of gases and solids led to the first real

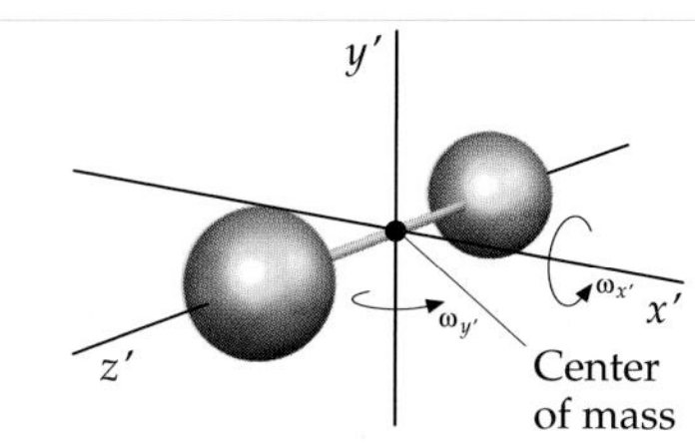

FIGURE 18-18 Rigid-dumbbell model of a diatomic molecule.

understanding of molecular structure in the nineteenth century, whereas its failures played an important role in the development of quantum mechanics in the twentieth century.

ROTATIONAL ENERGY OF THE HYDROGEN ATOM **EXAMPLE 18-11**

(*a*) Estimate the lowest (nonzero) rotational energy for the hydrogen atom and compare it to kT at room temperature, $T = 300$ K. (*b*) Calculate the critical temperature T_c.

PICTURE THE PROBLEM From Equation 18-30, the lowest rotational energy is for $\ell = 1$. We use Equation 18-30 to determine the energy in terms of the moment of inertia. We can neglect the moment of inertia of the nucleus because its radius is 100,000 times smaller than the radius of the atom. Therefore, the moment of inertia for the atom is essentially the moment of inertia of the electron about the nucleus. Then $I = m_e r^2$, where $r \approx 5 \times 10^{-11}$ m is the distance from the nucleus to the electron.

(*a*) 1. The lowest energy greater than zero occurs for $\ell = 1$:

$$E_\ell = \frac{\ell(\ell + 1)\hbar^2}{2I}, \qquad \ell = 0, 1, 2, \ldots$$

so

$$E_1 = \frac{1(1 + 1)\hbar^2}{2m_e r^2} = \frac{\hbar^2}{m_e r^2}$$

2. The numerical values are:

$$\hbar = 1.05 \times 10^{-34}\ \text{J}\cdot\text{s}$$

$$m_e = 9.11 \times 10^{-31}\ \text{kg}$$

$$r = 5 \times 10^{-11}\ \text{m}$$

3. Substitute the numerical values:

$$E_1 = \frac{\hbar^2}{m_e r^2} = \frac{(1.05 \times 10^{-34}\ \text{J}\cdot\text{s})^2}{(9.11 \times 10^{-31}\ \text{kg})(5 \times 10^{-11}\ \text{m})^2}$$

$$= \boxed{4.8 \times 10^{-18}\ \text{J}}$$

4. The value of kT at $T = 300$ K is:

$$kT = (1.38 \times 10^{-23}\ \text{J/K})(300\ \text{K}) = 4.1 \times 10^{-21}\ \text{J}$$

5. Compare E_1 and kT:

$$\frac{E_1}{kT} = \frac{4.8 \times 10^{-18}\ \text{J}}{4.1 \times 10^{-21}\ \text{J}} \approx 10^3$$

$\boxed{E_1 \text{ is about three orders of magnitude larger than } kT.}$

(*b*) Set $kT_c = E_1$ and solve for T_c:

$$kT_c = E_1$$

$$T_c = \frac{E_1}{k} = \frac{4.8 \times 10^{-18}\ \text{J}}{1.38 \times 10^{-23}\ \text{J/K}} = \boxed{3.48 \times 10^5\ \text{K}}$$

REMARKS The critical temperature of a hydrogen atom is so high that the atom would be ionized well before the critical temperature could be reached.

18-9 The Quasi-Static Adiabatic Compression of a Gas

A process in which no heat flows into or out of a system is called an **adiabatic process.** Such a process can occur when the system is extremely well insulated, or when the process happens very quickly. Consider the quasi-static adiabatic compression of a gas in which the gas in a thermally insulated container is slowly compressed by a piston, which is thereby doing work on the gas. Because no heat enters or leaves the gas, the work done on the gas equals the increase in the internal energy of the gas, and the temperature of the gas increases. The curve representing this process on a PV diagram is shown in Figure 18-19.

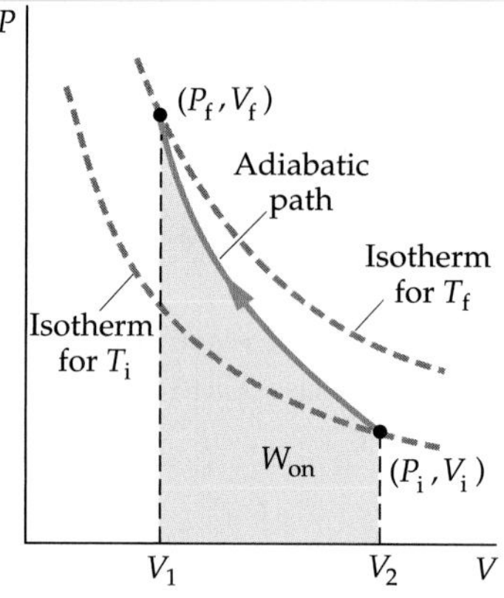

FIGURE 18-19 Quasi-static adiabatic compression of an ideal gas. The dashed lines are the isotherms for the initial and final temperatures. The curve connecting the initial and final states of the adiabatic compression is steeper than the isotherms because the temperature increases during the compression.

We can find the equation for the adiabatic curve for an ideal gas by using the equation of state ($PV = nRT$) and the first law of thermodynamics ($dE_{int} = dQ_m + dW_{on}$). We have

$$C_v dT = 0 + (-PdV) \qquad \text{18-33}$$

where we have used $dE_{int} = C_v dT$ (Equation 18-18*a*), $dQ_m = 0$ (the process is adiabatic), and $dW_{on} = -PdV$ (Equation 18-15). Then, substituting for P using $P = nRT/V$, we obtain

$$C_v\, dT = -nRT\frac{dV}{V}$$

Rearranging,

$$\frac{dT}{T} + \frac{nR}{C_v}\frac{dV}{V} = 0$$

Integration gives

$$\ln T + \frac{nR}{C_v}\ln V = \text{constant}$$

Simplifying,

$$\ln T + \frac{nR}{C_v}\ln V = \ln T + \ln V^{nR/C_v} = \ln TV^{nR/C_v} = \text{constant}$$

Thus,

$$TV^{nR/C_v} = \text{constant} \qquad \text{18-34}$$

where the constants in the two preceding equations are not the same. Equation 18-34 can be rewritten by noting that $C_p - C_v = nR$, so

$$\frac{nR}{C_v} = \frac{C_p - C_v}{C_v} = \frac{C_p}{C_v} - 1 = \gamma - 1 \qquad \text{18-35}$$

where γ is the ratio of the heat capacities:

$$\gamma = \frac{C_p}{C_v} \qquad \text{18-36}$$

Therefore,

$$TV^{\gamma-1} = \text{constant} \qquad \text{18-37}$$

We can eliminate T from Equation 18-37 using $PV = nRT$. We then have

$$\frac{PV}{nR}V^{\gamma-1} = \text{constant}$$

Clouds form if rising moist air cools due to adiabatic expansion of the air. Cooling causes water vapor to condense into liquid droplets.

or

$$PV^{\gamma} = \text{constant} \qquad \text{18-38}$$

QUASI-STATIC ADIABATIC PROCESS

Equation 18-38 relates P and V for adiabatic expansions and compressions.

EXERCISE Show that for quasi-static adiabatic process $T^{\gamma}/P^{\gamma-1} = \text{constant}$.

The work done on the gas in an adiabatic compression can be calculated from the first law of thermodynamics:

$$dE_{\text{int}} = dQ_{\text{in}} + dW_{\text{on}} \quad \text{or} \quad dW_{\text{on}} = dE_{\text{int}} - dQ_{\text{in}}$$

Since $dE_{\text{int}} = C_{\text{v}}\,dT$ and $dQ_{\text{in}} = 0$, we have

$$dW_{\text{on}} = C_{\text{v}}\,dT$$

Then

$$W_{\text{adiabatic}} = \int dW_{\text{on}} = \int C_{\text{v}}dT = C_{\text{v}}\Delta T \qquad \text{18-39}$$

ADIABATIC WORK ON GAS

where we have assumed that C_{v} is constant.† We note that the work done on the gas depends only on the change in the temperature of the gas. In an adiabatic compression, work is done on the gas, and its internal energy and temperature increase. In a quasi-static adiabatic *expansion*, work is done *by* the gas, and the internal energy and temperature decrease.

We can use the ideal-gas law to write Equation 18-39 in terms of the initial and final values of the pressure and volume. If T_{i} is the initial temperature and T_{f} is the final temperature, we have for the work done

$$W_{\text{adiabatic}} = C_{\text{v}}\Delta T = C_{\text{v}}(T_{\text{f}} - T_{\text{i}})$$

Using $PV = nRT$, we obtain

$$W_{\text{adiabatic}} = C_{\text{v}}\left(\frac{P_{\text{f}}V_{\text{f}}}{nR} - \frac{P_{\text{i}}V_{\text{i}}}{nR}\right) = \frac{C_{\text{v}}}{nR}(P_{\text{f}}V_{\text{f}} - P_{\text{i}}V_{\text{i}})$$

Using Equation 18-35 to simplify this expression, we have

$$W_{\text{adiabatic}} = \frac{P_{\text{f}}V_{\text{f}} - P_{\text{i}}V_{\text{i}}}{\gamma - 1} \qquad \text{18-40}$$

ADIABATIC WORK ON GAS

QUASI-STATIC ADIABATIC COMPRESSION OF AIR **EXAMPLE 18-12**

A quantity of air is compressed adiabatically and quasi-statically from an initial pressure of 1 atm and a volume of 4 L at temperature 20°C to half its original volume. Find (*a*) the final pressure, (*b*) the final temperature, and (*c*) the work done on the gas.

† For an ideal gas, E_{int} is proportional to the absolute temperature, and therefore $C_{\text{v}} = dE_{\text{int}}/dT$ is a constant.

PICTURE THE PROBLEM Because the process is both quasi-static and adiabatic, we know that PV^{γ} = constant, and $TV^{\gamma-1}$ = constant. These relations yield the final pressure and final temperature, respectively. Find γ using Equations 18-36, 18-20, and 18-25. The work done is found from $W_{\text{adiabatic}} = C_v\Delta T$ (Equation 18-39). For a diatomic gas, $C_v = \frac{5}{2}nR$. Let subscript 1 refer to initial values, and subscript 2 to final values. Then $P_1 = 1$ atm, $V_1 = 4$ L, $V_2 = 2$ L, $T_1 = 20°\text{C} = 293$ K.

(*a*) 1. Write PV^{γ} = constant in terms of initial and final values:

$$P_1V_1^{\gamma} = P_2V_2^{\gamma}$$

2. Find γ for a diatomic gas using Equations 18-36, 18-20, and 18-25:

$$\gamma = \frac{C_p}{C_v} = \frac{C_v + nR}{C_v} = 1 + \frac{nR}{C_v} = 1 + \frac{nR}{\frac{5}{2}nR}$$

3. Solve for P_2:

$$P_2 = P_1\left(\frac{V_1}{V_2}\right)^{\gamma} = (1\text{ atm})\left(\frac{4\text{ L}}{2\text{ L}}\right)^{1.4} = \boxed{2.64\text{ atm}}$$

(*b*) 1. Write $TV^{\gamma-1}$ = constant in terms of initial and final values:

$$T_1V_1^{\gamma-1} = T_2V_2^{\gamma-1}$$

2. Solve for T_2:

$$T_2 = T_1\left(\frac{V_1}{V_2}\right)^{\gamma-1} = (293\text{ K})\left(\frac{4\text{ L}}{2\text{ L}}\right)^{0.4}$$

$$= 387\text{ K} = \boxed{114°\text{C}}$$

(*c*) 1. Equation 18-39 gives the work done:

$$W_{\text{adiabatic}} = C_v\Delta T = \tfrac{5}{2}nR\Delta T$$

2. Using the ideal-gas law for the initial conditions, express nR in terms of P_1, V_1, and T_1:

$$W_{\text{adiabatic}} = \frac{5}{2}nR\Delta T = \frac{5}{2}\frac{P_1V_1}{T_1}(T_2 - T_1)$$

$$= \frac{5}{2}\frac{(1\text{ atm})(4\text{ L})}{293\text{ K}}(387\text{ K} - 293\text{ K})$$

$$= 3.20\text{ L·atm} = \boxed{324\text{ J}}$$

REMARKS The work can also be calculated using Equation 18-40, but using $W_{\text{adiabatic}} = C_v\Delta T$ is preferable because it is more directly connected to a principle, the first law of thermodynamics, and thus is easier to recall.

Speed of Sound Waves

We can use Equation 18-38 to calculate the adiabatic bulk modulus of an ideal gas, which is related to the speed of sound waves in air. We first compute the differential of both sides of PV^{γ} = constant (Equation 18-38):

$$Pd(V^{\gamma}) + V^{\gamma}dP = 0$$

or

$$\gamma PV^{\gamma-1}dV + V^{\gamma}dP = 0$$

Then

$$dP = -\frac{\gamma P dV}{V}$$

Referring to Equation 13-6, the adiabatic bulk modulus† is then:

$$B_{\text{adiabatic}} = -\frac{dP}{dV/V} = \gamma P \qquad \text{18-41}$$

† The bulk modulus, discussed in Chapter 13, is the negative ratio of the pressure change to the fractional change in volume, $B = -\Delta P/(\Delta V/V)$. The isothermal bulk modulus, which describes changes that occur at constant temperature, differs from the adiabatic bulk modulus, which describes changes with no heat transfer. For sound waves at audible frequencies, the pressure changes occur too rapidly for appreciable heat flow, so the appropriate bulk modulus is the adiabatic bulk modulus.

The speed of sound (Equation 15-4) is given by

$$v = \sqrt{\frac{B_{\text{adiabatic}}}{\rho}}$$

where the mass density ρ is related to the number of moles n and the molecular mass M by $\rho = m/V = nM/V$. Using the ideal-gas law, $PV = nRT$, we can eliminate V from the density

$$\rho = \frac{nM}{V} = \frac{nM}{nRT/P} = \frac{MP}{RT}$$

Using this result and γP for $B_{\text{adiabatic}}$, we obtain

$$v = \sqrt{\frac{B_{\text{adiabatic}}}{\rho}} = \sqrt{\frac{\gamma P}{MP/(RT)}} = \sqrt{\frac{\gamma RT}{M}}$$

which is Equation 15-5, the speed of sound in a gas.

SUMMARY

1. The first law of thermodynamics, which is a statement of the conservation of energy, is a fundamental law of physics.
2. The equipartition theorem is a fundamental law of classical physics. It breaks down if the typical thermal energy kT is small compared to the spacing of quantized energy levels.

Topic	Relevant Equations and Remarks	
1. Heat	Heat is energy that is transferred from one object to another because of a temperature difference.	
Calorie	The calorie, originally defined as the heat necessary to raise the temperature of 1 g of water by 1°C, is now defined to be 4.184 joules.	
2. Heat Capacity	Heat capacity is the amount of heat needed to raise the temperature of a substance by one degree. $C = \frac{Q}{\Delta T}$	**18-1**
At constant volume	$C_v = \frac{Q_v}{\Delta T}$	
At constant pressure	$C_p = \frac{Q_p}{\Delta T}$	
Specific heat (heat capacity per unit mass)	$c = \frac{C}{m}$	**18-2**
Molar specific heat (heat capacity per mole)	$c' = \frac{C}{n}$	**18-6**

Heat capacity–internal energy relation	$C_v = \frac{dE_{int}}{dT}$	**18-18*a***
Ideal gas	$C_p = C_v + nR$	**18-20**
Monatomic ideal gas	$C_v = \frac{3}{2}nR$	**18-22**
Diatomic ideal gas	$C_v = \frac{5}{2}nR$	**18-25**
3. Fusion and Vaporization	Both melting and vaporization occur at a constant temperature.	
Latent heat of fusion	The heat needed to melt a substance is the product of the mass of the substance and its latent heat of fusion L_f: $Q_f = mL_f$	**18-8**
L_f of water	$L_f = 333.5$ kJ/kg	
Latent heat of vaporization	The heat needed to vaporize a liquid is the product of the mass of the liquid and its latent heat of vaporization L_v: $Q_v = mL_v$	**18-9**
L_v of water	$L_v = 2257$ kJ/kg	
4. First Law of Thermodynamics	The change in the internal energy of a system equals the heat transferred into the system plus the work done on the system: $\Delta E_{int} = Q_{in} + W_{on}$	**18-10**
5. Internal Energy E_{int}	The internal energy of a system is a property of the state of the system, as are the pressure, volume, and temperature. Heat and work are not properties of state.	
Ideal gas	E_{int} depends only on the temperature T.	
Monatomic ideal gas	$E_{int} = \frac{3}{2}nRT$	**18-12**
Internal energy related to heat capacity	$dE_{int} = C_v\,dT$	**18-18*b***
6. Quasi-Static Process	A quasi-static process is one that occurs slowly so that the system moves through a series of equilibrium states.	
Isobaric	$P =$ constant	
Isothermal	$T =$ constant	
Adiabatic	$Q = 0$	
Adiabatic, ideal gas	$TV^{\gamma-1} =$ constant	**18-37**
	or $PV^{\gamma} =$ constant	**18-38**
	where $\gamma = \frac{C_p}{C_v}$	**18-36**

7. **Work Done on a Gas**	$W_{\text{on}} = -\int_{V_i}^{V_f} P\,dV = C_v\Delta T - Q_{\text{in}}$	**18-10, 18-15, and 18-18**
Constant volume	$W_{\text{on}} = -\int_{V_i}^{V_f} P\,dV = 0 \qquad V_f = V_i$	
Isobaric	$W_{\text{on}} = -\int_{V_i}^{V_f} P\,dV = -P\int_{V_i}^{V_f} dV = -P\Delta V$	
Isothermal	$W_{\text{isothermal}} = -\int_{V_i}^{V_f} P\,dV = -nRT\int_{V_i}^{V_f}\frac{dV}{V} = nRT\ln\frac{V_i}{V_f}$	**18-17**
Adiabatic	$W_{\text{adiabatic}} = C_v\Delta T$	**18-39**
8. **Equipartition Theorem**	The equipartition theorem states that if a system is in equilibrium, there is an average energy of $\frac{1}{2}kT$ per molecule or $\frac{1}{2}RT$ per mole associated with each degree of freedom.	
Failure of the equipartition theorem	The equipartition theorem fails if the thermal energy ($\sim kT$) that can be transferred in collisions is smaller than the energy gap ΔE between quantized energy levels. For example, monatomic gas molecules cannot rotate because the first nonzero energy permitted is much greater than kT.	
9. **Dulong–Petit Law**	The molar specific heat of most solids is $3R$. This is predicted by the equipartition theorem, assuming a solid atom has six degrees of freedom.	

PROBLEMS

In a few problems, you are given more data than you actually need; in a few other problems, you are required to supply data from your general knowledge, outside sources, or informed estimates.

- • Single-concept, single-step, relatively easy
- •• Intermediate-level, may require synthesis of concepts
- ••• Challenging
- SSM Solution is in the *Student Solutions Manual*
- iSOLVE Problems available on iSOLVE online homework service
- iSOLVE✓ These "Checkpoint" online homework service problems ask students additional questions about their confidence level, and how they arrived at their answer

Use $v = 340$ m/s for the speed of sound in air unless otherwise indicated.

Conceptual Problems

1 • Body A has twice the mass and twice the specific heat of body B. If they are supplied with equal amounts of heat, how do the subsequent changes in their temperatures compare?

2 • SSM The temperature change of two blocks of masses M_A and M_B is the same when they absorb equal amounts of heat. It follows that the specific heats are related by (*a*) $c_A = (M_A/M_B)c_B$, (*b*) $c_A = (M_B/M_A)c_B$, (*c*) $c_A = c_B$, (*d*) none of the above.

3 • The specific heat of aluminum is more than twice that of copper. Identical masses of copper and aluminum, both at 20°C, are dropped into a calorimeter containing water at 40°C. When thermal equilibrium is reached, (*a*) the aluminum is at a higher temperature than the copper, (*b*) the aluminum has absorbed less energy than the copper, (*c*) the aluminum has absorbed more energy than the copper, (*d*) both (*a*) and (*c*) are correct statements.

4 • Joule's experiment establishing the mechanical equivalence of heat involved the conversion of mechanical energy into internal energy. Give some examples of the internal energy of a system being converted into mechanical energy.

5 • SSM Can a system absorb heat with no change in its internal energy?

6 • In the equation $\Delta E_{\text{int}} = Q + W$ (the formal statement of the first law of thermodynamics), the quantities Q and W represent (*a*) the heat supplied to the system and the work done by the system, (*b*) the heat supplied to the system and the work done on the system, (*c*) the heat released by the system and the work done by the system, (*d*) the heat released by the system and the work done on the system.

7 • A real gas cools during a free expansion, though an ideal gas does not. Explain.

8 • An ideal gas at one atmosphere pressure and 300 K is confined to half of an insulated container by a thin partition. The partition is then removed and equilibrium is established. At that point of equilibrium, which of the following is correct? (*a*) The pressure is half an atmosphere and the temperature is 150 K, (*b*) the pressure is one atmosphere and the temperature is 150 K, (*c*) the pressure is half an atmosphere and the temperature is 300 K, (*d*) none of the above.

9 • A certain gas consists of ions that repel each other. The gas undergoes a free expansion with no heat exchange and no work done. How does the temperature of the gas change? Why?

10 •• SSM Two gas-filled rubber balloons of (initially) equal volume are at the bottom of a dark, cold lake. The top of the lake is warmer than the bottom. One balloon rises rapidly and expands adiabatically as it rises. The other balloon rises more slowly and expands isothermally. Which balloon is larger when it reaches the surface of the lake?

11 • A gas changes its state reversibly from A to C (Figure 18-20). The work done by the gas is (*a*) greatest for path A→B→C, (*b*) least for path A→C, (*c*) greatest for path A→D→C, (*d*) the same for all three paths.

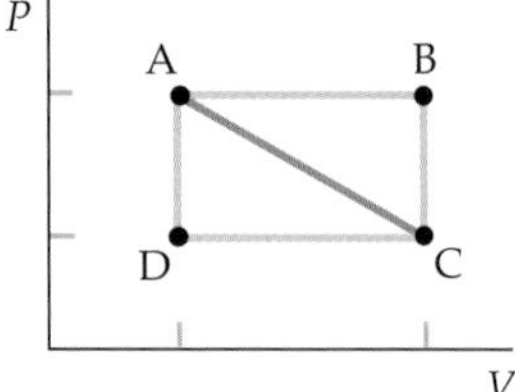

FIGURE 18-20 Problem 11

12 • When an ideal gas is subjected to an adiabatic process, (*a*) no work is done by the system, (*b*) no heat is supplied to the system, (*c*) the internal energy remains constant, (*d*) the heat supplied to the system equals the work done by the system.

13 • True or false:

(*a*) The heat capacity of a body is the amount of heat it can store at a given temperature.
(*b*) When a system goes from state 1 to state 2, the amount of heat added to the system is the same for all processes.
(*c*) When a system goes from state 1 to state 2, the work done on the system is the same for all processes.
(*d*) When a system goes from state 1 to state 2, the change in the internal energy of the system is the same for all processes.
(*e*) The internal energy of a given amount of an ideal gas depends only on its absolute temperature.
(*f*) A quasi-static process is one in which the system is never far from being in equilibrium.
(*g*) For any material that expands when heated, C_p is greater than C_v.

14 • SSM If a system's volume remains constant while undergoing changes in temperature and pressure, then (*a*) the internal energy of the system is unchanged, (*b*) the system does no work, (*c*) the system absorbs no heat, (*d*) the change in internal energy equals the heat absorbed by the system.

15 • When an ideal gas is subjected to an isothermal process, (*a*) no work is done by the system, (*b*) no heat is supplied to the system, (*c*) the heat supplied to the system equals the change in internal energy, (*d*) the heat supplied to the system equals the work done by the system.

16 •• The 1-L fuel tank of a gas grill contains 600 g of propane (C_3H_8) at a pressure of 2 MPa. What can you say about the phase state of the propane?

17 •• An ideal gas undergoes a process during which $P\sqrt{V}$ = constant and the volume of the gas decreases. What happens to the temperature?

18 •• SSM Which would you expect to have a higher heat capacity per *unit mass*, lead or copper? Why? (Don't look up the heat capacities to answer this question.)

19 •• Calculating the heat capacity of a liquid is very difficult, because of the strong intermolecular interactions and the random positions of the molecules in a liquid. However, simply based on counting degrees of freedom, would you expect a monatomic liquid to have a higher or lower heat capacity than the solid phase of the same substance it melted from? Assume that the melting temperature is high enough that you don't have to take quantum effects into consideration when calculating the heat capacity of the solid.

Estimation and Approximation

20 •• SSM A simple demonstration to show that heat is a form of energy is to take a bag of lead shot and drop it repeatedly onto a very rigid surface from a small height. The bag's temperature will increase, allowing an estimate of the heat capacity of lead. (*a*) Estimate the temperature increase of a bag filled with 1 kg of lead shot dropped 50 times from a height of 1 m. (*b*) In principle, the change in temperature is independent of the mass of the shot in the bag; in practice, it's better to use a larger mass than a smaller one. Why might this be true?

21 •• A "typical" microwave oven has a power consumption of about 1200 W. Estimate how long it should take to boil a cup of water in the microwave assuming that 50% of the power consumption goes into heating the water. Does this estimate mesh with everyday experience?

22 • A demonstration of the heating of a gas under adiabatic compression involves putting a few scraps of paper into a large glass test tube, which is then sealed off with a piston. If the piston compresses the trapped air very rapidly, the paper will catch fire. Assuming that the burning point of paper is 451°F, estimate the factor by which the volume of the air trapped by the piston must be reduced for this demonstration to work.

23 •• There is a small change in the volume of a liquid on heating at constant pressure. Use the following data for water to estimate the contribution this makes to the heat capacity of water between 4°C and 100°C:

Density of water at 4°C and 1 atm pressure: 1.000 g/cm^3

Density of liquid water at 100°C and 1 atm pressure: 0.9584 g/cm^3

24 •• SSM A certain molecule has vibrational energy levels that are equally spaced by 0.15 eV. Find the critical temperature T_c so for $T >> T_c$ you would expect the equipartition theorem to hold and for $T << T_c$ you would expect the equipartition theorem to fail.

Heat Capacity; Specific Heat; Latent Heat

25 • SSM A "typical" adult male consumes about 2500 kcal of food in a day. (*a*) How many joules is this? (*b*) If this consumed energy is dissipated over the course of 24 hours, what is his average output power in watts?

26 • ISOLVE A solar home contains 10^5 kg of concrete (specific heat = 1.00 kJ/kg·K). How much heat is given off by the concrete when it cools from 25 to 20°C?

27 • ISOLVE How many calories must be supplied to 60 g of ice at −10°C to melt it and raise the temperature of the water to 40°C?

28 •• ISOLVE How much heat must be removed when 100 g of steam at 150°C is cooled and frozen into 100 g of ice at 0°C? (Take the specific heat of steam to be 2.01 kJ/kg·K.)

29 •• ISOLVE A 50-g piece of aluminum at 20°C is cooled to −196°C by placing it in a large container of liquid nitrogen at that temperature. How much nitrogen is vaporized? (Assume that the specific heat of aluminum is constant and is equal to 0.90 J/kg·K.)

30 •• ISOLVE If 500 g of molten lead at 327°C is poured into a cavity in a large block of ice at 0°C, how much of the ice melts?

31 •• SSM ISOLVE A 30-g lead bullet initially at 20°C comes to rest in the block of a ballistic pendulum. Assume that half the initial kinetic energy of the bullet is converted into thermal energy within the bullet. If the speed of the bullet was 420 m/s, what is the temperature of the bullet immediately after coming to rest in the block?

32 •• ISOLVE A 1400-kg car traveling at 80 km/h is brought to rest by applying the brakes. If the specific heat of steel is 0.11 cal/g·K, what total mass of steel must be contained in the steel brake drums if the temperature of the brake drums is not to rise by more than 120 C°?

Calorimetry

33 • ISOLVE A 200-g piece of lead is heated to 90°C and is then dropped into a calorimeter containing 500 g of water that is initially at 20°C. Neglecting the heat capacity of the container, find the final temperature of the lead and water.

34 • SSM ISOLVE The specific heat of a certain metal can be determined by measuring the temperature change that occurs when a piece of the metal is heated and then placed in an insulated container made of the same material and containing water. Suppose a piece of metal has a mass of 100 g and is initially at 100°C. The container has a mass of 200 g and contains 500 g of water at an initial temperature of 20.0°C. The final temperature is 21.4°C. What is the specific heat of the metal?

35 •• In the 2002 Tour de France, champion bicyclist Lance Armstrong expended an average power of about 400 W, 5 hours a day for 20 days. What quantity of water, initially at 24°C, could be brought to the boiling point by harnessing all of that energy?

36 •• A 25-g glass tumbler contains 200 mL of water at 24°C. If two 15-g ice cubes, each at a temperature of −3°C, are dropped into the tumbler, what is the final temperature of the drink? Neglect thermal conduction between the tumbler and the room.

37 •• A 200-g piece of ice at 0°C is placed in 500 g of water at 20°C. The system is in a container of negligible heat capacity and is insulated from its surroundings. (*a*) What is the final equilibrium temperature of the system? (*b*) How much of the ice melts?

38 •• ISOLVE✓ A 3.5-kg block of copper at a temperature of 80°C is dropped into a bucket containing a mixture of ice and water whose total mass is 1.2 kg. When thermal equilibrium is reached, the temperature of the water is 8°C. How much ice was in the bucket before the copper block was placed in it? (Neglect the heat capacity of the bucket.)

39 •• ISOLVE✓ A well-insulated bucket contains 150 g of ice at 0°C. (*a*) If 20 g of steam at 100°C is injected into the bucket, what is the final equilibrium temperature of the system? (*b*) Is any ice left afterward?

40 •• ISOLVE✓ A calorimeter of negligible mass contains 1 kg of water at 303 K and 50 g of ice at 273 K. Find the final temperature *T*. Solve the same problem if the mass of ice is 500 g.

41 •• SSM A 200-g aluminum calorimeter contains 500 g of water at 20°C. A 100-g piece of ice cooled to −20°C is placed in the calorimeter. (*a*) Find the final temperature of the system, assuming no heat loss. (Assume that the specific heat of ice is 2.0 kJ/kg·K.) (*b*) A second 200-g piece of ice at −20°C is added. How much ice remains in the system after it reaches equilibrium? (*c*) Would you give a different answer for (*b*) if both pieces of ice were added at the same time?

42 •• ISOLVE The specific heat of a 100-g block of material is to be determined. The block is placed in a 25-g copper calorimeter that also holds 60 g of water. The system is initially at 20°C. Then 120 mL of water at 80°C are added to the calorimeter vessel. When thermal equilibrium is attained, the temperature of the water is 54°C. Determine the specific heat of the block.

43 •• A 100-g piece of copper is heated in a furnace to a temperature *t*. The copper is then inserted into a 150-g copper calorimeter containing 200 g of water. The initial temperature of the water and calorimeter is 16°C, and the final temperature after equilibrium is established is 38°C. When the calorimeter and its contents are weighed, 1.2 g of water are found to have evaporated. What was the temperature *t*?

44 •• A 200-g aluminum calorimeter contains 500 g of water at 20°C. Aluminum shot with a mass 300 g is heated to 100°C and is then placed in the calorimeter. (*a*) Using the value of the specific heat of aluminum given in Table 18-1, find the final temperature of the system, assuming that no heat is lost to the surroundings. (*b*) The error due to heat transfer between the system and its surroundings can be minimized if the initial temperature of the water and calorimeter is chosen to be below room temperature, where Δt_w is the temperature change of the calorimeter and water during the measurement. Then the final temperature is $\frac{1}{2}\Delta t_w$ above room temperature. What should the initial temperature of the water and container be if the room temperature is 20°C?

First Law of Thermodynamics

45 • ISOLVE A diatomic gas does 300 J of work and also absorbs 600 cal of heat. What is the change in internal energy of the gas?

46 • SSM ISOLVE If 400 kcal is added to a gas that expands and does 800 kJ of work, what is the change in the internal energy of the gas?

47 • ISOLVE A lead bullet moving at 200 m/s is stopped in a block of wood. Assuming that all of the energy change goes into heating the bullet, find the final temperature of the bullet if its initial temperature is 20°C.

48 • (*a*) At Niagara Falls, the water drops 50 m. If the change in potential energy goes into the internal energy of the water, compute the increase in its temperature. (*b*) Do the same for Yosemite Falls, where the water drops 740 m. (These temperature rises are not observed because the water cools by evaporation as it falls.)

49 • When 20 cal of heat are absorbed by a gas, the system performs 30 J of work. What is the change in the internal energy of the gas?

50 •• ISOLVE A lead bullet initially at 30°C just melts upon striking a target. Assuming that all of the initial kinetic energy of the bullet goes into the internal energy of the bullet to raise its temperature and melt it, calculate the speed of the bullet upon impact.

51 •• SSM On a cold day you can warm your hands by rubbing them together. (*a*) Assume that the coefficient of friction between your hands is 0.5, that the normal force between your hands is 35 N, and that you rub them together at an average speed of 35 cm/s. What is the rate at which heat is generated? (*b*) Assume further that the mass of each of your hands is approximately 350 g, that the specific heat of your hands is about 4 kJ/kg·K, and that all the heat generated goes into raising the temperature of your hands. How long must you rub your hands together to produce a 5 C° increase in their temperature?

Work and the *PV* Diagram for a Gas

In Problems 52 through 55, the initial state of 1 mol of an ideal gas is $P_1 = 3$ atm, $V_1 = 1$ L, and $E_{int,1} = 456$ J, and its final state is $P_2 = 2$ atm, $V_2 = 3$ L, and $E_{int,2} = 912$ J.

52 • ISOLVE✓ The gas is allowed to expand at constant pressure to a volume of 3 L. It is then cooled at constant volume until its pressure is 2 atm. (*a*) Show this process on a *PV* diagram, and calculate the work done by the gas. (*b*) Find the heat added during this process.

53 • ISOLVE✓ The gas is first cooled at constant volume until its pressure is 2 atm. It is then allowed to expand at constant pressure until its volume is 3 L. (*a*) Show this process on a *PV* diagram, and calculate the work done by the gas. (*b*) Find the heat added during this process.

54 •• SSM The gas is allowed to expand isothermally until its volume is 3 L and its pressure is 1 atm. It is then heated at constant volume until its pressure is 2 atm. (*a*) Show this process on a *PV* diagram, and calculate the work done by the gas. (*b*) Find the heat added during this process.

55 •• The gas is heated and is allowed to expand such that it follows a straight-line path on a *PV* diagram from its initial state to its final state. (*a*) Show this process on a *PV* diagram, and calculate the work done by the gas. (*b*) Find the heat added during this process.

56 •• ISOLVE✓ One mole of the ideal gas is initially in the state $P_0 = 1$ atm, $V_0 = 25$ L. As the gas is slowly heated, the plot of its state on a *PV* diagram moves in a straight line to the state $P = 3$ atm, $V = 75$ L. Find the work done by the gas.

57 •• One mole of the ideal gas is heated while its volume changes, so that $T = AP^2$, where A is a constant. The temperature changes from T_0 to $4T_0$. Find the work done by the gas.

58 • SSM An *isobaric* expansion is one carried out at constant pressure. Draw several isobars for an ideal gas on a diagram showing volume as a function of temperature.

59 •• ISOLVE✓ An ideal gas initially at 20°C and 200 kPa has a volume of 4 L. It undergoes a quasi-static, isothermal expansion until its pressure is reduced to 100 kPa. Find (*a*) the work done by the gas, and (*b*) the heat added to the gas during the expansion.

Heat Capacities of Gases and the Equipartition Theorem

60 • The heat capacity at constant volume of a certain amount of a monatomic gas is 49.8 J/K. (*a*) Find the number of moles of the gas. (*b*) What is the internal energy of the gas at $T = 300$ K? (*c*) What is the heat capacity of the gas at constant pressure?

61 • The Dulong–Petit law was originally used to determine the molecular mass of a substance from its measured heat capacity. The specific heat of a certain solid is measured to be 0.447 kJ/kg·K. (*a*) Find the molecular mass of the substance. (*b*) What element is this?

62 •• SSM (*a*) Calculate the specific heats per unit mass of air at constant volume and constant pressure. Assume a temperature of 300 K and a pressure of 10^5 N/m^2. Assume that air is composed of 74% N_2 (molecular weight 28 g/mole) molecules and 26% O_2 molecules (molecular weight 32 g/mole) and that both components are ideal gases. (*b*) Compare your answer to the value listed in the *Handbook of Chemistry and Physics* for the heat capacity at constant pressure of 1.032 J/g·K.

63 •• One mole of an ideal diatomic gas is heated at constant volume from 300 to 600 K. (*a*) Find the increase in internal energy, the work done, and the heat added. (*b*) Find the same quantities if this gas is heated from 300 to 600 K at constant pressure. Use the first law of thermodynamics and your results for (*a*) to calculate the work done. (*c*) Calculate the work done in (*b*) directly from $dW = P\,dV$.

64 •• ISOLVE✓ A diatomic gas (molar mass M) is confined to a closed container of volume V at a pressure P_0. What amount of heat Q should be transferred to the gas in order to triple the pressure? (Express your answer in terms of P_0 and V.)

65 •• One mole of air ($c_v = 5R/2$) is confined at atmospheric pressure in a cylinder with a piston at 0°C. The initial volume, occupied by gas, is V. Find the volume of gas V' after the equivalent of 13,200 J of heat is transferred to it.

66 •• The heat capacity of a certain amount of a particular gas at constant pressure is greater than that at constant volume by 29.1 J/K. (*a*) How many moles of the gas are there? (*b*) If the gas is monatomic, what are C_v and C_p? (*c*) If the gas consists of diatomic molecules that rotate but do not vibrate, what are C_v and C_p?

67 •• SSM Carbon dioxide (CO_2) at 1 atm of pressure and a temperature of −78.5°C sublimates directly from a solid to a gaseous state, without going through a liquid phase. What is the change in the heat capacity (at constant pressure) per mole of CO_2 when it undergoes sublimation? Assume that the gas molecules can rotate but do not vibrate. Is the change in the heat capacity positive or negative on sublimation? The CO_2 molecule is pictured in Figure 18-21.

FIGURE 18-21 Problem 67

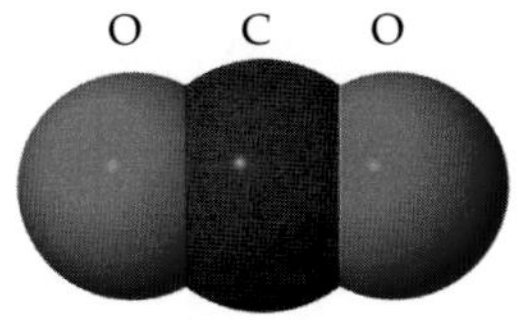

68 •• One mole of a monatomic ideal gas is initially at 273 K and 1 atm. (*a*) What is its initial internal energy? (*b*) Find its final internal energy and the work done by the gas when 500 J of heat are added at constant pressure. (*c*) Find the same quantities when 500 J of heat are added at constant volume.

69 •• List all of the degrees of freedom possible for a water molecule and estimate the heat capacity of water at a temperature very far above its boiling point. (Ignore the fact the molecule might dissociate at high temperatures.) Think carefully about all of the different ways in which a water molecule can vibrate.

Quasi-Static Adiabatic Expansion of a Gas

70 •• One mole of an ideal gas ($\gamma = \frac{5}{3}$) expands adiabatically and quasi-statically from a pressure of 10 atm and a temperature of 0°C to a pressure of 2 atm. Find (*a*) the initial and final volumes, (*b*) the final temperature, and (*c*) the work done by the gas.

71 • An ideal gas at a temperature of 20°C is compressed quasi-statically and adiabatically to half its original volume. Find its final temperature if (*a*) $C_v = \frac{3}{2}nR$ and (*b*) $C_v = \frac{5}{2}nR$.

72 • ISOLVE✓ Two moles of neon gas initially at 20°C and a pressure of 1 atm are compressed adiabatically to one-fourth of their initial volume. Determine the temperature and pressure following compression.

73 •• SSM Half a mole of an ideal monatomic gas at a pressure of 400 kPa and a temperature of 300 K expands until the pressure has diminished to 160 kPa. Find the final temperature and volume, the work done, and the heat absorbed by the gas if the expansion is (*a*) isothermal and (*b*) adiabatic.

74 •• Repeat Problem 73 for a diatomic gas.

75 •• One-half mole of helium is expanded adiabatically and quasi-statically from an initial pressure of 5 atm and temperature of 500 K to a final pressure of 1 atm. Find (*a*) the final temperature, (*b*) the final volume, (*c*) the work done by the gas, and (*d*) the change in the internal energy of the gas.

76 ••• SSM A hand pump is used to inflate a bicycle tire to a gauge pressure of 482 kPa (about 70 lb/in.^{2}). How much work must be done if each stroke of the pump is an adiabatic process? Atmospheric pressure is 1 atm, the air temperature is initially 20°C, and the volume of the air in the tire remains constant at 1 L.

77 ••• An ideal gas at initial volume V_1 and pressure P_1 expands quasi-statically and adiabatically to volume V_2 and pressure P_2. Calculate the work done by the gas directly by integrating $P\,dV$, and show that your result is the same as that given by Equation 18-39.

Cyclic Processes

78 •• One mole of N_2 ($C_v = \frac{5}{2}nR$) gas is originally at room temperature (20°C) and a pressure of 5 atm. It is allowed to expand adiabatically and quasi-statically until its pressure equals the room pressure of 1 atm. It is then heated at constant pressure until its temperature is again 20°C. During this heating, the gas expands. After it reaches room temperature, it is heated at constant volume until its pressure is 5 atm. It is then compressed at constant pressure until it is back to its original state. (*a*) Construct an accurate PV diagram showing each process in the cycle. (*b*) From your graph, determine the work done by the gas during the complete cycle. (*c*) How much heat is added or subtracted from the gas during the complete cycle? (*d*) Check your graphical determination of the work done by the gas in (*b*) by calculating the work done during each part of the cycle.

79 •• SSM One mole of an ideal diatomic gas is allowed to expand along the straight line from 1 to 2 in the PV diagram (Figure 18-22). It is then compressed back isothermally from 2 to 1. Calculate the total work done on the gas during this cycle.

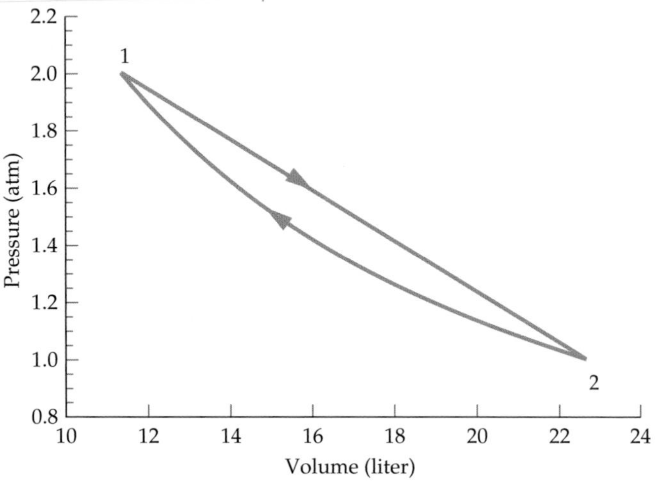

FIGURE 18-22 Problem 79

80 •• Two moles of an ideal monatomic gas have an initial pressure $P_1 = 2$ atm and an initial volume $V_1 = 2$ L. The gas is taken through the following quasi-static cycle: It is expanded isothermally until it has a volume $V_2 = 4$ L. It is then heated at constant volume until it has a pressure $P_3 = 2$ atm. It is then cooled at constant pressure until it is back to its initial state. (*a*) Show this cycle on a *PV* diagram. (*b*) Calculate the heat added and the work done by the gas during each part of the cycle. (*c*) Find the temperatures T_1, T_2, and T_3.

81 ••• At point D in Figure 18-23 the pressure and temperature of 2 mol of an ideal monatomic gas are 2 atm and 360 K. The volume of the gas at point B on the *PV* diagram is three times that at point D and its pressure is twice that at point C. Paths AB and CD represent isothermal processes. The gas is carried through a complete cycle along the path DABCD. Determine the total amount of work done by the gas and the heat supplied to the gas along each portion of the cycle.

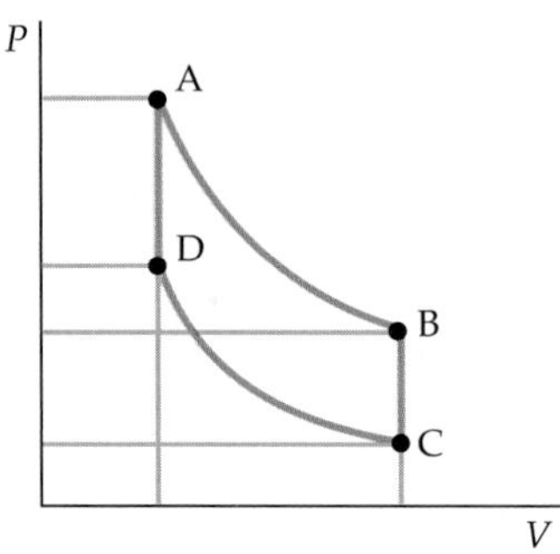

FIGURE 18-23 Problems 81 and 82

82 ••• SSM Repeat Problem 81 with a diatomic gas.

83 ••• An ideal gas of n mol is initially at pressure P_1, volume V_1, and temperature T_h. It expands isothermally until its pressure and volume are P_2 and V_2. It then expands adiabatically until its temperature is T_c and its pressure and volume are P_3 and V_3. It is then compressed isothermally until it is at a pressure P_4 and a volume V_4, which is related to its initial volume V_1 by $T_c V_4^{\gamma-1} = T_h V_1^{\gamma-1}$. The gas is then compressed adiabatically until it is back in its original state. (*a*) Assuming that each process is quasi-static, plot this cycle on a *PV* diagram. (This cycle is known as the Carnot cycle for an ideal gas.) (*b*) Show that the heat Q_h absorbed during the isothermal expansion at T_h is $Q_h = nRT_h \ln(V_2/V_1)$. (*c*) Show that the heat Q_c given off by the gas during the isothermal compression at T_c is $Q_c = nRT_c \ln(V_3/V_4)$. (*d*) Using the result that $TV^{\gamma-1}$ is constant for an adiabatic expansion, show that $V_2/V_1 = V_3/V_4$. (*e*) The efficiency of a Carnot cycle is defined to be the net work done divided by the heat absorbed Q_h. Using the first law of thermodynamics, show that the efficiency is $1 - Q_c/Q_h$. (*f*) Using your results from the previous parts of this problem, show that $Q_c/Q_h = T_c/T_h$.

General Problems

84 • ISOLVE ✓ The volume of three moles of a monatomic gas is increased from 50 L to 200 L at constant pressure. The initial temperature of the gas is 300 K. How much heat must be supplied to the gas?

85 • In the process of compressing n moles of an ideal diatomic gas to one-fifth of its initial volume, 180 kJ of work is done on the gas. If this is accomplished isothermally at room temperature (293 K), how many calories of heat are removed from the gas?

86 • SSM What is the number of moles n of the gas in Problem 85?

87 • The *PV* diagram in Figure 18-24 represents 3 mol of an ideal monatomic gas. The gas is initially at point A. The paths AD and BC represent isothermal changes. If the system is brought to point C along the path AEC, find (*a*) the initial and final temperatures, (*b*) the work done by the gas, and (*c*) the heat absorbed by the gas.

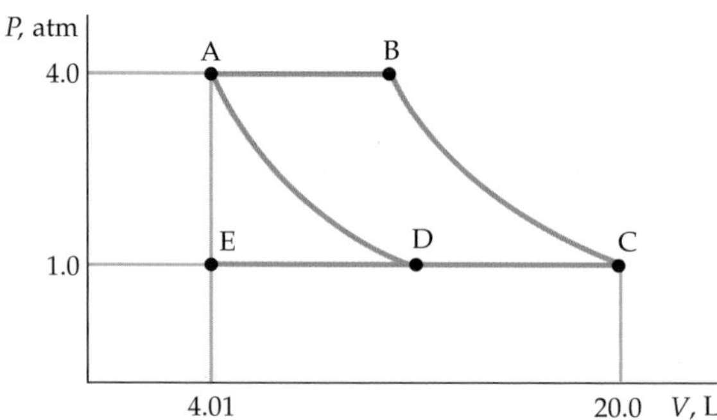

FIGURE 18-24 Problems 87-90

88 •• Repeat Problem 87 with the gas following path ABC.

89 •• SSM Repeat Problem 87 with the gas following path ADC.

90 •• Suppose that the paths AD and BC represent adiabatic processes. What then are the work done by the gas and the heat absorbed by the gas in following the path ABC?

91 •• At very low temperatures, the specific heat of a metal is given by $c = aT + bT^3$. For the metal copper, $a = 0.0108$ J/kg·K^2 and $b = 7.62 \times 10^{-4}$ J/kg·K^4. (*a*) What is the specific heat of copper at 4 K? (*b*) How much heat is required to heat copper from 1 to 3 K?

92 •• Two moles of a diatomic ideal gas are compressed isothermally from 18 L to 8 L. In the process, 170 calories escape from the system. Determine the amount of work done by the gas, the change in internal energy, and the initial and final temperatures of the gas.

93 •• Suppose the two moles of a diatomic ideal gas in Problem 92 are compressed from 18 L to 8 L adiabatically. The work done on the gas is 820 J. Find the initial temperature and the initial and final pressures.

94 •• How much work must be done to 30 grams of CO at standard temperature and pressure to compress it to one-fifth of its initial volume if the process is (*a*) isothermal; (*b*) adiabatic?

95 •• Repeat Problem 94 if the gas is CO_2.

96 •• Repeat Problem 94 if the gas is argon.

97 •• ISOLVE A thermally insulated system consists of 1 mol of a diatomic ideal gas at 100 K and 2 mol of a solid at 200 K that are separated by a rigid insulating wall. Find the equilibrium temperature of the system after the insulating wall is removed, assuming that the solid obeys the Dulong–Petit law.

98 •• SSM When an ideal gas undergoes a temperature change at constant volume, its energy changes by $\Delta E_{int} = C_v \Delta T$. (*a*) Explain why this result holds for an ideal gas for any temperature change independent of the process. (*b*) Show explicitly that this result holds for the expansion of an ideal gas at constant pressure by first calculating the work done and showing that it can be written as $W = nR\Delta T$, and then by using $\Delta E_{int} = Q - W$, where $Q = C_p \Delta T$.

99 •• One mole of an ideal monatomic gas is heated at constant volume from 300 to 600 K. (*a*) Find the heat added, the work done by the gas, and the change in its internal energy. (*b*) Find these same quantities if the gas is heated from 300 to 600 K at constant pressure.

100 •• SSM Heat in the amount of 500 J is supplied to 2 mol of an ideal diatomic gas. (*a*) Find the change in temperature if the pressure is kept constant. (*b*) Find the work done by the gas. (*c*) Find the ratio of the final volume of the gas to the initial volume if the initial temperature is 20°C.

101 •• An insulated cylinder is fitted with a movable piston to maintain constant pressure. The cylinder initially contains 100 g of ice at −10°C. Heat is supplied to the contents at a constant rate by a 100-W heater. Make a graph showing the temperature of the cylinder contents as a function of time starting at $t = 0$, when the temperature is −10°C, and ending when the temperature is 110°C. (Use $c = 2.0$ kJ/kg·K for the average specific heat of ice from −10 to 0°C and of steam from 100 to 110°C.)

102 •• SSM Two moles of a diatomic ideal gas expand adiabatically. The initial temperature of the gas is 300 K. The work done by the gas during the expansion is 3.5 kJ. What is the final temperature of the gas?

103 •• One mole of monatomic gas, initially at temperature T, undergoes a process in which its temperature is quadrupled and its volume is halved. Find the amount of heat Q transferred to the gas. It is known that in this process the pressure was never less than the initial pressure, and the work done on the gas was the minimum possible.

104 •• If a small amount of a substance is dissolved into a liquid, the liquid pressure will rise slightly. For a dilute solution, the change in pressure follows the ideal-gas law: $PV = NkT$, where N is the number of solute molecules dissolved in the liquid, V is the liquid volume, and P is the increase in the liquid pressure. Calculate the increase in pressure when 20 g of table salt (NaCl) are dissolved in 1 L of water at a temperature of 24°C.

105 •• A vertical heat-insulated cylinder is divided into two parts by a movable piston of mass m. Initially the piston is held at rest. The top part is evacuated and the bottom part is filled with 1 mole of diatomic ideal gas at temperature 300 K. After the piston is released and the system comes to equilibrium, the volume, occupied by gas, is halved. Find the final temperature of the gas.

106 •• According to the Einstein model of a crystalline solid, the internal energy per mole is given by

$$U = \frac{3N_A kT_E}{e^{T_E/T} - 1}$$

where T_E is a characteristic temperature called the Einstein temperature, and T is the temperature of the solid in kelvins. Calculate the molar internal energy of diamond ($T_E = 1060$ K) at 300 K and 600 K, and thereby the increase in internal energy as diamond is heated from 300 K to 600 K.

107 ••• SSM In an isothermal expansion, an ideal gas at an initial pressure P_0 expands until its volume is twice its initial volume. (*a*) Find its pressure after the expansion. (*b*) The gas is then compressed adiabatically and quasi-statically back to its original volume, at which point its pressure is $1.32P_0$. Is the gas monatomic, diatomic, or polyatomic? (*c*) How does the translational kinetic energy of the gas change in these processes?

Note: Problems 108 and 109 involve nonquasi-static processes. Nevertheless, assuming that the gases participating in these processes approximate ideal gases, one can calculate the state functions of the end products of the reactions using the first law of thermodynamics and the ideal-gas law. For $T > 2000$ K, vibration of the atoms contributes to C_p of H_2O and CO_2 so that C_p of these gases is 7.5R at high temperatures. Also, assume the gases do not dissociate.

108 ••• The combustion of benzene is represented by the chemical reaction $2(C_6H_6) + 15(O_2) \rightarrow 12(CO_2) + 6(H_2O)$. The amount of energy released in the combustion of 2 mol of benzene is 1516 kcal. One mol of benzene and 7.5 mol of oxygen at 300 K are confined in an insulated enclosure at a pressure of 1 atm. (*a*) Find the temperature and volume following combustion if the pressure is maintained at 1 atm. (*b*) If, following combustion, the thermal insulation about the container is removed and the system is cooled to 300 K, what is the final pressure?

109 ••• SSM Carbon monoxide and oxygen combine to form carbon dioxide with an energy release of 280 kJ/mol of CO according to the reaction $2(CO) + O_2 \rightarrow 2(CO_2)$. Two mol of CO and one mol of O_2 at 300 K are confined in an 80-L container; the combustion reaction is initiated with a spark. (*a*) What is the pressure in the container prior to the reaction? (*b*) If the reaction proceeds adiabatically, what are the final temperature and pressure? (*c*) If the resulting CO_2 gas is cooled to 0°C, what is the pressure in the container?

110 ••• Use the expression given in Problem 106 for the internal energy per mole of a solid according to the Einstein model to show that the molar heat capacity at constant volume is given by

$$c'_{\mathrm{v}} = 3R\left(\frac{T_{\mathrm{E}}}{T}\right)^2 \frac{e^{T_{\mathrm{E}}/T}}{(e^{T_{\mathrm{E}}/T} - 1)^2}$$

111 ••• (*a*) Use the results of Problem 110 to show that the Dulong–Petit law, $c'_{\mathrm{v}} \approx 3R$ holds for the Einstein model when $T > T_{\mathrm{E}}$. (*b*) For diamond, T_{E} is approximately 1060 K. Numerically integrate $dE_{\mathrm{int}} = c'_{\mathrm{v}}\, dT$ to find the increase in the internal energy if 1 mol of diamond is heated from 300 to 600 K. Compare your result to that obtained in Problem 106.

112 ••• If a hole is punctured in a tire, the gas inside will gradually leak out of it. Let's assume the following: the area of the hole is A; the tire volume is V; and the time, τ, it takes for most of the air to leak out of the tire can be expressed in terms of the ratio A/V, the temperature T, the Boltzmann constant k, and the mass of the gas molecules inside the tire, m. (*a*) Under these assumptions, use dimensional analysis to find an estimate for τ. (*b*) Use the result of Part (*a*) to estimate the time it takes for a car tire with a nail hole punched in it to go flat.

CHAPTER

The Second Law of Thermodynamics

THIS OLD-FASHIONED TRAIN ENGINE PRODUCES STEAM WHICH DOES WORK ON A PISTON THAT CAUSES THE TRAIN'S WHEELS TO MOVE. THE STEAM ENGINE'S EFFICIENCY IS LIMITED BY THE SECOND LAW OF THERMODYNAMICS.

? What is the maximum possible efficiency of this engine? (See Example 19-4.)

We are often asked to conserve energy. But according to the first law of thermodynamics, energy is always conserved. What then does it mean to conserve energy if the total amount of energy in the universe does not change regardless of what we do? The first law of thermodynamics does not tell the whole story. Energy is always conserved, but some forms of energy are more useful than others. The possibility or impossibility of putting energy to *use* is the subject of the **second law of thermodynamics.** For example, it is easy to convert work into thermal energy, but it is impossible to remove energy as heat from a single reservoir and convert it entirely into work with no other changes. This experimental fact is one statement of the second law of thermodynamics.

Solar energy is directed toward the solar oven at the center by this circular array of reflectors at Barstow, California.

No system can take energy as heat from a single reservoir and convert it entirely into work without additional net changes in the system or its surroundings.

SECOND LAW OF THERMODYNAMICS: KELVIN STATEMENT

➢ **In this chapter, we will encounter several other formulations of this law.**

A common example of the conversion of work into heat is movement with friction. For example, suppose you spend two minutes pushing a block this way and that way along a tabletop in a closed path, leaving the block in its initial position. Also, suppose that the block-table system is initially in thermal equilibrium with its surroundings. The work you do on the system is converted into internal energy of the system, and as a result the block-table system becomes warmer. Consequently, the system is no longer in thermal equilibrium with its surroundings. However, the system will transfer energy as heat to its surroundings until it returns to thermal equilibrium with those surroundings. Because the final and initial states of the system are the same, the first law of thermodynamics dictates that the energy transferred to the environment as heat equals the work done by you on the system. The reverse process never occurs—a block and table that are warm will never spontaneously cool by converting their internal energy into work that causes the block to push your hand around the table! Yet such an amazing occurrence would not violate the first law of thermodynamics or any other physical laws we have encountered so far. It does, however, violate the second law of thermodynamics. Thus, there is a lack of symmetry in the roles played by heat and work that is not evident from the first law. This lack of symmetry is related to the fact that some processes are *irreversible.*

There are many other irreversible processes, seemingly quite different from one another, but all related to the second law. For example, heat conduction is an irreversible process. If we place a hot body in contact with a cold body, heat will flow from the hot body to the cold body until they are at the same temperature. However, the reverse does not occur. Two bodies in contact at the same temperature remain at the same temperature; heat does not flow from one to the other leaving one colder and the other warmer. This experimental fact gives us a second statement of the second law of thermodynamics.

A process whose only net result is to transfer energy as heat from a cooler object to a hotter one is impossible.

SECOND LAW OF THERMODYNAMICS: CLAUSIUS STATEMENT

We will show in this chapter that the Kelvin and Clausius statements of the second law are equivalent.

19-1 Heat Engines and the Second Law of Thermodynamics

The study of the efficiency of heat engines gave rise to the first clear statements of the second law. A **heat engine** is a cyclic device whose purpose is to convert as much heat input into work as possible. Heat engines contain a **working substance** (water in a steam engine, air and gasoline vapor in an internal-combustion engine) that absorbs a quantity of heat Q_h from a high temperature reservoir, does work W on its surroundings, and gives off heat Q_c as it returns to its initial state, where

Q_h, W, and Q_c represent magnitudes and are positive.

The earliest heat engines were steam engines, invented in the eighteenth century for pumping water from coal mines. Today steam engines are used to generate electricity. In a typical steam engine, liquid water is heated under several hundred atmospheres of pressure until it vaporizes at about 500°C (Figure 19-1). This steam expands against a piston (or turbine blades), doing work, then exits at a much lower temperature and is cooled further in the condenser where heat is transferred from it, causing it to condense. The water is then pumped back into the boiler and heated again.

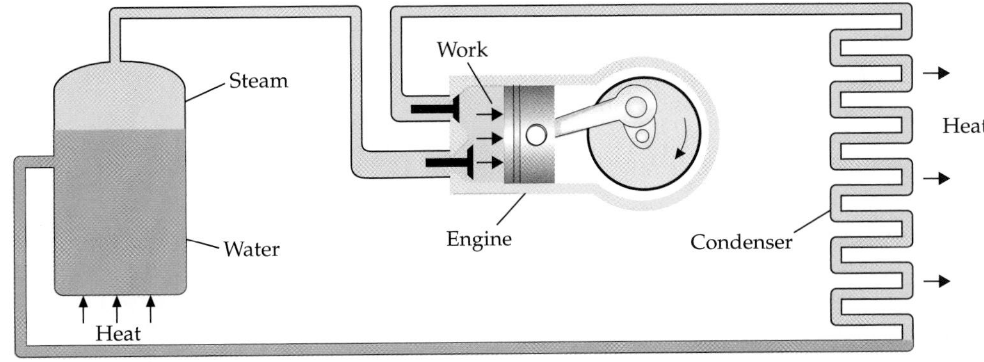

FIGURE 19-1 **Schematic drawing of a steam engine. High-pressure steam does work on the piston.**

Figure 19-2 is a schematic diagram of the heat engine used in many automobiles —the internal-combustion engine. With the exhaust valve closed, a mixture of gasoline vapor and air enters the combustion chamber as the piston moves down

Intake valve open
Gas vapor and air mixture
Intake valve
Exhaust valve
A mixture of gasoline vapor and air enters the combustion chamber as the piston moves down.
Intake stroke (1)
Spark plug
Both valves closed
Cylinder
Piston
Connecting rod
Crankshaft
The piston then moves up, compressing the gas for ignition.
Compression stroke (2)
Both valves closed
When the gas ignites, it expands.
Ignition (3)
Both valves closed
The expanding gas moves the piston down, a stage called the power stroke.
Power stroke (4)
Exhaust valve open
To exhaust pipe
The piston moves up again to exhaust the burned gases.
Exhaust stroke (5)

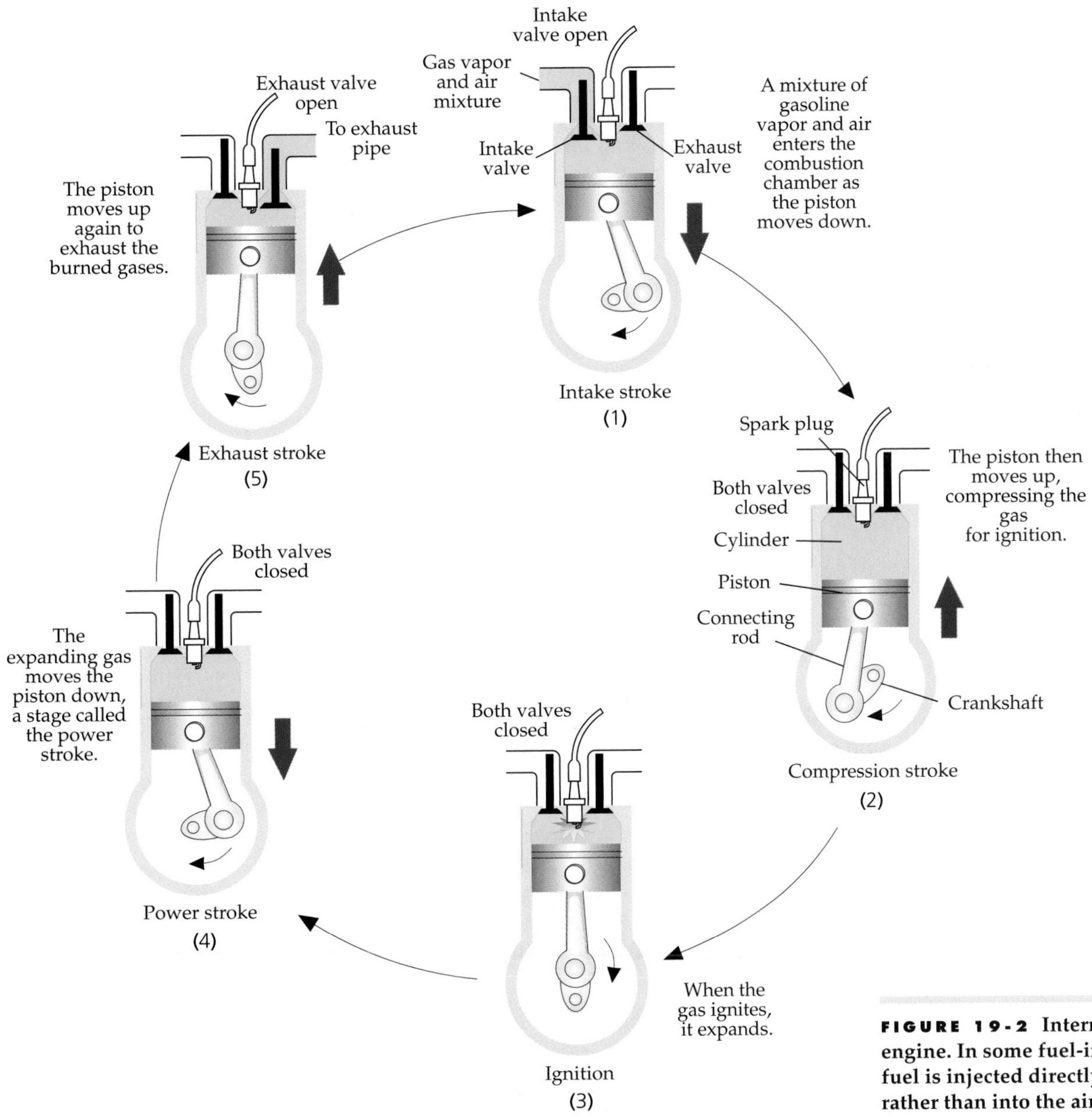

FIGURE 19-2 **Internal-combustion engine. In some fuel-injected engines, the fuel is injected directly into the cylinder rather than into the air stream.**

during the intake stroke. The mixture is then compressed, after which it is ignited by a spark from the spark plug. The hot gases then expand, driving the piston down and doing work on it in the stage called the power stroke. The gases are then exhausted through the exhaust valve, and the cycle repeats. An idealized model of the processes in the internal combustion engine is called the **Otto cycle** and is shown in Figure 19-3.

Figure 19-4 shows a schematic representation of a basic heat engine. The heat input is represented as coming from a **hot heat reservoir** at temperature T_h, and the exhaust goes into a **cold heat reservoir** at a lower temperature T_c. A hot or cold heat reservoir is an idealized body or system that has a very large heat capacity so that it can absorb or give off energy as heat with no noticeable change in its temperature. In practice, burning fossil fuel often acts as the high-temperature reservoir, and the surrounding atmosphere or a lake often acts as the low-temperature reservoir. Applying the first law of thermodynamics ($\Delta E_{int} = Q_{in} + W_{on}$) to the heat engine gives

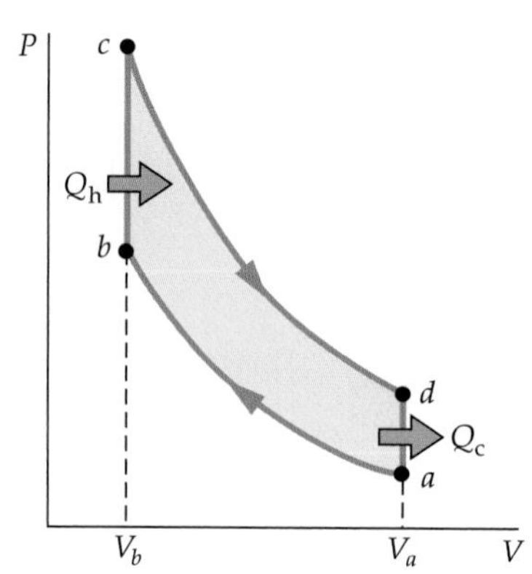

FIGURE 19-3 Otto cycle, representing the internal-combustion engine. The fuel-air mixture enters at *a* and is adiabatically compressed to *b*. It is then heated (by ignition from the spark plug) at constant volume to *c*. The power stroke is represented by the adiabatic expansion from *c* to *d*. The cooling at constant volume from *d* to *a* represents the exhausting of the combustion products and the intake of a fresh fuel-air mixture.

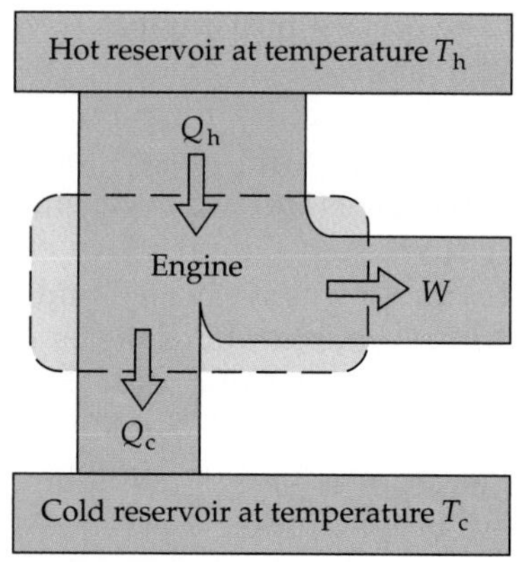

FIGURE 19-4 Schematic representation of a heat engine. The engine removes heat energy Q_h from a hot reservoir at a temperature T_h, does work W, and gives off heat Q_c to a cold reservoir at a temperature T_c.

$$W = Q_h - Q_c \qquad 19\text{-}1$$

where W is the work done *by* the engine during one complete cycle, $Q_h - Q_c$ is the total energy transferred to the engine as heat during one cycle, and ΔE_{int} is the change in internal energy of the engine (including the working substance) during one cycle. Since the initial and final states of the engine for a complete cycle are the same, the initial and final internal energies of the engine are equal. Thus, $\Delta E_{int} = 0$.

The **efficiency** ε of a heat engine is defined as the ratio of the work done by the engine to the heat absorbed from the high temperature reservoir:

$$\varepsilon = \frac{W}{Q_h} = \frac{Q_h - Q_c}{Q_h} = 1 - \frac{Q_c}{Q_h} \qquad 19\text{-}2$$

DEFINITION—EFFICIENCY OF A HEAT ENGINE

The heat Q_h is usually produced by burning some fuel like coal or oil that must be paid for, so it is desirable to get the most efficient use of the fuel as possible. The best steam engines operate near 40 percent efficiency; the best internal-combustion engines operate near 50 percent efficiency. At 100 percent efficiency ($\varepsilon = 1$), all the thermal energy absorbed from the hot reservoir would be converted into work and no thermal energy would be given off to the cold reservoir. However, *it is impossible to make a heat engine with an efficiency of 100 percent.* This experimental result is the **heat-engine statement of the second law of thermodynamics.** It is another way of expressing the Kelvin statement given earlier:

An exhaust manifold feeds the header pipes seen on this top-fuel dragster in order to carry heat away from the engine to reduce its temperature.

It is impossible for a heat engine working in a cycle to produce *only the effect* of extracting heat from a single reservoir and performing an equivalent amount of work.

SECOND LAW OF THERMODYNAMICS: HEAT-ENGINE STATEMENT

The word *cycle* in this statement is important because it *is* possible to convert heat completely into work in a noncyclic process. An ideal gas undergoing an isothermal expansion does just this. But after the expansion, the gas is not in its original state. To bring the gas back to its original state, work must be done on the gas, and some heat will be exhausted.

The second law tells us that to do work with energy extracted from a heat reservoir, we must have a colder reservoir available to receive part of the energy as exhaust. If this were not true, we could design a ship with a heat engine that was powered by simply extracting energy as heat from the ocean. Unfortunately, the lack of a colder reservoir for exhaust makes this enormous reservoir of energy unavailable for such use. (It is theoretically possible to run a heat engine between the warmer surface water of the ocean and the colder water at greater depths, but no practical scheme for using this temperature difference has yet emerged.) In order to convert completely disordered thermal energy at a single temperature into the completely ordered energy associated with work (with no other changes in the source or object), a separate cold reservoir must be used.

EFFICIENCY OF A HEAT ENGINE — EXAMPLE 19-1

During each cycle a heat engine absorbs 200 J of heat from a hot reservoir, does work, and exhausts 160 J to a cold reservoir. What is the efficiency of the engine?

PICTURE THE PROBLEM We use the definition of the efficiency of a heat engine (Equation 19-2).

1. The efficiency is the work done divided by the heat absorbed: $\varepsilon = \dfrac{W}{Q_h}$

2. The heat absorbed is given: $Q_h = 200\text{ J}$

3. The work is found from the first law: $W = Q_h - Q_c = 200\text{ J} - 160\text{ J} = 40\text{ J}$

4. Substitute the values of Q_h and W to calculate the efficiency: $\varepsilon = \dfrac{W}{Q_h} = \dfrac{40\text{ J}}{200\text{ J}} = 0.20 = \boxed{20\%}$

EXERCISE A heat engine has an efficiency of 35%. (*a*) How much work does it perform in a cycle if it extracts 150 J of energy as heat from a hot reservoir per cycle? (*b*) How much energy as heat is exhausted to the cold reservoir per cycle? (*Answer* (*a*) 52.5 J (*b*) 97.5 J)

EFFICIENCY OF AN IDEAL INTERNAL COMBUSTION ENGINE—THE OTTO CYCLE — EXAMPLE 19-2 Try It Yourself

(*a*) Find the efficiency of the Otto cycle shown in Figure 19-3. (*b*) Express your answer in terms of the ratio of the volumes $r = V_a/V_b = V_d/V_c$.

PICTURE THE PROBLEM (*a*) To find ε, you need to find Q_h and Q_c. Heat transfer occurs only during the two constant-volume processes, b to c and d to a. You can thus find Q_h and Q_c and therefore ε in terms of the temperatures T_a, T_b, T_c, and T_d. (*b*) The temperatures can be related to the volumes using $TV^{\gamma-1} =$ constant for adiabatic processes.

Cover the column to the right and try these on your own before looking at the answers.

Steps	**Answers**
(*a*) 1. Write the efficiency in terms of Q_h and Q_c.	$\varepsilon = 1 - \dfrac{Q_{cold}}{Q_{hot}} = 1 - \dfrac{Q_c}{Q_h}$
2. The heat out occurs at constant volume from d to a. Write Q_c in terms of C_v and the temperatures T_a and T_d.	$Q_c = \lvert Q_{d\to a}\rvert = C_v \lvert T_a - T_d \rvert = C_v(T_d - T_a)$
3. The heat in occurs at constant volume from b to c. Write Q_h in terms of C_v and the temperatures T_c and T_b.	$Q_h = Q_{b\to c} = C_v(T_c - T_b)$
4. Substitute these values of Q_c and Q_h to find the efficiency in terms of the temperatures T_a, T_b, T_c, and T_d.	$\varepsilon = \boxed{1 - \dfrac{T_d - T_a}{T_c - T_b}}$
(*b*) 1. Relate T_c to T_d using $TV^{\gamma-1}$ = constant, and $V_a/V_c = r$.	$T_cV_c^{\gamma-1} = T_dV_d^{\gamma-1}$ $T_c = T_d\dfrac{V_d^{\gamma-1}}{V_c^{\gamma-1}} = T_d r^{\gamma-1}$
2. Relate T_b to T_a as in step 1.	$T_b = T_a r^{\gamma-1}$
3. Use these relations to eliminate T_c and T_b from ε in Part (*a*) so that ε is expressed in terms of r.	$\varepsilon = 1 - \dfrac{T_d - T_a}{T_d r^{\gamma-1} - T_a r^{\gamma-1}} = \boxed{1 - \dfrac{1}{r^{\gamma-1}}}$

REMARKS The ratio r (volume before compression/volume after compression) is called the compression ratio.

19-2 Refrigerators and the Second Law of Thermodynamics

A **refrigerator** is essentially a heat engine run backwards (Figure 19-5*a*.). The refrigerator's engine extracts thermal energy from the interior of the refrigerator (cold reservoir) and transfers it to the surroundings (hot reservoir) (Figure 19-5*b*). Experience shows that such a transfer always requires work—a result known as the **refrigerator statement of the second law of thermodynamics,** which is another way of expressing the Clausius statement:

Hot reservoir at temperature T_h

Q_h

Refrigerator

W

Q_c

Cold reservoir at temperature T_c

(*a*)

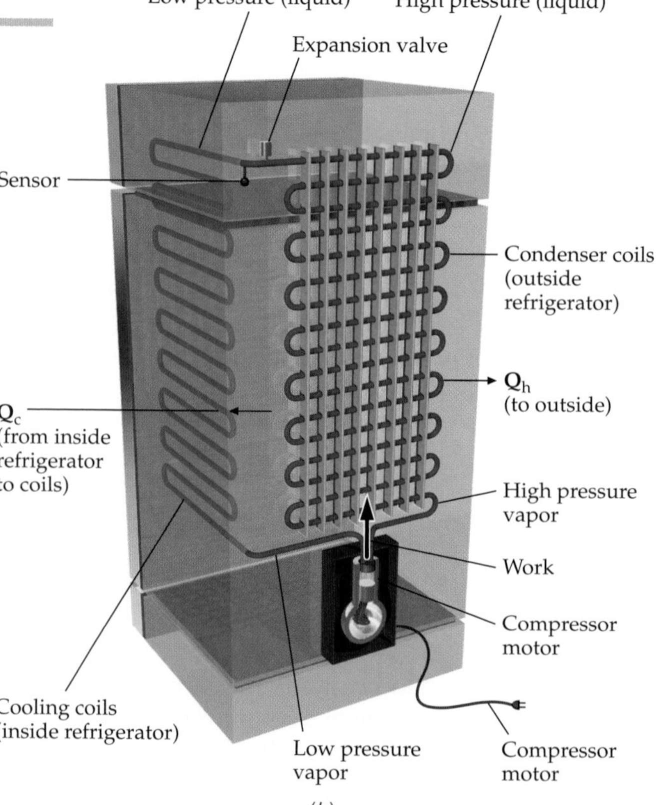

FIGURE 19-5 (*a*) Schematic representation of a refrigerator. Work W is done on the refrigerator and it removes heat energy Q_c from a cold reservoir and gives off heat Q_h. (*b*) An actual refrigerator.

It is impossible for a refrigerator working in a cycle to produce *only the effect* of extracting heat from a cold object and reject the same amount of heat to a hot object.

SECOND LAW OF THERMODYNAMICS: REFRIGERATOR STATEMENT

Were the above statement not true, we could cool our homes in the summer with refrigerators that pumped thermal energy to the outside without using any electricity or any other energy.

A measure of a refrigerator's performance is the ratio Q_c/W of the heat removed from the low temperature reservoir to the work done on the refrigerator. (This work equals the electrical energy that comes from the wall outlet.) The ratio Q_c/W is called the **coefficient of performance** (COP):

$$\text{COP} = \frac{Q_c}{W} \qquad 19\text{-}3$$

DEFINITION—COEFFICIENT OF PERFORMANCE (REFRIGERATOR)

The greater the COP, the better the refrigerator. Typical refrigerators have coefficients of performance of about 5 or 6. In terms of this ratio, the refrigerator statement of the second law says that the COP of a refrigerator cannot be infinite.

MAKING ICE CUBES **EXAMPLE 19-3** **Put It in Context**

You have one hour before guests start arriving for your party when you suddenly realize that you forgot to buy ice for drinks. You quickly put one liter of water at 10°C into your ice cube trays and pop them into the freezer. Will you have ice in time for your guests? The label on your refrigerator states that the appliance has a coefficient of performance of 5.5 and a power rating of 550 W. You estimate that only 10 percent of the power goes to freezing the ice.

PICTURE THE PROBLEM Work equals power times time. We are given the power, so we need to find the work to determine the time. The work is related to Q_c by Equation 19-3. To find Q_c we calculate how much heat must be extracted from the water.

1. The time needed is related to the power available and the work required:

$$P = W/t$$
$$t = W/P$$

2. The work is related to the coefficient of performance and the heat extracted:

$$W = \frac{Q_c}{\text{COP}}$$

3. The heat Q_c removed from inside of the refrigerator equals the heat Q_{cool} to be removed from the water to cool it plus the heat Q_{freeze} to be removed from the water to freeze it:

$$Q_c = Q_{cool} + Q_{freeze}$$

4. The heat needed to cool 1 L of water (mass 1 kg) by 10°C is:

$$Q_{cool} = mc\Delta T = (1\text{ kg})(4.18\text{ kJ}/(\text{kg}\cdot\text{K}))(10\text{ K}) = 41.8\text{ kJ}$$

5. The heat needed to freeze 1 L of water into ice cubes is:

$$Q_{freeze} = mL_f = (1\text{ kg})(333.5\text{ kJ/kg}) = 333.5\text{ kJ}$$

6. Add these heats to obtain Q_c: $Q_c = 41.8\text{ kJ} + 333.5\text{ kJ} = 375.3\text{ kJ} \approx 375\text{ kJ}$

7. Substitute Q_c into step 2 to find the work W: $W = \dfrac{Q_c}{\text{COP}} = \dfrac{375\text{ kJ}}{5.5} = 68.2\text{ kJ}$

8. Use this value of W and 55 W for the available power to find the time t: $t = \dfrac{W}{P} = \dfrac{68.2\text{ kJ}}{55\text{ J/s}} = 1.24\text{ ks} \times \dfrac{1\text{ min}}{60\text{ s}} = \boxed{20.7\text{ min}}$

REMARKS You won't make it in one hour. You should have done the calculation before you put the water into the refrigerator and then used only half as much water.

EXERCISE A refrigerator has a coefficient of performance of 4.0. How much heat is exhausted to the hot reservoir if 200 kJ of heat are removed from the cold reservoir? (*Answer* 250 kJ)

19-3 Equivalence of the Heat-Engine and Refrigerator Statements

The heat-engine and refrigerator statements (or the Kelvin and Clausius statements, respectively) of the second law of thermodynamics seem quite different, but they are actually equivalent. We can prove this by showing that if either statement is assumed to be false, then the other must also be false. We'll use a numerical example to show that if the heat-engine statement is false, then the refrigerator statement is false.

Figure 19-6*a* shows an ordinary refrigerator that uses 50 J of work to remove 100 J of energy as heat from a cold reservoir and rejects 150 J of energy as heat to a hot reservoir. Suppose the heat-engine statement of the second law were not true. Then a "perfect" heat engine could remove energy from the hot reservoir and convert it completely into work with 100 percent efficiency. We could use this perfect heat engine to remove 50 J of energy from the hot reservoir and do 50 J of work (Figure 19-6*b*) on the ordinary refrigerator. Then, the combination of the perfect heat engine and the ordinary refrigerator would be a perfect refrigerator, transferring 100 J of energy as heat from the cold reservoir to the hot reservoir without requiring any work, as illustrated in Figure 19-6*c*. This violates the refrigerator statement of the second law. Thus, if the heat-engine statement is false, the refrigerator statement is also false. Similarly, if a perfect refrigerator existed, it could be used in conjunction with an ordinary heat engine to construct a perfect heat engine. Thus, if the refrigerator statement is false, the heat-engine statement is also false. It then follows that if one statement is true, the other is also true. Therefore, the heat engine statement and the refrigerator statement are equivalent.

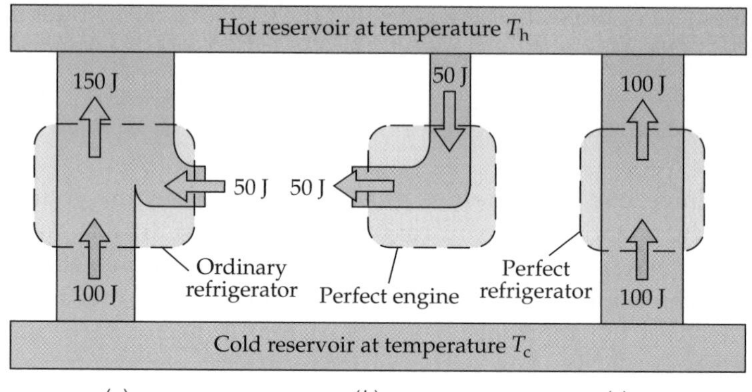

(*a*) An ordinary refrigerator removes 100 J from a cold reservoir, requiring the input of 50 J of work.

(*b*) A perfect heat engine violates the heat engine statement of the second law by removing 50 J from the hot reservoir and converting it completely into work.

(*c*) Putting the two together makes a perfect refrigerator that violates the refrigerator statement of the second law by transferring 100 J from the cold reservoir to the hot reservoir with no other effect.

FIGURE 19-6 Demonstration of the equivalence of the heat-engine and refrigerator statements of the second law of thermodynamics.

19-4 The Carnot Engine

According to the second law of thermodynamics, it is impossible for a heat engine working between two heat reservoirs to be 100% efficient. What, then, is the maximum possible efficiency for such an engine? A young French engineer,

Sadi Carnot answered this question in 1824, before either the first or the second law of thermodynamics had been established. Carnot found that a *reversible engine* is the most efficient engine that can operate between any two given reservoirs. This result is known as the Carnot theorem:

> No engine working between two given heat reservoirs can be more efficient than a reversible engine working between those two reservoirs.

CARNOT THEOREM

A reversible engine working in a cycle between two heat reservoirs is called a **Carnot engine,** and its cycle is called a **Carnot cycle.** Figure 19-7 illustrates the Carnot theorem with a numerical example.

If no engine can have a greater efficiency than a Carnot engine, it follows that all Carnot engines working between the same two reservoirs have the same efficiency. This efficiency, called the **Carnot efficiency,** must be independent of the working substance of the engine and thus can depend only on the temperatures of the reservoirs.

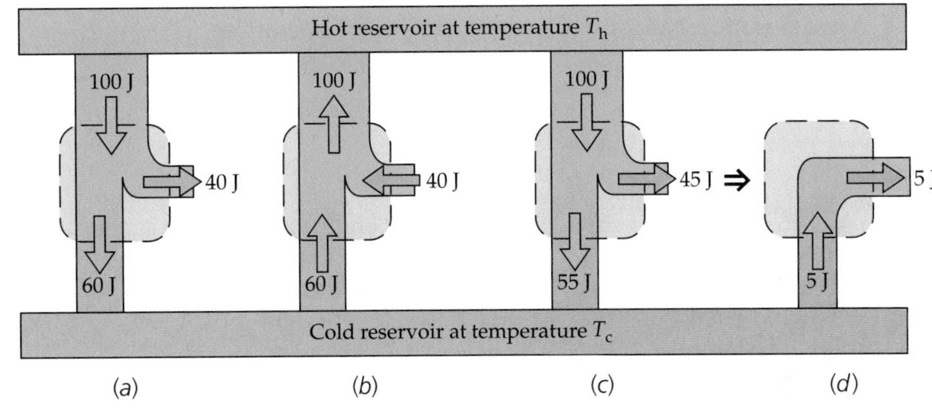

FIGURE 19-7 Illustration of the Carnot theorem. (*a*) A reversible heat engine with 40 percent efficiency removes 100 J from a hot reservoir, does 40 J work, and exhausts 60 J to the cold reservoir. (*b*) If the same engine runs backwards as a refrigerator, 40 J of work are done to remove 60 J from the cold reservoir and exhaust 100 J to the hot reservoir. (*c*) An assumed heat engine working between the same two reservoirs with an efficiency of 45 percent which is greater than that of the reversible engine in (*a*). (*d*) The net effect of running the engine in (*c*) in conjunction with the refrigerator in (*b*) is the same as that of a perfect heat engine that removes 5 J from the cold reservoir and converts it completely into work with no other effect, violating the second law of thermodynamics. Thus, the reversible engine in (*a*) is the most efficient engine that can operate between these two reservoirs.

Let us look at what makes a process reversible or irreversible. According to the second law, heat flows from hot objects to cold objects and never the other way around. Thus, the conduction of heat from a hot object to a cold one is *not* reversible. Also, friction can transform work into heat, but friction can never transform heat into work. The conversion of work into heat via friction is *not* reversible. Friction and other dissipative forces irreversibly transform mechanical energy into thermal energy. A third type of irreversibility occurs when a system passes through nonequilibrium states, such as when there is turbulence in a gas or when a gas explodes. For a process to be reversible, we must be able to move the system back through the same equilibrium states in the reverse order.

From these considerations and our statements of the second law of thermodynamics, we can list some conditions that are necessary for a process to be reversible:

1. No mechanical energy is transformed into thermal energy by friction, viscous forces, or other dissipative forces.
2. Energy transfer as heat can only occur between objects at the same temperature (or infinitesimally near the same temperature).
3. The process must be quasi-static so that the system is always in an equilibrium state (or infinitesimally near an equilibrium state).

CONDITIONS FOR REVERSIBILITY

Any process that violates any of the above conditions is irreversible. Most processes in nature are irreversible. To have a reversible process, great care must be taken to eliminate frictional and other dissipative forces and to make the process quasi-static. Because this can never be completely accomplished, a reversible process is an idealization similar to the idealization of motion without friction in mechanics problems. Reversibility can, nevertheless, be closely approximated in practice.

We can now understand the features of a Carnot cycle, which is a reversible cycle between two reservoirs only. Because all heat transfer must be done isothermally in order for the process to be reversible, the heat absorbed from the hot reservoir must be absorbed isothermally. The next step is a quasi-static adiabatic expansion to the lower temperature of the cold reservoir. Next, heat is given off isothermally to the cold reservoir. Finally, there is a quasi-static, adiabatic compression to the higher temperature of the hot reservoir. The Carnot cycle thus consists of four reversible steps:

1. A quasi-static isothermal absorption of heat from a hot reservoir
2. A quasi-static adiabatic expansion to a lower temperature
3. A quasi-static isothermal exhaustion of heat to a cold reservoir
4. A quasi-static adiabatic compression back to the original state

STEPS IN A CARNOT CYCLE

One way to calculate the efficiency of a Carnot engine is to choose as the working substance a material of which we have some knowledge—an ideal gas, and then explicitly calculate the work done on it over a Carnot cycle (Figures 19-8*a* and 8*b*). Since all Carnot cycles have the same efficiency independent of the working substance, our result will be valid in general.

The efficiency of the Carnot cycle (Equation 19-2) is

$$\varepsilon = 1 - \frac{Q_c}{Q_h}$$

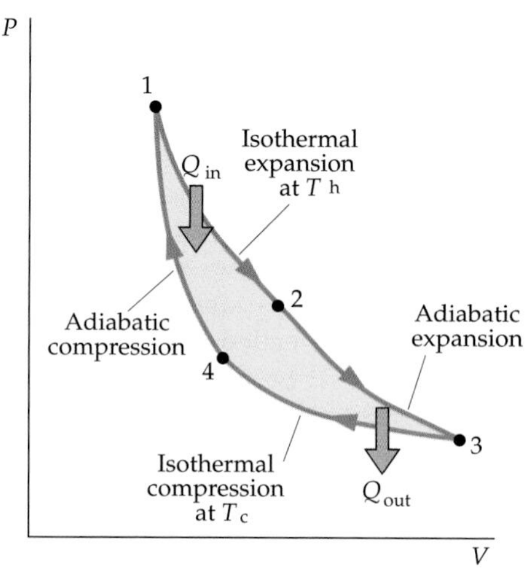

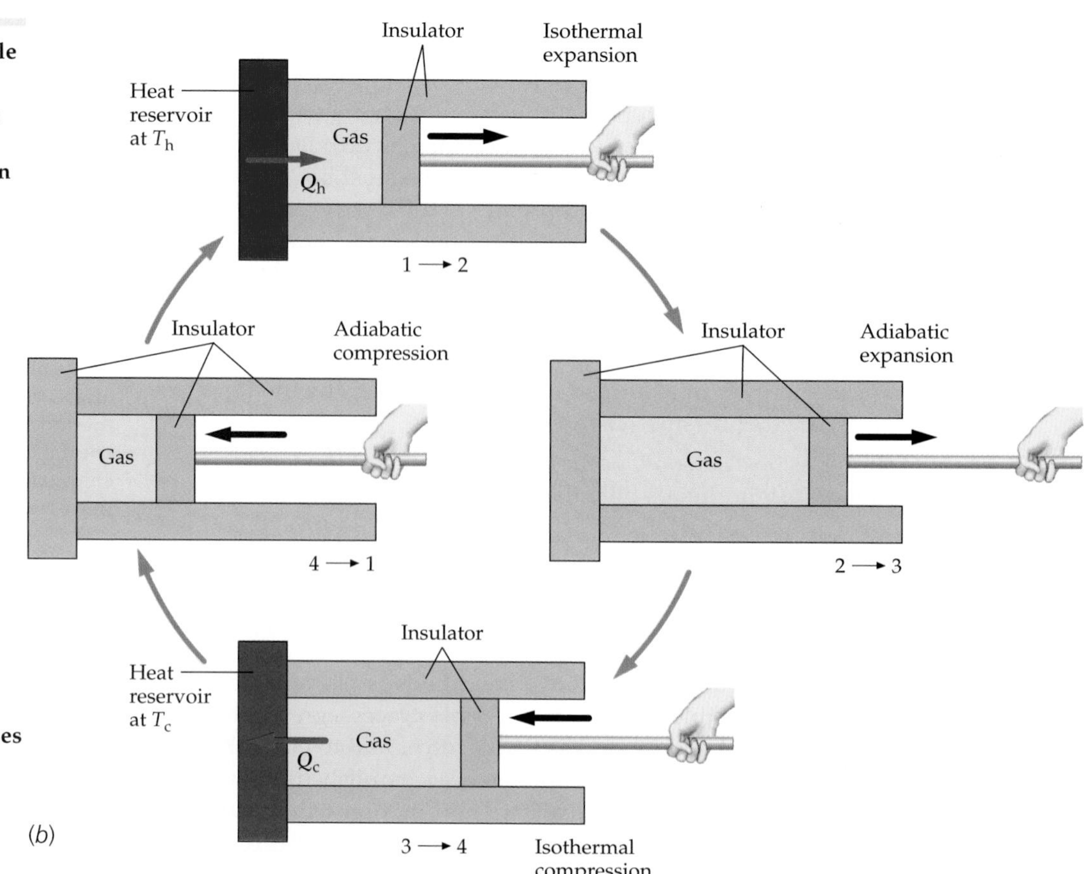

FIGURE 19-8 (*a*) Carnot cycle for an ideal gas:

Step 1: Heat is absorbed from a hot reservoir at temperature T_h during an isothermal expansion from state 1 to state 2.

Step 2: The gas expands adiabatically from state 2 to state 3, reducing its temperature to T_c.

Step 3: The gas gives off heat to the cold reservoir as it is compressed isothermally at T_c from state 3 to state 4.

Step 4: The gas is compressed adiabatically until its temperature is again T_h.

(*b*) Work is done on the gas or by the gas during each step. The net work done during the cycle is represented by the shaded area. All processes are reversible. All steps are quasi-static.

The heat Q_h is absorbed during the isothermal expansion from state 1 to state 2. The first law of thermodynamics is $\Delta E_{int} = Q_{in} + W_{on}$. For an isothermal expansion of an ideal gas $\Delta E_{int} = 0$. Applying the first law to the isothermal expansion from state 1 to state 2 we have $Q_h = Q_{in}$, so Q_h equals the work done by the gas.

$$Q_h = W_{\text{by gas}} = \int_{V_1}^{V_2} P\,dV = \int_{V_1}^{V_2} \frac{nRT_h}{V}\,dV = nRT_h \int_{V_1}^{V_2} \frac{dV}{V} = nRT_h \ln \frac{V_2}{V_1}$$

Similarly, the heat given off to the cold reservoir equals the work done on the gas during the isothermal compression at temperature T_c from state 3 to state 4. This work has the same magnitude as that done by the gas if it expands from state 4 to state 3. The heat rejected is thus

$$Q_c = W_{\text{on gas}} = nRT_c \ln \frac{V_3}{V_4}$$

The ratio of these heats is

$$\frac{Q_c}{Q_h} = \frac{T_c \ln \dfrac{V_3}{V_4}}{T_h \ln \dfrac{V_2}{V_1}} \qquad 19\text{-}4$$

We can relate the ratios V_2/V_1 and V_3/V_4 using Equation 18-37 for a quasi-static adiabatic expansion. For the expansion from state 2 to state 3, we have

$$T_h V_2^{\gamma-1} = T_c V_3^{\gamma-1}$$

Similarly, for the adiabatic compression from state 4 to state 1, we have

$$T_h V_1^{\gamma-1} = T_c V_4^{\gamma-1}$$

Dividing these two equations, we obtain

$$\left(\frac{V_2}{V_1}\right)^{\gamma-1} = \left(\frac{V_3}{V_4}\right)^{\gamma-1} \Rightarrow \frac{V_2}{V_1} = \frac{V_3}{V_4}$$

Coal-fueled electric generating plant at Four Corners, New Mexico.

Power plant at Wairakei, New Zealand, that converts geothermal energy into electricity.

Solar energy is focused and collected individually to produce electricity by these heliostats that are being tested at Sandia National Laboratory.

Control rods are inserted into this nuclear reactor at Tihange, Belgium.

Thus, Equation 19-4 gives

$$\frac{Q_c}{Q_h} = \frac{T_c \ln \frac{V_2}{V_1}}{T_h \ln \frac{V_2}{V_1}} = \frac{T_c}{T_h} \qquad 19\text{-}5$$

The Carnot efficiency ε_c is thus

$$\varepsilon_C = 1 - \frac{T_c}{T_h} \qquad 19\text{-}6$$

CARNOT EFFICIENCY

Equation 19-6 demonstrates that because the Carnot efficiency must be independent of the working substance of any particular engine, it depends only on the temperatures of the two reservoirs.

An experimental wind-powered electric generator at Sandia National Laboratory. The propeller is designed for optimum transfer of wind energy to mechanical energy.

EFFICIENCY OF A STEAM ENGINE **EXAMPLE 19-4**

A steam engine works between a hot reservoir at 100°C (373 K) and a cold reservoir at 0°C (273 K). (*a*) What is the maximum possible efficiency of this engine? (*b*) If the engine is run backwards as a refrigerator, what is its maximum coefficient of performance?

PICTURE THE PROBLEM The maximum efficiency is the Carnot efficiency given by Equation 19-6. To find the maximum COP, we use the definition of efficiency ($\varepsilon = W/Q_h$), the definition of COP (COP = Q_c/W), and Equation 19-5.

(*a*) The maximum efficiency is the Carnot efficiency:

$$\varepsilon_{max} = \varepsilon_C = 1 - \frac{T_c}{T_h} = 1 - \frac{273\text{ K}}{373\text{ K}}$$

$$= 0.268 = \boxed{26.8\%}$$

(*b*) 1. Write the expression for the COP if the engine is run in reverse for a single cycle:

$$\text{COP} = \frac{Q_c}{W}$$

2. Write the expression for the efficiency if the engine is run forward for a single cycle. (Since for maximum possible performance the engine is reversible, the values for Q_h, Q_c, and W are the same whether the engine is run backward or forward.):

$$\varepsilon = \frac{W}{Q_h}$$

3. Solve the step 2 result for the work and substitute it into the step 1 result:

$$\text{COP} = \frac{Q_c}{W} = \frac{Q_c}{\varepsilon Q_h}$$

4. Using Equation 19-5 and the Part (*a*) result, solve for the COP:

$$\text{COP} = \frac{Q_c}{\varepsilon Q_h} = \frac{T_c}{\varepsilon T_h} = \frac{273\text{ K}}{0.268(373\text{ K})}$$

$$\text{COP} = \boxed{2.73}$$

REMARKS Even though this maximum efficiency seems to be quite low, it is the greatest efficiency possible for any engine working between these temperatures. Real engines will have lower efficiencies because of friction, heat conduction, and other irreversible processes. Real refrigerators will have a lower coefficient of performance. It can be shown that the coefficient of performance of a Carnot refrigerator is $T_c/\Delta T$.

The Carnot efficiency gives us an upper limit on possible efficiencies, and is therefore useful to know. For example, we calculated in Example 19-4 that the Carnot efficiency is 26.8 percent. This means that, no matter how much we reduce friction and other irreversible losses, the best efficiency obtained between reservoirs at 373 K and 273 K is 26.8 percent. We would know, then, that an engine working between those two temperatures with an efficiency of 25 percent is a very good engine!

WORK LOST BY AN ENGINE — EXAMPLE 19-5

An engine removes 200 J from a hot reservoir at 373 K, does 48 J of work, and exhausts 152 J to a cold reservoir at 273 K. How much work is "lost" per cycle due to irreversible processes in this engine?

PICTURE THE PROBLEM The difference between maximum amount of work that could be done using a Carnot engine and 48 J is the work lost.

1. The work lost is the maximum amount of work that could be done minus the work actually done:

$$W_{lost} = W_{max} - W$$

2. The maximum amount of work that could be done is the work done using a Carnot engine:

$$W_{max} = \varepsilon_C Q_h$$

3. The work lost is then:

$$W_{lost} = \varepsilon_C Q_h - W$$

4. The Carnot efficiency can be expressed in terms of the temperatures:

$$\varepsilon_C = 1 - \frac{T_c}{T_h}$$

5. Substituting for ε_C gives:

$$W_{lost} = \left(1 - \frac{T_c}{T_h}\right)Q_h - W$$
$$= \left(1 - \frac{273\text{ K}}{373\text{ K}}\right)(200\text{ J}) - 48\text{ J}$$
$$= \boxed{5.6\text{ J}}$$

REMARKS The 5.6 J of energy in the answer is not "lost" to the universe—total energy is conserved. That 5.6 J of energy exhausted into the cold reservoir by the non-ideal engine of the problem is only lost in that it would have been converted into useful work if an ideal (reversible) engine had been used.

Work Lost Between Heat Reservoirs — EXAMPLE 19-6

If 200 J of heat are conducted from a heat reservoir at 373 K to one at 273 K, how much work capability is "lost" in this process?

We saw in the previous example that a Carnot engine working between these two reservoirs could do 53.6 J of work if it extracted 200 J from the 373-K reservoir and exhausted to a 273-K reservoir. Thus, if 200 J is conducted directly from the hot reservoir to the cold reservoir without any work being done, 53.6 J of this energy has been "lost" in the sense that it could have been converted into useful work.

EXERCISE A Carnot engine works between heat reservoirs at 500 K and 300 K. (*a*) What is its efficiency? (*b*) If it removes 200 kJ of heat from the hot reservoir, how much work does it do? (*Answer* (*a*) 40% (*b*) 80 kJ)

EXERCISE A real engine works between heat reservoirs at 500 K and 300 K. It removes 500 kJ of heat from the hot reservoir and does 150 kJ of work during each cycle. What is its efficiency? (*Answer* 30%)

The Thermodynamic or Absolute Temperature Scale

In Chapter 17, the ideal-gas temperature scale was defined in terms of the properties of gases at low densities. Because the Carnot efficiency depends only on the temperatures of the two heat reservoirs, it can be used to define the ratio of the temperatures of the reservoirs independent of the properties of any substance. We *define* the ratio of the thermodynamic temperatures of the hot and cold reservoirs to be

$$\frac{T_c}{T_h} = \frac{Q_c}{Q_h} \qquad 19\text{-}7$$

DEFINITION OF THERMODYNAMIC TEMPERATURE

where Q_h is the energy removed from the hot reservoir and Q_c is the energy exhausted to the cold reservoir by a Carnot engine working between the two reservoirs. Thus, to find the ratio of two reservoir temperatures, we set up a reversible engine operating between them and measure the energy transferred as heat to or from each reservoir during one cycle. The **thermodynamic temperature** is completely specified by Equation 19-7 *and* the choice of one fixed point. If the fixed point is defined to be 273.16 K for the triple point of water, then the thermodynamic temperature scale matches the ideal-gas temperature scale for the range of temperatures over which a gas thermometer can be used. Any temperature that reads zero at absolute zero is called an *absolute temperature scale.*

*19-5 Heat Pumps

A **heat pump** is a refrigerator with a different objective. Typically, the objective of a refrigerator is to cool an object or region of interest. The objective of a heat pump, however, is to heat an object or region of interest. For example, if you use a heat pump to heat your house you transfer heat from the cold air outside the house to the warmer air inside it. Your objective is to heat the region inside your house. If work W is done on a heat pump to remove heat Q_c from the cold reservoir and reject heat Q_h to the hot reservoir, the coefficient of performance for a heat pump is defined as

$$\mathrm{COP}_{\mathrm{HP}} = \frac{Q_h}{W} \qquad 19\text{-}8$$

DEFINITION—COEFFICIENT OF PERFORMANCE (HEAT PUMP)

This coefficient of performance differs from that for the refrigerator, which is Q_c/W (Equation 19-3). Using $W = Q_h - Q_c$, this can be written

$$\mathrm{COP}_{\mathrm{HP}} = \frac{Q_h}{Q_h - Q_c} = \frac{1}{1 - \frac{Q_c}{Q_h}} \qquad 19\text{-}9$$

The maximum coefficient of performance is obtained using a Carnot heat pump. Then Q_c and Q_h are related by Equation 19-5. Substituting $Q_c/Q_h = T_c/T_h$ into Equation 19-9, we obtain for the maximum coefficient of performance

$$\mathrm{COP}_{\mathrm{HP\,max}} = \frac{1}{1 - \frac{T_c}{T_h}} = \frac{T_h}{T_h - T_c} = \frac{T_h}{\Delta T} \qquad 19\text{-}10$$

where ΔT is the difference in temperature between the hot and cold reservoirs. Real heat pumps have coefficients of performance less than the $\mathrm{COP}_{\mathrm{HP\,max}}$ because of friction, heat conduction, and other irreversible processes.

The two coefficients are related. Using $Q_h = Q_c + W$, we can relate Equations 19-3 and 19-10:

$$\mathrm{COP}_{\mathrm{HP}} = \frac{Q_h}{W} = \frac{Q_c + W}{W} = 1 + \frac{Q_c}{W} = 1 + \mathrm{COP} \qquad 19\text{-}11$$

AN IDEAL HEAT PUMP — EXAMPLE 19-7 Try It Yourself

An ideal heat pump is used to pump heat from the outside air at −5°C to the hot-air supply for the heating fan in a house, which is at 40°C. How much work is required to pump 1 kJ of heat into the house?

PICTURE THE PROBLEM Use Equation 19-11 with $\mathrm{COP}_{\mathrm{HP\,max}}$ calculated from Equation 19-10 for $T_c = -5°\mathrm{C} = 268$ K and $\Delta T = 45$ K.

Steps	Answers
1. Calculate the work from Equation 19-8:	$W = \dfrac{Q_h}{\text{COP}_{HP}}$
2. Calculate the COP_{HP} from Equation 19-10:	$\text{COP}_{HP} = \text{COP}_{HP\,max} = \dfrac{T_h}{\Delta T}$
3. Solve for the work:	$W = \dfrac{Q_h}{\text{COP}_{HP}} = Q_h \dfrac{\Delta T}{T_h} = (1\text{ kJ}) \dfrac{45\text{ K}}{313\text{ K}}$
	$W = \boxed{0.144\text{ kJ}}$

REMARKS The $\text{COP}_{HP\,max} = T_h/\Delta T = 6.96$. That is, the energy transferred inside the house as heat is 6.96 times larger than the work done. (Only 0.144 kJ of work is needed to pump 1 kJ of heat into the hot-air supply in the house.)

19-6 Irreversibility and Disorder

There are many irreversible processes that cannot be described by the heat-engine or refrigerator statements of the second law, such as a glass falling to the floor and breaking or a balloon popping. However, all irreversible processes have one thing in common—the system plus its surroundings moves toward a less ordered state.

Suppose a box containing a gas of mass M at a temperature T is moving along a frictionless table with a velocity v_{cm} (Figure 19-9a). The total kinetic energy of the gas has two components: that associated with the movement of the center of mass $\frac{1}{2}Mv_{cm}^2$, and the energy of the motion of its molecules relative to its center of mass. The center of mass energy $\frac{1}{2}Mv_{cm}^2$ is ordered mechanical energy that could be converted entirely into work. (For example, if a weight were attached to the moving box by a string passing over a pulley, this energy could be used to lift the weight.) The relative energy is the internal thermal energy of the gas, which is related to its temperature T. It is random, non-ordered energy that cannot be converted entirely into work.

Now, suppose that the box hits a fixed wall and stops (Figure 19-9b). This inelastic collision is clearly an irreversible process. The ordered mechanical energy of the gas is converted into random internal energy and the temperature of the gas rises. The gas still has the same total energy, but now all of that energy is associated with the random motion of the gas molecules about the center of mass of the gas, which is now at rest. Thus, the gas has become less ordered (more disordered), and has lost some of its ability to do work.

19-7 Entropy

There is a thermodynamic function called **entropy** S that is a measure of the disorder of a system. Entropy S, like pressure P, volume V, temperature T, and

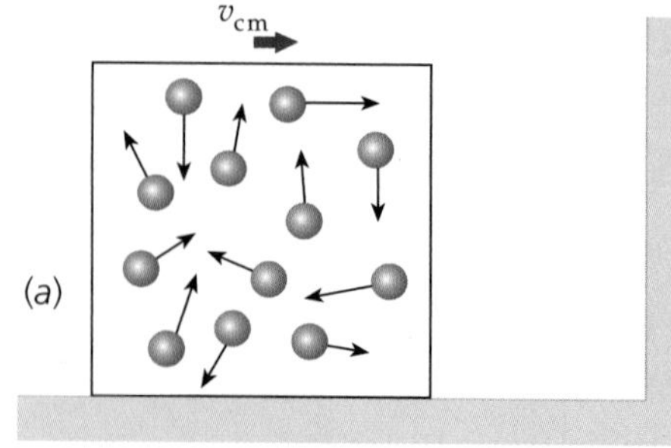

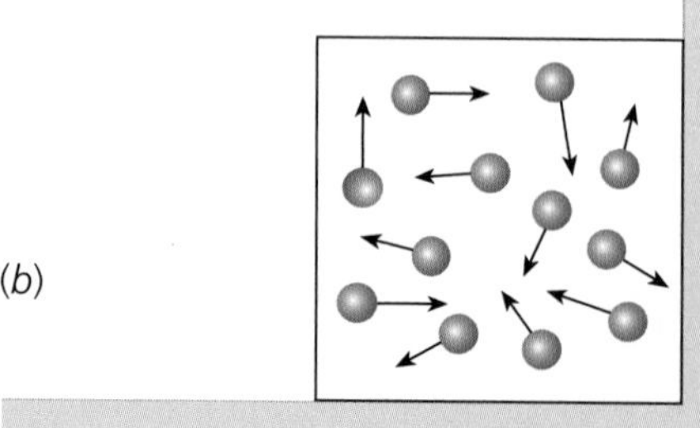

FIGURE 19-9

internal energy U, is a function of the state of a system. As with potential energy, it is the *change* in entropy that is important. The change in entropy dS of a system as it goes from one state to another is defined as

$$dS = \frac{dQ_{rev}}{T} \qquad \text{19-12}$$

DEFINITION—ENTROPY CHANGE

where dQ_{rev} is the energy that must be transferred to the system as heat in a *reversible* process that brings the system from the initial state to the final state. If dQ_{rev} is negative, then the entropy change of the system is negative.

The term dQ_{rev} does not mean that a reversible heat transfer must take place in order for the entropy of a system to change. Indeed, there are many situations in which the entropy of a system changes when there is no transfer of heat whatsoever, for example, the box of gas colliding with the wall in Figure 19-9. Equation 19-12 simply gives us a method for *calculating* the entropy difference between two states of a system. Because entropy is a state function, the change in entropy when the system moves from one state to another depends only on the system's initial and final states, not on the process by which the change occurs.

Entropy of an Ideal Gas

We can illustrate that dQ_{rev}/T is in fact the differential of a state function for an ideal gas (even though dQ_{rev} is not). Consider an arbitrary reversible quasi-static process in which a system consisting of an ideal gas absorbs an amount of heat dQ_{rev}. According to the first law, dQ_{rev} is related to the change in the internal energy dE_{int} of the gas and the work done on the gas ($dW_{on} = -PdV$) by

$$dE_{int} = dQ_{rev} + dW_{on} = dQ_{rev} - PdV$$

For an ideal gas, we can write dE_{int} in terms of the heat capacity, $dE_{int} = C_v dT$, and we can substitute nRT/V for P from the equation of state. Then

$$C_v dT = dQ_{rev} - nRT\frac{dV}{V} \qquad \text{19-13}$$

Equation 19-13 cannot be integrated unless we know how T depends on V. This is just another way of saying that dQ_{rev} is not a differential of a state function Q_{rev}. But if we divide each term by T, we obtain

$$C_v\frac{dT}{T} = \frac{dQ_{rev}}{T} - nR\frac{dV}{V} \qquad \text{19-14}$$

Since C_v depends only on T, the term on the left can be integrated as can the second term on the right.† Thus, dQ_{rev}/T is the differential of a function, the entropy function S.

$$dS = \frac{dQ_{rev}}{T} = \left(C_v \times \frac{dT}{T}\right) + nR\frac{dV}{V} \qquad \text{19-15}$$

For simplicity, we will assume that C_v is constant. Integrating Equation 19-15, we obtain

$$\Delta S = \int\frac{dQ}{T} = C_v \ln\frac{T_2}{T_1} + nR\ln\frac{V_2}{V_1} \qquad \text{19-16}$$

Equation 19-16 gives the entropy change of an ideal gas that undergoes a reversible expansion from an initial state of volume V_1 and temperature T_1 to a final state of volume V_2 and temperature T_2.

† Mathematically, the factor $1/T$ is called an integrating factor for Equation 19-13.

Entropy Changes for Various Processes

ΔS for an Isothermal Expansion of an Ideal Gas If an ideal gas undergoes an isothermal expansion, then $T_2 = T_1$ and its entropy change is

$$\Delta S = \int \frac{dQ}{T} = nR \ln \frac{V_2}{V_1} \qquad 19\text{-}17$$

The entropy change of the gas is positive because V_2 is greater than V_1. In this process, an amount of energy Q is transferred as heat from the reservoir to the gas. This heat equals the work done by the gas:

$$Q = W_{\text{by}} = \int_{V_1}^{V_2} PdV = nRT \int_{V_1}^{V_2} \frac{dV}{V} = nRT \ln \frac{V_2}{V_1} \qquad 19\text{-}18$$

The entropy change of the gas is $+Q/T$. Because the same amount of heat leaves the reservoir at temperature T, the entropy change of the reservoir is $-Q/T$. The net entropy change of the gas plus the reservoir is zero. We will refer to the system under consideration plus its surroundings as the "universe." This example illustrates a general result:

> In a reversible process, the entropy change of the universe is zero.

ΔS for a Free Expansion of an Ideal Gas In the free expansion of a gas discussed in Section 18-4, a gas is initially confined in one compartment of a container, which is connected by a stopcock to another compartment that is evacuated. The whole system has rigid walls and is thermally insulated from its surroundings so that no heat can flow in or out, and no work can be done on (or by) the system (Figure 19-10). When the stopcock is opened, the gas rushes into the evacuated chamber. Eventually, the gas reaches thermal equilibrium with itself. Since there is no work done and no heat transferred, the final internal energy of the gas must equal its initial internal energy. If we assume that the gas is ideal, the final temperature T equals the initial temperature.

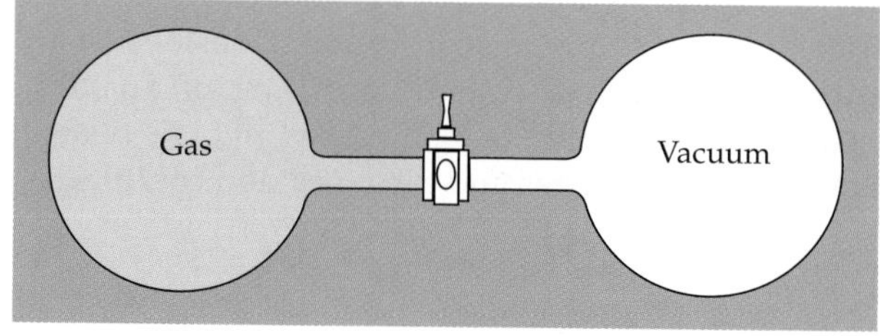

FIGURE 19-10 Free expansion of a gas. When the stopcock is opened, the gas expands rapdily into the evacuated chamber. Since no work is done on the gas and the whole system is thermally insulated, the initial and final internal energies of the gas are equal.

We might think that there is no entropy change of the gas because there is no heat transfer. But this process is not reversible, so we cannot use $\int dQ/T$ to find the change in entropy of the gas. However, the initial and final states of the gas in the free expansion are the same as those of the gas in the isothermal expansion just discussed. *Because the change in the entropy of a system for any process depends only on the initial and final states of the system, the entropy change of the gas for the free expansion is the same as that for the isothermal expansion.* If V_1 is the initial volume of the gas and V_2 is its final volume, the entropy change of the gas is given by Equation 19-17, or

$$\Delta S_{\text{gas}} = nR \ln \frac{V_2}{V_1}$$

In this case, there is no change in the surroundings, so the entropy change of the gas is also the entropy change of the universe:

$$\Delta S_{\text{u}} = nR \ln \frac{V_2}{V_1} \qquad 19\text{-}19$$

Note that because V_2 is greater than V_1, the change in entropy of the universe for this irreversible process is positive; that is, the entropy of the universe increases. This is also a general result:

In an irreversible process, the entropy of the universe increases.

If the final volume in the free expansion were less than the initial volume, then the entropy of the universe would decrease—but this does not happen. A gas does not freely contract by itself into a smaller volume. This leads us to yet another statement of the second law of thermodynamics:

For any process, the entropy of the universe never decreases.

FREE EXPANSION OF AN IDEAL GAS — **EXAMPLE 19-8**

Find the entropy change for the free expansion of 0.75 mol of an ideal gas from $V_1 = 1.5$ L to $V_2 = 3$ L.

PICTURE THE PROBLEM For a free expansion of an ideal gas the initial and final temperatures are the same. Thus, the entropy change ΔS for a free expansion from V_1 to V_2 is the same as ΔS for an isothermal process from V_1 to V_2. For the isothermal process $\Delta E_{int} = 0$, so $Q = W_{by}$. First we calculate Q, then we set $\Delta S = Q/T$.

1. The entropy change is the same as for an isothermal expansion from V_1 to V_2:

$$\Delta S = \Delta S_{isothermal} = \frac{Q}{T}$$

2. The heat Q that would enter the gas during an isothermal expansion at temperature T equals the work done by the gas during the expansion:

$$Q = W_{by} = nRT \ln \frac{V_2}{V_1}$$

3. Substitute this value of Q to calculate ΔS:

$$\Delta S = \frac{Q}{T} = nR \ln \frac{V_2}{V_1}$$

$$\Delta S = (0.75 \text{ mol})(8.31 \text{ J/mol·K}) \ln 2$$

$$\Delta S = \boxed{4.32 \text{ J/K}}$$

ΔS for Constant-Pressure Processes If a substance is heated from temperature T_1 to temperature T_2 at constant pressure, the heat absorbed dQ is related to its temperature change dT by

$$dQ = C_p \, dT$$

We can approximate reversible heat conduction if we have a large number of heat reservoirs with temperatures ranging from just slightly greater than T_1 to T_2 in very small steps. We could place the substance, with initial temperature T_1, in contact with the first reservoir at a temperature just slightly greater than T_1 and let the substance absorb a small amount of heat. Because the heat transfer is approximately isothermal, the process will be approximately reversible. We then place the substance in contact with the next reservoir at a slightly higher temperature, and so on, until the final temperature T_2 is reached. If heat dQ is absorbed reversibly, the entropy change of the substance is

$$dS = \frac{dQ}{T} = C_p \frac{dT}{T}$$

Integrating from T_1 to T_2, we obtain the total entropy change of the substance:

$$\Delta S = C_p \int_{T_1}^{T_2} \frac{dT}{T} = C_p \ln \frac{T_2}{T_1} \tag{19-20}$$

This result gives the entropy change of a substance that is heated from T_1 to T_2 by any process, reversible or irreversible, as long as the final pressure equals the initial pressure. It also gives the entropy change of a substance that is cooled. In the case of cooling, T_2 is less than T_1, and $\ln (T_2/T_1)$ is negative, giving a negative entropy change.

EXERCISE Find the change in entropy of 1 kg of water that is heated at constant pressure from 0°C to 100°C. (*Answer* $\Delta S = 1.31$ kJ/K)

ENTROPY CHANGES DURING HEAT TRANSFER — EXAMPLE 19-9 Try It Yourself

Suppose 1 kg of water at temperature $T_1 = 30°C$ is added to 2 kg of water at $T_2 = 90°C$ in a calorimeter of negligible heat capacity at a constant pressure of 1 atm. (*a*) Find the change in entropy of the system. (*b*) Find the change in entropy of the universe.

PICTURE THE PROBLEM When the two amounts of water are combined, they eventually come to a final equilibrium temperature, T_f, that can be found by setting the heat lost equal to the heat gained. To calculate the entropy change of each mass of water, we consider a reversible isobaric heating of the 1-kg mass of water from 30°C to T_f and a reversible isobaric cooling of the 2-kg mass from 90°C to T_f using Equation 19-18. The entropy change of the system is the sum of the entropy changes of each part. The entropy change of the universe is the entropy change of the system plus the entropy change of its surroundings. To find the entropy change of the surroundings, assume no heat leaves the calorimeter during the time it takes the water to reach its final temperature.

Cover the column to the right and try these on your own before looking at the answers.

Steps	Answers
(*a*) 1. Calculate T_f by setting the heat lost equal to the heat gained:	$T_f = 70°C = 343$ K
2. Use your result for T_f and the data given to calculate ΔS_1 and ΔS_2:	$\Delta S_1 = 0.519$ kJ/K $\Delta S_2 = -0.474$ kJ/K
3. Add ΔS_1 and ΔS_2 to find the total entropy change of the system:	$\Delta S_{system} = \boxed{+0.0453 \text{ kJ/K}}$
(*b*) 1. Assuming no heat leaves the calorimeter, find the entropy change of the surroundings:	$\Delta S_{surroundings} = 0$
2. Add ΔS_{system} and $\Delta S_{surroundings}$ to find the entropy change of the universe:	$\Delta S_u = \boxed{+0.0453 \text{ kJ/K}}$

REMARKS Note that we had to convert the temperatures to the absolute scale to calculate the entropy changes. The entropy change of the universe is positive, as expected.

ΔS for an Inelastic Collision Because mechanical energy is converted into thermal energy in an inelastic collision, such a process is clearly irreversible. The entropy of the universe must therefore increase. Consider a block of mass m

falling from a height h and making an inelastic collision with the ground. Let the block, ground, and atmosphere all be at a temperature T, which is not significantly changed by the process. If we consider the block, ground, and atmosphere as our isolated system, there is no heat conducted into or out of the system. The state of the system has been changed because its internal energy has been increased by an amount mgh. This change is the same as if we added heat $Q = mgh$ to the system at constant temperature T. To calculate the change in entropy of the system, we thus consider a reversible process in which heat $Q_{\text{rev}} = mgh$ is added at a constant temperature T. According to Equation 19-12, the change in entropy is then

$$\Delta S = \frac{Q_{\text{rev}}}{T} = \frac{mgh}{T}$$

This positive entropy change is also the entropy change of the universe.

ΔS for Heat Conduction from One Reservoir to Another Heat conduction is also an irreversible process, and so we expect the entropy of the universe to increase when this occurs. Consider the simple case of heat Q conducted from a hot reservoir at a temperature T_{h} to a cold reservoir at a temperature T_{c}. The state of a heat reservoir is determined by its temperature and its internal energy only. The change in entropy of a heat reservoir due to a heat exchange is the same whether the heat exchange is reversible or not. If heat Q is put into a reservoir at temperature T, then the entropy of the reservoir increases by Q/T. If the heat is removed, then the entropy of the reservoir decreases by $-Q/T$. In the case of heat conduction, the hot reservoir loses heat, so its entropy change is

$$\Delta S_{\text{h}} = -\frac{Q}{T_{\text{h}}}$$

The cold reservoir absorbs heat, so its entropy change is

$$\Delta S_{\text{c}} = +\frac{Q}{T_{\text{c}}}$$

The net entropy change of the universe is

$$\Delta S_{\text{u}} = \Delta S_{\text{c}} + \Delta S_{\text{h}} = \frac{Q}{T_{\text{c}}} - \frac{Q}{T_{\text{h}}} \qquad \text{19-21}$$

Note that, because heat flows from a hot reservoir to a cold reservoir, the change in entropy of the universe is positive.

ΔS for a Carnot Cycle Because a Carnot cycle is by definition reversible, the entropy change of the universe after a cycle must be zero. We demonstrate this by showing that the entropy change of the reservoirs in a Carnot engine is zero. (Since a Carnot engine works in a cycle, the entropy change of the engine itself is zero, so the entropy change of the universe is just the sum of the entropy changes of the reservoirs.) The entropy change of the hot reservoir is $\Delta S_{\text{h}} = -\frac{Q_{\text{h}}}{T_{\text{h}}}$ and the entropy change of the cold reservoir is $\Delta S_{\text{c}} = +\frac{Q_{\text{c}}}{T_{\text{c}}}$. These heats are related to the temperatures by the definition of thermodynamic temperature (Equation 19-7)

$$\frac{T_{\text{c}}}{T_{\text{h}}} = \frac{Q_{\text{c}}}{Q_{\text{h}}} \left(\text{or} \quad \frac{Q_{\text{h}}}{T_{\text{h}}} = \frac{Q_{\text{c}}}{T_{\text{c}}} \right)$$

The entropy change of the universe is thus

$$\Delta S_u = \Delta S_h + \Delta S_c = -\frac{Q_h}{T_h} + \frac{Q_c}{T_c} = -\frac{Q_h}{T_h} + \frac{Q_h}{T_h} = 0$$

The entropy change of the universe is zero as expected.

Notice that we have ignored any entropy change associated with the energy transferred via work from the Carnot engine to its surroundings. If this work is used to raise a weight, or some other ordered process, then there is no entropy change. However, if this work is used to push a block across a table top where friction is involved, then there is an additional entropy increase associated with this work.

ENTROPY CHANGES IN A CARNOT CYCLE — EXAMPLE 19-10

During each cycle, a Carnot engine removes 100 J of energy from a reservoir at 400 K, does work, and exhausts heat to a reservoir at 300 K. Compute the entropy change of each reservoir for each cycle, and show explicitly that the entropy change of the universe is zero for this reversible process.

PICTURE THE PROBLEM Since the engine works in a cycle, its entropy change is zero. We therefore compute the entropy change of each reservoir and add them to obtain the entropy change of the universe.

1. The entropy change of the universe equals the sum of the entropy changes of the reservoirs:

$$\Delta S_u = \Delta S_{400} + \Delta S_{300}$$

2. Calculate the entropy change of the hot reservoir:

$$\Delta S_{400} = -\frac{Q_h}{T_h} = -\frac{100\text{ J}}{400\text{ K}} = \boxed{-0.250\text{ J/K}}$$

3. The entropy change of the cold reservoir is Q_c divided by T_c, where $Q_c = Q_h - W$:

$$\Delta S_{300} = \frac{Q_c}{T_c} = \frac{Q_h - W}{T_c}$$

4. We use $W = \varepsilon_C Q_h$ (Equation 19-2) to relate W to Q_h. The efficiency is the Carnot efficiency (Equation 19-6):

$$W = \varepsilon Q_h, \text{ where } \varepsilon = \varepsilon_C = 1 - \frac{T_c}{T_h}$$

so

$$W = \left(1 - \frac{T_c}{T_h}\right)Q_h$$

5. Calculate the entropy change of the cold reservoir:

$$\Delta S_{300} = \frac{Q_h - W}{T_c} = \frac{Q_h - Q_h\left(1 - \frac{T_c}{T_h}\right)}{T_c} = \frac{Q_h}{T_h}$$

$$= \frac{100\text{ J}}{400\text{ K}} = \boxed{0.250\text{ J/K}}$$

6. Substitute these results into step 1 to find the entropy change of the universe:

$$\Delta S_u = \Delta S_{400} + \Delta S_{300}$$

$$\Delta S_u = -0.250\text{ J/K} + 0.250\text{ J/K} = \boxed{0}$$

REMARKS Suppose that an ordinary, nonreversible engine removed 100 J from the hot reservoir. Because its efficiency must be less than that of a Carnot engine, it would do less work and exhaust more heat to the cold reservoir. Then the entropy increase of the cold reservoir would be greater than the entropy decrease of the hot reservoir, and the entropy change of the universe would be positive.

THE *ST* PLOT

EXAMPLE 19-11

Because entropy is a state function, thermodynamic processes can be represented as *ST*, *SV*, or *SP* diagrams instead of the *PV* diagrams we have used so far. Make a sketch of the Carnot cycle on an *ST* plot.

PICTURE THE PROBLEM The Carnot cycle consists of a reversible isothermal expansion followed by a reversible adiabatic expansion, then a reversible isothermal compression followed by a reversible adiabatic compression. During the isothermal processes, heat is absorbed or expelled at constant temperature, so S increases or decreases at constant T. During the adiabatic processes, the temperature changes, but since $\Delta Q_{rev} = 0$, S is constant.

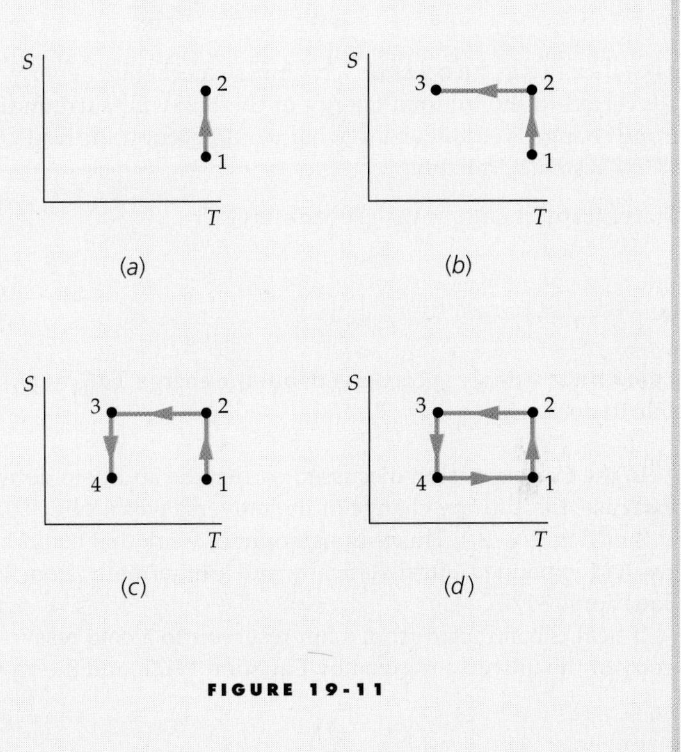

FIGURE 19-11

1. During the isothermal expansion (1 to 2 in Figure 19-11*a*), heat is absorbed reversibly so, S increases at constant T:

2. During the reversible adiabatic expansion (2 to 3 in Figure 19-11*b*), the temperature decreases while S is constant:

3. During the isothermal compression (3 to 4 in Figure 19-11*c*) heat is rejected reversibly, so S decreases at constant T:

4. During the reversible adiabatic compression (4 to 1 in Figure 19-11*d*) the temperature increases while S is constant:

REMARKS The Carnot cycle is a rectangle if plotted on an S versus T diagram.

19-8 Entropy and the Availability of Energy

If an irreversible process occurs, energy is conserved, but some of the energy becomes unavailable to do work and is "wasted." Consider a block falling to the ground. The entropy change of the universe for this process is mgh/T. When the block was at a height h, its potential energy mgh could have been used to do useful work. But after the inelastic collision of the block with the ground, this energy is no longer available because it has become the disordered internal energy of the block and its surroundings. The energy that has become unavailable (wasted) is equal to $mgh = T\Delta S_u$. This is a general result:

In an irreversible process, energy equal to $T\Delta S_u$ becomes unavailable to do work, where T is the temperature of the coldest available reservoir.

For simplicity, we will call the energy that becomes unavailable to do work the "work lost":

$$W_{lost} = T\Delta S_u \qquad 19\text{-}22$$

A Sliding Box Revisited **EXAMPLE 19-12**

Suppose that the box shown in Figure 19-9*a* and *b* has a mass of 2.4 kg and slides with a speed of $v = 3$ m/s before crashing into a fixed wall and stopping. The temperature T of the box, table, and surroundings is 293 K and does not change appreciably as the box comes to rest. Find the entropy change of the universe.

PICTURE THE PROBLEM The initial mechanical energy of the box $\frac{1}{2}Mv^2$ is converted to the internal energy of the box-wall-surroundings system. The entropy change is equivalent to what would occur if the heat $Q = \frac{1}{2}Mv^2$ were added to the system reversibly.

The entropy change of the universe is Q/T:

$$\Delta S_u = \frac{Q}{T} = \frac{\frac{1}{2}Mv^2}{T} = \frac{\frac{1}{2}(2.4\text{ kg})(3\text{ m/s})^2}{293\text{ K}}$$

$$\Delta S_u = \boxed{0.0369\text{ J/K}}$$

REMARKS Energy is conserved, but the energy $T\Delta S_u = \frac{1}{2}Mv^2$ is no longer available to do work.

In the free expansion discussed earlier, the ability to do work was also lost. In that case, the entropy change of the universe was $nR \ln (V_2/V_1)$, so the work lost was $nRT \ln (V_2/V_1)$. This is the amount of work that could have been done if the gas had expanded quasi-statically and isothermally from V_1 to V_2, as given by Equation 19-17.

If heat is conducted from a hot reservoir to a cold reservoir, the change in entropy of the universe is given by Equation 19-21, and the work lost is

$$W_{\text{lost}} = T_c\,\Delta S_u = T_c\left(\frac{Q}{T_c} - \frac{Q}{T_h}\right) = Q\left(1 - \frac{T_c}{T_h}\right)$$

We can see that this is just the work that could have been done by a Carnot engine running between these reservoirs, removing heat Q from the hot reservoir and doing work $W = \varepsilon_C Q$, where $\varepsilon_C = 1 - T_c/T_h$.

19-9 Entropy and Probability

Entropy, which is a measure of the disorder of a system, is related to probability. Essentially, a state of high order has a low probability, whereas a state of low order has a high probability. Thus, in an irreversible process, the universe moves from a state of low probability to one of high probability.

Let us consider a free expansion in which a gas expands from an initial volume V_1 to a final volume $V_2 = 2V_1$. The entropy change of the universe for this process is given by Equation 19-19:

$$\Delta S = nR \ln\frac{V_2}{V_1} = nR \ln 2$$

Why is this process irreversible? Why can't the gas spontaneously compress back into its original volume? Such a compression would not violate the first law of thermodynamics, as there is no energy change involved. The reason that the gas does not compress to its original volume is merely that such a compression is extremely *improbable.* To see this, let's assume that the gas consists of only

10 molecules and that, initially, these molecules occupy the entire volume of their container. Then the chance that any one particular molecule will be in the left half of the container at any given time is $\frac{1}{2}$. The chance that any two particular molecules will both be in the left half is $\frac{1}{2} \times \frac{1}{2} = \frac{1}{4}$. (This is the same as the chance that a coin flipped twice will come up heads both times.) The chance that three particular molecules will be in the left half is $\frac{1}{2} \times \frac{1}{2} \times \frac{1}{2} = (\frac{1}{2})^3 = \frac{1}{8}$. The chance that all 10 molecules will be in the left half is $(\frac{1}{2})^{10} = \frac{1}{1024}$. That is, there is 1 chance in 1024 that all 10 molecules will be in the left half of the container at any given time.

Though the probability of all 10 molecules being on one side of the container is small, we would not be completely surprised to see it occur. If we look at the gas once each second, we could expect to see it happen once in every 1024 sec, or about once every 17 min. If we started with the 10 molecules randomly distributed and then found them all in the left half of the original volume, the entropy of the universe would have *decreased* by $nR \ln 2$. However, this decrease is extremely small, since the number of moles n corresponding to 10 molecules is only about 10^{-23}. Still, it would violate the entropy statement of the second law of thermodynamics, which says that for any process, the entropy of the universe never decreases. Therefore, if we wish to apply the second law of thermodynamics to microscopic systems such as a small number of molecules, we should consider the second law to be a statement of *probability.*

We can relate the probability of a gas spontaneously compressing itself into a smaller volume to the change in its entropy. If the original volume is V_1, the probability p of finding N molecules in a smaller volume V_2 is

$$p = \left(\frac{V_2}{V_1}\right)^N$$

Taking the natural logarithm of both sides of this equation, we obtain

$$\ln p = N \ln\frac{V_2}{V_1} = nN_A \ln\frac{V_2}{V_1} \qquad 19\text{-}23$$

where n is the number of moles and N_A is Avogadro's number. The entropy change of the gas is

$$\Delta S = nR \ln\frac{V_2}{V_1} \qquad 19\text{-}24$$

Comparing Equations 19-23 and 19-24, we see that

$$\Delta S = \frac{R}{N_A} \ln p = k \ln p \qquad 19\text{-}25$$

where k is Boltzmann's constant.

It may be disturbing to learn that irreversible processes, such as the spontaneous compression of a gas or the spontaneous conduction of heat from a cold body to a hot body, are not impossible—they are just improbable. As we have just seen, there is a reasonable chance that an irreversible process will occur in a system consisting of a very small number of molecules; however, *thermodynamics itself is applicable only to macroscopic systems,* that is, to systems that have a very large number of molecules. Consider trying to measure the pressure of a gas consisting of only 10 molecules. The pressure would vary wildly depending on whether no molecule, 2 molecules, or 10 molecules were colliding with the wall of the container at the time of measurement. The macroscopic variables of pressure and temperature are not applicable to a microscopic system with only 10 molecules.

As we increase the number of molecules in a system, the chance of an irreversible process occurring decreases dramatically. For example, if we have 50 molecules in a container, the chance that they will all be in the left half of the container is $(\frac{1}{2})^{50} \approx 10^{-15}$. Thus, if we look at the gas once each second, we could expect to see all 50 molecules in the left half of the volume about once in every 10^{15} seconds or once in every 36 million years! For 1 mole (6×10^{23} molecules), the chance that all will wind up in half of the volume is vanishingly small, essentially zero. For macroscopic systems, then, the probability of a process resulting in a decrease in the entropy of the universe is so extremely small that the distinction between improbable and impossible becomes blurred.

SUMMARY

The second law of thermodynamics is a fundamental law of nature.

Topic	Relevant Equations and Remarks	
1. Efficiency of a Heat Engine	If the engine removes Q_h from a hot reservoir, does work W, and exhausts heat Q_c to a cold reservoir, its efficiency is $\varepsilon = \frac{W}{Q_h} = \frac{Q_h - Q_c}{Q_h} = 1 - \frac{Q_c}{Q_h}$	**19-2**
2. Coefficient of Performance of a Refrigerator	$\text{COP} = \frac{Q_c}{W}$	**19-3**
3. Coefficient of Performance of a Heat Pump	$\text{COP}_{\text{HP}} = \frac{Q_h}{W}$	**19-8**
4. Equivalent Statements of the Second Law of Thermodynamics		
The Kelvin statement	No system can take energy as heat from a single reservoir and convert it entirely into work without additional net changes in the system or its surroundings.	
The heat-engine statement	It is impossible for a heat engine working in a cycle to produce *only the effect* of extracting heat from a single reservoir and performing an equivalent amount of work.	
The Clausius statement	A process whose only net result is to transfer energy as heat from a cooler object to a hotter one is impossible.	
The refrigerator statement	It is impossible for a refrigerator working in a cycle to produce *only the effect* of extracting heat from a cold object and rejecting the same amount of heat to a hot object.	
The entropy statement	The entropy of the universe (system plus surroundings) can never decrease.	
5. Conditions for a Reversible Process	1. No mechanical energy is transformed into thermal energy by friction, viscous forces, or other dissipative forces. 2. Energy transfer as heat can only occur between objects at the same temperature (or infinitesimally near the same temperature). 3. The process must be quasi-static so that the system is always in an equilibrium state (or infinitesimally near an equilibrium state).	

6. Carnot Engine	A Carnot engine is a reversible engine that works between two reservoirs. It uses a Carnot cycle, which consists of
Carnot cycle	1. A quasi-static isothermal absorption of heat at temperature T_h 2. A quasi-static adiabatic expansion 3. A quasi-static isothermal exhaustion of heat at temperature T_c 4. A quasi-static adiabatic compression back to the original state
Carnot efficiency	$\varepsilon_C = 1 - \frac{Q_c}{Q_h} = 1 - \frac{T_c}{T_h}$ **19-6**
7. Thermodynamic Temperature	The ratio of the thermodynamic temperatures of two reservoirs is defined to be the ratio of the heat exhausted to the heat intake of a Carnot engine running between the reservoirs. $\frac{T_c}{T_h} = \frac{Q_c}{Q_h}$ **19-7**
8. Entropy	Entropy is a measure of the disorder of a system. The difference in entropy between two nearby states is given by $dS = \frac{dQ_{rev}}{T}$ **19-12** where dQ_{rev} is the heat added in a reversible process connecting the states. The entropy change of a system can be positive or negative.
Entropy and loss of work capability	During an irreversible process, the entropy of the universe S_u increases and an amount of energy $W_{lost} = T\Delta S_u$ **19-22** becomes unavailable for doing work.
Entropy and probability	Entropy is related to probability. A highly ordered system is one of low probability and low entropy. An isolated system moves towards a state of high probability, low order, and high entropy.

PROBLEMS

In a few problems, you are given more data than you actually need; in a few other problems, you are required to supply data from your general knowledge, outside sources, or informed estimates.

- • Single-concept, single-step, relatively easy
- •• Intermediate-level, may require synthesis of concepts
- ••• Challenging
- SSM Solution is in the *Student Solutions Manual*
- iSOLVE Problems available on iSOLVE online homework service
- iSOLVE ✓ These "Checkpoint" online homework service problems ask students additional questions about their confidence level and how they arrived at their answer

Conceptual Problems

1 • How does kinetic friction in an engine affect its efficiency?

2 • SSM Explain why you can't just open your refrigerator to cool your kitchen on a hot day. Why is it that turning on a room air conditioner will cool down the room but opening a refrigerator door will not?

3 • Why do power-plant designers try to increase the temperature of the steam fed to engines as much as possible?

4 •• On a humid day, water vapor condenses on a cold surface. During condensation, the entropy of the water (*a*) increases, (*b*) remains constant, (*c*) decreases, (*d*) may decrease or remain unchanged.

5 • SSM In a reversible adiabatic process, (*a*) the internal energy of the system remains constant, (*b*) no work is done by the system, (*c*) the entropy of the system remains constant, (*d*) the temperature of the system remains constant.

6 •• True or false:

(*a*) Work can never be converted completely into heat.
(*b*) Heat can never be converted completely into work.
(*c*) All heat engines have the same efficiency.
(*d*) It is impossible to transfer a given quantity of heat from a cold reservoir to a hot reservoir.
(*e*) The coefficient of performance of a refrigerator cannot be greater than 1.
(*f*) All Carnot engines are reversible.
(*g*) The entropy of a system can never decrease.
(*h*) The entropy of the universe can never decrease.

7 •• An ideal gas is taken reversibly from an initial state P_i, V_i, T_i to the final state P_f, V_f, T_f. Two possible paths are (A) an isothermal expansion followed by an adiabatic compression, and (B) an adiabatic compression followed by an isothermal expansion. For these two paths, (*a*) $\Delta E_{\text{int A}} > \Delta E_{\text{int B}}$, (*b*) $\Delta S_A > \Delta S_B$, (*c*) $\Delta S_A < \Delta S_B$, (*d*) none of the above is correct.

8 •• SSM Figure 19-12 shows a thermodynamic cycle on an *ST* diagram. Identify this cycle and sketch it on a *PV* diagram.

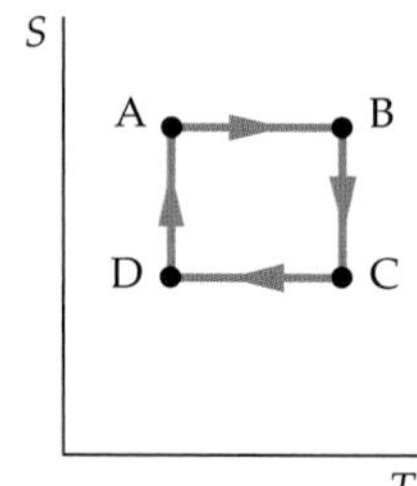

FIGURE 19-12
Problems 8 and 68

9 •• Figure 19-13 shows a thermodynamic cycle on an *SV* diagram. Identify the type of engine represented by this diagram.

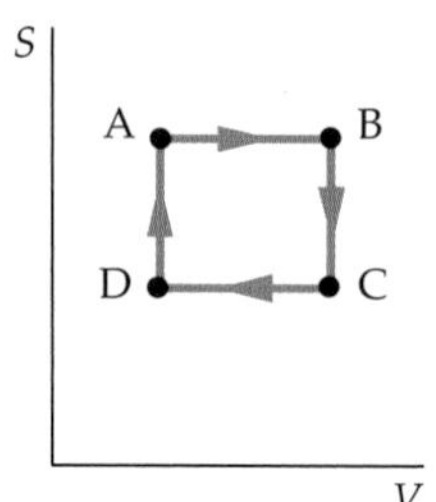

FIGURE 19-13
Problem 9

10 •• Sketch an *ST* diagram of the Otto cycle.

11 •• Sketch an *SV* diagram of the Carnot cycle.

12 •• Sketch an *SV* diagram of the Otto cycle.

13 •• Figure 19-14 shows a thermodynamic cycle on an *SP* diagram. Make a sketch of this cycle on a *PV* diagram.

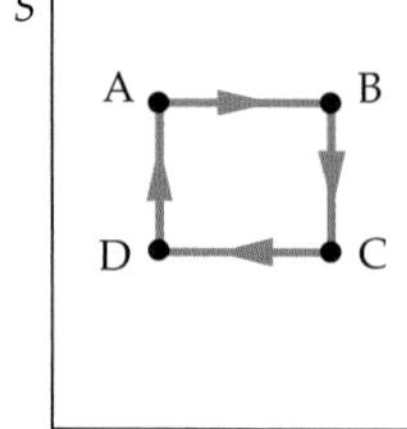

FIGURE 19-14 **Problem 13**

14 • SSM Which has a greater effect on increasing the efficiency of a Carnot engine, a 5-K increase in the temperature of the hot reservoir or a 5-K decrease in the temperature of the cold reservoir?

Estimation and Approximation

15 •• Estimate the maximum efficiency of an automobile engine with a compression ratio of 8:1. Assume the Otto cycle and assume $\gamma = 1.4$.

16 •• SSM (*a*) Estimate the highest COP possible for a "typical" household refrigerator. (*b*) If the refrigerator draws 600 W of electrical power, estimate the rate at which heat is being drawn from the refrigerator compartment.

17 •• The temperature of the sun is about 5400 K, the Earth's average temperature is about 290 K, and the solar constant (the intensity of sunlight reaching the Earth's orbit) is about 1.3 kW/m^2. (*a*) Calculate the total power of sunlight hitting the Earth. (*b*) Calculate the net rate at which the Earth's entropy is increasing due to the flow of solar radiation. (*c*) Calculate the net rate at which the sun's entropy is decreasing just due to the outflow of solar radiation *hitting the Earth.*

18 •• (*a*) Using the information given in Problem 17 and the known distance from the Earth to the sun (1.5×10^{11} m), calculate the total power that the sun radiates into space. (*b*) There are about 10^{11} stars like the sun in the Milky Way galaxy, and 10^{11} galaxies in the universe. Use this information to estimate the rate at which the entropy of the universe is increasing, assuming that the average temperature of the universe is 2.73 K.

19 •• A typical human body produces about 100 W of heat. Estimate the increase in entropy of the universe produced by a single human body over the course of a spring day where the temperature is 70°F during the day and 55°F at night.

20 ••• SSM How long, on average, should we have to wait until all of the air molecules in a room rush to one half of the room? (As a friend of mine put it, "Don't hold your breath. . . .") Assume that the air molecules are contained in a 1 m × 1 m × 1 m box and that they reshuffle their positions 100 times per second. Calculate the average time it should take for all the molecules to occupy only one half of the box if there are (*a*) 10 molecules, (*b*) 100 molecules, (*c*) 1000 molecules, and (*d*) 1 mole of molecules in the box. (*e*) The highest vacuums that have been created to date have pressures of about 10^{-12} torr. If a typical vacuum chamber has a capacity of about 1 liter, how long will a physicist have to wait before all of the gas molecules in the vacuum chamber occupy only one half of it? Compare that to the expected lifetime of the universe, which is about 10^{10} years.

Heat Engines and Refrigerators

21 • ISOLVE✓ An engine with 20% efficiency does 100 J of work in each cycle. (*a*) How much heat is absorbed in each cycle? (*b*) How much heat is rejected in each cycle?

22 • ISOLVE✓ An engine absorbs 400 J of heat and does 120 J of work in each cycle. (*a*) What is its efficiency? (*b*) How much heat is rejected in each cycle?

23 • ISOLVE✓ An engine absorbs 100 J and rejects 60 J in each cycle. (*a*) What is its efficiency? (*b*) If each cycle takes 0.5 s, find the power output of this engine in watts.

24 • SSM ISOLVE A refrigerator absorbs 5 kJ of energy from a cold reservoir and rejects 8 kJ to a hot reservoir. (*a*) Find the coefficient of performance of the refrigerator. (*b*) The refrigerator is reversible and is run backward as a heat engine between the same two reservoirs. What is its efficiency?

25 •• An engine operates with 1 mol of an ideal gas, for which $C_v = \frac{3}{2}R$ and $C_p = \frac{5}{2}R$, as its working substance. The cycle begins at $P_1 = 1$ atm and $V_1 = 24.6$ L. The gas is heated at constant volume to $P_2 = 2$ atm. It then expands at constant pressure until $V_2 = 49.2$ L. During these two steps, heat is absorbed by the gas. The gas is then cooled at constant volume until its pressure is again 1 atm. It is then compressed at constant pressure to its original state. During the last two steps, heat is rejected by the gas. All the steps are quasi-static and reversible. (*a*) Show this cycle on a *PV* diagram. Find the work done, the heat added, and the change in the internal energy of the gas for each step of the cycle. (*b*) Find the efficiency of the cycle.

26 •• An engine using 1 mol of a diatomic ideal gas performs a cycle consisting of three steps: (1) an adiabatic expansion from an initial pressure of 2.64 atm and an initial volume of 10 L to a pressure of 1 atm and a volume of 20 L, (2) a compression at constant pressure to its original volume of 10 L, and (3) heating at constant volume to its original pressure of 2.64 atm. Find the efficiency of this cycle.

27 •• ISOLVE An engine using 1 mol of an ideal gas initially at $V_1 = 24.6$ L and $T = 400$ K performs a cycle consisting of four steps: (1) an isothermal expansion at $T = 400$ K to twice its initial volume, (2) cooling at constant volume to $T = 300$ K, (3) an isothermal compression to its original volume, and (4) heating at constant volume to its original temperature of 400 K. Assume that $C_v = 21$ J/K. Sketch the cycle on a *PV* diagram and find its efficiency.

28 •• SSM One mole of an ideal monatomic gas at an initial volume $V_1 = 25$ L follows the cycle shown in Figure 19-15. All the processes are quasi-static. Find (*a*) the temperature of each state of the cycle, (*b*) the heat flow for each part of the cycle, and (*c*) the efficiency of the cycle.

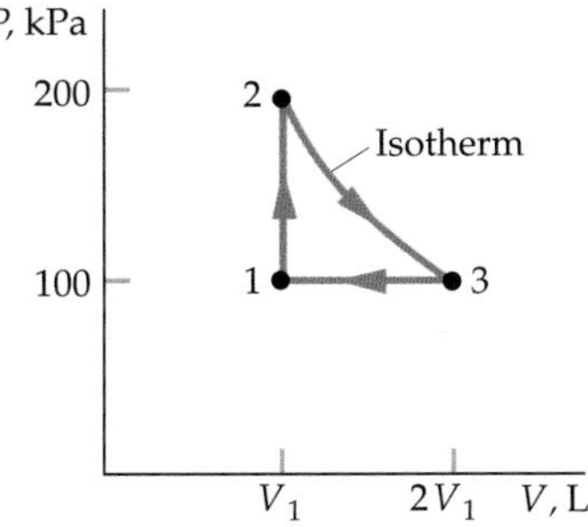

FIGURE 19-15 Problem 28

29 •• An ideal gas ($\gamma = 1.4$) follows the cycle shown in Figure 19-16. The temperature of state 1 is 200 K. Find (*a*) the temperatures of the other three states of the cycle and (*b*) the efficiency of the cycle.

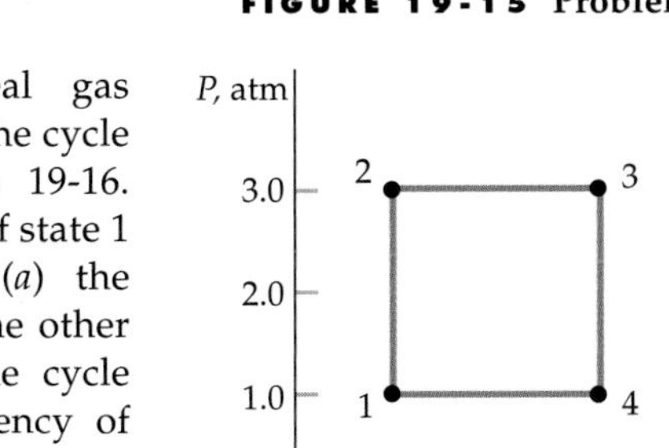

FIGURE 19-16 Problem 29

30 ••• The *diesel cycle* shown in Figure 19-17 approximates the behavior of a diesel engine. Process *ab* is an adiabatic compression, process *bc* is an expansion at constant pressure, process *cd* is an adiabatic expansion, and process *da* is cooling at constant volume. Find the efficiency of this cycle in terms of the volumes V_a, V_b, V_c, and V_d.

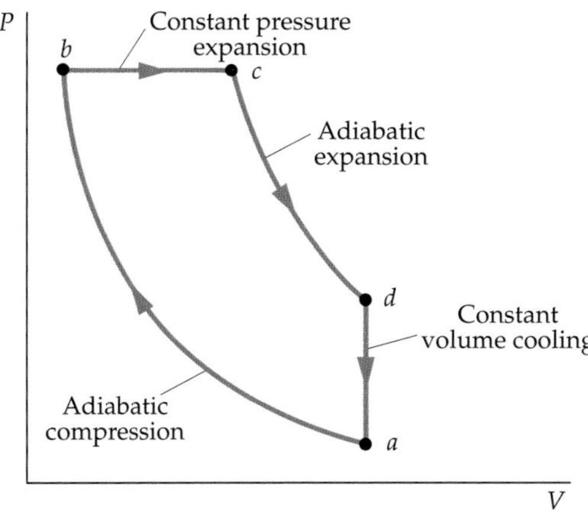

FIGURE 19-17 Diesel cycle for Problem 30

31 •• SSM "As far as we know, Nature has never evolved a heat engine"—Steven Vogel, *Life's Devices,* Princeton University Press (1988). (*a*) Calculate the efficiency of a heat engine operating between body temperature (98.6°F) and a typical outdoor temperature (70°F), and compare this to the human body's efficiency for converting chemical energy into work (approximately 20%). Does this contradict the Second Law of Thermodynamics? (*b*) From the result of Part (*a*), and a general knowledge of the conditions under which most warm-blooded animal life exists, explain why no warm-blooded animals have evolved heat engines to supply their internal energy.

32 ••• The Clausius equation of state is $P(V - bn) = nRT$, where b is a constant. Show that the efficiency of a Carnot cycle is the same for a gas that obeys this equation of state as it is for one that obeys the ideal-gas equation of state, $PV = nRT$.

Second Law of Thermodynamics

33 •• A refrigerator takes in 500 J of heat from a cold reservoir and gives off 800 J to a hot reservoir. Assume that the heat-engine statement of the second law of thermodynamics is false, and show how a perfect engine working with this refrigerator can violate the refrigerator statement of the second law.

34 •• SSM If two adiabatic curves intersect on a *PV* diagram, a cycle could be completed by an isothermal path between the two adiabatic curves shown in Figure 19-18. Show that such a cycle could violate the second law of thermodynamics.

P
Adiabatic
Isotherm
Adiabatic
V

FIGURE 19-18 Problem 34

Carnot Engines

35 • ISOLVE ✓ A Carnot engine works between two heat reservoirs at temperatures $T_h = 300$ K and $T_c = 200$ K. (*a*) What is its efficiency? (*b*) If it absorbs 100 J from the hot reservoir during each cycle, how much work does it do? (*c*) How much heat does it give off during each cycle? (*d*) What is the COP of this engine when it works as a refrigerator between the same two reservoirs?

36 • ISOLVE An engine removes 250 J from a reservoir at 300 K and exhausts 200 J to a reservoir at 200 K. (*a*) What is its efficiency? (*b*) How much more work could be done if the engine were reversible?

37 •• A reversible engine working between two reservoirs at temperatures T_h and T_c has an efficiency of 30%. Working as a heat engine, it gives off 140 J of heat to the cold reservoir. A second engine working between the same two reservoirs also gives off 140 J to the cold reservoir. Show that if the second engine has an efficiency greater than 30%, the two engines working together would violate the heat-engine statement of the second law.

38 •• A reversible engine working between two reservoirs at temperatures T_h and T_c has an efficiency of 20%. Working as a heat engine, it does 100 J of work in each cycle. A second engine working between the same two reservoirs also does 100 J of work in each cycle. Show that if the efficiency of the second engine is greater than 20%, the two engines working together would violate the refrigerator statement of the second law.

39 •• SSM A Carnot engine works between two heat reservoirs as a refrigerator. It does 50 J of work to remove 100 J from the cold reservoir and gives off 150 J to the hot reservoir during each cycle. Its coefficient of performance $\text{COP} = Q_c/W = (100\text{ J})/(50\text{ J}) = 2$. (*a*) What is the efficiency of the Carnot engine when it works as a heat engine between the same two reservoirs? (*b*) Show that no other engine working as a refrigerator between the same two reservoirs can have a COP greater than 2.

40 •• ISOLVE A Carnot engine works between two heat reservoirs at temperatures $T_h = 300$ K and $T_c = 77$ K. (*a*) What is its efficiency? (*b*) If it absorbs 100 J from the hot reservoir during each cycle, how much work does it do? (*c*) How much heat does it give off in each cycle? (*d*) What is the coefficient of performance of this engine when it works as a refrigerator between these two reservoirs?

41 •• In the cycle shown in Figure 19-19, 1 mol of an ideal gas ($\gamma = 1.4$) is initially at a pressure of 1 atm and a temperature of 0°C. The gas is heated at constant volume to $T_2 = 150$°C and is then expanded adiabatically until its pressure is again 1 atm. It is then compressed at constant pressure back to its original state. Find (*a*) the temperature T_3 after the adiabatic expansion, (*b*) the heat entering or leaving the system during each process, (*c*) the efficiency of this cycle, and (*d*) the efficiency of a Carnot cycle operating between the temperature extremes of this cycle.

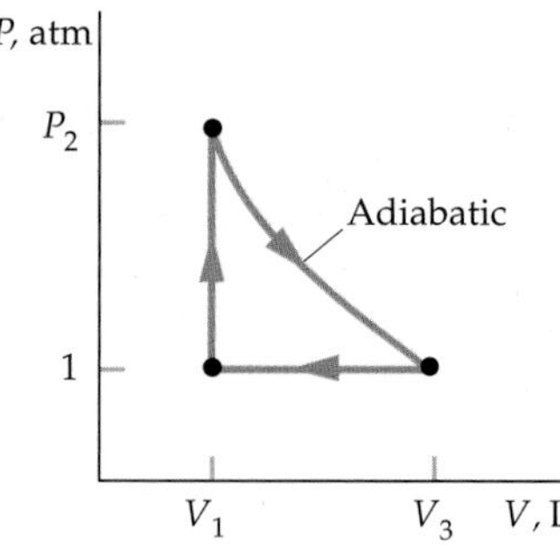

FIGURE 19-19
Problem 41

42 •• ISOLVE A steam engine takes in superheated steam at 270°C and discharges condensed steam from its cylinder at 50°C. Its efficiency is 30%. (*a*) How does this efficiency compare with the maximum possible efficiency for these temperatures? (*b*) If the useful power output of the engine is 200 kW, how much heat does the engine discharge to its surroundings in 1 h?

*Heat Pumps

43 • SSM ISOLVE ✓ A heat pump delivers 20 kW to heat a house. The outside temperature is −10°C and the inside temperature of the hot-air supply for the heating fan is 40°C. (*a*) What is the coefficient of performance of a Carnot heat pump operating between these temperatures? (*b*) What must be the minimum power of the engine needed to run the heat pump? (*c*) If the COP of the heat pump is 60% of the efficiency of an ideal pump, what must be the minimum power of the engine?

44 • ISOLVE ✓ A refrigerator is rated at 370 W. (*a*) What is the maximum amount of heat it can remove in 1 min if the inside temperature of the refrigerator is 0°C and it exhausts into a room at 20°C? (*b*) If the COP of the refrigerator is 70% of that of an ideal pump, how much heat can it remove in 1 min?

45 • Rework Problem 44 for a room temperature of 35°C.

Entropy Changes

46 • What is the change in entropy of 1 mol of water at 0°C that freezes?

47 •• Consider the freezing of 50 g of water by placing it in the freezer compartment of a refrigerator. Assume the walls of the freezer are maintained at −10°C. The water, initially liquid at 0°C, is frozen into ice and cooled to −10°C. Show that even though the entropy of the ice decreases, the net entropy of the universe increases.

48 • ISOLVE Two moles of an ideal gas at $T = 400$ K expand quasi-statically and isothermally from an initial volume of 40 L to a final volume of 80 L. (*a*) What is the entropy change of the gas? (*b*) What is the entropy change of the universe for this process?

49 • The gas in Problem 48 is taken from the same initial state ($T = 400$ K, $V_1 = 40$ L) to the same final state ($T = 400$ K, $V_2 = 80$ L) by a process that is not quasi-static. (*a*) What is the entropy change of the gas? (*b*) What can be said about the entropy change of the universe?

50 • What is the change in entropy of 1.0 kg of water when it changes to steam at 100°C and a pressure of 1 atm?

51 • ISOLVE ✓ What is the change in entropy of 1.0 kg of ice when it changes to water at 0°C and a pressure of 1 atm?

52 •• A system absorbs 200 J of heat reversibly from a reservoir at 300 K and gives off 100 J reversibly to a reservoir at 200 K as it moves from state A to state B. During this process, the system does 50 J of work. (*a*) What is the change in the internal energy of the system? (*b*) What is the change in entropy of the system? (*c*) What is the change in entropy of the universe? (*d*) If the system goes from state A to state B by a nonreversible process, how would your answers for Parts (*a*), (*b*), and (*c*) differ?

53 •• SSM ISOLVE A system absorbs 300 J from a reservoir at 300 K and 200 J from a reservoir at 400 K. It then returns to its original state, doing 100 J of work and rejecting 400 J of heat to a reservoir at a temperature T. (*a*) What is the entropy change of the system for the complete cycle? (*b*) If the cycle is reversible, what is the temperature T?

54 •• ISOLVE Two moles of an ideal gas originally at $T = 400$ K and $V = 40$ L undergo a free expansion to twice their initial volume. What is (*a*) the entropy change of the gas and (*b*) the entropy change of the universe?

55 •• A 200-kg block of ice at 0°C is placed in a large lake. The temperature of the lake is just slightly higher than 0°C, and the ice melts. (*a*) What is the entropy change of the ice? (*b*) What is the entropy change of the lake? (*c*) What is the entropy change of the universe (the ice plus the lake)?

56 •• A 100-g piece of ice at 0°C is placed in an insulated container with 100 g of water at 100°C. (*a*) When thermal equilibrium is established, what is the final temperature of the water (Ignore the heat capacity of the container.) (*b*) Find the entropy change of the universe for this process.

57 •• SSM A 1-kg block of copper at 100°C is placed in a calorimeter of negligible heat capacity containing 4 L of water at 0°C. Find the entropy change of (*a*) the copper block, (*b*) the water, and (*c*) the universe.

58 •• If a 2-kg piece of lead at 100°C is dropped into a lake at 10°C, find the entropy change of the universe.

59 •• ISOLVE A 1500-kg car traveling at 100 km/h crashes into a concrete wall. If the temperature of the air is 20°C, calculate the entropy change of the universe.

60 •• SSM A box is divided into two identical halves by an impermeable partition through its middle. On one side is 1 mole of ideal gas A; on the other, 1 mole of ideal gas B (which is different from A). (*a*) Calculate the change in entropy when the partition is lifted, and the two gases mix together. (*b*) If we repeat the process with the same type of gas in each side, should the entropy change when the partition is lifted? Explain. (Think carefully about this question!)

Entropy and Work Lost

61 •• SSM ISOLVE If 500 J of heat is conducted from a reservoir at 400 K to one at 300 K, (*a*) what is the change in entropy of the universe, and (*b*) how much of the 500 J of heat conducted could have been converted into work using a cold reservoir at 300 K?

62 •• One mole of an ideal gas undergoes a free expansion from $V_1 = 12.3$ L and $T_1 = 300$ K to $V_2 = 24.6$ L and $T_2 = 300$ K. It is then compressed isothermally and quasi-statically back to its original state. (*a*) What is the entropy change of the universe for the complete cycle? (*b*) How much work is wasted in this cycle? (*c*) Show that the work wasted is $T\Delta S_u$.

General Problems

63 • ISOLVE ✓ An engine with an output of 200 W has an efficiency of 30%. It works at 10 cycles/s. (*a*) How much work is done in each cycle? (*b*) How much heat is absorbed and how much is given off in each cycle?

64 • ISOLVE ✓ In each cycle, an engine removes 150 J from a reservoir at 100°C and gives off 125 J to a reservoir at 20°C. (*a*) What is the efficiency of this engine? (*b*) What is the ratio of its efficiency to that of a Carnot engine working between the same reservoirs? (This ratio is called the *second law efficiency*.)

65 • An engine removes 200 kJ of heat from a hot reservoir at 500 K in each cycle and exhausts heat to a cold reservoir at 200 K. Its efficiency is 85% of a Carnot engine working between the same reservoirs. (*a*) What is the efficiency of this engine? (*b*) How much work is done in each cycle? (*c*) How much heat is exhausted in each cycle?

66 •• (*a*) Calvin Cliffs Nuclear Power Plant, located on the Hobbes River, generates 1 GW of power. In this plant liquid sodium circulates between the reactor core and a heat exchanger located in the superheated steam that drives the turbine. Heat is transferred into the liquid sodium in the core, and out of the liquid sodium (and into the superheated steam) in the heat exchanger. The temperature of the superheated steam is 500 K. Waste heat is dumped into the river, which flows by at a temperature of 25°C. (*a*) What is the highest efficiency that this plant can have? (*b*) How much waste heat is dumped into the river every second? (*c*) How much heat must be generated to supply 1 GW of power? (*d*) Assume that new, tough environmental laws have been passed (to preserve the unique wildlife of the river). Because of this, the plant is not allowed to heat the river by more than 0.5°C. What is the minimum flow rate that the Hobbes river must have (in L/sec)?

67 • ISOLVE To maintain the temperature inside a house at 20°C, the power consumption of the electric baseboard heaters is 30 kW on a day when the outside temperature is −7°C. At what rate does this house contribute to the increase in the entropy of the universe?

68 •• The system represented in Figure 19-12 (Problem 8) is 1 mol of an ideal monatomic gas. The temperatures at points A and B are 300 and 750 K, respectively. What is the thermodynamic efficiency of the cyclic process ABCDA?

69 •• (*a*) Which process is more wasteful: (1) a block moving with 500 J of kinetic energy being slowed to rest by friction when the temperature of the atmosphere is 300 K, or (2) 1 kJ of heat being conducted from a reservoir at 400 K to one at 300 K? *Hint: How much of the 1 kJ of heat could be converted into work in an ideal situation?* (*b*) What is the change in entropy of the universe for each process?

70 •• Helium gas ($\gamma = 1.67$) is initially at a pressure of 16 atm, a volume of 1 L, and a temperature of 600 K. It is expanded isothermally until its volume is 4 L and is then compressed at constant pressure until its volume and temperature are such that an adiabatic compression will return the gas to its original state. (*a*) Sketch this cycle on a *PV* diagram. (*b*) Find the volume and temperature after the isobaric compression. (*c*) Find the work done during each cycle. (*d*) Find the efficiency of the cycle.

71 •• SSM A heat engine that does the work of blowing up a balloon at a pressure of 1 atm extracts 4 kJ from a hot reservoir at 120°C. The volume of the balloon increases by 4 L, and heat is exhausted to a cold reservoir at a temperature T_c. If the efficiency of the heat engine is 50% of the efficiency of a Carnot engine working between the same reservoirs, find the temperature T_c.

72 •• Show that the COP of a Carnot refrigerator is related to the efficiency of a Carnot engine by COP $= T_c/(\varepsilon_C T_h)$.

73 •• ISOLVE✓ A freezer has a temperature $T_c = -23°C$. The air in the kitchen has a temperature $T_h = 27°C$. Since the heat insulation is not perfect, some heat flows into the freezer at a rate of 50 W. Find the power of the motor that is needed to maintain the temperature in the freezer.

74 •• Two moles of a diatomic gas are taken through the cycle ABCA as shown on the *PV* diagram in Figure 19-20. At A the pressure and temperature are 5 atm and 600 K. The volume at B is twice that at A. The segment BC is an adiabatic expansion and the segment CA is an isothermal compression. (*a*) What is the volume of the gas at A? (*b*) What are the volume and temperature of the gas at B? (*c*) What is the temperature of the gas at C? (*d*) What is the volume of the gas at C? (*e*) How much work is done by the gas in each of the three segments of the cycle? (*f*) How much heat is absorbed by the gas in each segment of the cycle? (*g*) What is the thermodynamic efficiency of this cycle?

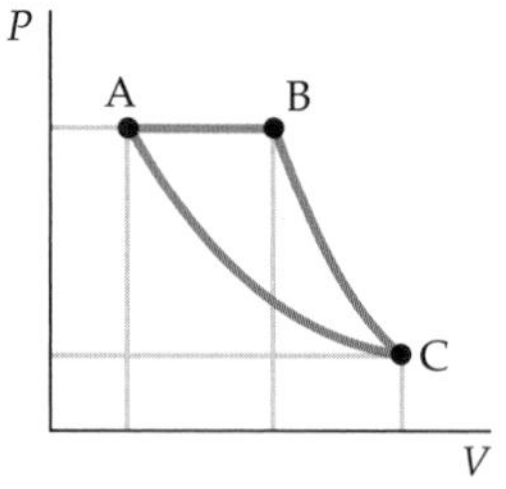

FIGURE 19-20
Problems 74 and 76

75 •• Two moles of a diatomic gas are carried through the cycle ABCDA shown in the *PV* diagram in Figure 19-21. The segment AB represents an isothermal expansion, the segment BC an adiabatic expansion. The pressure and temperature at A are 5 atm and 600 K. The volume at B is twice that at A. The pressure at D is 1 atm. (*a*) What is the pressure at B? (*b*) What is the temperature at C? (*c*) Find the work done by the gas in one cycle and the thermodynamic efficiency of this cycle.

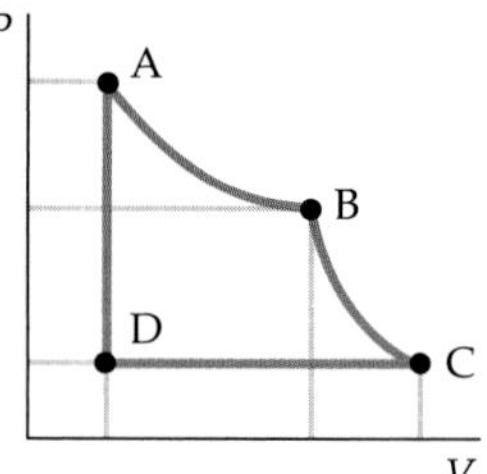

FIGURE 19-21
Problems 75 and 77

76 •• Repeat Problem 74 for a monatomic gas.

77 •• Repeat Problem 75 for a monatomic gas.

78 •• Compare the efficiency of the Otto engine and the Carnot engine operating between the same maximum and minimum temperatures.

79 ••• SSM Using the equation for the entropy change of an ideal gas when the volume and temperature change and $TV^{\gamma-1}$ is a constant, show explicitly that the entropy change is zero for a quasi-static adiabatic expansion from state (V_1, T_1) to state (V_2, T_2).

80 ••• (*a*) Show that if the refrigerator statement of the second law of thermodynamics were not true, then the entropy of the universe could decrease. (*b*) Show that if the heat-engine statement of the second law were not true, then the entropy of the universe could decrease. (*c*) An alternative statement of the second law is that the entropy of the universe cannot decrease. Have you just proved that this statement is equivalent to the refrigerator and heat-engine statements?

81 ••• Suppose that two heat engines are connected in series, such that the heat exhaust of the first engine is used as the heat input of the second engine as shown in Figure 19-22. The efficiencies of the engines are ε_1 and ε_2, respectively. Show that the net efficiency of the combination is given by

$$\varepsilon_{net} = \varepsilon_1 + (1 - \varepsilon_1)\varepsilon_2.$$

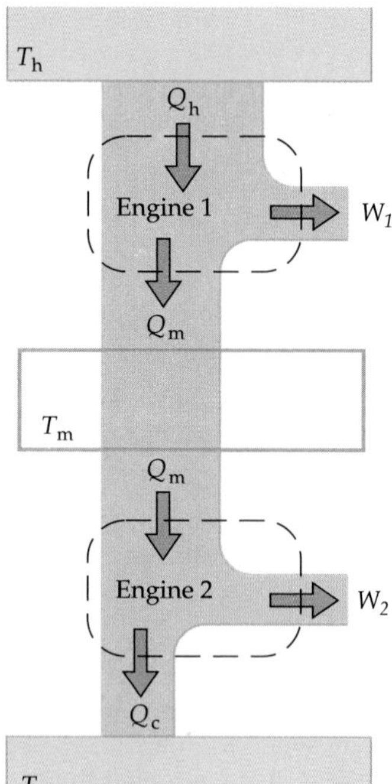

FIGURE 19-22
Problems 81 and 82

82 ••• SSM Suppose that each engine in Figure 19-22 is an ideal reversible heat engine. Engine 1 operates between temperatures T_h and T_m and Engine 2 operates between T_m and T_c, where $T_h > T_m > T_c$. Show that

$$\varepsilon_{net} = 1 - \frac{T_c}{T_h}$$

This means that two reversible heat engines in series are equivalent to one reversible heat engine operating between the hottest and coldest reservoirs.

83 ••• Bertrand Russell once said that if a million monkeys were given a million typewriters and typed away at random for a million years, they would produce all of Shakespeare's works. Let's limit ourselves to the following fragment of Shakespeare (*Julius Caesar* III:ii):

Friends, Romans, countrymen! Lend me your ears.
I come to bury Caesar, not to praise him.
The evil that men do lives on after them,
The good is oft interred with the bones.
So let it be with Caesar.
The noble Brutus hath told you that Caesar was ambitious,
And, if so, it were a grievous fault,
And grievously hath Caesar answered it . . .

Even with this small fragment, it will take a lot longer than a million years! By what factor (roughly speaking) was Russell in error? Make any reasonable assumptions you want. (You may even assume that the monkeys are immortal.)

*CHAPTER

Thermal Properties and Processes

EXPANSION JOINTS LIKE THIS ONE ARE INCLUDED ON STEEL BRIDGES TO ALLOW THE BRIDGE TO EXPAND AND CONTRACT WITH CHANGES IN TEMPERATURE WITHOUT PUTTING STRESS ON THE STEEL STRUCTURE.

What might eventually happen to this bridge if it did not have expansion joints? (See Example 20-1.)

When an object absorbs thermal energy, various changes may occur in the physical properties of the object. For example, its temperature may rise, accompanied by an expansion or contraction of the object, or the object may liquefy or vaporize, during which its temperature remains constant.

➢ **In this chapter, we examine some of the thermal properties of matter and some important processes involving thermal energy.**

20-1 Thermal Expansion

When the temperature of an object increases, the object typically expands. (Consider that on concrete highways, expansion joints appear every 10 to 15 m, allowing the road to expand without cracking.) Suppose that we have a long rod of length L at a temperature T. When the temperature changes by ΔT, the fractional change in length ΔL is proportional to ΔT:

$$\frac{\Delta L}{L} = \alpha \Delta T \tag{20-1}$$

where α, called the **coefficient of linear expansion,** is the ratio of the fractional change in length to the change in temperature:

$$\alpha = \frac{\Delta L/L}{\Delta T} \qquad \text{20-2}$$

The units for the coefficient of linear expansion are reciprocal Celsius degrees (1/°C), which are the same as reciprocal kelvins (1/K). The value of α for a solid or liquid doesn't vary much with pressure, but it may vary significantly with temperature. Equation 20-2 gives the average value over the temperature interval ΔT. The coefficient of linear expansion at a particular temperature T is found by taking the limit as ΔT approaches zero:

$$\alpha = \lim_{\Delta T \to 0} \frac{\Delta L/L}{\Delta T} = \frac{1}{L}\frac{dL}{dT} \qquad \text{20-3}$$

The accuracy obtained by using the average value of α over a wide temperature range is sufficient for most purposes.

The **coefficient of volume expansion** β is similarly defined as the ratio of the fractional change in volume to the change in temperature (at constant pressure):

$$\beta = \lim_{\Delta T \to 0} \frac{\Delta V/V}{\Delta T} = \frac{1}{V}\frac{dV}{dT} \qquad \text{20-4}$$

Like α, β does not usually vary with pressure for solids and liquids, but may vary with temperature. Average values for α and β for various substances are given in Table 20-1.

For a given material, $\beta = 3\alpha$. We can show this by considering a box of dimensions L_1, L_2, and L_3. Its volume at a temperature T is

$$V = L_1L_2L_3$$

The rate of change of the volume with respect to temperature is

$$\frac{dV}{dT} = L_1L_2\frac{dL_3}{dT} + L_1L_3\frac{dL_2}{dT} + L_2L_3\frac{dL_1}{dT}$$

Dividing each side of the equation by the volume, we obtain

$$\beta = \frac{1}{V}\frac{dV}{dT} = \frac{1}{L_3}\frac{dL_3}{dT} + \frac{1}{L_2}\frac{dL_2}{dT} + \frac{1}{L_1}\frac{dL_1}{dT}$$

We can see that each term on the right side of the above equation equals α, and so we have

$$\beta = 3\alpha \qquad \text{20-5}$$

Similarly, the coefficient of area expansion is twice that of linear expansion.

The increase in size of any part of an object for a given temperature change is proportional to the original size of that part of the body. For example, if we increase the temperature of a steel ruler, the effect will be similar to that of a (very slight) photographic enlargement. That is, the dimensions of the ruler itself will be larger, as will the distance between the equally spaced lines. **If the ruler has a 1-cm-diameter hole in it, say between the 3-cm and 4-cm lines, the hole will get larger, just as the distance between the 3-cm and 4-cm lines does.**

TABLE 20-1

Approximate Values of the Coefficients of Thermal Expansion for Various Substances

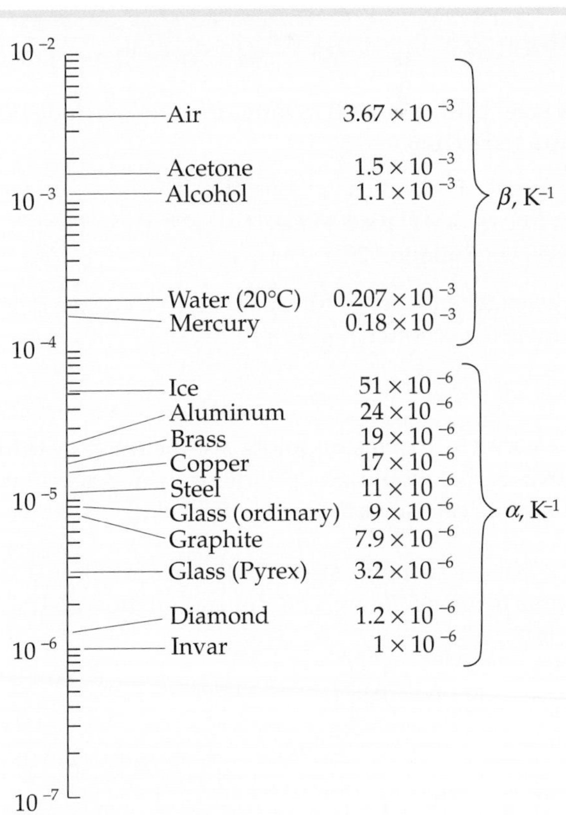

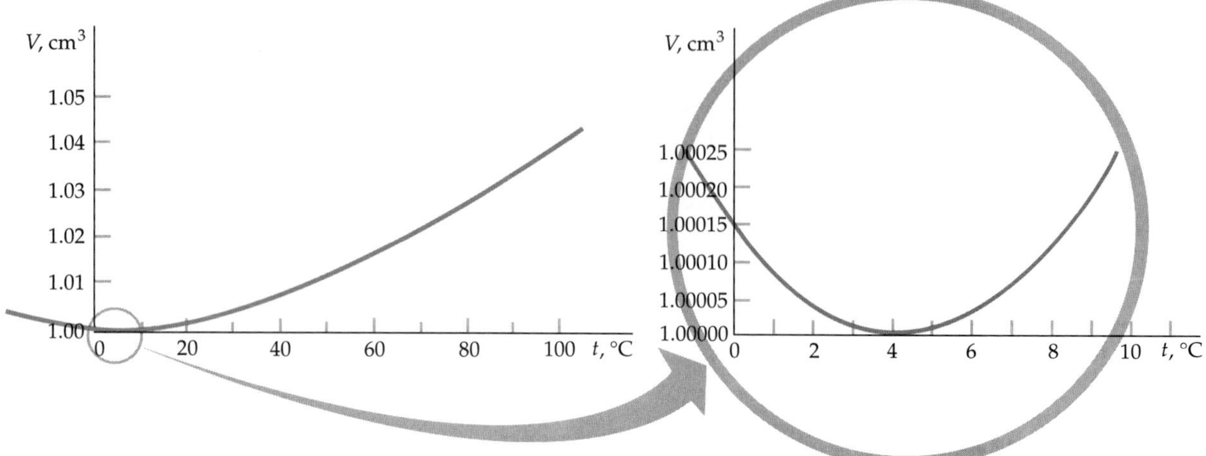

FIGURE 20-1 Volume of 1 g of water at atmospheric pressure versus temperature. The minimum volume, which corresponds to the maximum density, occurs at 4°C. At temperatures below 0°C, the curve shown is for supercooled water. (Supercooled water is water that is cooled below the normal freezing point without solidifying.)

Most materials expand when heated and contract when cooled. Water, however, presents an important exception. Figure 20-1 shows the volume occupied by 1 g of water as a function of temperature. The minimum volume, and therefore the maximum density, is at 4°C. Thus, when water below 4°C is cooled, it expands rather than contracts and vice versa. This property of water has important consequences for the ecology of lakes. At temperatures above 4°C, the water in a lake becomes denser as it cools and therefore sinks to the bottom. But as the water cools below 4°C, it becomes less dense and rises to the surface. This is the reason that ice forms first on the surface of a lake. Since ice is less dense than liquid water, it remains at the surface and acts as a thermal insulator for the water below. If water behaved like most substances and contracted when it froze, then ice would sink and expose more water at the surface that would then freeze. Lakes would fill with ice from the bottom up and would be much more likely to freeze completely in the winter, killing fish and other aquatic life.

AN EXPANDING BRIDGE **EXAMPLE 20-1**

A steel bridge is 1000 m long. By how much does it expand when the temperature rises from 0 to 30°C?

PICTURE THE PROBLEM Use $\alpha = 11 \times 10^{-6}\ \text{K}^{-1}$ from Table 20-1 and calculate ΔL from Equation 20-1.

The change in length for a 30 C° (30 K) change in temperature is the product of α, L, and ΔT:

$$\Delta L = \alpha L \Delta T = (11 \times 10^{-6}\ \text{K}^{-1})(1000\ \text{m})(30\ \text{K})$$

$$= \boxed{0.33\ \text{m} = 33\ \text{cm}}$$

REMARKS Expansion joints are included in bridges to relieve the enormous stresses that would occur without them. Excessive stress caused by temperature increases can cause the bridge to buckle.

We can calculate the stress that would result in a steel bridge without expansion joints by using Young's modulus (Equation 12-1):

$$Y = \frac{\text{stress}}{\text{strain}} = \frac{F/A}{\Delta L/L}$$

Then

$$\frac{F}{A} = Y\frac{\Delta L}{L} = Y\alpha\Delta T$$

For $\Delta T = 30$ K, $\Delta L/L = 0.33$ m/1000 m as found in Example 20-1. Then using $Y = 2 \times 10^{11}$ N/m^2 (from Table 12-1),

$$\frac{F}{A} = Y\frac{\Delta L}{L} = (2 \times 10^{11}\ \text{N/m}^2)\frac{0.33\ \text{m}}{1000\ \text{m}} = 6.6 \times 10^7\ \text{N/m}^2$$

This stress is about one-third of the breaking stress for steel under compression. A compression stress of this magnitude would cause a steel bridge to buckle and become permanently deformed.

A COMPLETELY FILLED GLASS — EXAMPLE 20-2

While working in the laboratory, you fill a 1-L glass flask to the brim with water at 10°C. You heat the flask, raising the temperature of the water and flask to 30°C. How much water spills out of the flask?

PICTURE THE PROBLEM The glass flask and the water both expand when heated, but the water expands more, so some spills out. We calculate the amount spilled by finding the changes in volume for $\Delta T = 20$ K using $\Delta V_a = \beta V \Delta T$ with $\beta = 1.1 \times 10^{-3}$ K^{-1} for water (from Table 20-1), and $\Delta V_g = \beta V \Delta T = 3\alpha V \Delta T$ with $\alpha = 9 \times 10^{-6}$ K^{-1} for glass. The difference in these volume changes equals the volume spilled.

1. The volume of water spilled ΔV_s is the difference in the changes in volume of the water and glass: $\quad \Delta V_s = \Delta V_a - \Delta V_g$

2. Find the increase in the volume of the water: $\quad \Delta V_a = \beta_a V \Delta T$

3. Find the increase in the volume of the glass flask: $\quad \Delta V_g = \beta_g V \Delta T = 3\alpha_g V \Delta T$

4. Subtract to find the amount of water spilled:

$$\begin{aligned}\Delta V_s &= \Delta V_a - \Delta V_g = \beta_a V \Delta T - \beta_g V \Delta T \\ &= (\beta_a - \beta_g)\, V \Delta T = (\beta_a - 3\alpha_g)\, V \Delta T \\ &= [0.207 \times 10^{-3}\ \text{K}^{-1} - 3(9 \times 10^{-6}\ \text{K}^{-1})](1\ \text{L})(20\ \text{K}) \\ &= 3.6 \times 10^{-3}\ \text{L} = \boxed{3.6\ \text{mL}}\end{aligned}$$

BREAKING COPPER — EXAMPLE 20-3

A copper bar is heated to 300°C. Then it is clamped rigidly between two fixed points so that it can neither expand nor contract. If the breaking stress of copper is 230 MN/m², at what temperature will the bar break as it cools?

PICTURE THE PROBLEM As the bar cools, the change in length ΔL that *would* occur if the bar contracted is offset by an equal stretching due to tensile stress in the bar. The stress F/A is related to the stretching ΔL by $Y = (F/A)/(\Delta L/L)$, where Young's modulus for copper is $Y = 110$ GN/m^2 (from Table 12-1). The maximum allowable stretching occurs when F/A equals 230 MN/m^2. Thus, we find the temperature change that would produce this maximum contraction.

1. Calculate the change in length ΔL_1 that would occur if the bar were unclamped and cooled by ΔT: $\quad \Delta L_1 = \alpha L \Delta T$

2. A tensile stress F/A stretches the bar by ΔL_2, where $L_1 + \Delta L_2 = 0$: $\quad Y = \dfrac{F/A}{\Delta L_2/L}$, so $\Delta L_2 = L\dfrac{F/A}{Y}$

3. Substitute the step 1 and step 2 results into $\Delta L_1 + \Delta L_2 = 0$ and solve for ΔT with the stress equal to the breaking value:

$$\Delta L_1 + \Delta L_2 = 0$$

$$\alpha L \Delta T + L\frac{F/A}{Y} = 0$$

so

$$\Delta T = -\frac{F/A}{\alpha Y}$$

$$= -\frac{230 \times 10^6\ \text{N/m}^2}{(17 \times 10^{-6}\ \text{K}^{-1})(110 \times 10^9\ \text{N/m}^2)}$$

$$= -123\ \text{K} = -123\ \text{C}°$$

4. Add this result to the original temperature to find the final temperature at which the bar breaks:

$$T_f = T_1 + \Delta T = 300°\text{C} - 123\ \text{C}° = \boxed{177°\text{C}}$$

20-2 The van der Waals Equation and Liquid–Vapor Isotherms

At ordinary pressures most gases behave like an ideal gas. However, this ideal behavior breaks down when the pressure is high enough or the temperature is low enough such that the density of the gas is high and the molecules are, on average, closer together. An equation of state called the **van der Waals equation** describes the behavior of many real gases over a wide range of pressures more accurately than does the ideal-gas equation of state ($PV = nRT$). The van der Waals equation for n moles of gas is

$$\left(P + \frac{an^2}{V^2}\right)(V - bn) = nRT \qquad \text{20-6}$$

THE VAN DER WAALS EQUATION OF STATE

The constant b in this equation arises because the gas molecules are not point particles but objects that have a finite size; therefore, the volume available to each molecule is reduced. The magnitude of b is the volume of one mole of gas molecules. The term an^2/V^2 arises from the attraction of the gas molecules to each other. As a molecule approaches the wall of the container, it is pulled back by the molecules surrounding it with a force that is proportional to the density of those molecules n/V. Because the number of molecules that hit the wall in a given time is also proportional to the density of the molecules, the decrease in pressure due to the attraction of the molecules is proportional to the square of the density and therefore to n^2/V^2. The constant a depends on the gas and is small for inert gases, which have very weak chemical interactions. The terms bn and an^2/V^2 are both negligible when the volume V is large, so at low densities the van der Waals equation approaches the ideal-gas law. At high densities the van der Waals equation provides a much better description of the behavior of real gases than does the ideal-gas law.

Figure 20-2 shows PV isothermal curves for a substance at various temperatures. Except for the region where the liquid and vapor coexist, these curves are described quite accurately by the van der Waals equation and can be used to determine the constants a and b. For example, the values of these constants that give the best fit to the experimental curves for nitrogen are $a = 0.14\ \text{Pa·m}^6/\text{mol}^2$ and $b = 39.1\ \text{mL/mol}$. This volume of 39.1 mL per mole is about 0.2 percent of the

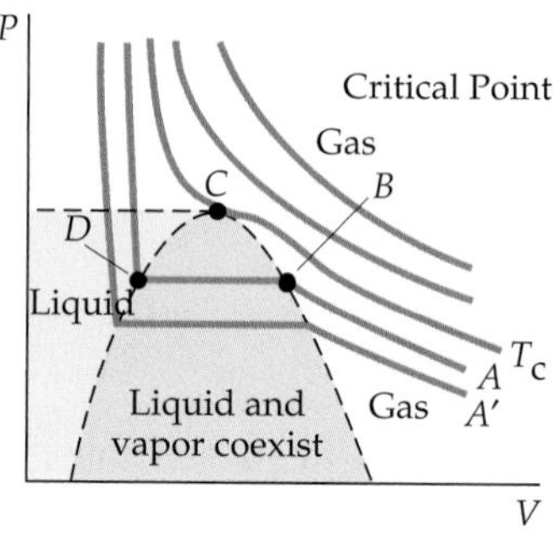

FIGURE 20-2 **Isotherms on the PV diagram for a substance. For temperatures above the critical temperature T_c, the substance remains a gas at all pressures. Except for the region where the liquid and vapor coexist, these curves are described quite well by the van der Waals equation. The pressure for the horizontal portions of the curves in the shaded region is the vapor pressure, which is the pressure at which the vapor and liquid are in equilibrium. To the left of the shaded region for temperatures below the critical temperature, the substance is a liquid and is nearly incompressible.**

volume of 22.4 L occupied by 1 mol of nitrogen under standard conditions. Since the molar mass of nitrogen is 28 g/mol, if 1 mol of nitrogen molecules were packed into a volume of 39.1 mL, then the density would be

$$\rho = \frac{M}{V} = \frac{28 \text{ g}}{39.1 \text{ mL}} = 0.72 \text{ g/mL} = 0.72 \text{ kg/L}$$

which is almost the same as the density of liquid nitrogen, 0.80 kg/L.

The value of the constant b can be used to estimate the size of a molecule. Since 1 mol (N_A molecules) of nitrogen has a volume of 39.1 cm^3, the volume of one nitrogen molecule is

$$V = \frac{b}{N_A} = \frac{39.1 \text{ cm}^3/\text{mol}}{6.02 \times 10^{23} \text{ molecules/mol}} = 6.50 \times 10^{-23} \text{ cm}^3/\text{molecule}$$

If we assume that each molecule occupies a cube of side d, we obtain

$$d^3 = 6.50 \times 10^{-23} \text{ cm}^3$$

or

$$d = 4.0 \times 10^{-8} \text{ cm} = 0.4 \text{ nm}$$

which is a plausible estimate for the "diameter" of a nitrogen molecule.

At temperatures below T_c, the van der Waals equation describes those portions of the isotherms outside the shaded region in Figure 20-2 but not those portions inside the shaded region. Suppose we have a gas at a temperature below T_c that initially has a low pressure and a large volume. We begin to compress the gas while holding the temperature constant (isotherm A in the figure). At first the pressure rises, but when we reach point B on the dashed curve, the pressure ceases to rise and the gas begins to liquefy at constant pressure. Along the horizontal line BD in the figure, the gas and liquid are in equilibrium. As we continue to compress the gas, more and more gas liquefies until point D on the dashed curve, at which point we have only liquid. Then, if we try to compress the substance further, the pressure rises sharply because a liquid is nearly incompressible.

Now consider injecting a liquid such as water into a sealed evacuated container. As some of the water evaporates, water-vapor molecules fill the previously empty space in the container. Some of these molecules will hit the liquid surface and rejoin the liquid water in a process called condensation. Initially, the rate of evaporation will be greater than the rate of condensation, but eventually equilibrium will be reached. The pressure at which a liquid is in equilibrium with its own vapor is called the **vapor pressure.** If we now heat the container slightly, the liquid boils, more liquid evaporates, and a new equilibrium is established at a higher vapor pressure. Vapor pressure thus depends on the temperature. We can see this from Figure 20-2. If we had started compressing the gas at a lower temperature, as with isotherm A' in Figure 20-2, the vapor pressure would be lower, as is indicated by the horizontal constant-pressure line for A' at a lower value of pressure. The temperature for which the vapor pressure for a substance equals 1 atm is the **normal boiling point** of that substance. For example, the temperature at which the vapor pressure of water is 1 atm is 373 K (= 100°C), so this temperature is the normal boiling point of water. At high altitudes, such

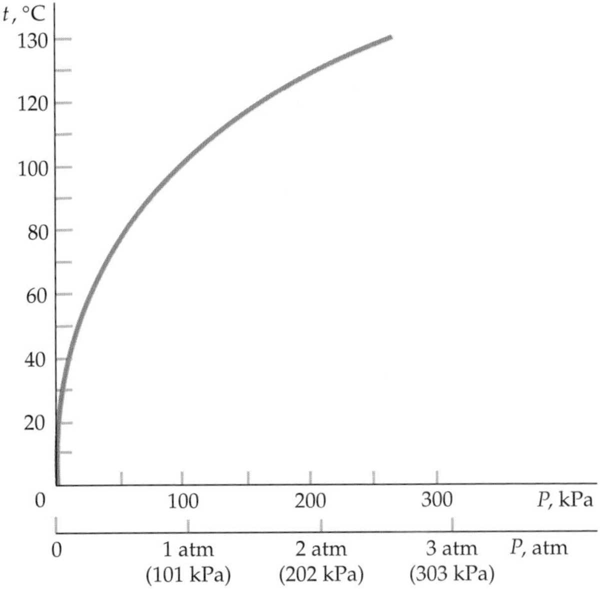

FIGURE 20-3 Boiling point of water versus temperature.

as on the top of a mountain, the pressure is less than 1 atm, therefore, water boils at a temperature lower than 373 K. Figure 20-3 gives the vapor pressures of water at various temperatures.

At temperatures greater than the critical temperature T_c, a gas will not liquefy at any pressure. The critical temperature for water vapor is 647 K (= 374°C). The point at which the critical isotherm intersects the dashed curve in Figure 20-2 (point C) is called the **critical point.**

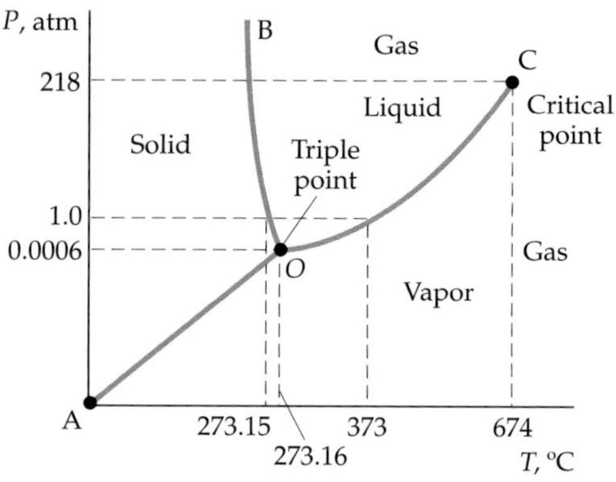

FIGURE 20-4 Phase diagram for water. The pressure and temperature scales are not linear but are compressed to show the interesting points. Curve *OC* is the curve of vapor pressure versus temperature. Curve *OB* is the melting curve, and curve *OA* is the sublimation curve.

20-3 Phase Diagrams

Figure 20-4 is a plot of pressure versus temperature at a constant volume for water. Such a plot is called a **phase diagram.** The portion of the diagram between points *O* and *C* shows vapor pressure versus temperature. As we continue to heat the container, the density of the liquid decreases and the density of the vapor increases. At point C on the diagram, these densities are equal. Point C is called the **critical point.** At this point and above it, there is no distinction between the liquid and the gas. Critical-point temperatures T_c for various substances are listed in Table 20-2. At temperatures greater than the critical temperature a gas will not liquefy at any pressure.

If we now cool our container, some of the vapor condenses into a liquid as we move back down the curve *OC* until the substance reaches point *O* in Figure 20-4. At this point, the liquid begins to solidify. Point *O* is the **triple point,** that one point at which the vapor, liquid, and solid phases of a substance can coexist in equilibrium. Every substance has a unique triple point at a specific temperature and pressure. The triple-point temperature for water is 273.16 K (= 0.01°C) and the triple-point pressure is 4.58 mmHg.

At temperatures and pressures below the triple point, the liquid cannot exist. The curve *OA* in the phase diagram of Figure 20-4 is the locus of pressures and temperatures for which the solid and vapor coexist in equilibrium. The direct change from a solid to a vapor is called **sublimation.** You can observe sublimation by putting a few loose ice cubes in the freezer compartment of a no-frost (self-defrosting) refrigerator. Over time, the ice cubes will shrink and eventually disappear due to sublimation. This happens because the atmospheric pressure is well above the triple-point pressure of water, and therefore, equilibrium is never established between the ice and water vapor. The triple-point temperature and pressure of carbon dioxide (CO_2) are 216.55 K and 3880 mmHg, which means that liquid CO_2 can only exist at pressures above 3880 mmHG (= 5.1 atm). Thus, at ordinary atmospheric pressures, liquid carbon dioxide cannot exist at any temperature. When solid carbon dioxide "melts," it sublimates directly into gaseous CO_2 without going through the liquid phase, hence the name "dry ice."

The curve *OB* in Figure 20-4 is the melting curve separating the liquid and solid phases. For a substance like water for which the melting temperature decreases as the pressure increases, curve *OB* slopes upward to the left from the triple point, as in this figure. For most other substances, the melting temperature increases as the pressure increases. For such a substance, curve *OB* slopes upward to the right from the triple point.

For a molecule to escape (evaporate) from a substance in the liquid state, energy is required to break the molecular bonds at the liquid's surface. Vaporization cools the liquid left behind. If water is brought to a boil over heat, this cooling effect keeps the temperature of the liquid constant at the boiling point. This is the reason that the boiling point of a substance can be used to calibrate thermometers. However, water can also be made to boil without adding heat by evacuating the air above it, thereby lowering the applied pressure. The energy needed for vaporization is then taken from the water left behind. As a result, the water will cool down, even to the point that ice forms on top of the boiling water!

TABLE 20-2

Critical Temperatures T_c for Various Substances

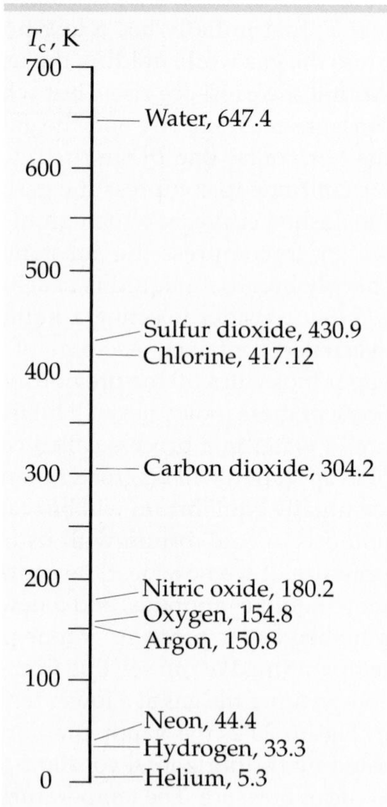

20-4 The Transfer of Thermal Energy

Thermal energy is transferred from one place to another by three processes: conduction, convection, and radiation.

In **conduction,** energy is transferred as heat by interactions among atoms or molecules, although there is no transport of the atoms or molecules themselves. For example, if one end of a solid bar is heated, the atoms in the heated end vibrate with greater energy than do those at the cooler end. The interaction of the more energetic atoms with their neighbors causes this energy to be transported along the bar.†

In **convection,** energy is transported as heat by direct mass transport. For example, warm air in a region of a room expands, its density decreases, and the buoyant force on it due to the surrounding air causes it to rise. Energy is thus transported upward along with the mass of warm air.

In **radiation,** energy is transported as heat through space in the form of electromagnetic waves that move at the speed of light. Thermal radiation, light waves, radio waves, television waves, and X rays are all forms of electromagnetic radiation that differ from one another in their wavelengths and frequencies.

In all mechanisms of heat transfer, the rate of cooling of a body is approximately proportional to the temperature difference between the body and its surroundings. This result is known as **Newton's law of cooling.**

In many real situations, all three mechanisms for heat transfer occur simultaneously, though one may be more dominant than the others. For example, an ordinary space heater uses both radiation and convection. If the heating element is quartz, then the main mechanism of heat transference is radiation. If the heating element is metal (which does not radiate as efficiently as quartz), then convection is the main mechanism by which heat is transmitted, with the heated air rising to be replaced by cooler air. Fans are often included in heaters to speed the convection process.

Conduction

Figure 20-5a shows an insulated uniform solid bar of cross-sectional area A. If we keep one end of the bar at a high temperature and the other end at a low temperature, energy is conducted down the bar from the hot end to the cold end. In the steady state, the temperature varies linearly from the hot end to the cold end. The rate of change of the temperature along the bar dT/dx is called the **temperature gradient.**

Let ΔT be the temperature difference across a small segment of length Δx (Figure 20-5*b*). If ΔQ is the amount of heat conducted through the segment in some time Δt, then the rate of conduction of heat $\Delta Q/\Delta t$ is called the thermal current I. Experimentally, it is found that the thermal current is proportional to the temperature gradient and to the cross-sectional area A:

$$I = \frac{\Delta Q}{\Delta t} = kA\frac{\Delta T}{\Delta x} \tag{20-7}$$

DEFINITION—THERMAL CURRENT

The proportionality constant k called the *thermal conductivity,* depends on the composition of the bar.‡ In SI units, thermal current is expressed in watts, and the thermal conductivity has units of W/(m·K).§ In practical calculations in the

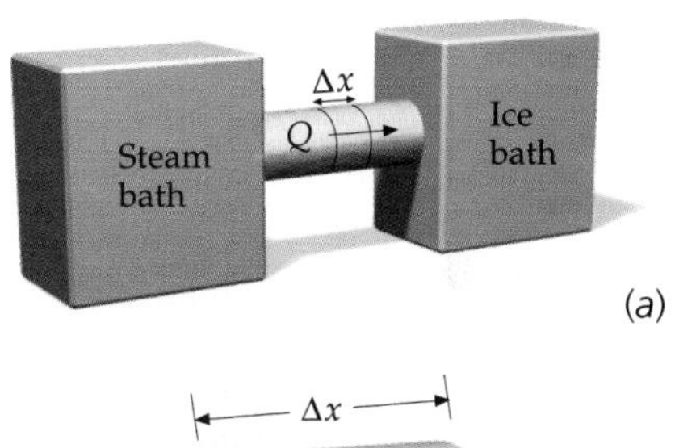

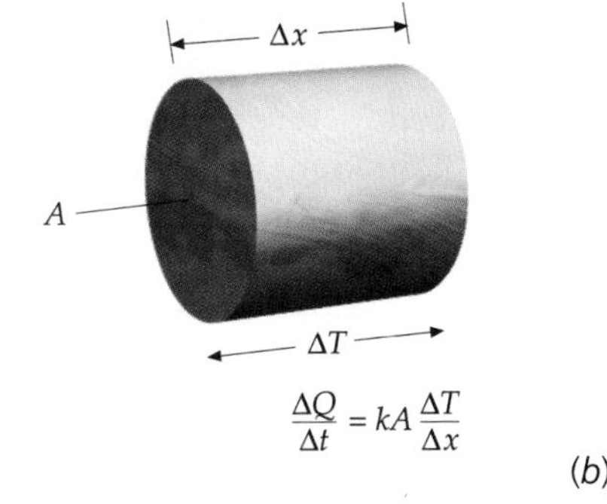

FIGURE 20-5 *(a)* **An insulated conducting bar with its ends at two different temperatures.** *(b)* **A segment of the bar of length Δx. The rate at which thermal energy is conducted across the segment is proportional to the cross-sectional area of the bar and the temperature difference across the segment, and it is inversely proportional to the length of the segment.**

† If the solid is a metal, the transport of thermal energy is helped by free electrons, which move throughout the metal.
‡ Don't confuse the thermal conductivity with Boltzmann's constant, which is also designated by k.
§ In some tables, the energy may be given in calories or kilocalories and the thickness in centimeters.

United States, the thermal current is usually expressed in Btu per hour, the area in square feet, the length (or thickness) in inches, and the temperature in degrees Fahrenheit. The thermal conductivity is then given in Btu·in./(h·ft²·F°). Table 20-3 gives the thermal conductivities of various materials.

If we solve Equation 20-7 for the temperature difference, we obtain

$$\Delta T = I\frac{\Delta x}{kA} \tag{20-8}$$

or

$$\Delta T = IR \tag{20-9}$$

TEMPERATURE CHANGE VERSUS CURRENT

where $\Delta x/(kA)$ is the **thermal resistance** R:

$$R = \frac{\Delta x}{kA} \tag{20-10}$$

DEFINITION—THERMAL RESISTANCE

EXERCISE Calculate the thermal resistance of an aluminum slab of cross-sectional area 15 cm² and thickness 2 cm. (*Answer* 0.0563 K/W = 56.3 mK/W)

EXERCISE What thickness of silver would be required to give the same thermal resistance as a 1-cm thickness of air of the same area? (*Answer* Δx = (1 cm)(429)/(0.026) = 16,500 cm = 165 m)

TABLE 20-3

Thermal Conductivities k for Various Materials

k, Btu·in./(h·ft²·F°)		k, W/(m·K)
(2980)	Silver	(429)
(2780)	Copper	(401)
(2450)	Lead	(353)
(2200)	Gold	(318)
(1644)	Aluminum	(237)
(558)	Iron	(80.4)
(319)	Steel	(46)
(6–9)	Concrete	(0.19–1.3)
(5–6)	Glass	(0.7–0.9)
(4.22)	Water at 27°C	(0.609)
(4.11)	Ice	(0.592)
(1.02)	Oak	(0.15)
(0.78)	White Pine	(0.11)
(0.18)	Air at 27°C	(0.026)

(Left scale: 10^{-1} to 10^{4} Btu·in./(h·ft²·F°); right scale: 10^{-2} to 10^{3} W/(m·K).)

In many practical problems, we are interested in the flow of heat through two or more conductors (or insulators) in series. For example, we may want to know the effect of adding insulating material of a certain thickness and thermal conductivity to the space between two layers of wallboard. Figure 20-6 shows two thermally conducting slabs of the same cross-sectional area but of different materials and of different thickness. Let T_1 be the temperature on the warm side, T_2 be the temperature at the interface between the slabs, and T_3 be the temperature on the cool side. Under the conditions of steady-state heat flow, the thermal current I must be the same through both slabs. This follows from energy conservation; for steady-state flow, the rate at which energy enters any region must equal the rate at which it exits that region.

If R_1 and R_2 are the thermal resistances of the two slabs, we have from Equation 20-9 for each slab

$$T_1 - T_2 = IR_1$$

and

$$T_2 - T_3 = IR_2$$

Adding these equations gives

$$\Delta T = T_1 - T_3 = I(R_1 + R_2) = IR_{eq}$$

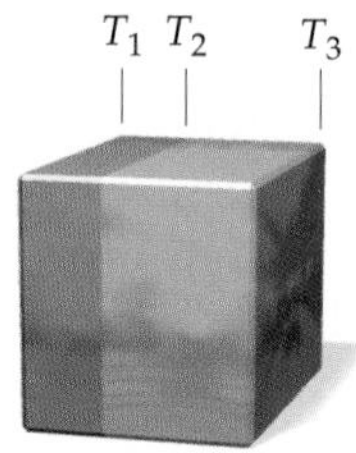

FIGURE 20-6 Two thermally conducting slabs of different materials in series. The equivalent thermal resistance of the slabs in series is the sum of their individual thermal resistances. The thermal current is the same through both slabs.

or

$$I = \frac{\Delta T}{R_{eq}} \qquad 20\text{-}11$$

where R_{eq} is the **equivalent resistance.** Thus, for thermal resistances in series, the equivalent resistance is the sum of the individual resistances:

$$R_{eq} = R_1 + R_2 + \cdots \qquad 20\text{-}12$$

THERMAL RESISTANCES IN SERIES

This thermogram of a house shows the heat energy being radiated to its surroundings.

This result can be applied to any number of resistances in series. In Chapter 25, we will find that the same formula applies to electrical resistances in series.

To calculate the amount of heat leaving a room by conduction in a given time, we need to know how much heat leaves through the walls, the windows, the floor, and the ceiling. For this type of problem, in which there are several paths for heat flow, the resistances are said to be in parallel. The temperature difference is the same for each path, but the thermal current is different. The total thermal current is the sum of the thermal currents through each of the parallel paths:

$$I_{total} = I_1 + I_2 + \cdots = \frac{\Delta T}{R_1} + \frac{\Delta T}{R_2} + \cdots = \Delta T\left(\frac{1}{R_1} + \frac{1}{R_2} + \cdots\right)$$

or

$$I_{total} = \frac{\Delta T}{R_{eq}} \qquad 20\text{-}13$$

where the equivalent thermal resistance is given by

$$\frac{1}{R_{eq}} = \frac{1}{R_1} + \frac{1}{R_2} + \cdots \qquad 20\text{-}14$$

THERMAL RESISTANCES IN PARALLEL

We will encounter this equation again in Chapter 25 when we study electric conduction through parallel resistances. Note that for both resistors in series (Equation 20-11) and resistors in parallel (Equation 20-13) I is proportional to ΔT, which is in agreement with Newton's law of cooling.

THERMAL CURRENT BETWEEN TWO METAL BARS **EXAMPLE 20-4** Try It Yourself

Two insulated metal bars, each of length 5 cm and rectangular cross section with sides 2 cm and 3 cm, are wedged between two walls, one held at 100°C and the other at 0°C. (Figure 20-7). The bars are lead and silver. Find (*a*) the total thermal current through the two-bar combination, and (*b*) the temperature at the interface.

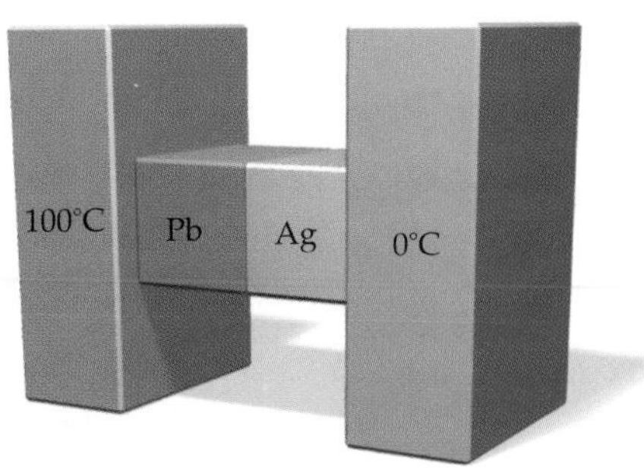

FIGURE 20-7 **Two thermally conducting slabs of different materials in parallel.**

PICTURE THE PROBLEM The bars are thermal resistors connected in series. (*a*) You can find the total thermal current from $I = R_{eq}/\Delta T$, where the equivalent resistance R_{eq} is the sum of the individual resistances. Using Equation 20-10 and the thermal conductivities given in Table 20-3, the individual resistances can be determined. (*b*) You can find the temperature at the interface by applying $I = R_1/\Delta T$ to the lead bar only, and solving for ΔT in terms of the value for I found in Part (*a*).

Cover the column to the right and try these on your own before looking at the answers.

Steps	Answers
(*a*) 1. Write the equivalent thermal resistance in terms of the thermal resistances of the two bars.	$R_{eq} = R_{Pb} + R_{Ag}$
2. Using Equation 20-10, write each resistance in terms of the individual thermal conductivities and geometric parameters:	$R_{Pb} = \dfrac{\Delta x_{Pb}}{k_{Pb}A_{Pb}},\ R_{Ag} = \dfrac{\Delta x_{Ag}}{k_{Ag}A_{Ag}}$
3. Use Equation 20-13 to find the thermal current.	$I = \Delta T/R_{eq} = \boxed{232\text{ W}}$
(*b*) 1. Calculate the temperature difference across the lead bar using the current and thermal resistance found in Part (*a*).	$\Delta T_{Pb} = IR_{Pb} = 54.9\text{ K} = 54.9^\circ\text{C}$
2. Use your result from the previous step to find the temperature at the interface.	$T_{if} = 100^\circ\text{C} - \Delta T_{Pb} = \boxed{45.1^\circ\text{C}}$
3. Check your result by finding the temperature difference across the silver bar.	$\Delta T_{Ag} = IR_{Ag} = 45.1^\circ\text{C}$

THE METAL BARS REARRANGED — EXAMPLE 20-5

FIGURE 20-8

The metal bars in Example 20-4 are rearranged as shown in Figure 20-8. Find (*a*) the thermal current in each bar, (*b*) the total thermal current, and (*c*) the equivalent thermal resistance of the two-bar system.

PICTURE THE PROBLEM The current in each bar is found from $I = \Delta T/R$, where R is the thermal resistance of the bar (found in Example 20-4). The total current is the sum of the currents. The equivalent resistance can be found from Equation 20-14 or from $I_{total} = \Delta T/R_{eq}$.

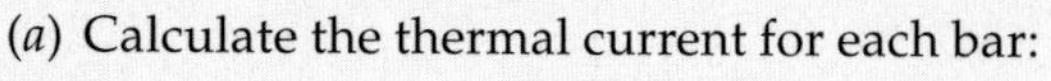

(*a*) Calculate the thermal current for each bar:

$$I_{Pb} = \frac{\Delta T}{R_{Pb}} = \frac{100\text{ K}}{0.236\text{ K/W}} = \boxed{424\text{ W}}$$

$$I_{Ag} = \frac{\Delta T}{R_{Ag}} = \frac{100\text{ K}}{0.194\text{ K/W}} = \boxed{515\text{ W}}$$

(*b*) Add your results to find the total thermal current:

$$I_{total} = I_{Pb} + I_{Ag}$$

$$= 424\text{ W} + 515\text{ W} = \boxed{938\text{ W}}$$

(*c*) 1. Use Equation 20-14 to calculate the equivalent resistance of the two bars in parallel:

$$\frac{1}{R_{eq}} = \frac{1}{R_{Pb}} + \frac{1}{R_{Ag}} = \frac{1}{0.236\text{ W}} + \frac{1}{0.194\text{ W}}, \quad \text{so}$$

$$R_{eq} = \boxed{0.107\text{ K/W}}$$

2. Check your result using, $I_{total} = \Delta T/R_{eq}$:

$$I_{total} = \frac{\Delta T}{R_{eq}};$$

$$R_{eq} = \frac{\Delta T}{I_{total}} = \frac{100\text{ K}}{938\text{ W}} = 0.107\text{ K/W}$$

REMARKS Note that the equivalent resistance is less than either of the individual resistances. This is always the case for parallel resistors.

In the building industry, the thermal resistance of a square foot of cross-sectional area of a material is called its **R factor** R_f. Consider a 32 ft² sheet of insulating material with thickness Δx and R factor R_f of 7.2. That is, each square foot (Figure 20-9) has a thermal resistance of 7.2 F°/(Btu/h). The 32 square feet are in parallel, so the net resistance R_{net} is calculated using Equation 20-14 giving

$$\frac{1}{R_{net}} = \frac{1}{R_f} + \frac{1}{R_f} + \cdots = \frac{32}{R_f}, \quad \text{so} \quad R_{total} = \frac{R_f}{32}$$

Thus, the total thermal resistance R in F°/(Btu/h) equals the R factor divided by the area A in square feet. That is

$$R_{net} = \frac{R_f}{A}$$

FIGURE 20-9 For a 1-in. thickness of this material, the $R_f = 7.2$.

Since the total resistance R_{total} is related to the conductivity by $R_{net} = \Delta x/(kA)$ (Equation 20-10), we can express the R factor by

$$R_f = R_{net}A = \frac{\Delta x}{k} \qquad \text{20-15}$$

DEFINITION—R FACTOR

where Δx is the thickness in inches and k is the conductivity in Btu·in./(h·ft²·F°). Table 20-4 lists R factors for several materials. In terms of the R factor, Equation 20-9 for the thermal current is

$$\Delta T = IR_{net} = \frac{I}{A}R_f \qquad \text{20-16}$$

For slabs of insulating material of the same area in series, R_f is replaced by the equivalent R factor $R_{f,eq}$

$$R_{f,eq} = R_{f1} + R_{f2} + \ldots$$

For parallel slabs, we calculate the thermal current through each slab and add all these currents together in order to obtain the total current.

TABLE 20-4

R Factors $\Delta x/k$ for Various Building Materials

Material	Thickness, in.	R_f, h·ft²·F°/Btu
Building board		
Gypsum or plasterboard	0.375	0.32
Plywood (Douglas fir)	0.5	0.62
Plywood or wood panels	0.75	0.93
Particle board, medium density	1.0	1.06
Finish flooring materials		
Carpet and fibrous pad	1.0	2.08
Tile		0.5
Wood, hardwood finish	0.75	0.68
Roof insulation	1.0	2.8
Roofing		
Asphalt roll roofing		0.15
Asphalt shingles		0.44
Windows		
Single-pane		0.9
Double pane		1.8

Heat Loss Through a Roof **EXAMPLE 20-6** **Put It in Context**

You are helping your friend's family put new asphalt shingles on the roof of their winter cabin. The 60 ft × 20 ft roof is made of 1-in. pine board covered with asphalt shingles. There is room for 2 in. of roof insulation, and your friend's family is wondering how much of a difference it would make to their energy bill if they were to install the two inches of insulation. Knowing that you are studying physics, they ask for your opinion.

PICTURE THE PROBLEM To assess the situation, you first calculate the R factor for each layer of the roof. Since the layers are in series, the equivalent R factor is just the sum of the individual R factors. The aim is to calculate the equivalent R factor of the roof with and without the insulation. The R factors for asphalt shingles and for roof insulation are found in Table 20-4. The R factor for the pine board is calculated from its thermal conductivity, which is found in Table 20-3. Note that when you shingle a roof you have to overlap the shingles, so there are two layers of asphalt shingling on the roof.

1. The equivalent R factor is the sum of the individual R factors:

$$R_{\text{f,eq}} = R_{\text{f,pine}} + R_{\text{f,asph}} + R_{\text{f,insul}}$$

2. The R factor for the double layer of shingles is twice the R factor for one layer:

$$R_{\text{f,asph}} = 2(0.44\ \text{h}\cdot\text{ft}^2\cdot\text{F}^\circ/\text{Btu}) = 0.88\ \text{h}\cdot\text{ft}^2\cdot\text{F}^\circ/\text{Btu}$$

3. The R factor for 2 in. of roof insulation is twice that for 1 in.:

$$R_{\text{f,insul}} = 2(2.8\ \text{h}\cdot\text{ft}^2\cdot\text{F}^\circ/\text{Btu}) = 5.6\ \text{h}\cdot\text{ft}^2\cdot\text{F}^\circ/\text{Btu}$$

4. The R factor for 1-in.-thick pine is obtained from the conductivity:

$$R_{\text{f,p}} = \frac{\Delta x_{\text{p}}}{k_{\text{p}}} = \frac{1\ \text{in.}}{0.78\ \text{Btu}\cdot\text{in.}/(\text{h}\cdot\text{ft}^2\cdot\text{F}^\circ)} = 1.28\ \text{h}\cdot\text{ft}^2\cdot\text{F}^\circ/\text{Btu}$$

5. The equivalent R factor without the insulation is:

$$R'_{\text{f,eq}} = R_{\text{f,pine}} + R_{\text{f,asph}} = 1.28\ \text{h}\cdot\text{ft}^2\cdot\text{F}^\circ/\text{Btu} + 0.88\ \text{h}\cdot\text{ft}^2\cdot\text{F}^\circ/\text{Btu} = 2.16\ \text{h}\cdot\text{ft}^2\cdot\text{F}^\circ/\text{Btu}$$

6. The equivalent R factor with insulation is:

$$R_{\text{f,eq}} = R_{\text{f,pine}} + R_{\text{f,asph}} + R_{\text{f,insul}} = R'_{\text{f,eq}} + R_{\text{f,insul}} = 2.16\ \text{h}\cdot\text{ft}^2\cdot\text{F}^\circ/\text{Btu} + 5.6\ \text{h}\cdot\text{ft}^2\cdot\text{F}^\circ/\text{Btu} = 7.76\ \text{h}\cdot\text{ft}^2\cdot\text{F}^\circ/\text{Btu}$$

7. One comparison of the two equivalent R factors is their ratio:

$$\frac{R'_{\text{f,eq}}}{R_{\text{f,eq}}} = \frac{2.16}{7.76} = 0.28$$

8. By adding the insulation the heat loss rate per square foot is reduced by 78%. Is it 78% of a large heat loss or a small heat loss? Using Equation 20-16 we calculate the thermal current I through the entire roof.

$$\Delta T = IR_{\text{net}} = \frac{I}{A}\text{R}_{\text{f}}$$

$$I' = \frac{A}{R'_{\text{f,eq}}}\Delta T = \frac{(60\ \text{ft})(20\ \text{ft})}{2.16\ \text{h}\cdot\text{ft}^2\cdot\text{F}^\circ/\text{Btu}}\Delta T = [556\ (\text{Btu/h})/\text{F}^\circ]\,\Delta T$$

9. To complete the calculation we estimate that the temperature inside the cabin is maintained at 70°F and the temperature outside the cabin during the winter is typically 40°F colder.

$$I' = [556\ (\text{Btu/h})/\text{F}^\circ]\Delta T = [556\ (\text{Btu/h})/\text{F}^\circ](40^\circ\text{F}) = 22{,}200\ \text{Btu/h}$$

and

$$I = 0.28I' = 0.28(22{,}200\ \text{Btu/h}) = 6200\ \text{Btu/h}$$

so the reduction due to the insulation is

$$I - I' = 22{,}200 \text{ Btu/h} - 6200 \text{ Btu/h}$$

$$= 16{,}000 \text{ Btu/h}$$

10. Estimate the savings that would result from adding the 2 in. of insulation.

See the following Remarks for an estimate of the cost.

REMARKS Installing 2 in. of roof insulation reduces the heat loss through the roof by 22,200 Btu/h. The cabin is heated with propane, and the energy content of propane is about 92,000 Btu/gal. Insulating the roof reduces consumption by approximately 6 gal of propane every 24 h of use. Propane costs about $1.40/gal, so this amounts to a savings of approximately $8.40 per day, or $252 per month. Your friend's family is impressed by the potential savings (and by the benefits of your physics knowledge). They decide to install the 2 in. of roofing insulation.

EXERCISE How much additional savings can be had by adding even more insulation to the roof? (*Answer* The maximum additional savings is 6200 Btu/h which would save $68 per month.)

REMARKS These cost estimates do not include the cost of purchasing and installing the insulation.

The thermal conductivity of air is very small compared with that of solid materials, which makes air a very good insulator. However, when there is a large air gap—say, between a storm window and the inside window—the insulating efficiency of air is greatly reduced because of convection. Whenever there is a temperature difference between different parts of the air space, convection currents act quickly to equalize the temperature, so the effective conductivity is greatly increased. For storm windows, air gaps of about 1 to 2 cm are optimal. Wider air gaps actually reduce the thermal resistance of a double-pane window due to convection.

The insulating properties of air are most effectively used when the air is trapped in small pockets that prevent convection from taking place. This is the principle underlying the excellent insulating properties of both goose down and Styrofoam.

If you touch the inside surface of a glass window when it is cold outside, you will observe that the surface is considerably colder than the inside air. The thermal resistance of windows is mainly due to thin films of insulating air that adhere to either side of the glass surface. The thickness of the glass has little effect on the overall thermal resistance. The air film on each side typically adds an R factor of about 0.45 per side. Thus, the R factor of a window with N separated glass layers is approximately $0.9N$ because of the two sides of each layer. Under windy conditions, the outside air film may be greatly decreased, leading to a smaller R factor for the window.

Convection

Convection is the transport of energy as heat by the transport of the material medium itself. This thermal property is responsible for the great ocean currents as well as the global circulation of the atmosphere. In the simplest case, convection arises when a fluid (gas or liquid) is heated from below. The warm fluid then expands and rises as the cooler fluid sinks. The mathematical description of convection is very complex because the flow depends on the temperature difference in different parts of the fluid, and this temperature difference is affected by the flow itself.

The heat transferred from an object to its surroundings by convection is approximately proportional to the area of the object and to the difference in temperature between the object and the surrounding fluid. It is possible to write an equation for the energy transported as heat by convection and to define a coefficient of convection, but the analyses of practical problems involving convection is quite complex and will not be discussed here.

Radiation

All objects emit and absorb electromagnetic radiation. When an object is in thermal equilibrium with its surroundings, it emits heat and absorbs heat at the same rate. The rate at which an object radiates energy is proportional to both the area of the object and to the fourth power of its absolute temperature. This result, found empirically by Josef Stefan in 1879 and derived theoretically by Ludwig Boltzmann about five years later, is called the **Stefan-Boltzmann law:**

$$P_r = e\sigma AT^4 \qquad \text{20-17}$$

STEFAN–BOLTZMANN LAW

where P_r is the power radiated, A is the area, σ is a universal constant called Stefan's constant, which has the value

$$\sigma = 5.6703 \times 10^{-8}\ \mathrm{W/(m^2{\cdot}K^4)} \qquad \text{20-18}$$

and e is the **emissivity** of the object, a fractional quantity between 0 and 1 that is dependent upon the composition of the surface of the object.

When electromagnetic radiation falls on an opaque object, part of the radiation is reflected and part is absorbed. Light-colored objects reflect most visible radiation, whereas dark objects absorb most of it. The rate at which an object absorbs radiation is given by

$$P_a = e\sigma AT_0^4 \qquad \text{20-19}$$

where T_0 is the temperature of the source of the radiation.

If an object emits more radiation than it absorbs, then it cools, while the object's surroundings absorb radiation from the object and become warmer. If the object absorbs more radiation than it emits, then the object warms and its surroundings cool. The net power radiated by an object at temperature T in an environment at temperature T_0 is

$$P_{net} = e\sigma A(T^4 - T_0^4) \qquad \text{20-20}$$

When an object is in thermal equilibrium with its surroundings, $T = T_0$, and the object emits and absorbs radiation at the same rate.

An object that absorbs all the radiation incident upon it has an emissivity equal to 1, and it is called a **blackbody.** A blackbody is also an ideal radiator. The concept of a blackbody is important because the characteristics of the radiation emitted by such an ideal object can be calculated theoretically. Materials such as black velvet come close to being ideal blackbodies. The best practical approximation of an ideal blackbody is a small hole leading into a cavity, such as a keyhole in a closet door (Figure 20-10). Radiation incident on the hole has little chance of being reflected out the hole before the walls of the cavity absorb it. Thus, the radiation emitted out of the hole is characteristic of the temperature of the walls of the cavity.

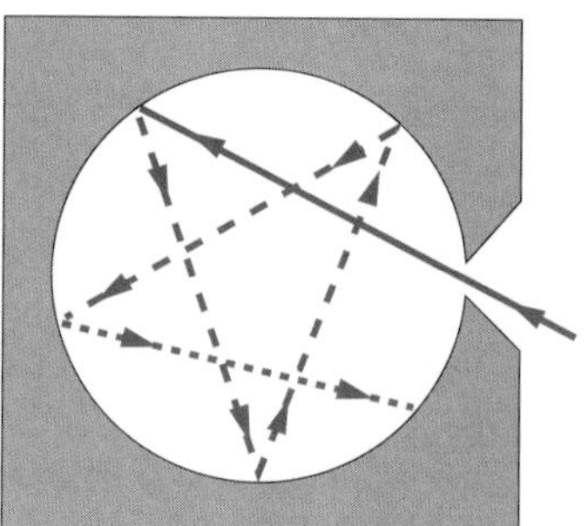

FIGURE 20-10 A hole in a cavity approximates an ideal blackbody. Radiation entering the cavity has little chance of leaving the cavity before it is completely absorbed. The radiation emitted through the hole (not shown) is therefore characteristic of the temperature of the walls of the cavity.

The radiation emitted by an object at temperatures below approximately 600°C is not visible. Most radiation emissions are concentrated at wavelengths much longer than those of visible light.† As an object is heated, the rate of energy

† When we study light, we will see that visible light is electromagnetic radiation with wavelengths between about 400 and 700 nm.

emission increases, and the energy radiated extends to higher frequencies (and shorter wavelengths). Between about 600 and 700°C, enough of the radiated energy is in the visible spectrum for the object to glow a dull red. At higher temperatures, it may become bright red or even "white hot." Figure 20-11 shows the power radiated by a blackbody as a function of wavelength for several different temperatures. The wavelength at which the power is a maximum varies inversely with the temperature, a result known as Wien's displacement law:

$$\lambda_{max} = \frac{2.898\ \text{mm}\cdot\text{K}}{T} \qquad 20\text{-}21$$

WIEN'S DISPLACEMENT LAW

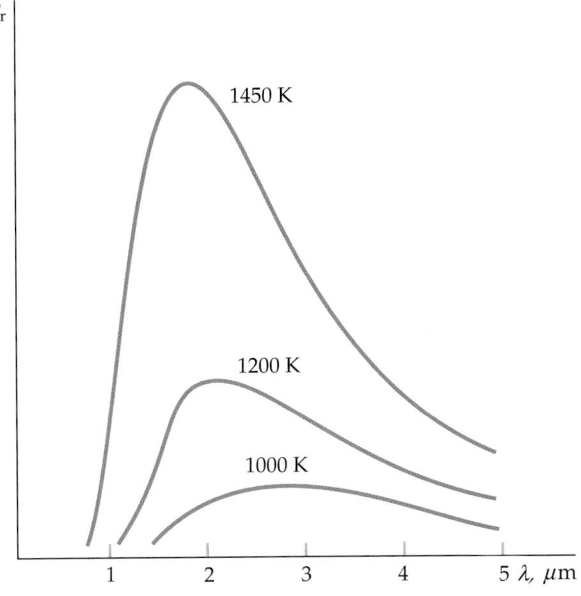

FIGURE 20-11 Radiated power versus wavelength for radiation emitted by a blackbody. The wavelength of the maximum power varies inversely with the absolute temperature of the blackbody.

This law is used to determine the surface temperatures of stars by analyzing their radiation. It can also be used to map out the variation in temperature over different regions of the surface of an object. Such a map is called a thermograph. Thermographs can be used to detect cancer because cancerous tissue results in increased circulation which produces a slight increase in skin temperature.

The spectral-distribution curves shown in Figure 20-11 played an important role in the history of physics. It was the discrepancy between theoretical calculations (using classical thermodynamics) of what the blackbody spectral distribution should be and the actual experimental measurements of spectral distributions that led to Max Planck's first ideas about the quantization of energy in 1900.

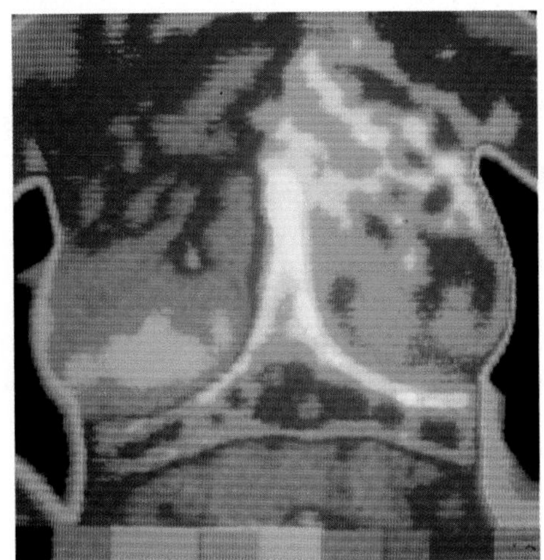

A thermograph was used to detect this cancerous tumor.

RADIATION FROM THE SUN **EXAMPLE 20-7**

(*a*) The radiation emitted by the surface of the sun emits maximum power at a wavelength of about 500 nm. Assuming the sun to be a blackbody emitter, what is its surface temperature? (*b*) Calculate λ_{max} for a blackbody at room temperature, $T = 300$ K.

(*a*) We can find T given λ_{max} using Wien's displacement law:

$$\lambda_{max} = \frac{2.898\ \text{mm}\cdot\text{K}}{T}$$

so

$$T = \frac{2.898\ \text{mm}\cdot\text{K}}{\lambda_{max}} = \frac{2.898\ \text{mm}\cdot\text{K}}{500\ \text{nm}}$$

$$= \boxed{5800\ \text{K}}$$

(*b*) We can find λ_{max} from Wien's displacement law for $T = 300$ K:

$$\lambda_{max} = \frac{2.898\ \text{mm}\cdot\text{K}}{300\ \text{K}} = 9.66 \times 10^{-3}\ \text{mm}$$

$$= \boxed{9.66\ \mu\text{m}}$$

REMARKS The peak wavelength from the sun is in the visible spectrum. The blackbody radiation spectrum describes the sun's radiation fairly well, so the sun is indeed a good example of a blackbody.

For $T = 300$ K, the spectrum peaks in the infrared at wavelengths much longer than the wavelengths visible to the eye. Surfaces that are not black to our eyes may act as blackbodies for infrared radiation and absorption. For example, it has been found experimentally that the skin of human beings of all races is black to infrared radiation; hence, the emissivity of skin is 1.00 for its own radiation process.

RADIATION FROM THE HUMAN BODY — EXAMPLE 20-8 Try It Yourself

Calculate the net rate of heat loss in radiated energy for a naked person in a room at 20°C, assuming the person to be a blackbody with a surface area of 1.4 m² and a surface temperature of 33°C (= 306 K). (The surface temperature of the human body is slightly less than the internal temperature of 37°C because of the thermal resistance of the skin.)

Cover the column to the right and try these on your own before looking at the answers.

Steps	Answer
Use $P_{net} = e\sigma A(T^4 - T_0^4)$ with $e = 1$, $T = 306$ K, and $T_0 = 293$ K.	$P_{net} = 111$ W

REMARKS This large energy loss is approximately equal to the basal metabolic rate of about 120 W. We protect ourselves from this great loss of energy by wearing clothing, which, because of its low thermal conductivity, has a much lower outside temperature and therefore a much lower rate of thermal radiation.

When the temperature of an object T is not too different from the surrounding temperature T_0, a radiating object obeys Newton's law of cooling. We can see this by writing Equation 20-20 as

$$P_{net} = e\sigma A(T^4 - T_0^4) = e\sigma A(T^2 + T_0^2)(T^2 - T_0^2)$$
$$= e\sigma A(T^2 + T_0^2)(T + T_0)(T - T_0)$$

When $T - T_0$ is small, we can replace T by T_0 in the sums with little change in the result. Then

$$P_{net} = e\sigma A(T^4 - T_0^4) \approx e\sigma A(T_0^2 + T_0^2)(T_0 + T_0)(T - T_0) = 4e\sigma AT_0^3\,\Delta T$$

The net power radiated is approximately proportional to the temperature difference, in agreement with Newton's law of cooling. This result can also be obtained by using the differential approximation.

$$\Delta P_r \approx \left.\frac{dP_r}{dT}\right|_{T=T_0}(T - T_0)$$

where $P_r = e\sigma A(T^4 - T_0^4)$. For a small temperature difference $T - T_0$ we have

$$\Delta P_r \approx e\sigma A\, 4T^3|_{T=T_0}(T - T_0) = 4e\sigma AT_0^3\,\Delta T$$

SUMMARY

Topic	Relevant Equations and Remarks	
1. Thermal Expansion		
Coefficient of linear expansion	$\alpha = \frac{\Delta L/L}{\Delta T}$	**20-2**
Coefficient of volume expansion	$\beta = \frac{\Delta V/V}{\Delta T} = 3\alpha$	**20-4, 20-5**
2. The van der Waals Equation of State	The van der Waals equation of state describes the behavior of real gases over a wide range of temperatures and pressures, taking into account the space occupied by the gas molecules themselves and the attraction of the molecules to one another.	
	$\left(P + \frac{an^2}{V^2}\right)(V - bn) = nRT$	**20-6**
3. Vapor Pressure	Vapor pressure is the pressure at which the liquid and gas phases of a substance are in equilibrium at a given temperature. The liquid boils at that temperature for which the external pressure equals the vapor pressure.	
4. The Triple Point	The triple point is the unique temperature and pressure at which the gas, liquid, and solid phases of a substance can coexist in equilibrium. At temperatures and pressures below the triple point, the liquid phase of a substance cannot exist.	
5. Heat Transfer	The three mechanisms by which thermal energy is transferred are radiation, conduction, and convection.	
Newton's law of cooling	For all mechanisms of heat transfer, if the temperature difference between the body and its surroundings is small, the rate of cooling of a body is approximately proportional to the temperature difference.	
6. Heat Conduction		
Current	The rate of conduction of thermal energy is given by	
	$I = \frac{\Delta Q}{\Delta t} = kA\frac{\Delta T}{\Delta x}$	**20-7**
	where I is the thermal current, k is the coefficient of thermal conductivity, and $\Delta T/\Delta x$ is the temperature gradient.	
Thermal resistance	$\Delta T = IR$	**20-9**
	where R is the thermal resistance:	
	$R = \frac{\Delta x}{kA}$	**20-10**
Equivalent resistance:		
series	$R_{eq} = R_1 + R_2 + \ldots$	**20-12**
parallel	$\frac{1}{R_{eq}} = \frac{1}{R_1} + \frac{1}{R_2} + \ldots$	**20-14**

R factor	The R factor is the thermal resistance in in.·ft²·°F/(Btu/h) for a square foot of a slab of material $R_f = R_{net}A = \dfrac{\Delta x}{k}$	**20-15**
7. Thermal Radiation		
Rate of power radiated	$P_r = e\sigma AT^4$ where $\sigma = 5.6703 \times 10^{-8}$ W/m²·K⁴ is Stefan's constant and e is the emissivity, which varies between 0 and 1 (depending on the composition of the surface of the object). Materials that are good heat absorbers are also good heat radiators.	**20-17**
Net power radiated by an object at T to its environment at T_0	$P_{net} = e\sigma A(T^4 - T_0^4)$	**20-20**
Blackbody	A blackbody has an emissivity of 1. It is a perfect radiator, and it absorbs all radiation incident upon it.	
Wein's law	The power spectrum of electromagnetic energy radiated by a blackbody has a maximum at a wavelength λ_{max}, which varies inversely with the absolute temperature of the body: $\lambda_{max} = \dfrac{2.898 \text{ mm·K}}{T}$	**20-21**

PROBLEMS

- • Single-concept, single-step, relatively easy
- •• Intermediate-level, may require synthesis of concepts
- ••• Challenging
- SSM Solution is in the *Student Solutions Manual*
- iSOLVE Problems available on iSOLVE online homework service
- iSOLVE✓ These "Checkpoint" online homework service problems ask students additional questions about their confidence level, and how they arrived at their answer

In a few problems, you are given more data than you actually need; in a few other problems, you are required to supply data from your general knowledge, outside sources, or informed estimates.

Conceptual Problems

1 • SSM Why does the mercury level first decrease slightly when a thermometer is placed in warm water?

2 • A large sheet of metal has a hole cut in the middle of it. When the sheet is heated, the area of the hole will (*a*) not change, (*b*) always increase, (*c*) always decrease, (*d*) increase if the hole is not in the exact center of the sheet, (*e*) decrease only if the hole is in the exact center of the sheet.

3 • Mountaineers say that you cannot hard boil an egg on the top of Mount Rainier. This is true because (*a*) the air is too cold to boil water, (*b*) the air pressure is too low for stoves to burn, (*c*) boiling water is not hot enough to hard boil the egg, (*d*) the oxygen content of the air is too low, (*e*) eggs always break in their backpacks.

4 • Which gases in Table 20-2 cannot be liquefied by applying pressure at 20°C?

5 •• SSM The phase diagram in Figure 20-12 can be interpreted to yield information on how the boiling and melting points of water change with altitude. (*a*) Explain how this information can be obtained. (*b*) How might this information affect cooking procedures in the mountains?

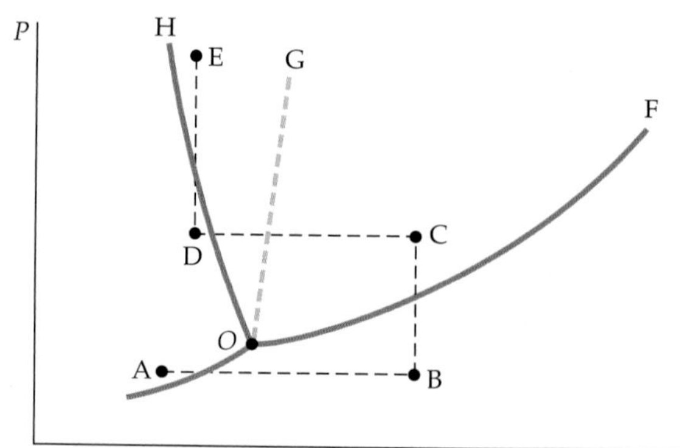

FIGURE 20-12 Problem 5

6 • If the absolute temperature of an object is tripled, the rate at which it radiates thermal energy (*a*) triples, (*b*) increases by a factor of 9, (*c*) increases by a factor of 27, (*d*) increases by a factor of 81, (*e*) depends on whether the absolute temperature is above or below zero.

7 • SSM In a cool room, a metal or marble table top feels much colder to the touch than does a wood surface even though they are at the same temperature. Why?

8 • True or false:

(*a*) During a phase change, the temperature of a substance remains constant.
(*b*) The rate of conduction of thermal energy is proportional to the temperature gradient.
(*c*) The rate at which an object radiates energy is proportional to the square of its absolute temperature.
(*d*) All materials expand when they are heated.
(*e*) The vapor pressure of a liquid depends on the temperature.

9 • The earth loses heat by (*a*) conduction, (*b*) convection, (*c*) radiation, (*d*) all of the above.

10 • Which heat-transfer mechanisms are the most important in the warming effect of a fire in a fireplace?

11 • Which heat-transfer mechanism is important in the transfer of energy from the sun to the earth?

12 •• Explain why turning down the temperature of a house at night in winter can save money on heating costs. Why doesn't the cost of the fuel consumed to heat the house back up in the morning equal the savings realized by cooling it down?

13 •• Two cylinders made of materials A and B have the same lengths; their diameters are related by $d_A = 2d_B$. When the same temperature difference is maintained between the ends of the cylinders they conduct heat at the same rate. Their thermal conductivities are related by (*a*) $k_A = k_B/4$, (*b*) $k_A = k_B/2$, (*c*) $k_A = k_B$, (*d*) $k_A = 2k_B$, (*e*) $k_A = 4k_B$.

14 • Infrared light is sometimes referred to as "heat waves." Explain why infrared light has received this label, and why the label is inaccurate.

15 • SSM In artistic nomenclature, blue is often referred to as a "cool" color, while red is referred to as a "warm" color. In physics, however, red is considered a "cooler" color than blue. Explain why.

Estimation and Approximation

16 ••• Liquid helium is stored in containers fitted with 7-cm-thick "superinsulation" consisting of numerous layers of very thin aluminized Mylar sheets. The rate of evaporation of liquid in a 200-L container is about 0.7 L per day. Assume the container is spherical and that the external temperature is 20°C. The specific gravity of liquid helium is 0.125 and the latent heat of vaporization is 21 kJ/kg. Estimate the thermal conductivity of superinsulation.

17 •• Estimate the thermal conductivity of the skin, given that the body of an "average" adult male has about 1.8 m^2 of skin area, and produces about 130 W of heat when resting. Use an internal temperature of 37°C (98.6°F) and an external skin temperature of 33°C. Assume that the skin has an average thickness of about 1 mm.

18 •• SSM Estimate the effective emissivity of the earth, given the following information: the solar constant (the intensity of light incident on the earth from the sun) is 1370 W/m^2, 70 percent of this light is absorbed by the earth, and the earth's average temperature is 288 K. (Assume that the effective area that is absorbing the light is πR^2, where R is the earth's radius, while the blackbody-emission area is $4\pi R^2$.)

19 •• Black holes in orbit around a normal star are detected from earth due to the frictional heating of infalling gas into the black hole, which can reach temperatures greater than 10^6 K. Assuming that the infalling gas can be modeled as a blackbody radiator, estimate λ_{max} for use in an astronomical detection of a black hole. (Remark: This is in the x-ray region of the electromagnetic spectrum.)

Thermal Expansion

20 • ISOLVE A steel ruler has a length of 30 cm at 20°C. What is its length at 100°C?

21 •• (*a*) Define a coefficient of area expansion. (*b*) Calculate it for a square and a circle, and show that it is two times the coefficient of linear expansion.

22 •• ISOLVE The density of aluminum is 2.70×10^3 kg/m^3 at 0°C. What is the density of aluminum at 200°C?

23 •• ISOLVE A copper collar is to fit tightly about a steel shaft that has a diameter of 6.0000 cm at 20°C. The inside diameter of the copper collar at that temperature is 5.9800 cm. To what temperature must the copper collar be raised so that it will just slip on the steel shaft, assuming the steel shaft remains at 20°C?

24 •• SSM ISOLVE Repeat Problem 23 when the temperature of both the steel shaft and copper collar are raised simultaneously.

25 •• A container is filled to the brim with 1.4 L of mercury at 20°C. When the temperature of container and mercury is raised to 60°C, 7.5 mL of mercury spill over the brim of the container. Determine the linear expansion coefficient of the container.

26 •• ISOLVE✓ A hole is drilled in an aluminum plate with a steel drill bit whose diameter at 20°C is 6.245 cm. In the process of drilling, the temperature of the drill bit and of the aluminum plate rise to 168°C. What is the diameter of the hole in the aluminum plate when it has cooled to room temperature?

27 •• SSM A rookie crew was left to put in the final 1 km of rail for a stretch of railroad track. When they finished, the temperature was 20°C, and they headed to town for some refreshments. After an hour or two, one of the old-timers noticed that the temperature had gone up to 25°C, so he said, "I hope you left some gaps to allow for expansion." From the look on their faces, he knew that they had not, and they all rushed back to the work site. The rail had buckled into an isosceles triangle. How high was the buckle?

28 •• ISOLVE A car has a 60-L steel gas tank filled to the top with gasoline when the temperature is 10°C. The coefficient of volume expansion of gasoline is $\beta = 0.900 \times 10^{-3}\ \mathrm{K}^{-1}$. Taking the expansion of the steel tank into account, how much gasoline spills out of the tank when the car is parked in the sun and its temperature rises to 25°C?

29 •• A thermometer has an ordinary glass bulb and thin glass tube filled with 1 mL of mercury. A temperature change of 1°C changes the level of mercury in the thin tube by 3.0 mm. Find the inside diameter of the thin glass tube.

30 •• ISOLVE✓ A mercury thermometer consists of a 0.4-mm capillary tube connected to a glass bulb. The mercury level rises 7.5 cm as the temperature of the thermometer increases from 35°C to 43°C. Find the volume of the thermometer bulb.

31 ••• A grandfather's clock is calibrated at a temperature of 20°C. (*a*) On a hot day, when the temperature is 30°C, does the clock run fast or slow? (*b*) How much does it gain or lose in a 24-h period? Assume that the pendulum is a thin brass rod of negligible mass with a heavy bob attached to the end.

32 ••• ISOLVE✓ A steel tube has an outside diameter of 3.000 cm at room temperature (20°C). A brass tube has an inside diameter of 2.997 cm at the same temperature. To what temperature must the ends of the tubes be heated if the steel tube is to be inserted into the brass tube?

33 ••• SSM What is the tensile stress in the copper collar of Problem 23 when its temperature returns to 20°C?

The van der Waals Equation, Liquid–Vapor Isotherms, and Phase Diagrams

34 • ISOLVE✓ (*a*) Calculate the volume of 1 mol of steam at 100°C and a pressure of 1 atm, assuming that it is an ideal gas. (*b*) Find the temperature at which the steam will occupy the volume found in Part (*a*) if it obeys the van der Waals equation with $a = 0.55\ \text{Pa·m}^6/\text{mol}^2$ and $b = 30\ \text{cm}^3/\text{mol}$.

35 •• From Figure 20-3, find (*a*) the temperature at which water boils on a mountain where the atmospheric pressure is 70 kPa, (*b*) the temperature at which water will boil in a container in which the pressure has been reduced to 0.5 atm, and (*c*) the pressure at which water will boil at 115°C.

36 •• SSM The van der Waals constants for helium are $a = 0.03412\ \text{L}^2\text{·atm/mol}^2$ and $b = 0.0237$ L/mol. Use these data to find the volume in cubic centimeters occupied by one helium atom and to estimate the radius of the atom.

37 ••• (*a*) For a van der Waals gas, show that the critical temperature is $8a/27Rb$ and the critical pressure is $a/27b^2$. (*b*) Rewrite the van der Waals equation of state in terms of the reduced variable $V_r = V/V_c$, $P_r = P/P_c$, and $T_r = T/T_c$.

Heat Conduction

38 • ISOLVE A copper bar 2 m long has a circular cross section of radius 1 cm. One end is kept at 100°C and the other end is kept at 0°C. The surface of the bar is insulated so that there is negligible heat loss through it. Find (*a*) the thermal resistance of the bar, (*b*) the thermal current *I*, (*c*) the temperature gradient $\Delta T/\Delta x$, and (*d*) the temperature of the bar 25 cm from the hot end.

39 • ISOLVE✓ A 20 × 30-ft slab of insulation has an *R* factor of 11. How much heat (in Btu/h) is conducted through the slab if the temperature on one side is 68°F and on the other side it is 30°F?

40 •• ISOLVE✓ Two metal cubes with 3-cm edges, one copper (Cu) and one aluminum (Al), are arranged as shown in Figure 20-13. Find (*a*) the thermal resistance of each cube, (*b*) the thermal resistance of the two-cube system, (*c*) the thermal current *I*, and (*d*) the temperature at the interface of the two cubes.

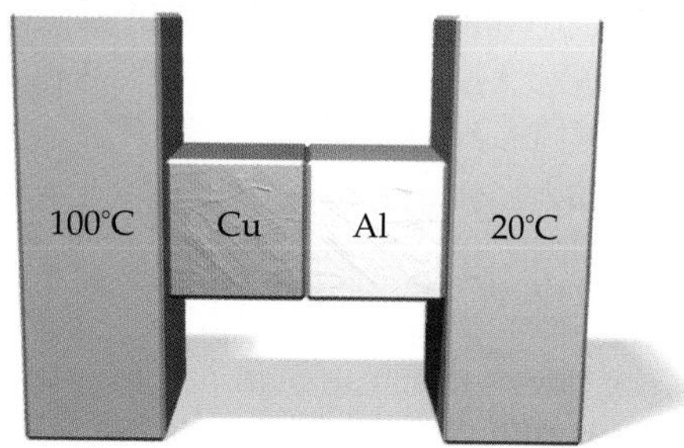

FIGURE 20-13 Problem 40

41 •• The cubes in Problem 40 are rearranged in parallel as shown in Figure 20-14. Find (*a*) the thermal current carried by each cube from one side to the other, (*b*) the total thermal current, and (*c*) the equivalent thermal resistance of the two-cube system.

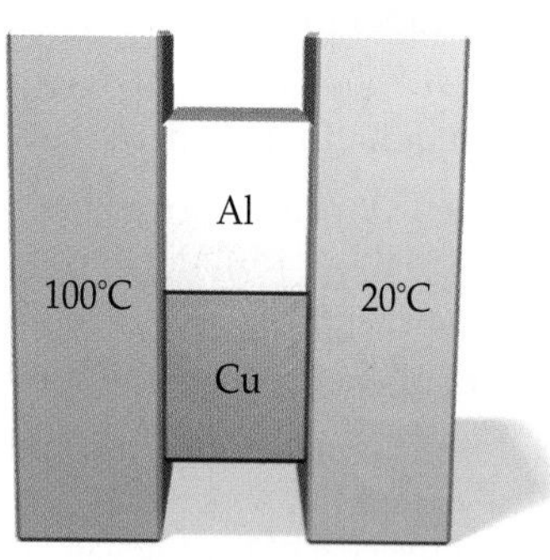

FIGURE 20-14 Problem 41

42 •• SSM The cost of air conditioning a house is proportional to the rate at which heat flows from the house and is inversely proportional to the coefficient of performance (COP) of the air conditioner. We denote the temperature difference between the house and its surroundings as ΔT. Assuming that the rate at which heat flows from a house is proportional to ΔT and that the air conditioner is operating ideally, show that the cost of air conditioning is proportional to $(\Delta T)^2$.

43 ••• A spherical shell of thermal conductivity *k* has inside radius r_1 and outside radius r_2 (Figure 20-15). The inside of the shell is held at a temperature T_1, and the outside at temperature T_2. In this problem, you are to show that the thermal current through the shell is given by

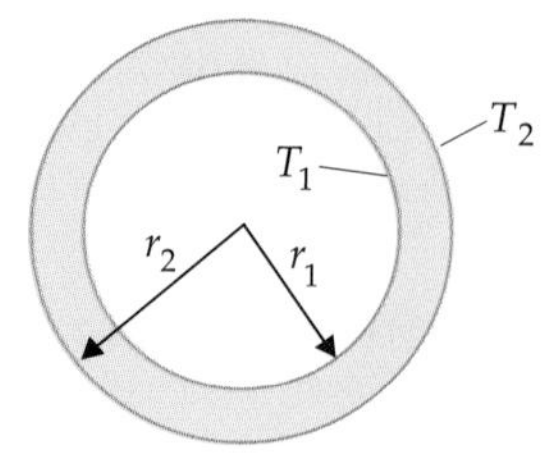

FIGURE 20-15 Problem 43

$$I = \frac{4\pi k r_1 r_2}{r_2 - r_1}(T_2 - T_1) \qquad 20\text{-}22$$

Consider a spherical element of the shell of radius *r* and thickness *dr*. (*a*) Why must the thermal current through each such element be the same? (*b*) Write the thermal current *I* through such a shell element in terms of the area $A = 4\pi r^2$, the thickness *dr*, and the temperature difference *dT* across the element. (*c*) Solve for *dT* in terms of *dr* and integrate from $r = r_1$ to $r = r_2$. (*d*) Show that when r_1 and r_2 are much larger than $r_2 - r_1$, Equation 20-22 (shown above) is the same as Equation 20-7.

44 •• SSM ISOLVE For a boiler at a power station, heat must be transferred to boiling water at the rate of 3 GW. The boiling water passes through copper pipes having a wall thickness of 4.0 mm and a surface area of 0.12 m^2 per meter length of pipe. Find the total length of pipe (actually there are many pipes in parallel) that must pass through the furnace if the steam temperature is 225°C and the external temperature of the pipes is 600°C.

45 ••• A steam pipe of length L is insulated with a layer of material of thermal conductivity k. Find the rate of heat transfer if the temperature outside the insulation is T_1, the temperature inside is T_2, the outside radius of the insulation is r_1, and the inside radius is r_2.

Radiation

46 • SSM ISOLVE✓ Calculate λ_{max} for a human blackbody radiator, assuming the surface temperature of the skin to be 33°C.

47 • ISOLVE✓ The heating wires of a 1-kW electric heater are red hot at a temperature of 900°C. Assuming that 100% of the heat output is due to radiation and that the wires act as blackbody radiators, what is the effective area of the radiating surface? (Assume a room temperature of 20°C.)

48 •• ISOLVE✓ A blackened, solid copper sphere of radius 4.0 cm hangs in a vacuum in an enclosure whose walls have a temperature of 20°C. If the sphere is initially at 0°C, find the rate at which its temperature changes, assuming that heat is transferred by radiation only.

49 •• ISOLVE The surface temperature of the filament of an incandescent lamp is 1300°C. If the electric power input is doubled, what will the temperature become? *Hint: Show that you can neglect the temperature of the surroundings.*

50 •• Liquid helium is stored at its boiling point (4.2 K) in a spherical can that is separated by a vacuum space from a surrounding shield that is maintained at the temperature of liquid nitrogen (77 K). If the can is 30 cm in diameter and is blackened on the outside so that it acts as a blackbody radiator, how much helium boils away per hour?

General Problems

51 • SSM ISOLVE A steel tape is placed around the earth at the equator when the temperature is 0°C. What will the clearance between the tape and the ground (assumed to be uniform) be if the temperature of the tape rises to 30°C? Neglect the expansion of the earth.

52 •• Show that change in the density of an isotropic material due to an increase in temperature ΔT is given by $\Delta\rho = -\beta\rho\Delta T$.

53 •• ISOLVE The solar constant is the power received from the sun per unit area perpendicular to the sun's rays at the mean distance of the earth from the sun. Its value at the upper atmosphere of the earth is about 1.35 kW/m^2. Calculate the effective temperature of the sun if it radiates like a blackbody. (The radius of the sun is 6.96×10^8 m.)

54 •• ISOLVE✓ To determine the R value of insulating material that comes in sheets of $\frac{1}{2}$-in. thickness, you construct a cubical box of 12 in. per side and place a thermometer and a 100-W heater inside the box. After thermal equilibrium has been attained, the temperature inside the box is 90°C when the external temperature is 20°C. Determine the R value of this material.

55 •• A 2-cm-thick copper sheet is pressed against a sheet of aluminum. What should be the thickness of the aluminum sheet so that the temperature of the copper–aluminum interface is $(T_1 + T_2)/2$, where T_1 and T_2 are the temperatures at the copper–air and aluminum–air interfaces?

56 •• At a temperature of 20°C, a steel bar of radius 2.2 cm and length 60 cm is jammed horizontally perpendicular between two vertical concrete walls. With a blowtorch, the temperature of the bar is raised to 60°C. Find the force exerted by the bar on each wall.

57 •• (*a*) From the definition of β, the coefficient of volume expansion (at constant pressure), show that $\beta = 1/T$ for an ideal gas. (*b*) The experimentally determined value of β for N_2 gas at 0°C is 0.003673 K^{-1}. Compare this value with the theoretical value $\beta = 1/T$, assuming that N_2 is an ideal gas.

58 •• One way to construct a device with two points whose separation remains the same in spite of temperature changes is to bolt together one end of two rods, both of which have different coefficients of linear expansion as in the arrangement shown in Figure 20-16. (*a*) Show that the distance L will not change with temperature if the lengths L_A and L_B are chosen such that $L_A/L_B = \alpha_B/\alpha_A$. (*b*) If material B is steel, material A is brass, and $L_A = 250$ cm at 0°C, what is the value of L?

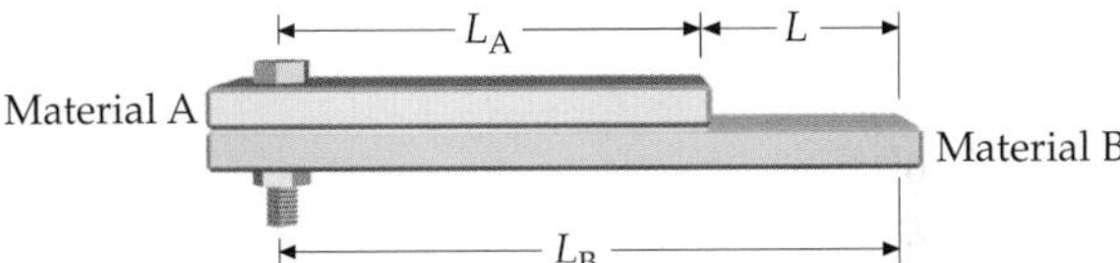

FIGURE 20-16 Problem 58

59 •• On the average, the temperature of the earth's crust increases 1.0 C° for every 30 m of depth. The average thermal conductivity of the earth's crust is 0.74 J/m·s·K. What is the heat loss of the earth per second due to conduction from the core? How does this heat loss compare with the average power received from the sun? (The solar constant is about 1.35 kW/m^2.)

60 •• ISOLVE A copper-bottomed saucepan containing 0.8 L of boiling water boils dry in 10 min. Assuming that all the heat flows through the flat copper bottom, which has a diameter of 15 cm and a thickness of 3.0 mm, calculate the temperature of the outside of the copper bottom while some water is still in the pan.

61 •• SSM A hot-water tank of cylindrical shape has an inside diameter of 0.55 m and inside height of 1.2 m. The tank is enclosed with a 5-cm-thick insulating layer of glass wool whose thermal conductivity is 0.035 W/m·K. The metallic interior and exterior walls of the container have thermal conductivities that are much greater than that of the glass wool. How much power must be supplied to this tank in order to maintain the water temperature at 75°C when the external temperature is 1°C?

62 ••• The diameter of a rod is given by $d = d_0(1 + ax)$, where a is a constant and x is the distance from one end. If the thermal conductivity of the material is k what is the thermal resistance of the rod if its length is L?

63 ••• A solid disk of radius r and mass m is spinning in a frictionless environment with angular velocity ω_1 at temperature T_1. The temperature of the disk is then changed to T_2. Express the angular velocity ω_2, rotational kinetic energy E_2, and angular momentum L_2 in terms of their values at the temperature T_1 and the linear expansion coefficient α of the disk.

64 ••• Write a spreadsheet program to graph the temperature of the earth as a function of emissivity, using the results of Problem 18. How much does the emissivity have to change in order for the average temperature to rise by 1 K? This can be thought of as a model for the effect of increasing concentrations of greenhouse gases like methane and CO_2 in the earth's atmosphere.

65 ••• A small pond has a layer of ice 1 cm thick floating on its surface. (*a*) If the air temperature is -10°C, find the rate in centimeters per hour at which ice is added to the bottom of the layer. The density of ice is 0.917 g/cm^3. (*b*) How long does it take for a 20-cm layer to be built up?

66 ••• **SSM** A 200-g copper container holding 0.7 L of water is thermally isolated from its surroundings—except for a 10-cm-long copper rod of cross-sectional area 1.5 cm^2 connecting it to a second copper container filled with an ice and water mixture so its temperature remains at 0°C. The initial temperature of the first container is T_0 = 60°C. (Assume the heat capacity of the rod to be negligible.) (*a*) Show that the temperature T of the first container changes over time t according to

$$T = T_0 e^{-t/RC}$$

where T is in degrees Celsius, R is the thermal resistance of the rod, and C is the total heat capacity of the container plus the water. (Neglect the heat capacity of the rod.) (*b*) Evaluate R, C, and the "time constant" RC. (*c*) Show that the total amount of heat Q conducted in time t is

$$Q = CT_0(1 - e^{-t/RC})$$

(*d*) Using a spreadsheet program, graph both $T(t)$ and $Q(t)$; from the graph, find the time it takes for the temperature of the first container to be reduced to 30°C.

67 ••• A blackened copper cube that is 1 cm along an edge is heated to a temperature of 300°C, and then is placed in a vacuum chamber whose walls are at a temperature of 0°C. In the vacuum chamber, the cube cools radiatively. (*a*) Show that the (absolute) temperature T of the cube follows the differential equation:

$$\frac{dT}{dt} = -\frac{e\sigma A}{C}(T^4 - T_0^4)$$

where C is the heat capacity of the cube, A is its surface area, e the emissivity, and T_0 the temperature of the vacuum chamber. (*b*) Using Euler's method, numerically solve the differential equation to find $T(t)$, and graph it. Assume e = 1. (The Euler method is discussed in Section 4 of Chapter 5.) How long does it take the cube to cool to a temperature of 15°C?

APPENDIX A

SI Units and Conversion Factors

Basic Units

Length	The *meter* (m) is the distance traveled by light in a vacuum in 1/299,792,458 s.
Time	The *second* (s) is the duration of 9,192,631,770 periods of the radiation corresponding to the transition between the two hyperfine levels of the ground state of the ^{133}Cs atom.
Mass	The *kilogram* (kg) is the mass of the international standard body preserved at Sèvres, France.
Current	The *ampere* (A) is that current in two very long parallel wires 1 m apart that gives rise to a magnetic force per unit length of 2×10^{-7} N/m.
Temperature	The *kelvin* (K) is 1/273.16 of the thermodynamic temperature of the triple point of water.
Luminous intensity	The *candela* (cd) is the luminous intensity, in the perpendicular direction, of a surface of area 1/600,000 m^2 of a blackbody at the temperature of freezing platinum at a pressure of 1 atm.

Derived Units

Force	newton (N)	$1\ N = 1\ kg{\cdot}m/s^2$
Work, energy	joule (J)	$1\ J = 1\ N{\cdot}m$
Power	watt (W)	$1\ W = 1\ J/s$
Frequency	hertz (Hz)	$1\ Hz = cy/s$
Charge	coulomb (C)	$1\ C = 1\ A{\cdot}s$
Potential	volt (V)	$1\ V = 1\ J/C$
Resistance	ohm (Ω)	$1\ \Omega = 1\ V/A$
Capacitance	farad (F)	$1\ F = 1\ C/V$
Magnetic field	tesla (T)	$1\ T = 1\ N/(A{\cdot}m)$
Magnetic flux	weber (Wb)	$1\ Wb = 1\ T{\cdot}m^2$
Inductance	henry (H)	$1\ H = 1\ J/A^2$

Conversion Factors

Conversion factors are written as equations for simplicity; relations marked with an asterisk are exact.

Length

1 km = 0.6215 mi
1 mi = 1.609 km
1 m = 1.0936 yd = 3.281 ft = 39.37 in.
*1 in. = 2.54 cm
*1 ft = 12 in. = 30.48 cm
*1 yd = 3 ft = 91.44 cm
1 lightyear = 1 c·y = 9.461×10^{15} m
*1 Å = 0.1 nm

Area

*1 m² = 10^4 cm²
1 km² = 0.3861 mi² = 247.1 acres
*1 in.² = 6.4516 cm²
1 ft² = 9.29×10^{-2} m²
1 m² = 10.76 ft²
*1 acre = 43,560 ft²
1 mi² = 640 acres = 2.590 km²

Volume

*1 m³ = 10^6 cm³
*1 L = 1000 cm³ = 10^{-3} m³
1 gal = 3.786 L
1 gal = 4 qt = 8 pt = 128 oz = 231 in³
1 in³ = 16.39 cm³
1 ft³ = 1728 in.³ = 28.32 L
= 2.832×10^4 cm³

Time

*1 h = 60 min = 3.6 ks
*1 d = 24 h = 1440 min = 86.4 ks
1 y = 365.24 d = 3.156×10^7 s

Speed

*1 m/s = 3.6 km/h
1 km/h = 0.2778 m/s = 0.6215 mi/h
1 mi/h = 0.4470 m/s = 1.609 km/h
1 mi/h = 1.467 ft/s

Angle and Angular Speed

*π rad = 180°
1 rad = 57.30°
1° = 1.745×10^{-2} rad
1 rev/min = 0.1047 rad/s
1 rad/s = 9.549 rev/min

Mass

*1 kg = 1000 g
*1 tonne = 1000 kg = 1 Mg
1 u = 1.6606×10^{-27} kg
1 kg = 6.022×10^{26} u
1 slug = 14.59 kg
1 kg = 6.852×10^{-2} slug
1 u = 931.50 MeV/c^2

Density

*1 g/cm³ = 1000 kg/m³ = 1 kg/L
(1 g/cm³)g = 62.4 lb/ft³

Force

1 N = 0.2248 lb = 10^5 dyn
*1 lb = 4.448222 N
(1 kg)g = 2.2046 lb

Pressure

*1 Pa = 1 N/m²
*1 atm = 101.325 kPa = 1.01325 bars
1 atm = 14.7 lb/in.² = 760 mmHg
= 29.9 in.Hg = 33.8 ftH_2O
1 lb/in.² = 6.895 kPa
1 torr = 1 mmHg = 133.32 Pa
1 bar = 100 kPa

Energy

*1 kW·h = 3.6 MJ
*1 cal = 4.1840 J
1 ft·lb = 1.356 J = 1.286×10^{-3} Btu
*1 L·atm = 101.325 J
1 L·atm = 24.217 cal
1 Btu = 778 ft·lb = 252 cal = 1054.35 J
1 eV = 1.602×10^{-19} J
1 u·c^2 = 931.50 MeV
*1 erg = 10^{-7} J

Power

1 horsepower = 550 ft·lb/s = 745.7 W
1 Btu/h = 1.055 kW
1 W = 1.341×10^{-3} horsepower
= 0.7376 ft·lb/s

Magnetic Field

*1 T = 10^4 G

Thermal Conductivity

1 W/(m·K) = 6.938 Btu·in./(h·ft²·F°)
1 Btu·in./(h·ft²·F°) = 0.1441 W/(m·K)

APPENDIX B

Numerical Data

Terrestrial Data

Free-fall acceleration g	$9.80665\ \text{m/s}^2$; $32.1740\ \text{ft/s}^2$
(Standard value at sea level at 45° latitude)†	
Standard value	
At sea level, at equator†	$9.7804\ \text{m/s}^2$
At sea level, at poles†	$9.8322\ \text{m/s}^2$
Mass of earth M_E	5.98×10^{24} kg
Radius of earth R_E, mean	6.37×10^6 m; 3960 mi
Escape speed $\sqrt{2R_E g}$	1.12×10^4 m/s; 6.95 mi/s
Solar constant‡	$1.35\ \text{kW/m}^2$
Standard temperature and pressure (STP):	
Temperature	273.15 K
Pressure	101.325 kPa (1.00 atm)
Molar mass of air	28.97 g/mol
Density of air (STP), ρ_{air}	$1.293\ \text{kg/m}^3$
Speed of sound (STP)	331 m/s
Heat of fusion of H_2O (0°C, 1 atm)	333.5 kJ/kg
Heat of vaporization of H_2O (100°C, 1 atm)	2.257 MJ/kg.

† Measured relative to the earth's surface.
‡ Average power incident normally on 1 m^2 outside the earth's atmosphere at the mean distance from the earth to the sun.

Astronomical Data†

Earth	
Distance to moon‡	3.844×10^8 m; 2.389×10^5 mi
Distance to sun, mean‡	1.496×10^{11} m; 9.30×10^7 mi; 1.00 AU
Orbital speed, mean	2.98×10^4 m/s
Moon	
Mass	7.35×10^{22} kg
Radius	1.738×10^6 m
Period	27.32 d
Acceleration of gravity at surface	$1.62\ \text{m/s}^2$
Sun	
Mass	1.99×10^{30} kg
Radius	6.96×10^8 m

† Additional solar-system data is available from NASA at <http://nssdc.gsfc.nasa.gov/planetary/planetfact.html>.
‡ Center to center.

Physical Constants†

Gravitational constant	G	$6.673(10) \times 10^{-11}$ N·m²/kg²
Speed of light	c	$2.997\,924\,58 \times 10^{8}$ m/s
Fundamental charge	e	$1.602\,1764\,62(63) \times 10^{-19}$ C
Avogadro's number	N_A	$6.022\,141\,99(47) \times 10^{23}$ particles/mol
Gas constant	R	8.314 472(15) J/(mol·K)
		1.987 2065(36) cal/(mol·K)
		$8.205\,746(15) \times 10^{-2}$ L·atm/(mol·K)
Boltzmann constant	$k = R/N_A$	$1.380\,6503(24) \times 10^{-23}$ J/K
		$8.617\,342(15) \times 10^{-5}$ eV/K
Stefan-Boltzmann constant	$\sigma = (\pi^2/60)k^4/(\hbar^3 c^2)$	$5.670\,400(40) \times 10^{-8}$ W/(m²k⁴)
Atomic mass constant	$m_u = \frac{1}{12}m(^{12}C)$	$1.660\,538\,73(13) \times 10^{-27}$ kg = 1u
Coulomb constant	$k = 1/(4\pi\epsilon_0)$	$8.987\,551\,788 \ldots \times 10^{9}$ N·m²/C²
Permittivity of free space	ϵ_0	$8.854\,187\,817 \ldots \times 10^{-12}$ C²/(N·m²)
Permeability of free space	μ_0	$4\,\pi \times 10^{-7}$ N/A²
		$1.256\,637 \times 10^{-6}$ N/A²
Planck's constant	h	$6.626\,068\,76(52) \times 10^{-34}$ J·s
		$4.135\,667\,27(16) \times 10^{-15}$ eV·s
	$\hbar = h/2\pi$	$1.054\,571\,596(82) \times 10^{-34}$ J·s
		$6.582\,118\,89(26) \times 10^{-16}$ eV·s
Mass of electron	m_e	$9.109\,381\,88(72) \times 10^{-31}$ kg
		0.510 998 902(21) MeV/c^2
Mass of proton	m_p	$1.672\,621\,58(13) \times 10^{-27}$ kg
		938.271 998(38) × MeV/c^2
Mass of neutron	m_n	$1.674\,927\,16(13) \times 10^{-27}$ kg
		939.565 330(38) MeV/c^2
Bohr magneton	$m_B = eh/2m_e$	$9.274\,0008\,99(37) \times 10^{-24}$ J/T
		$5.788\,381\,749(43) \times 10^{-5}$ eV/T
Nuclear magneton	$m_n = eh/2m_p$	$5.050\,783\,17(20) \times 10^{-27}$ J/T
		$3.152\,451\,238(24) \times 10^{-8}$ eV/T
Magnetic flux quantum	$\phi_0 = h/2e$	$2.067\,833\,636(81) \times 10^{-15}$ T·m²
Quantized Hall resistance	$R_K = h/e^2$	$2.581\,280\,7572(95) \times 10^{4}$ Ω
Rydberg constant	R_H	$1.097\,373\,156\,8549(83) \times 10^{7}$ m⁻¹
Josephson frequency-voltage quotient	$K_J = 2e/h$	$4.835\,978\,98(19) \times 10^{14}$ Hz/V
Compton wavelength	$\lambda_C = h/m_e c$	$2.426\,310\,215(18) \times 10^{-12}$ m

† The values for these and other constants may be found on the Internet at http://physics.nist.gov/cuu/Constants/index.html. The numbers in parentheses represent the uncertainties in the last two digits. (For example, 2.044 43(13) stands for 2.044 43 ± 0.000 13.) Values with without uncertainties are exact, including those values with ellipses (like the value of pi is exactly 3.1415. . .).

For additional data, see the following tables in the text.

APPENDIX C

Periodic Table of Elements

1	2	3	4	5	6	7	8	9	10	11	12	13	14	15	16	17	18
1 **H** 1.00797																	2 **He** 4.003
3 **Li** 6.941	4 **Be** 9.012											5 **B** 10.81	6 **C** 12.011	7 **N** 14.007	8 **O** 15.9994	9 **F** 19.00	10 **Ne** 20.179
11 **Na** 22.990	12 **Mg** 24.31											13 **Al** 26.98	14 **Si** 28.09	15 **P** 30.974	16 **S** 32.064	17 **Cl** 35.453	18 **Ar** 39.948
19 **K** 39.102	20 **Ca** 40.08	21 **Sc** 44.96	22 **Ti** 47.88	23 **V** 50.94	24 **Cr** 52.00	25 **Mn** 54.94	26 **Fe** 55.85	27 **Co** 58.93	28 **Ni** 58.69	29 **Cu** 63.55	30 **Zn** 65.38	31 **Ga** 69.72	32 **Ge** 72.59	33 **As** 74.92	34 **Se** 78.96	35 **Br** 79.90	36 **Kr** 83.80
37 **Rb** 85.47	38 **Sr** 87.62	39 **Y** 88.906	40 **Zr** 91.22	41 **Nb** 92.91	42 **Mo** 95.94	43 **Tc** (98)	44 **Ru** 101.1	45 **Rh** 102.905	46 **Pd** 106.4	47 **Ag** 107.870	48 **Cd** 112.41	49 **In** 114.82	50 **Sn** 118.69	51 **Sb** 121.75	52 **Te** 127.60	53 **I** 126.90	54 **Xe** 131.29
55 **Cs** 132.905	56 **Ba** 137.33	57–71 **Rare Earths**	72 **Hf** 178.49	73 **Ta** 180.95	74 **W** 183.85	75 **Re** 186.2	76 **Os** 190.2	77 **Ir** 192.2	78 **Pt** 195.09	79 **Au** 196.97	80 **Hg** 200.59	81 **Tl** 204.37	82 **Pb** 207.19	83 **Bi** 208.98	84 **Po** (210)	85 **At** (210)	86 **Rn** (222)
87 **Fr** (223)	88 **Ra** (226)	89–103 **Actinides**	104 **Rf** (261)	105 **Ha** (260)	106 (263)	107 (262)	108 (265)	109 (266)									

Rare Earths (Lanthanides)	57 **La** 138.91	58 **Ce** 140.12	59 **Pr** 140.91	60 **Nd** 144.24	61 **Pm** (147)	62 **Sm** 150.36	63 **Eu** 152.0	64 **Gd** 157.25	65 **Tb** 158.92	66 **Dy** 162.50	67 **Ho** 164.93	68 **Er** 167.26	69 **Tm** 168.93	70 **Yb** 173.04	71 **Lu** 174.97
Actinides	89 **Ac** 227.03	90 **Th** 232.04	91 **Pa** 231.04	92 **U** 238.03	93 **Np** 237.05	94 **Pu** (244)	95 **Am** (243)	96 **Cm** (247)	97 **Bk** (247)	98 **Cf** (251)	99 **Es** (252)	100 **Fm** (257)	101 **Md** (258)	102 **No** (259)	103 **Lr** (260)

The 1–18 group designation has been recommended by the International Union of Pure and Applied Chemistry (IUPAC).

Atomic Numbers and Atomic Masses†

Name	Symbol	Atomic Number	Mass	Name	Symbol	Atomic Number	Mass
Actinium	Ac	89	227.03	Mercury	Hg	80	200.59
Aluminum	Al	13	26.98	Molybdenum	Mo	42	95.94
Americium	Am	95	(243)	Neodymium	Nd	60	144.24
Antimony	Sb	51	121.75	Neon	Ne	10	20.179
Argon	Ar	18	39.948	Neptunium	Np	93	237.05
Arsenic	As	33	74.92	Nickel	Ni	28	58.69
Astatine	At	85	(210)	Niobium	Nb	41	92.91
Barium	Ba	56	137.3	Nitrogen	N	7	14.007
Berkelium	Bk	97	(247)	Nobelium	No	102	(259)
Beryllium	Be	4	9.012	Osmium	Os	76	190.2
Bismuth	Bi	83	208.98	Oxygen	O	8	15.9994
Boron	B	5	10.81	Palladium	Pd	46	106.4
Bromine	Br	35	79.90	Phosphorus	P	15	30.974
Cadmium	Cd	48	112.41	Platinum	Pt	78	195.09
Calcium	Ca	20	40.08	Plutonium	Pu	94	(244)
Californium	Cf	98	(251)	Polonium	Po	84	(210)
Carbon	C	6	12.011	Potassium	K	19	39.098
Cerium	Ce	58	140.12	Praseodymium	Pr	59	140.91
Cesium	Cs	55	132.905	Promethium	Pm	61	(147)
Chlorine	Cl	17	35.453	Protactinium	Pa	91	231.04
Chromium	Cr	24	52.00	Radium	Ra	88	(226)
Cobalt	Co	27	58.93	Radon	Rn	86	(222)
Copper	Cu	29	63.55	Rhenium	Re	75	186.2
Curium	Cm	96	(247)	Rhodium	Rh	45	102.905
Dysprosium	Dy	66	162.50	Rubidium	Rb	37	85.47
Einsteinium	Es	99	(252)	Ruthenium	Ru	44	101.1
Erbium	Er	68	167.26	Rutherfordium	Rf	104	(261)
Europium	Eu	63	152.0	Samarium	Sm	62	150.36
Fermium	Fm	100	(257)	Scandium	Sc	21	44.96
Fluorine	F	9	19.00	Selenium	Se	34	78.96
Francium	Fr	87	(223)	Silicon	Si	14	28.09
Gadolinium	Gd	64	157.25	Silver	Ag	47	107.870
Gallium	Ga	31	69.72	Sodium	Na	11	22.990
Germanium	Ge	32	72.59	Strontium	Sr	38	87.62
Gold	Au	79	196.97	Sulfur	S	16	32.064
Hafnium	Hf	72	178.49	Tantalum	Ta	73	180.95
Hahnium	Ha	105	(260)	Technetium	Tc	43	(98)
Helium	He	2	4.003	Tellurium	Te	52	127.60
Holmium	Ho	67	164.93	Terbium	Tb	65	158.92
Hydrogen	H	1	1.0079	Thallium	Tl	81	204.37
Indium	In	49	114.82	Thorium	Th	90	232.04
Iodine	I	53	126.90	Thulium	Tm	69	168.93
Iridium	Ir	77	192.2	Tin	Sn	50	118.69
Iron	Fe	26	55.85	Titanium	Ti	22	47.88
Krypton	Kr	36	83.80	Tungsten	W	74	183.85
Lanthanum	La	57	138.91	Uranium	U	92	238.03
Lawrencium	Lr	103	(260)	Vanadium	V	23	50.94
Lead	Pb	82	207.2	Xenon	Xe	54	131.29
Lithium	Li	3	6.941	Ytterbium	Yb	70	173.04
Lutetium	Lu	71	174.97	Yttrium	Y	39	88.906
Magnesium	Mg	12	24.31	Zinc	Zn	30	65.38
Manganese	Mn	25	54.94	Zirconium	Zr	40	91.22
Mendelevium	Md	101	(258)				

† More precise values for the atomic masses, along with the uncertainties in the masses, can be found at http://physics.nist.gov/PhysRefData/.

APPENDIX D

Review of Mathematics

In this appendix, we will review some of the basic results of algebra, geometry, trigonometry, and calculus. In many cases, we will merely state results without proof. Table D-1 lists some mathematical symbols.

Equations

The following operations can be performed on mathematical equations to facilitate their solution:

1. The same quantity can be added to or subtracted from each side of the equation.
2. Each side of the equation can be multiplied or divided by the same quantity.
3. Each side of the equation can be raised to the same power.

It is important to understand that the preceding rules apply to each *side* of the equation and not to each term in the equation.

EXAMPLE D-1

Solve the following equation for x: $(x - 3)^2 + 7 = 23$.

1. Subtract 7 from each side:	$(x-3)^2 = 16$
2. Take the square root of each side:	$x - 3 = \pm 4$
3. Add 3 to each side:	$x = 4 + 3 = 7$
	or
	$x = -4 + 3 = -1$

REMARKS Note that in step 2 we do not need to write $-(x - 3) = \pm 4$ because all possibilities are included in

$$x - 3 = \pm 4.$$

CHECK THE RESULT We check our result by substituting each value into the original equation: $(7 - 3)^2 + 7 = 16 + 7 = 23$ and $(-1 - 3)^2 + 7 = 16 + 7 = 23$.

TABLE D-1

Mathematical Symbols

$=$	is equal to
$\neq$	is not equal to
$\approx$	is approximately equal to
$\sim$	is of the order of
$\propto$	is proportional to
$>$	is greater than
$\geq$	is greater than or equal to
$>>$	is much greater than
$<$	is less than
$\leq$	is less than or equal to
$<<$	is much less than
Δx	change in x
$\lvert x \rvert$	absolute value of x
$n!$	$n(n-1)(n-2)\ldots 1$
Σ	sum
lim	limit
$\Delta t \to 0$	Δt approaches zero
$\frac{dx}{dt}$	derivative of x with respect to t
$\frac{\partial x}{\partial t}$	partial derivative of x with respect to t
$\int$	integral

EXAMPLE D-2

Solve the following equation for x:

$$\frac{1}{x} + \frac{1}{4} = \frac{1}{3}$$

1. Subtract from each side: $\frac{1}{x} = \frac{1}{3} - \frac{1}{4} = \frac{4}{12} - \frac{3}{12} = \frac{1}{12}$

2. Multiply each side by $12x$: $x = 12$

REMARKS This type of equation occurs both in geometric optics and in analyses of electric circuits. Although it is easy to solve, errors are often made. A typical mistake is to take the reciprocal of each *term,* obtaining $x + 4 = 3$. Taking the reciprocal of each term is not allowed; taking the reciprocal of each *side* of an equation is allowed. Note that multiplying each side by $12x$ in step 2 is equivalent to taking the reciprocal of each side of the equation.

Direct and Inverse Proportion

The relationships of direct proportion and inverse proportion are so important in physics that they deserve special consideration. Often much algebraic manipulation can be avoided through a simple knowledge of these relationships. Suppose, for example, that you work for 5 days at a certain pay rate and earn \$400. How much would you earn at the same pay rate if you worked 8 days? In this problem, the money earned is *directly proportional* to the time worked. We can write an equation relating the money earned M to the time worked t using a constant of proportionality R:

$$M = Rt$$

The constant of proportionality in this case is the pay rate. We can express R in dollars per day. Since \$400 was earned in 5 d, the value of R is \$400/(5 d) = \$80/d. In 8 d, the amount earned is therefore

$$M = (\$80/\text{d})(8\text{ d}) = \$640$$

However, we do not have to find the pay rate explicitly to work the problem. Since the amount earned in 8 d is $\frac{8}{5}$ times that earned in 5 d, this amount is

$$M = \tfrac{8}{5}(\$400) = \$640$$

We can use a similar example to illustrate inverse proportion. If you get a 25% raise, how long would you need to work to earn \$400? Here we consider R to be a variable and we wish to solve for t:

$$t = \frac{M}{R}$$

In this equation, the time t is *inversely proportional* to the pay rate R. Thus, if the new rate is $\frac{5}{4}$ times the old rate, the new time will be $\frac{4}{5}$ times the old time or 4 d.

There are some situations in which one quantity varies as the square or some other power of another quantity where the ideas of proportionality are also very useful. Suppose, for example, that a 10-in. diameter pizza costs \$8.50. How much would you expect a 12-in. diameter pizza to cost? We expect the cost of a pizza to be approximately proportional to the amount of its contents, which is proportional to the area of the pizza. Since the area is in turn proportional to the square of the diameter, the cost should be proportional to the square of the diameter. If we increase the diameter by a factor of 12/10, the area increases by a factor of $(12/10)^2 = 1.44$, so we should expect the cost to be (1.44)($8.50) = $12.24.

EXAMPLE D-3

The intensity of light from a point source varies inversely with the square of the distance from the source. If the intensity is 3.20 W/m² at 5 m from a source, what is it at 6 m from the source?

1. Write an equation expressing the fact that the intensity varies inversely with the square of the distance:
$$I = \frac{C}{r^2}$$
where C is some constant.

2. Let I_1 be the intensity at $r_1 = 5$ m and I_2 be the intensity at $r_2 = 6$ m, and express the ratio I_2/I_1 in terms of r_1 and r_2:
$$\frac{I_2}{I_1} = \frac{C/r_2^2}{C/r_1^2} = \frac{r_1^2}{r_2^2} = \left(\frac{r_1}{r_2}\right)^2 = \left(\frac{5}{6}\right)^2 = 0.694$$

3. Solve for I_2:
$$I_2 = 0.694 I_1 = (0.694)(3.20\ \text{W/m}^2) = 2.22\ \text{W/m}^2$$

Linear Equations

An equation in which the variables occur only to the first power is said to be linear. A linear equation relating y and x can always be put into the standard form

$$y = mx + b \qquad \text{D-1}$$

where m and b are constants that may be either positive or negative. Figure D-1 shows a graph of the values of x and y that satisfy (Equation D-1). The constant b, called the **intercept,** is the value of y at $x = 0$. The constant m is the **slope** of the line, which equals the ratio of the change in y to the corresponding change in x. In the figure, we have indicated two points on the line, x_1, y_1 and x_2, y_2, and the changes $\Delta x = x_2 - x_1$ and $\Delta y = y_2 - y_1$. The slope m is then

$$m = \frac{y_2 - y_1}{x_2 - x_1} = \frac{\Delta y}{\Delta x}$$

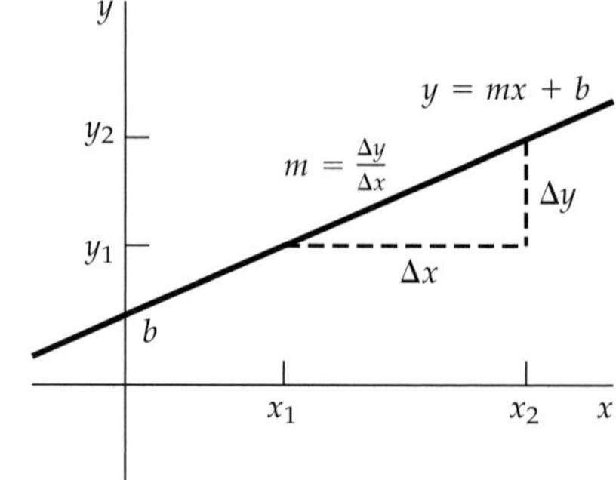

FIGURE D-1 Graph of the linear equation $y = mx + b$, where b is the intercept and $m = \Delta y/\Delta x$ is the slope.

If x and y are both unknown, there is no unique solution for their values. Any pair of values x_1, y_1 on the line in Figure D-1 will satisfy the equation. If we have two equations, each with the same two unknowns x and y, the equations can be solved simultaneously for the unknowns.

EXAMPLE D-4

Find the values of x and y that satisfy

$$3x - 2y = 8 \qquad \text{D-2}$$

and

$$y - x = 2 \qquad \text{D-3}$$

PICTURE THE PROBLEM Figure D-2 shows a graph of each of these equations. At the point where the lines intersect, the values of x and y satisfy both equations. We can solve two simultaneous equations by first solving either equation for one variable in terms of the other variable and then substituting the result into the other equation. An alternative method is to multiply one equation by a constant such that one of the unknown terms is eliminated when the equations are added or subtracted.

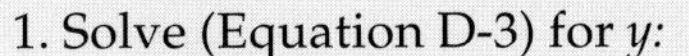

1. Solve (Equation D-3) for y: $\quad y = x + 2$
2. Substitute this value for y into (Equation D-2): $\quad 3x - 2(x + 2) = 8$
3. Simplify and solve for x: $\quad 3x - 2x - 4 = 8$

   $x = 12$
4. To solve these equations using the alternative method, we first multiply (Equation D-3) by 2: $\quad 2y - 2x = 4$
5. Add this equation to (Equation D-2):

$$\begin{array}{r} 2y - 2x = 4 \\ 3x - 2y = 8 \\ \hline 3x - 2x = 12 \end{array}$$

$$x = 12$$

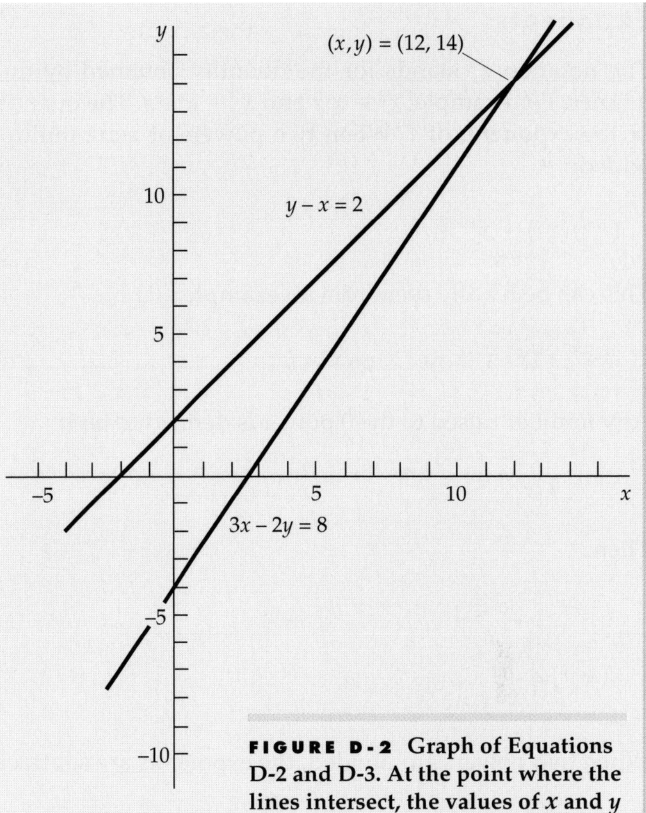

FIGURE D-2 Graph of Equations D-2 and D-3. At the point where the lines intersect, the values of x and y satisfy both equations.

Factoring

Equations can often be simplified by factoring. Three important examples are

1. Common factor: $\quad 2ax + 3ay = a(2x + 3y)$
2. Perfect square: $\quad x^2 - 2xy + y^2 = (x - y)^2$
3. Difference of squares: $\quad x^2 - y^2 = (x + y)(x - y)$

The Quadratic Formula

An equation that contains a variable to the second power is called a *quadratic equation.* The standard form for a quadratic equation is

$$ax^2 + bx + c = 0 \qquad \text{D-4}$$

where a, b, and c are constants. The general solution of this equation is

$$x = -\frac{b}{2a} \pm \frac{1}{2a}\sqrt{b^2 - 4ac} \qquad \text{D-5}$$

When b^2 is greater than $4ac$, there are two solutions corresponding to the + and − signs. Figure D-3 shows a graph of y versus x where $y = ax^2 + bx + c$. The curve, called a **parabola,** crosses the x axis twice. The values of x for which $y = 0$ are the solutions to (Equation D-4). When $b^2 < 4ac$, the graph of y versus x does not intersect the x axis, as is shown in Figure D-4, and there are no real solutions to (Equation D-4). When $b^2 = 4ac$, the graph of y versus x is tangent to the x axis at the point $x = -b/2a$.

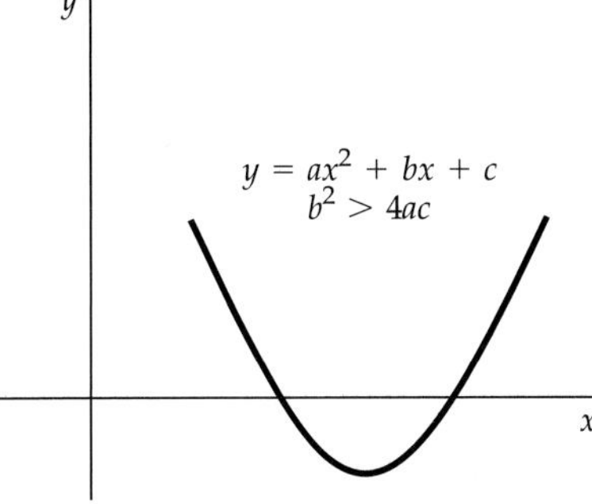

FIGURE D-3 Graph of y versus x when $y = ax^2 + bx + c$ for the case $b^2 > 4ac$. The two values of x for which $y = 0$ satisfy the quadratic equation (Equation D-4).

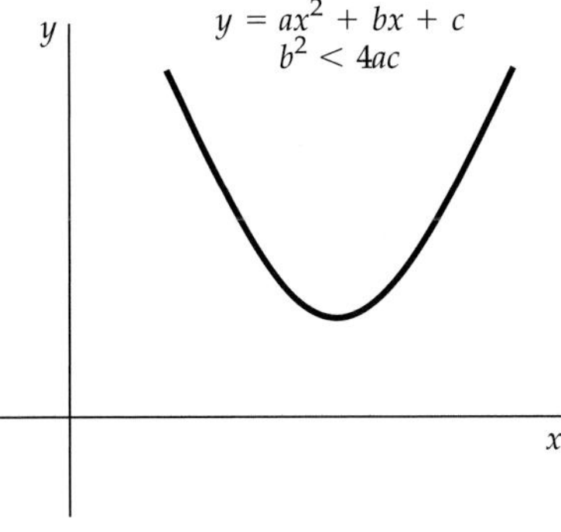

FIGURE D-4 Graph of y versus x when $y = ax^2 + bx + c$ for the case $b^2 < 4ac$. In this case, there are no (real) values of x for which $y = 0$.

Exponents

The notation x^n stands for the quantity obtained by multiplying x times itself n times. For example, $x^2 = x{\cdot}x$ and $x^3 = x{\cdot}x{\cdot}x$. The quantity n is called the **power,** or the **exponent,** of x. When two powers of x are multiplied, the exponents are added:

$$(x^m)(x^n) = x^{m+n} \qquad \text{D-6}$$

This can be readily seen from an example:

$$x^2x^3 = (x \cdot x)(x \cdot x \cdot x) = x^5$$

Any number raised to the 0 power is defined to be 1:

$$x^0 = 1 \qquad \text{D-7}$$

Then

$$x^n x^{-n} = x^0 = 1$$
$$x^{-n} = \frac{1}{x^n} \qquad \text{D-8}$$

When two powers are divided, the exponents are subtracted:

$$\frac{x^n}{x^m} = x^n x^{-m} = n^{n-m} \qquad \text{D-9}$$

Using these rules, we have

$$x^{1/2} \cdot x^{1/2} = x$$

so

$$x^{1/2} = \sqrt{x}$$

When a power is raised to another power, the exponents are multiplied:

$$(x^n)^m = x^{nm} \qquad \text{D-10}$$

Logarithms

If y is related to x by $y = a^x$, the number x is said to be the logarithm of y to the base a and is written

$$x = \log_a y$$

then if $x = 1$ then $y = a^1 = a$ and

$$\log_a a = 1 \qquad \text{D-11}$$

and, if $x = 0$ then $y = a^0 = 1$ and

$$\log_a 1 = 0 \qquad \text{D-12}$$

Also, if $y_1 = a^n$ and $y_2 = a^m$, then

$$y_1 y_2 = a^n a^m = a^{n+m}$$

and

$$\log_a y_1 = n, \log_a y_2 = m, \text{ and } \log_a y_1 y_2 = n + m$$

so

$$\log_a y_1 y_2 = \log_a y_1 + \log_a y_2 \quad \text{D-13}$$

It immediately follows that

$$\log_a y^n = n \log_a y \quad \text{D-14}$$

There are two bases in common use: base 10, called **common logarithms,** and base $e(e = 2.728 \ldots)$, called **natural logarithms.** When no base is specified, the base is usually understood to be 10. Thus, $\log 100 = \log_{10} 100 = 2$ since $100 = 10^2$.

The symbol ln is used for natural logarithms. Thus,

$$\log_e x = \ln x \quad \text{D-15}$$

and $y = \ln x$ implies

$$x = e^y \quad \text{D-16}$$

Logarithms can be changed from one base to another. Suppose that

$$z = \log x \quad \text{D-17}$$

Then

$$10^z = 10^{\log x} = x \quad \text{D-18}$$

Taking the natural logarithm of both sides of (Equation D-18), we obtain

$$z \ln 10 = \ln x$$

Substituting $\log x$ for z (see Equation D-17) gives

$$\ln x = (\ln 10) \log x \quad \text{D-19}$$

The Exponential Function

When the rate of change of a quantity is proportional to the quantity itself, the quantity increases or decreases exponentially. An example of *exponential decrease* is nuclear decay. If N is the number of radioactive nuclei at some time, then the change dN in some very small time interval dt will be proportional to N and to dt:

$$dN = -\lambda N \, dt$$

where the constant of proportionality λ is the decay rate. The function N satisfying this equation is

$$N = N_0 e^{-\lambda t} \qquad \text{D-20}$$

where N_0 is the number at time $t = 0$. Figure D-5 shows N versus t. A characteristic of exponential decay is that N decreases by a constant factor in a given time interval. The time interval for N to decrease to half its original value is its half-life $t_{1/2}$. The half life is obtained from (Equation D-20) by setting $N = \frac{1}{2}N_0$ and solving for the time. This gives

$$t_{1/2} = \frac{\ln 2}{\lambda} = \frac{0.693}{\lambda} \qquad \text{D-21}$$

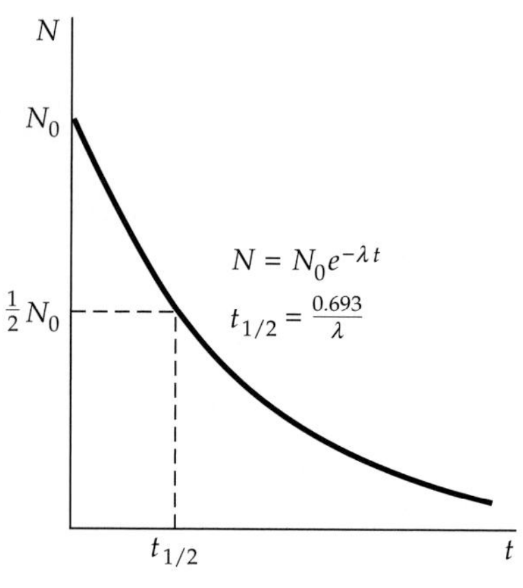

FIGURE D-5 **Graph of N versus t when N decreases exponentially. The time $t_{1/2}$ is the time it takes for N to decrease by one-half.**

An example of *exponential increase* is population growth. If the number of organisms is N, the change in N after a small time interval dt is given by

$$dN = +\lambda N\, dt$$

where λ is a constant that characterizes the rate of increase. The function N satisfying this equation is

$$N = N_0 e^{\lambda t} \qquad \text{D-22}$$

A graph of this function is shown in Figure D-6. An exponential increase is characterized by a doubling time T_2, which is related to λ by

$$T_2 = \frac{\ln 2}{\lambda} = \frac{0.693}{\lambda} \qquad \text{D-23}$$

If the rate of increase λ is expressed as a percentage, $r = \lambda/100\%$, the doubling time is

$$T_2 = \frac{69.3}{r} \qquad \text{D-24}$$

For example, if the population increases by 2 percent per year, the population will double every $69.3/2 \approx 35$ years. Table D-2 lists some useful relations for exponential and logarithmic functions.

TABLE D-2

Exponential and Logarithmic Functions

$e = 2.71828;$

$e^0 = 1$

If $y = e^x$, then $x = \ln y$.

$e^{\ln x} = x$

$e^x e^y = e^{(x+y)}$

$(e^x)^y = e^{xy} = (e^y)^x$

$\ln e = 1; \quad \ln 1 = 0$

$\ln xy = \ln x + \ln y$

$\ln \frac{x}{y} = \ln x - \ln y$

$\ln e^x = x; \quad \ln a^x = x \ln a$

$\ln x = (\ln 10) \log x$

$\quad = 2.3026 \log x$

$\log x = (\log e) \ln x = 0.43429 \ln x$

$e^x = 1 + x + \frac{x^2}{2!} + \frac{x^3}{3!} + \ldots$

$\ln(1 + x) = x \pm \frac{x^2}{2} + \frac{x^3}{3} \pm \frac{x^4}{4} + \ldots$

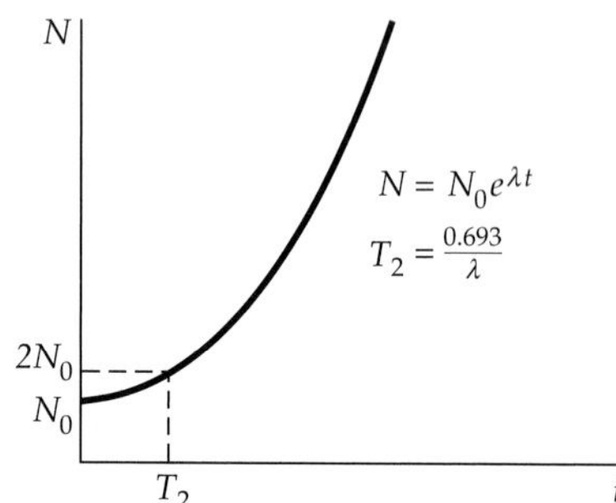

FIGURE D-6 **Graph of N versus t when N increases exponentially. The time T_2 is the time it takes for N to double.**

Geometry

The ratio of the circumference of a circle to its diameter is a natural number π, which has the approximate value

$$\pi = 3.141592$$

The circumference C of a circle is thus related to its diameter d and its radius r by

$$C = \pi d = 2\pi r \qquad \text{circumference of circle} \tag{D-25}$$

The area of a circle is

$$A = \pi r^2 \qquad \text{area of circle} \tag{D-26}$$

The area of a parallelogram is the base b times the height h (Figure D-7) and that of a triangle is one-half the base times the height (Figure D-8). A sphere of radius r (Figure D-9) has a surface area given by

$$A = 4\pi r^2 \qquad \text{spherical surface area} \tag{D-27}$$

and a volume given by

$$V = \tfrac{4}{3}\pi r^3 \qquad \text{spherical volume} \tag{D-28}$$

A cylinder of radius r and length L (Figure D-10) has surface area (not including the end faces) of

$$A = 2\pi rL \qquad \text{cylindrical surface} \tag{D-29}$$

and volume of

$$V = \pi r^2 L \qquad \text{cylindrical volume} \tag{D-30}$$

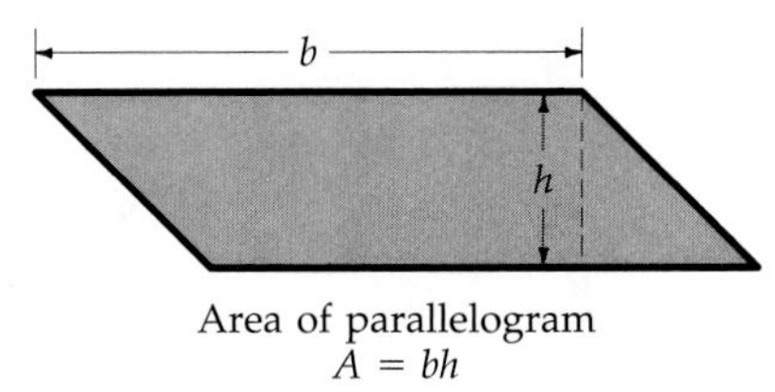

FIGURE D-7 **Area of a parallelogram.**

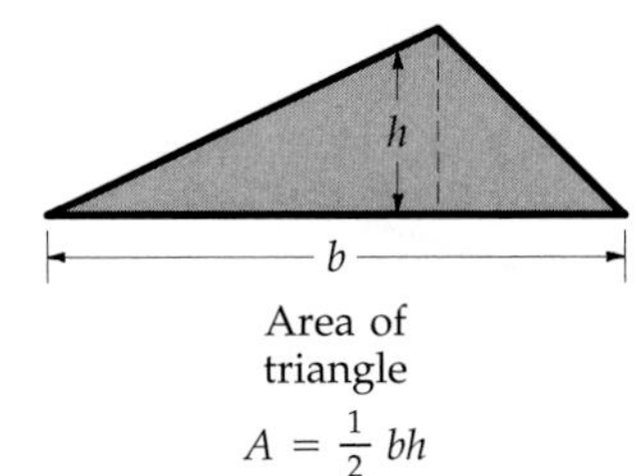

FIGURE D-8 **Area of a triangle.**

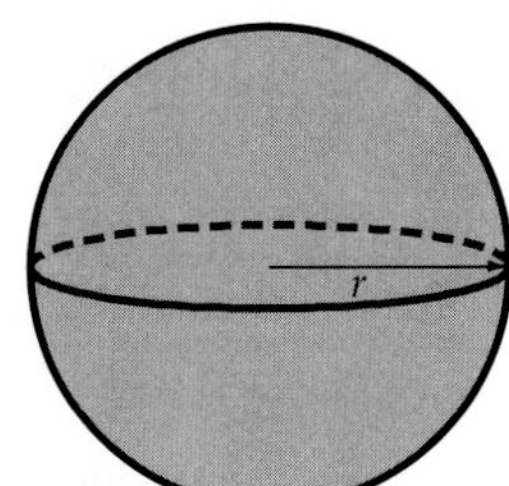

FIGURE D-9 **Surface area and volume of a sphere.**

Trigonometry

The angle between two intersecting straight lines is measured as follows. A circle is drawn with its center at the intersection of the lines, and the circular arc is divided into 360 parts called **degrees.** The number of degrees in the arc between the lines is the measure of the angle between the lines. For very small angles, the degree is divided into minutes (′) and seconds (″) with $1' = 1°/60$ and $1'' = 1'/60 = 1°/3600$. For scientific work, a more useful measure of an angle is the radian (rad). In radian measure, the angle between two intersecting straight lines is found by again drawing a circle with its center at the intersection of the lines. The measure of the angle in radians is then defined as the length of the circular arc between the lines divided by the radius of the circle (Figure D-11). If s is the arc length and r is the radius of the circle, the angle θ measured in radians is

$$\theta = \frac{s}{r} \tag{D-31}$$

Since the angle measured in radians is the ratio of two lengths, it is dimensionless. The relation between radians and degrees is

$$360° = 2\pi \text{ rad}$$

or

$$1 \text{ rad} = \frac{360°}{2\pi} = 57.3°$$

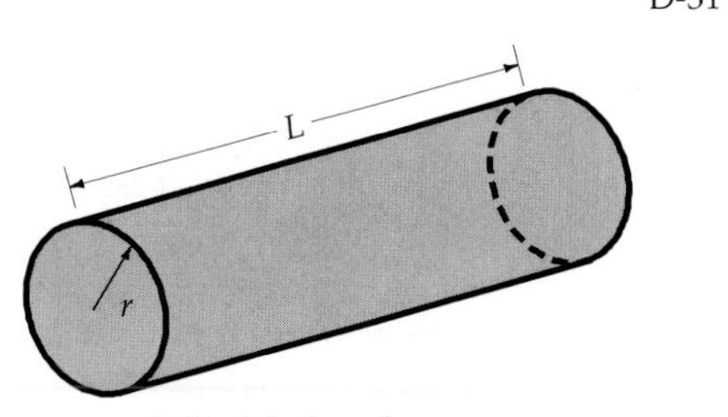

FIGURE D-10 **Surface area (not including the end faces) and volume of a cylinder.**

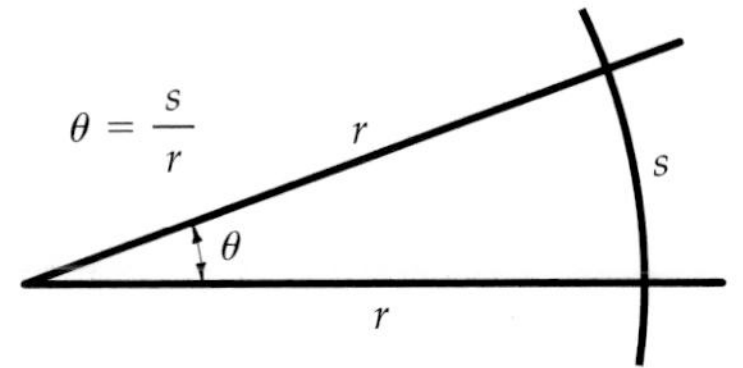

FIGURE D-11 **The angle θ in radians is defined to be the ratio s/r, where s is the arc length intercepted on a circle of radius r.**

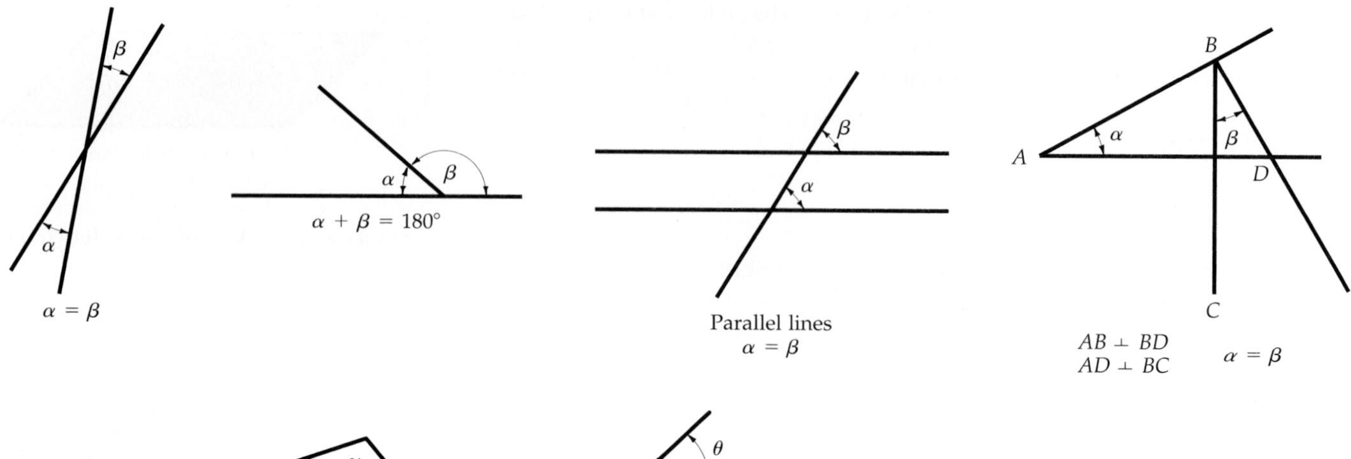

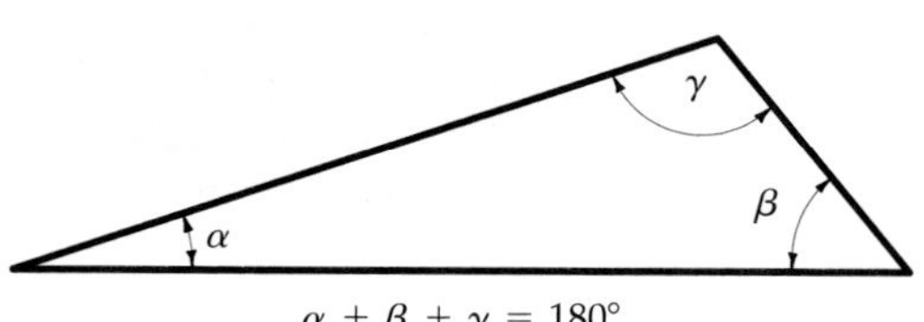

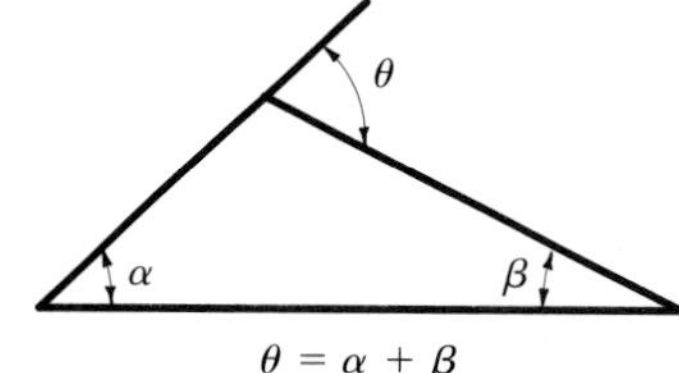

FIGURE D-12 Some useful relations for angles.

Figure D-12 shows some useful relations for angles.

Figure D-13 shows a right triangle formed by drawing the line BC perpendicular to AC. The lengths of the sides are labeled a, b, and c. The trigonometric functions sin θ, cos θ, and tan θ for an acute angle θ are defined as

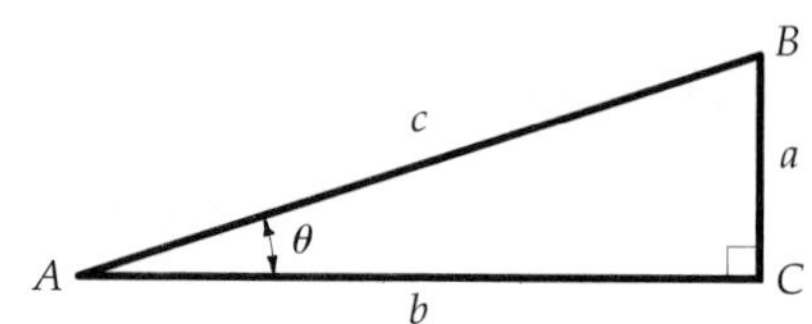

FIGURE D-13 A right triangle with sides of length a and b and a hypotenuse of length c.

$$\sin\theta = \frac{a}{c} = \frac{\text{opposite side}}{\text{hypotenuse}} \qquad \text{D-33}$$

$$\cos\theta = \frac{b}{c} = \frac{\text{adjacent side}}{\text{hypotenuse}} \qquad \text{D-34}$$

$$\tan\theta = \frac{a}{b} = \frac{\text{opposite side}}{\text{adjacent side}} = \frac{\sin\theta}{\cos\theta} \qquad \text{D-35}$$

Three other trigonometric functions, defined as the reciprocals of these functions, are

$$\sec\theta = \frac{c}{b} = \frac{1}{\cos\theta} \qquad \text{D-36}$$

$$\csc\theta = \frac{c}{a} = \frac{1}{\sin\theta} \qquad \text{D-37}$$

$$\cot\theta = \frac{b}{a} = \frac{1}{\tan\theta} = \frac{\cos\theta}{\sin\theta} \qquad \text{D-38}$$

The angle θ whose sine is x is called the arcsine of x, and is written $\sin^{-1} x$. That is, if

$$\sin\theta = x$$

then

$$\theta = \arcsin x = \sin^{-1} x \qquad \text{D-39}$$

The arcsine is the inverse of the sine. The inverse of the cosine and tangent are defined similarly. The angle whose cosine is y is the arccosine of y. That is, if

$$\cos\theta = y$$

then

$$\theta = \arccos y = \cos^{-1} y \qquad \text{D-40}$$

The angle whose tangent is z is the arctangent of z. That is, if

$$\tan\theta = z$$

$$\theta = \arctan z = \tan^{-1} z \qquad \text{D-41}$$

The Pythagorean theorem

$$a^2 + b^2 = c^2 \qquad \text{D-42}$$

gives some useful identities. If we divide each term in this equation by c^2, we obtain

$$\frac{a^2}{c^2} + \frac{b^2}{c^2} = 1$$

or, from the definitions of $\sin\theta$ and $\cos\theta$,

$$\sin^2\theta + \cos^2\theta = 1 \qquad \text{D-43}$$

Similarly, we can divide each term in (Equation D-42) by a^2 or b^2 and obtain

$$1 + \cot^2\theta = \csc^2\theta \qquad \text{D-44}$$

and

$$1 + \tan^2\theta = \sec^2\theta \qquad \text{D-45}$$

These and other useful trigonometric formulas are listed in Table D-3.

The following equation is actually two equations. For one of the equations the two upper signs are taken, and for the other equation the two lower signs are taken. The same sign rule applies to others of the following "equations":

$$\sin(A \pm B) = \sin A\cos B \pm \cos A\sin B$$

$$\cos(A \pm B) = \cos A\cos B \mp \sin A\sin B$$

$$\tan(A \pm B) = \frac{\tan A \pm \tan B}{1 \mp \tan A\tan B}$$

$$\sin A \pm \sin B = 2\sin\left[\tfrac{1}{2}(A \pm B)\right]\cos\left[\tfrac{1}{2}(A \mp B)\right]$$

$$\cos A + \cos B = 2\cos\left[\tfrac{1}{2}(A + B)\right]\cos\left[\tfrac{1}{2}(A \pm B)\right]$$

$$\cos A \pm \cos B = 2\sin\left[\tfrac{1}{2}(A + B)\right]\sin\left[\tfrac{1}{2}(B \pm A)\right]$$

$$\tan A \pm \tan B = \frac{\sin(A \pm B)}{\cos A\cos B}$$

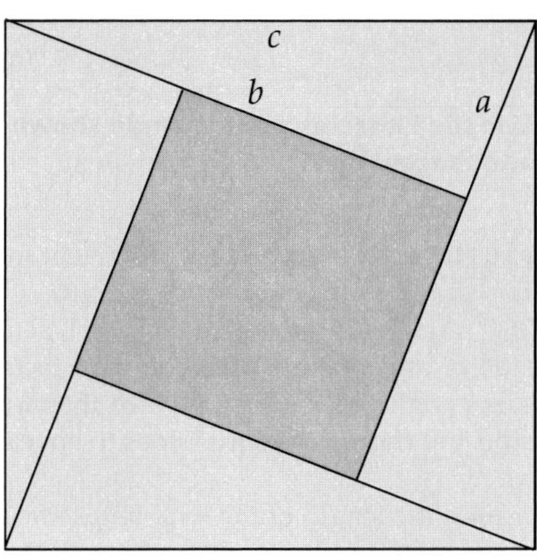

When this figure was first published the letters were absent and it was accompanied by the single word "Behold!" Using the drawing, establish the Pythagorean theorem ($a^2 + b^2 = c^2$).

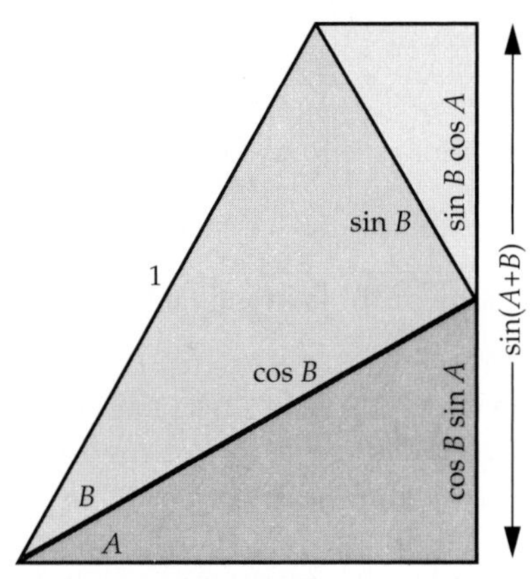

Using this drawing, establish the identity $\sin(A + B) = \sin A\cos B + \cos A\sin B$. You can also use it to establish the identity $\cos(A + B) = \cos A\cos B - \sin A\sin B$. Try it.

TABLE D-3

Trigonometric Formulas

$$\sin^2\theta + \cos^2\theta = 1; \quad \sec^2\theta - \tan^2\theta = 1; \quad \csc^2\theta - \cot^2\theta = 1$$

$$\sin 2\theta = 2\sin\theta\cos\theta$$

$$\cos 2\theta = \cos^2\theta - \sin^2\theta = 2\cos^2\theta - 1 = 1 - 2\sin^2\theta$$

$$\tan 2\theta = \frac{2\tan\theta}{1 - \tan^2\theta}$$

$$\sin\frac{1}{2}\theta = \sqrt{\frac{1 - \cos\theta}{2}}; \quad \cos\frac{1}{2}\theta = \sqrt{\frac{1 + \cos\theta}{2}}; \quad \tan\frac{1}{2}\theta = \sqrt{\frac{1 - \cos\theta}{1 + \cos\theta}}$$

EXAMPLE D-5

Use the isosceles right triangle shown in Figure D-14 to find the sine, cosine, and tangent of 45°.

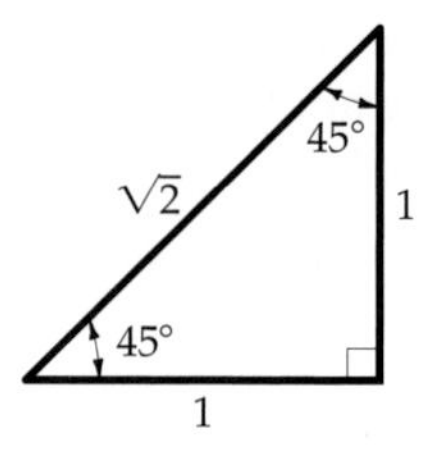

FIGURE D-14 An isosceles right triangle.

PICTURE THE PROBLEM It is clear from the figure that the two acute angles of this triangle are equal. Since the sum of the three angles in a triangle must equal 180°, and the right angle is 90°, each acute angle must be 45°. If we multiply each side of any triangle by a common factor, we obtain a similar triangle with the same angles as the first. We can therefore choose any convenient length for one side. We choose the equal sides to have a length of 1 unit.

1. Find the length of the hypotenuse from the Pythagorean theorem: $c = \sqrt{a^2 + b^2} = \sqrt{1^2 + 1^2} = \sqrt{2}$ units

2. Calculate sin 45° from its definition: $\sin 45° = \dfrac{a}{c} = \dfrac{1}{\sqrt{2}} = 0.707$

3. Calculate cos 45° from its definition: $\cos 45° = \dfrac{b}{c} = \dfrac{1}{\sqrt{2}} = 0.707$

4. Calculate tan 45° from its definition: $\tan 45° = \dfrac{a}{b} = \dfrac{1}{1} = 1$

EXAMPLE D-6

The sine of 30° is exactly 0.5. Find the ratios of the sides of a 30–60° right triangle.

PICTURE THE PROBLEM This common triangle is shown in Figure D-15. We choose a length of 1 unit for the side opposite the 30° angle.

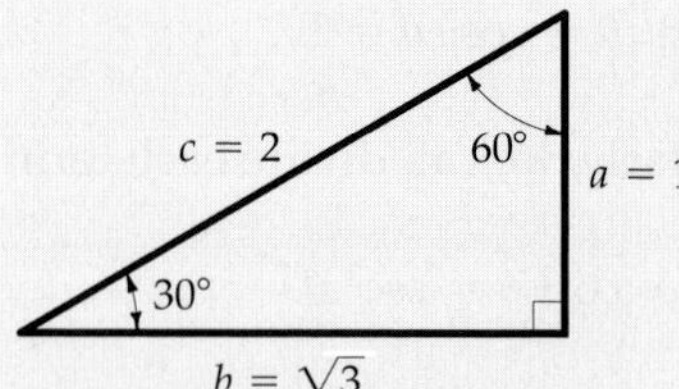

FIGURE D-15 A 30–60° right triangle.

1. Calculate the hypotenuse from the definition of the sine and the choice of 1 unit for the opposite side: $\sin 30° = \dfrac{a}{c} = \dfrac{1}{c} = 0.5$

2. Use the Pythagorean theorem to find the length b of the side opposite the 60° angle: $c = \dfrac{1}{0.5} = 2$

$$b = \sqrt{c^2 - a^2} = \sqrt{2^2 - 1^2} = \sqrt{3}$$

3. Use these results to calculate cos 30°, tan 30°, sin 60°, cos 60°, and tan 60°: $\cos 30° = \dfrac{b}{c} = \dfrac{\sqrt{3}}{2} = 0.866$

$$\tan 30° = \frac{a}{b} = \frac{1}{\sqrt{3}} = 0.577$$

$$\sin 60° = \frac{b}{c} = \cos 30° = 0.866$$

$$\cos 60° = \frac{a}{c} = \sin 30° = 0.500$$

$$\tan 60° = \frac{b}{a} = \frac{\sqrt{3}}{1} = 1.732$$

For small angles, the length a is nearly equal to the arc length s, as can be seen in Figure D-16. The angle $\theta = s/c$ is therefore nearly equal to $\sin\theta = a/c$:

$$\sin\theta \approx \theta \qquad \text{for small values of } \theta \qquad \text{D-46}$$

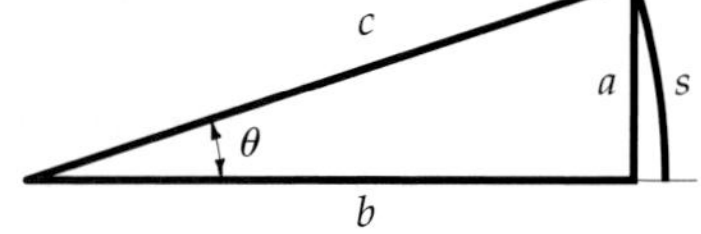

FIGURE D-16 For small angles, $\sin\theta = a/c$, $\tan\theta = a/b$, and the angle $\theta = s/c$ are all approximately equal.

Similarly, the lengths c and b are nearly equal, so $\tan\theta = a/b$ is nearly equal to both θ and $\sin\theta$ for small values of θ:

$$\tan\theta \approx \sin\theta \approx \theta \qquad \text{for small values of } \theta \qquad \text{D-47}$$

Equations D-46 and D-47 hold only if θ is measured in radians. Since $\cos\theta = b/c$ and these lengths are nearly equal for small values of θ, we have

$$\cos\theta \approx 1 \qquad \text{for small values of } \theta \qquad \text{D-48}$$

EXAMPLE D-7

By how much do $\sin\theta$, $\tan\theta$, and θ differ when $\theta = 15°$?

1. Convert 15° to radians: $\theta = 15°\,\dfrac{2\pi\text{ rad}}{360°} = 0.262\text{ rad}$

2. Find sin 15° and tan 15° using a calculator: $\sin 15° = 0.259$

   $\tan 15° = 0.268$

3. Compute the percentage difference between θ and $\sin\theta$: $\dfrac{|\sin\theta - \theta|}{\theta} = \dfrac{|0.259 - 0.262|}{0.262} = \dfrac{0.003}{0.262} = 0.011 \approx 1\%$

4. Compute the percentage difference between θ and $\tan\theta$: $\dfrac{|\tan\theta - \theta|}{\theta} = \dfrac{|0.268 - 0.262|}{0.262} = \dfrac{0.006}{0.262} = 0.023 \approx 2\%$

REMARKS For smaller angles, the approximation $\theta < \sin\theta < \tan\theta$ is even more accurate.

Example D-7 shows that if accuracy of a few percent is needed, small angle approximations can be used only for angles of about 15° or less. Figure D-17 shows graphs of θ, $\sin\theta$, and $\tan\theta$ versus θ, for small values of θ.

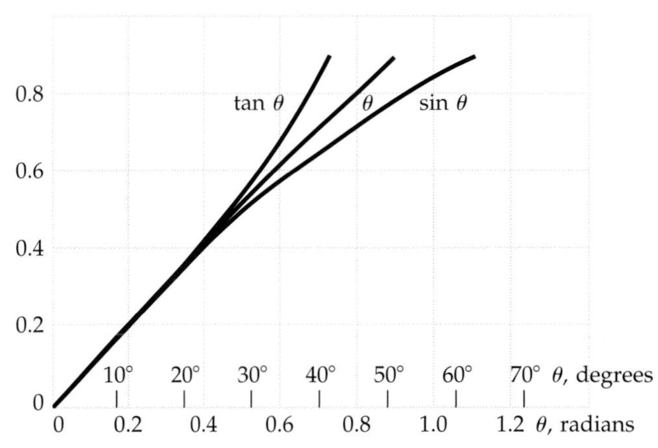

FIGURE D-17 Graphs of $\tan\theta$, θ, and $\sin\theta$ versus θ for small values of θ.

Figure D-18 shows an obtuse angle with its vertex at the origin and one side along the x axis. The trigonometric functions for a general angle such as this are defined by

$$\sin\theta = \frac{y}{c} \quad \text{D-49}$$

$$\cos\theta = \frac{x}{c} \quad \text{D-50}$$

$$\tan\theta = \frac{y}{x} \quad \text{D-51}$$

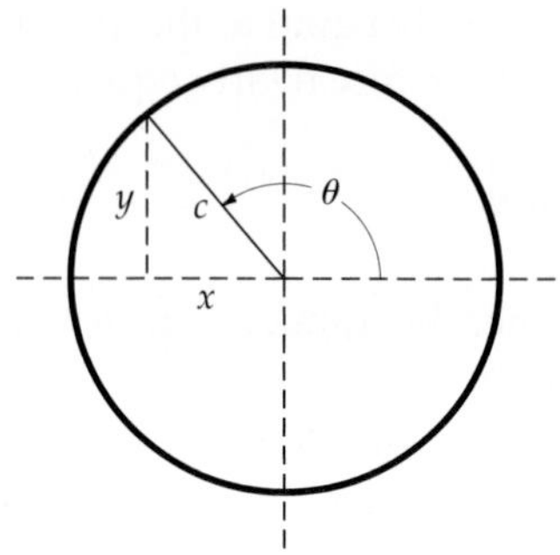

FIGURE D-18 Diagram for defining the trigonometric functions for an obtuse angle.

Figure D-19 shows plots of these functions versus θ. The sine function has a period of 2π rad. Thus, for any value of θ $\sin(\theta + 2\pi) = \sin\theta$ and so forth. That is, when an angle changes by 2π rad, the function returns to its original value. The tangent function has a period of π rad. Thus, $\sin(\theta + 2\pi) = \sin\theta$ and so forth. Some other useful relations are

$$\sin(\pi - \theta) = \sin\theta \quad \text{D-52}$$

$$\cos(\pi - \theta) = -\cos\theta \quad \text{D-53}$$

$$\sin(\pi/2 - \theta) = \cos\theta \quad \text{D-54}$$

$$\cos(\pi/2 - \theta) = \sin\theta \quad \text{D-55}$$

The trigonometric functions can be expressed as power series in θ. The series for $\sin\theta$ and $\cos\theta$ are

$$\sin\theta = \theta - \frac{\theta^3}{3!} + \frac{\theta^5}{5!} - \frac{\theta^7}{7!} + \dots \quad \text{D-56}$$

$$\cos\theta = 1 - \frac{\theta^2}{2!} + \frac{\theta^4}{4!} - \frac{\theta^6}{6!} + \dots \quad \text{D-57}$$

When θ is small, good approximations are obtained using only the first few terms in the series.

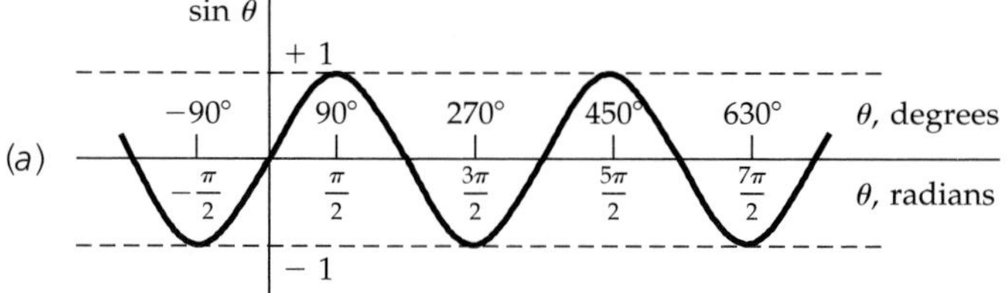

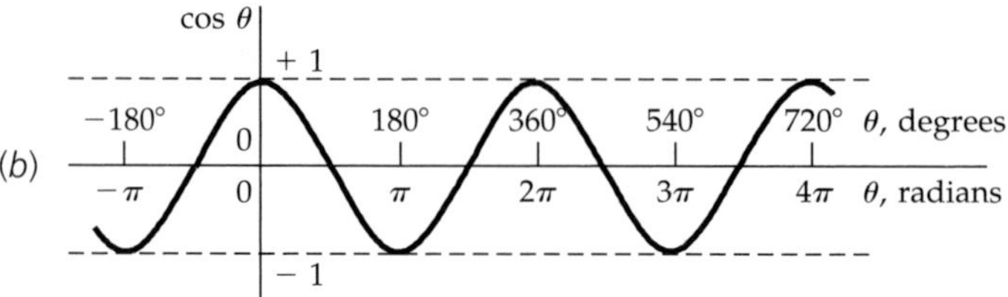

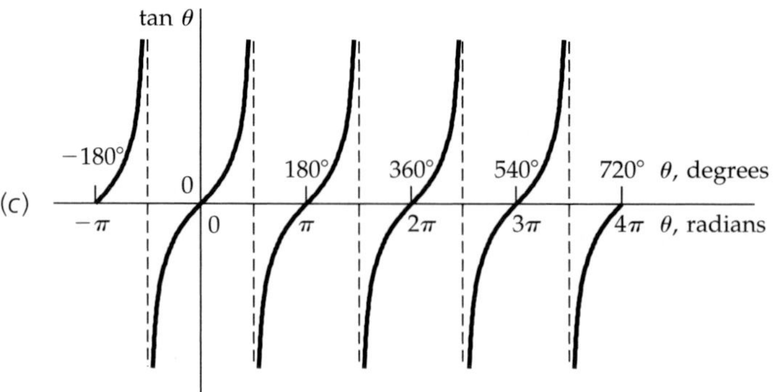

FIGURE D-19 The trigonometric functions sin θ, cos θ, and tan θ versus θ.

The Binomial Expansion

The binomial theorem is very useful for making approximations. One form of this theorem is

$$(1 + x)^n = 1 + nx + \frac{n(n-1)}{2!}x^2 + \frac{n(n-1)(n-2)}{3!}x^3 + \dots \quad \text{D-58}$$

If n is a positive integer, there are $n + 1$ terms in this series. If n is a real number other than a positive integer, there are an infinite number of terms. The series is valid for any value of n if x^2 is less than 1. It is also valid for $x^2 = 1$ if n is positive. The series is particularly useful if x^2 is small compared to 1. Then each term is much smaller than the previous term and we can drop all but the first two or three terms in the equation. If x^2 is much less compared to 1, we have

$$(1 + x)^n \approx 1 + nx, \qquad |x| << 1 \quad \text{D-59}$$

EXAMPLE D-8

Use (Equation D-59) to find an approximate value for the square root of 101.

1. Write $(101)^{1/2}$ so it is in the form $(1 + x)^n$ with x much less than 1: $(101)^{1/2} = (100 + 1)^{1/2} = (100)^{1/2}(1 + 0.01)^{1/2} = 10(1 + 0.01)^{1/2}$

2. Use Equation D-59 with $n = \frac{1}{2}$ and $x = 0.1$ to expand $(1 + 0.01)^{1/2}$. $(1 + 0.01)^{1/2} \approx 1 + \frac{1}{2}(0.01) = 1.005$

3. Substitute this result into the equation in step 1: $(101)^{1/2} = 10(1 + 0.01)^{1/2} \approx \boxed{10.05}$

REMARKS We can assess the accuracy of this result by computing the first term in Equation D-58 that was neglected. This term is

$$\frac{n(n-1)}{2}x^2 = \frac{\frac{1}{2}(-\frac{1}{2})}{2}(0.01)^2 = -\frac{0.0001}{8} \approx -0.00001 = -0.001$$

We therefore expect our answer to be correct to within about 0.001%. The value of $(101)^{1/2}$ to eight significant figures is 10.049875, which differs from 10.05 by 0.000124 or about 0.001% of 10.05.

Complex Numbers

A general complex number z can be written

$$z = a + bi \qquad \text{D-60}$$

where a and b are real numbers and $i = \sqrt{-1}$. The quantity a is called the real part and the quantity ib is called the imaginary part of z. We can represent a complex number in a plane as shown in Figure D-20, where the x axis is the real axis and the y axis is the imaginary axis. We can use the relations $a = r\cos\theta$ and $b = r\sin\theta$ from Figure D-20 to write the complex number z in polar coordinates:

$$z = r\cos\theta + ir\sin\theta \qquad \text{D-61}$$

where $r = \sqrt{a^2 + b^2}$ is called the magnitude of z.

When complex numbers are added or subtracted, the real and imaginary parts are added or subtracted separately:

$$z_1 + z_2 = (a_1 + ib_1) + (a_2 + ib_2) = (a_1 + a_2) + i(b_1 + b_2) \qquad \text{D-62}$$

However, when two complex numbers are multiplied, each part of one number is multiplied by each part of the other number:

$$z_1 z_2 = (a_1 + ib_1)(a_2 + ib_2) = a_1a_2 + i^2b_1b_2 + i(a_1b_2 + a_2b_1) \qquad \text{D-63}$$
$$= a_1a_2 - b_1b_2 + i(a_1b_2 + a_2b_1)$$

where we have used $i^2 = -1$.

The complex conjugate z^* of the complex number z is that number obtained by replacing i with $-i$. If $z = a + ib$ then

$$z^* = (a + ib)^* = a - ib \qquad \text{D-64}$$

The product of a complex number and its complex conjugate equals the square of the magnitude of the number:

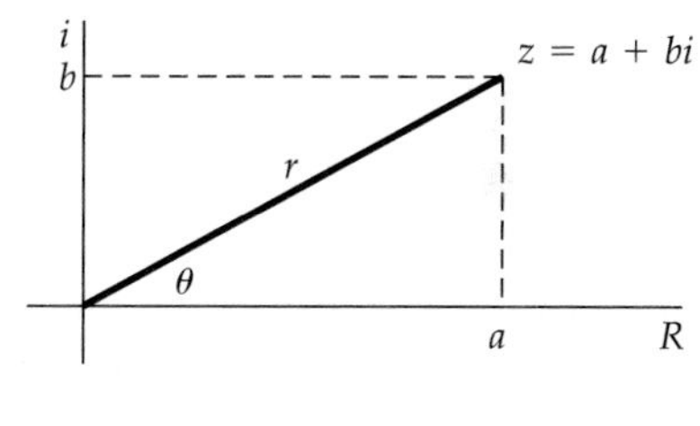

$z = a + bi$
$= r\cos\theta + (r\sin\theta)i$
$= r(\cos\theta + i\sin\theta)$

FIGURE D-20 Representation of a complex number in a plane. The real part of the complex number is plotted along the horizontal axis, and the imaginary part is plotted along the vertical axis.

$$zz^* = (a + ib)(a - ib) = a^2 + b^2 \tag{D-65}$$

A particularly useful function of a complex number is the exponential $e^{i\theta}$. Using the expansion for e^x given in Table D-2, we have

$$e^{i\theta} = 1 + i\theta + \frac{(i\theta)^2}{2!} + \frac{(i\theta)^3}{3!} + \frac{(i\theta)^4}{4!} + \ldots$$

Using $i^2 = -1$, $i^3 = -i$, $i^4 = +1$, and so forth and separating the real parts from the imaginary parts, this expansion can be written

$$e^{i\theta} = \left(1 - \frac{\theta^2}{2!} + \frac{\theta^4}{4!} - \ldots\right) + i\left(\theta - \frac{\theta^3}{3!} + \ldots\right)$$

Comparing this result with Equations D-56 and D-57, we can see that

$$e^{i\theta} = \cos\theta + i\sin\theta \tag{D-66}$$

Using this result, we can express a general complex number as an exponential:

$$z = a + ib = r\cos\theta + ir\sin\theta = re^{i\theta} \tag{D-67}$$

where $r = \sqrt{a^2 + b^2}$.

Solving differential equations using complex numbers Consider an equation of the form

$$a\frac{d^2x}{dt^2} + b\frac{dx}{dt} + cx = A\cos\omega t \tag{D-68}$$

that represents a physical process, such as a damped harmonic oscillator driven by a sinusoidal force, or a series *RLC* combination being driven by a sunusoidal potential drop. Each of the parameters in (Equation D-68) is a real number. We wish to obtain the steady-state solution to this equation using complex numbers. To do so, we first construct the equation:

$$a\frac{d^2y}{dt^2} + b\frac{dy}{dt} + cy = A\sin\omega t \tag{D-69}$$

Equation D-69 has no physical meaning and we have no interest in solving it. However, it is of use in solving Equation D-68. After multiplying both sides of (Equation D-69) by i we add it and (Equation D-68) to obtain

$$\left(a\frac{d^2x}{dt^2} + ai\frac{d^2y}{dt^2}\right) + \left(b\frac{dx}{dt} + bi\frac{dy}{dt}\right) + (cx + ciy) = A\cos\omega t + Ai\sin\omega t$$

We next combine terms to get

$$a\frac{d^2(x + iy)}{dt^2} + b\frac{d(x + iy)}{dt} + c(x + iy) = A(\cos\omega t + i\sin\omega t) \tag{D-70}$$

whose validity depends on the derivative of a sum being equal to the sum of the derivatives. We simplify our result by defining $z = x + iy$ and by using the identity $e^{i\omega t} = \cos\omega t + i\,\sin\omega t$. Substituting these into (Equation D-70) we obtain

$$a\frac{d^2z}{dt^2} + b\frac{dz}{dt} + cz = Ae^{i\omega t} \qquad \text{D-71}$$

which we now solve for z. Once z is obtained we can solve for x using $x = \text{Re}(z)$.

Since we are looking only for the steady state solution for (Equation D-68) we can assume its solution is of the form $x = x_0 \cos(\omega t - \phi)$, where ϕ is a constant. This is equivalent to assuming that the solution to (Equation D-71) is of the form $z = z_0 e^{i\omega t,}$ where z_0 is a complex number. Then $dz/dt = i\omega z\ d^2z/dt^2 = -\omega^2 z,$ and $e^{i\omega t} = z/z_0$.

Substituting these into (Equation D-70) gives $-a\omega^2 z + i\omega bz + cz = A\frac{z}{z_0}$

Dividing both sides of this equation by z and solving for z_0 gives

$$z_0 = \frac{A}{-a\omega^2 + i\omega b + c}$$

Expressing the denominator in polar form gives $(-a\omega^2 + c) + i\omega b = \sqrt{(-a\omega^2 + c)^2 + \omega^2 b^2}e^{i\phi}$, where $\tan \phi = \omega^2 b^2/(-a\omega^2 + c)$. Thus,

$$z_0 = \frac{A}{\sqrt{(-a\omega^2 + c)^2 + \omega^2 b^2}}e^{-i\phi}$$

so

$$z = z_0 e^{i\omega t} = \frac{A}{\sqrt{(-a\omega^2 + c)^2 + \omega^2 b^2}}e^{i(\omega t - \phi)}$$

$$= \frac{A}{\sqrt{(-a\omega^2 + c)^2 + \omega^2 b^2}}(\cos(\omega t - \delta) + i \sin(\omega t - \delta)) \qquad \text{D-72}$$

It follows that

$$x = \text{Re}(z) = \frac{A}{\sqrt{(-a\omega^2 + c)^2 + \omega^2 b^2}}\cos(\omega t - \phi) \qquad \text{D-73}$$

Differential Calculus

When we say that x is a function of t, we mean that for each value of t there is a single corresponding value of x. An example is $x = At^2$, where A is a constant. To indicate that x is a function of t, we sometimes write $x(t)$ for x. Figure D-21 is a graph of x versus t for a typical function $x(t)$. At a particular value $t = t_1$, x has the value of x_1 as indicated. At another value t_2, x has the value x_2. The change in t, $t_2 - t_1$, is written $\Delta t = t_2 - t_1$ and the corresponding change in x is written $\Delta x = x_2 - x_1$. The ratio $\Delta x/\Delta t$ is the slope of the straight line connecting (x_1, t_1) and (x_2, t_2). If we make Δt smaller and smaller, the line connecting (x_1, t_1) and (x_2, t_2) approaches the line that is tangent to the curve at the point (x_1, t_1). The slope of this tangent line is called the derivative of x with respect to t and is written dx/dt:

$$\frac{dx}{dt} = \lim_{\Delta t \to 0} \frac{\Delta x}{\Delta t} \qquad \text{D-74}$$

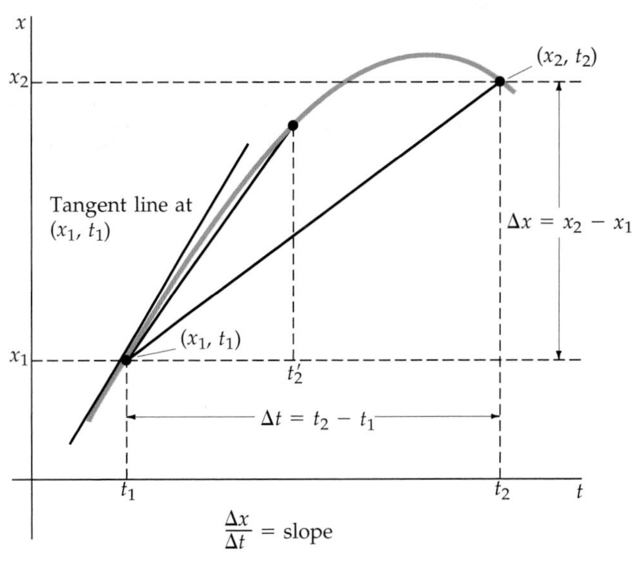

FIGURE D-21 Graph of a typical function $x(t)$. The points (x_1, t_1) and (x_2, t_2) are connected by a straight line. The slope of this line is $\Delta x/\Delta t$. As the time interval beginning at t_1 is decreased, the slope for that interval approaches the slope of the line tangent to the curve at time t_1, which is the derivative of x with respect to t.

The derivative of a function of t is generally another function of t. If x is a constant, the graph of x versus t is a horizontal line with zero slope. The derivative of a constant is thus zero. In Figure D-22, x is proportional to t:

$$x = Ct$$

This function has a constant slope equal to C. Thus the derivative of Ct is C. Table D-4 lists some properties of derivatives and the derivatives of some particular functions that occur often in physics. It is followed by comments aimed at making these properties and rules clearer. More detailed discussion can be found in most calculus books.

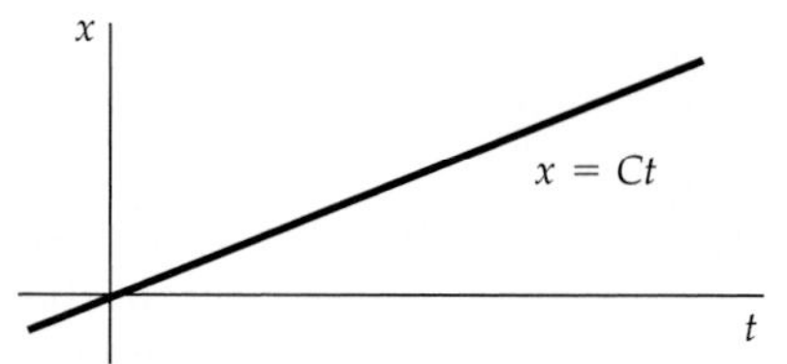

FIGURE D-22 Graph of the linear function $x = Ct$. This function has a constant slope C.

TABLE D-4

Properties of Derivatives and Derivatives of Particular Functions

Linearity

1. The derivative of a constant times a function equals the constant times the derivative of the function:

$$\frac{d}{dt}[Cf(t)] = C\frac{df(t)}{dt}$$

2. The derivative of a sum of functions equals the sum of the derivatives of the functions:

$$\frac{d}{dt}[f(t) + g(t)] = \frac{df(t)}{dt} + \frac{dg(t)}{dt}$$

Chain rule

3. If f is a function of x and x is in turn a function of t, the derivative of f with respect to t equals the product of the derivative of f with respect to x and the derivative of x with respect to t:

$$\frac{d}{dt}f(t) = \frac{df}{dx}\frac{dx}{dt}$$

Derivative of a product

4. The derivative of a product of functions $f(t)g(t)$ equals the first function times the derivative of the second plus the second function times the derivative of the first:

$$\frac{d}{dt}[f(t)g(t)] = f(t)\frac{dg(t)}{dt} + \frac{df(t)}{dt}g(t)$$

Reciprocal derivative

5. The derivative of t with respect to x is the reciprocal of the derivative of x with respect to t, assuming that neither derivative is zero:

$$\frac{dx}{dt} = \left(\frac{dt}{dx}\right)^{-1} \quad \text{if} \quad \frac{dt}{dx} \neq 0$$

Derivatives of particular functions

6. $\frac{dC}{dt} = 0$ where C is a constant

7. $\frac{d(t^n)}{dt} = nt^{n-1}$

8. $\frac{d}{dt}\sin \omega t = \omega \cos \omega t$

9. $\frac{d}{dt}\cos \omega t = -\omega \sin \omega t$

10. $\frac{d}{dt}\tan \omega t = \omega \sec^2 \omega t$

11. $\frac{d}{dt}e^{bt} = be^{bt}$

12. $\frac{d}{dt}\ln bt = \frac{1}{t}$

EXAMPLE D-9

Find the derivative of $x = at^2 + bt + c$, where a, b, and c are constants.

PICTURE THE PROBLEM From rule 2, we can differentiate each term separately and add the results.

1. Use rules 1 and 7 to find the derivative of the first term: $\dfrac{d(at^2)}{dt} = 2at^1 = 2at$

2. Compute the derivatives of the second and third terms: $\dfrac{d(bt)}{dt} = b,\ \dfrac{d(c)}{dt} = 0$

3. Add these results: $\dfrac{dx}{dt} = 2at + b$

Comments on Rules 1 Through 5

Rules 1 and 2 follow from the fact that the limiting process is linear. We can understand rule 3, the chain rule, by multiplying $\Delta f/\Delta t$ by $\Delta x/\Delta x$ and noting that, since x is a function of t, both Δx and Δf approach zero as Δt approaches zero. Since the limit of a product of two functions equals the product of their limits, we have

$$\lim_{\Delta t\to 0}\frac{\Delta f}{\Delta t} = \lim_{\Delta t\to 0}\frac{\Delta f}{\Delta x}\frac{\Delta x}{\Delta t} = \left(\lim_{\Delta t\to 0}\frac{\Delta f}{\Delta x}\right)\left(\lim_{\Delta t\to 0}\frac{\Delta x}{\Delta t}\right) = \frac{df}{dx}\frac{dx}{dt}$$

Rule 4 is not immediately apparent. The derivative of a product of functions is the limit of the ratio

$$\frac{f(t+\Delta t)g(t+\Delta t) - f(t)g(t)}{\Delta t}$$

If we add and subtract the quantity $f(t + \Delta t)g(t)$ in the numerator, we can write this ratio as

$$\frac{f(t+\Delta t)g(t+\Delta t) - f(t+\Delta t)g(t) + f(t+\Delta t)g(t) - f(t)g(t)}{\Delta t}$$

$$= f(t+\Delta t)\left[\frac{g(t+\Delta t) - g(t)}{\Delta t}\right] + g(t)\left[\frac{f(t+\Delta t) - f(t)}{\Delta t}\right]$$

As Δt approaches zero, the terms in square brackets become $dg(t)/dt$ and $df(t)/dt$, respectively, and the limit of the expression is

$$f(t)\frac{dg(t)}{dt} + g(t)\frac{df(t)}{dt}$$

Rule 5 follows directly from the definition:

$$\frac{dx}{dt} = \lim_{\Delta t\to 0}\frac{\Delta x}{\Delta t} = \lim_{\Delta x\to 0}\left(\frac{\Delta t}{\Delta x}\right)^{-1} = \left(\frac{dt}{dx}\right)^{-1}$$

Comments on Rule 7

We can obtain this important result using the binomial expansion. We have

$$f(t) = t^n$$

$$f(t + \Delta t) = (t + \Delta t)^n = t^n\left(1 + \frac{\Delta t}{t}\right)^n$$

$$= t^n\left[1 + n\frac{\Delta t}{t} + \frac{n(n-1)}{2!}\left(\frac{\Delta t}{t}\right)^2 + \frac{n(n-1)(n-2)}{3!}\left(\frac{\Delta t}{t}\right)^3 + \cdots\right]$$

Then

$$f(t - \Delta t) - f(t) = t^n\left[n\frac{\Delta t}{t} + \frac{n(n-1)}{2!}\left(\frac{\Delta t}{t}\right)^2 + \cdots\right]$$

and

$$\frac{f(t - \Delta t) - f(t)}{\Delta t} = nt^{n-1} + \frac{n(n-1)}{2!}t^{n-2}\,\Delta t + \cdots$$

The next term omitted from the last sum is proportional to $(\Delta t)^2$, the following to $(\Delta t)^3$, and so on. Each term except the first approaches zero as Δt approaches zero. Thus

$$\frac{df}{dt} = \lim_{\Delta x \to 0}\frac{f(t + \Delta t) - f(t)}{\Delta t} = nt^{n\pm 1}$$

Comments on Rules 8 to 10

We first write $\sin \omega t = \sin \theta$ with $\theta = \omega t$ and use the chain rule,

$$\frac{d \sin \theta}{dt} = \frac{d \sin \theta}{d\theta}\frac{d\theta}{dt} = \omega\frac{d \sin \theta}{d\theta}$$

We then use the trigonometric formula for the sine of the sum of two angles θ and $\Delta\theta$:

$$\sin(\theta + \Delta\theta) = \sin \Delta\theta \cos \theta + \cos \Delta\theta \sin \theta$$

Since $\Delta\theta$ is to approach zero, we can use the small-angle approximations

$$\sin \Delta\theta \approx \Delta\theta \qquad \text{and} \qquad \cos \Delta\theta \approx 1$$

Then

$$\sin(\theta + \Delta\theta) \approx \Delta\theta \cos \theta + \sin \theta$$

and

$$\frac{\sin(\theta + \Delta\theta) - \sin \theta}{\Delta\theta} \approx \cos \theta$$

Similar reasoning can be applied to the cosine function to obtain rule 9.

Rule 10 is obtained by writing $\tan \theta = \sin \theta/\cos \theta$ and applying rule 4 along with rules 8 and 9.

$$\frac{d}{dt}(\tan\theta) = \frac{d}{dt}(\sin\theta)(\cos\theta)^{-1} = \sin\theta\frac{d}{dt}(\cos\theta)^{-1} + \frac{d(\sin\theta)}{dt}(\cos\theta)^{-1}$$

$$= \sin\theta(-1)(\cos\theta)^{-2}(-\sin\theta) + (\cos\theta)(\cos\theta)^{-1}$$

$$= \frac{\sin^2\theta}{\cos^2\theta} + 1 = \tan^2\theta + 1 = \sec^2\theta$$

Comments on Rule 11

Again we use the chain rule

$$\frac{de^\theta}{dt} = \frac{bde^\theta}{bdt} = b\frac{de^\theta}{d(bt)} = b\frac{de^\theta}{d\theta} \qquad \text{with} \qquad \theta = bt$$

and the series expansion for the exponential function:

$$e^{\theta+\Delta\theta} = e^\theta e^{\Delta\theta} = e^\theta\left[1 + \Delta\theta + \frac{(\Delta\theta)^2}{2!} + \frac{(\Delta\theta)^3}{3!} + \cdots\right]$$

Then

$$\frac{e^{\theta+\Delta\theta} - e^\theta}{\Delta\theta} = e^\theta + e^\theta\frac{\Delta\theta}{2!} + e^\theta\frac{(\Delta\theta)^2}{3!} + \cdots$$

As $\Delta\theta$ approaches zero, the right side of the equation above approaches e^θ.

Comments on Rule 12

Let

$$y = \ln bt$$

Then

$$e^y = bt \qquad \text{and} \qquad \frac{dt}{dy} = \frac{1}{b}e^y = t$$

Then using rule 5, we obtain

$$\frac{dy}{dt} = \left(\frac{dt}{dy}\right)^{-1} = \frac{1}{t}$$

Integral Calculus

Integration is related to the problem of finding the area under a curve. It is also the inverse of differentiation. Figure D-23 shows a function $f(t)$. The area of the shaded element is approximately $f_i\Delta t_i$, where f_i is evaluated anywhere in the interval Δt_i. This approximation improves if Δt_i is very small. The total area from t_1 to t_2 is found by summing all the area elements from t_1 to t_2 and taking the limit as each Δt_i approaches zero. This limit is called the integral of f over t and is written

$$\int_{t_1}^{t_2} f\,dt = \text{Area} = \lim_{\Delta t_i \to 0} \sum_i f_i\,\Delta t_i$$

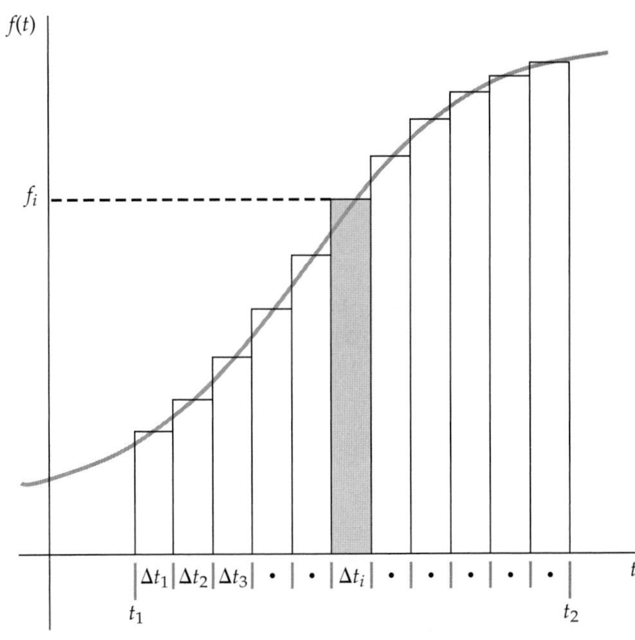

FIGURE D-23 A general function $f(t)$. The area of the shaded element is approximately $f_i\Delta t_i$, where f_i is evaluated anywhere in the interval.

If we integrate some function $f(t)$ from t_1 to some general value of t, we obtain another function of t. Let us call this function y:

$$y = \int_{t_1}^{t} f\,dt$$

The function y is the area under the f-versus-t curve from t_1 to a general value t. For a small interval Δt, the change in the area Δy is approximately $f\Delta t$.

$$\Delta y \approx f\,\Delta t$$

$$f \approx \frac{\Delta y}{\Delta t}$$

If we take the limit as Δt approaches 0, we can see that f is the derivative of y:

$$f = \frac{dy}{dt}$$

The relation between y and f is often written

$$y = \int f\,dt$$

where $\int f\,dt$ is called an **indefinite integral.** To evaluate an indefinite integral, we find the function y whose derivative is f. The definite integral of f from t_1 to t_2 is $y(t_1) - y(t_2)$, where $df/dt = y$:

$$\int_{t_1}^{t_2} f\,dt = y(t_2) - y(t_1)$$

EXAMPLE D-10

Find the indefinite integral of $f(t) = t$.

The function whose derivative is t is $\frac{1}{2}t^2$ plus any constant:

$$\int t\,dt = \tfrac{1}{2}t^2 + C$$

where C is any constant

Table D-5 lists some important integration formulas. More extensive lists of differentiation and integration formulas can be found on the Internet. They can be found by using a search engine and searching for *table of integrals,* and also in handbooks such as Herbert Dwight's *Tables of Integrals and Other Mathematical Data,* fourth edition, Macmillan Publishing Company, Inc., New York, 1961.

TABLE D-5

Integration Formulas†

1. $\int A\,dt = At$
2. $\int At\,dt = \tfrac{1}{2}At^2$
3. $\int At^n\,dt = A\dfrac{t^{n+1}}{n+1} \quad n \neq -1$
4. $\int At^{-1}\,dt = A\ln t$
5. $\int e^{bt}\,dt = \dfrac{1}{b}e^{bt}$
6. $\int \cos\omega t\,dt = \dfrac{1}{\omega}\sin\omega t$
7. $\int \sin\omega t\,dt = -\dfrac{1}{\omega}\cos\omega t$
8. $\int_0^\infty e^{-ax}\,dx = \dfrac{1}{a}$
9. $\int_0^\infty e^{-ax^2}\,dx = \dfrac{1}{2}\sqrt{\dfrac{\pi}{a}}$
10. $\int_0^\infty xe^{-ax^2}\,dx = \dfrac{2}{a}$
11. $\int_0^\infty x^2e^{-ax^2}\,dx = \dfrac{1}{4}\sqrt{\dfrac{\pi}{a^3}}$
12. $\int_0^\infty x^3e^{-ax^2}\,dx = \dfrac{4}{a^2}$
13. $\int_0^\infty x^4e^{-ax^2}\,dx = \dfrac{3}{8}\sqrt{\dfrac{\pi}{a^5}}$

† In these formulas, A, b, and ω are constants. In formulas 1 through 7 an arbitrary constant C can be added to the right side of each equation. The constant a is greater than zero.

ILLUSTRATION CREDITS

Chapter 1

Opener p. 1 Jeff Divine/FPG/Getty; **p. 4 (a)** The Granger Collection; **(b)** © 1999 Geoffrey Wheeler; **p. 6 (a)** McDonald Observatory; **(b)** Bruce Coleman; **p. 7** Eunice Harris/Photo Researchers, Inc.; **p. 11 (a)** IBM Almaden Research Center; **(b)** From *The Body Victorius,* The Delacourt Press/Lennart Nilsson; **(c)** Kent and Donnan Dannen/Photo Researchers; **(d)** NASA; **(e)** Smithsonian Institution.

Chapter 2

Opener p. 17 Corbis/Stock Market; **p. 23** Novastock/Dembinsky Photo Associates; **p. 24** Estate of Harold E. Edgerton/Palm Press Inc.; **p. 25** Gunter Ziesler/Peter Arnold Inc.; **p. 27** © Sydney Harris; **p. 28 Figure 2-9** James Sugar/Black Star; **p. 31** ©1994 General Motors Corporation. All rights reserved GM Media Archives; **p. 32 (left)** Stanford Linear Accelerator, U.S. Department of Energy; **(right)** Stanford Linear Accelerator Center, U.S. Department of Energy; **p. 40** Courtesy of Gene Mosca; **p. 46** Courtesy of Chuck Adler.

Chapter 3

Opener p. 53 Kevin Miller/Getty.

Chapter 4

Opener p. 85 John Neubauer/FPG/Getty; **p. 86** Jose Dupont/Explorer/Photo Researchers; **p. 88** NASA/Science Source/Photo Researchers; **p. 92 (a)** Cotton Coulson/Woodfin Camp and Assoc.; **(b)** Gary Ladd; **(c)** Los Alamos National Lab; **(d)** Science Photo Library/Photo Researchers; **p. 94 (b)** Fundamental Photographs; **Figure 4-7** David J. Phillip/AP Wide World.

Chapter 5

Opener p. 117 Courtesy of BMW; **p. 119 (top)** F.P. Bowden and D. Tabor, *Friction and Lubrication of Solids,* Oxford University Press, 2000; **(bottom)** Uzi Landman and David W. Leudtke/Georgia Institute of Technology; **p. 121** Nicole Villamora; **p. 122** Jean-Claude LeJeune/Stock, Boston; **p. 132** Courtesy of BMW; **p. 133** Sandia National Laboratory; **p. 134** NASA; **p. 135 (top)** Joe McBride/Stone; **(bottom)** Stuart Williams/Dembinsky Photo Associates; **p. 147 Figure 5-57** David de Lossy/The Image Bank.

Chapter 6

Opener p. 151 Getty; **p. 156** Courtesy of Dr. Roger Craig; **p. 168** Bill Bacjman/Photo Researchers; **p. 171 Figure 6-25** David J. Phillip/AP Wide World.

Chapter 7

Opener p. 183 Mark E. Gibson/Dembinsky Photo Associates; **p. 185** Loren Winters/Visuals Unlimited; **p. 194 (top left)** Visual Horizons/FPG International; **(top right)** New York State Commerce Department; **p. 195** Courtesy of Blyth Offshore Wind Limited; **p. 197** The Photo Works/Photo Researchers, Inc.; **p. 198** Stan Sholik/FPG International; **p. 205** Leicester University/Science Photo Library/Photo Researchers; **p. 215** Courtesy of PASCO.

Chapter 8

Opener p. 217 Jerry Wachter/Photo Researchers, Inc.; **p. 223 Figure 8-14** Estate of Harold E. Edgerton/Palm Press Inc.; **p. 228** Courtesy of Daedalon Corporation; **p. 229** NASA; **p. 234 (bottom)** Robert R. Edwards/BOB-E Photography; **p. 235 Figure 8-26** Romilly Lockyer/The Image Bank; **p. 237** Courtesy of Mercedes-Benz of N.A., Montvale, NJ; **p. 240** Estate of Harold E. Edgerton/Palm Press Inc.; **p. 244 (top)** Joe Strunk/Visuals Unlimited; **(bottom)** M. Hans/Vandystadtl/Photo Researchers, Inc.; **p. 246 (bottom right)** Brookhaven National Laboratory; **p. 251 Figure 8-47** NASA/Superstock.

Chapter 9

Opener p. 267 Photodisk; **p. 268** Fred Habegger/Grant Heilman Photography, Inc.; **p. 269** David Malin/Anglo-Australian Telescope Board; **p. 271** David Malin/Anglo-Australian Telescope Board; **p. 278** D.S. Kerr/Visuals Unlimited, Inc.; **p. 280** Richard Menga/Fundamental Photographs, Inc.; **p. 284** Fundamental Photographs, Inc.; **p. 288** Loren Winters/Visuals Unlimited; **p. 290** Scott Goldsmith/Stone/Getty; **p. 298 Figure 9-41** ©Treë.

Chapter 10

Opener p. 309 © Michael Newman/PhotoEdit; **p. 314** Dick Luria/Science Source/Photo Researchers; **p. 317** Courtesy of Segway; **p. 318 (left)** © The Harold E. Edgerton 1992 Trust; **(right)** Mike Powell/Getty; **p. 319** Dick Luria/FPG International; **p. 321** NASA/Goddard Space Flight Center; **p. 330** Chris Sorenson/The Stock Market; **p. 331** Chris Trotman/DUOMO/Corbis; **p. 337 Figure 10-56** Courtesy of Tangent Toy Co.

Relativity

Opener p. R-1 Courtesy of NASA.

Chapter 11

Opener p. 339 Stocktrek/Corbis; **p. 340** Collection of Historical Scientific Instruments, Harvard University; **p. 344 (top)** NASA; **(bottom)** NASA; **p. 345** Courtesy Central Scientific Company; **p. 353** NASA.

Chapter 12

Opener p. 370 Courtesy of Department of Physics, Purdue University; **p. 372** © 2002 Estate of Alexander Calder/Artists Rights Society (ARS), New York; **p. 379** Photodisk.

Chapter 13

Opener p. 395 Andy Pernick/Bureau of Reclamation; **p. 401** Vanessa Vick/Photo Researchers, Inc.; **p. 403** Chuck O'Rear/Woodfin Camp and Assoc.; **p. 405** David Burnett/Woodfin Camp and Assoc.; **p. 407 (top)** Estate of Harold E. Edgerton; **(bottom)** Takeski Takahara/Photo Researchers, Inc.; **p. 408** P. Motta/Photo Researchers, Inc.; **p. 412** Michael Dunn/The Stock Market; **p. 415** Picker International.

Chapter 14

Opener p. 425 Barry Slaven/Visuals Unlimited; **p. 427** Citibank; **p. 429** NASA; **p. 433** Institute for Marine Dynamics; **p. 442** Richard Menga/Fundamental Photographers; **p. 447 (top)** Monroe Auto Equipment; **(bottom)** David Wrobel/Visuals Unlimited; **p. 449** Eye Wire/Getty; **p. 451** Royal Swedish Academy of Music.

Chapter 15

Opener p. 465 John Cetrino/Check Six/Picture Quest; **p. 466 Figure 15-1** Richard Menga/Fundamental Photographs; **Figure 15-2** Richard Menga/Fundamental Photographs; **p. 478 (top right)** David Sacks/The Image Bank/Getty; **(left)** Maynard and Boucher/Visuals Unlimited; **p. 480** From Winston E. Cock, *Lasers and Holography,* Dover Publications, New York, 1981; **p. 484** Courtesy of Davies Symphony Hall; **p. 485 (top and bottom)** Fundamental Photographs; **p. 486 (top)** Bernard Benoit/Photo Researchers, Inc.; **(a)** Education Development Center; **p. 491 (top, a)** Sandia National Laboratory; **(top, b)** Robert de Gast/Photo Researchers; **(c)** Estate of Harold E. Edgerton/Palm Press Inc.; **(bottom, b)** Education Development Center; **p. 496 Figure 15-30** Estate of Harold E. Edgerton/Palm Press Inc.

Chapter 16

Opener p. 503 David Yost/Steinway & Sons; **p. 507** Rubberball Productions; **p. 510 (a)** Berenice Abbott (8J 1328)/Photo Researchers; **p. 513 (left)** University of Washington; **(center)** University of Washington; **(right)** University of Washington; **p. 519** Professor Thomas D. Rossing, Northern Illinois University, DeKalb; **p. 525** Courtesy of Chuck Adler.

Chapter 17

Opener p. 532 Hoby Finn/PhotoDisk/Getty; **p. 535 (a)** Courtesy of Taylor Precision Products; **(b)** Courtesy Honeywell, Inc.; **p. 536** Richard Menga/Fundamental Photographs; **p. 538** NASA; **p. 554** Jet Propulsion Laboratory/NASA.

Chapter 18

Opener p. 558 Donna Day/PhotoDisk/Getty; **p. 559** Phoenix Pipe & Tube/Lana Berkovich; **p. 561** From Frank Press and Raymond Sievert, *Understanding Earth,* 3rd ed., W.H. Freeman and Co., 2001; **p. 562** From Donald Wink, Sharon Gislason, and Sheila McNicholas, *The Practice of Chemistry,* W.H. Freeman and Co., 2002; **p. 582** Will and Deni McIntyre/Photo Researchers.

Chapter 19

Opener p. 595 (top) Paul Chesley/National Geographic/Getty; **p. 595 (bottom)** Sandia National Laboratory; **p. 598** © 2002 Robert Briggs; **p. 601** Anderson Ross/PhotoDisk/Getty; **p. 605 (right)** Michael Collier/Stock, Boston; **(left)** Jean-Pierre Horlin/The Image Bank; **p. 606 (top left)** Sandia National Laboratory; **(top right)** Peter Miller/The Image Bank; **(bottom)** Sandia National Laboratory.

Chapter 20

Opener p. 628 Frank Siteman/Stock Boston, Inc./PictureQuest; **p. 637** Alfred Pasieka/Photo Researchers, Inc.; **p. 639** Courtesy of Eugene Mosca; **p. 643** Science Photo Library/Photo Researchers, Inc.

ANSWERS

Problem answers are calculated using $g = 9.81 \text{ m/s}^2$ unless otherwise specified in the Problem. Differences in the last figure can easily result from differences in rounding the input data and are not important.

Chapter 14

1. $0; 4\pi^2 f^2 A$
3. *(a)* False
   (b) True
   (c) True
5. *(a)*
7. False
9. Assume that the first cart is given an initial velocity v by the blow. After the initial blow, there are no external forces acting on the carts, so their center of mass moves at a constant velocity $v/2$. The two carts will oscillate about their center of mass in simple harmonic motion where the amplitude of their velocity is $v/2$. Therefore, when one cart has velocity $v/2$ with respect to the center of mass, the other will have velocity $-v/2$. So, the velocity with respect to the laboratory frame of reference will be $+v$ and 0, respectively. Half a period later, the situation is reversed; so, one will appear to move as the other stops, and vice-versa.
11. True
13. Examples of driven oscillators include the pendulum of a clock, a bowed violin string, and the membrane of any loudspeaker.
15. Because f' varies inversely with the square root of m, taking into account the effective mass of the spring predicts that the frequency will be reduced.
17. *(d)*
19. *(b)*
21. 8π
23. *(a)* 3.00 Hz
    (b) 0.333 s
    (c) 7.00 cm
    (d) 0.0833 s; Because $v < 0$, the particle is moving in the negative direction at $t = 0.0833$ s.
25. *(a)* $x = (25\text{ cm})\cos[(4.19\text{ s}^{-1})t]$
    (b) $v = -(105\text{ cm/s})\sin[(4.19\text{ s}^{-1})t]$
    (c) $a = -(439\text{ cm/s}^2)\cos[(4.19\text{ s}^{-1})t]$
27. *(a)* $x = (27.7\text{ cm})\cos[(4.19\text{ s}^{-1})t - 0.445]$
    (b) $v = -(116\text{ cm/s})\sin[(4.19\text{ s}^{-1})t - 0.445]$
    (c) $a = -(486\text{ cm/s}^2)\cos[(4.19\text{ s}^{-1})t - 0.445]$
29. *(a)*

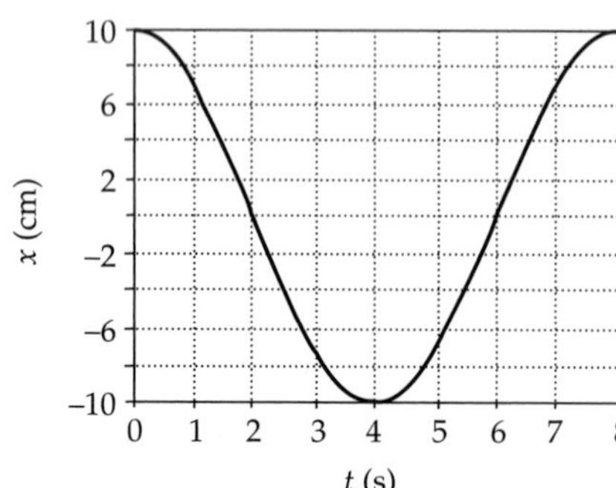

(b)

t_f	t_i	Δx
(s)	(s)	(cm)
1	0	2.93
2	1	7.07
3	2	7.07
4	3	2.93

31. *(a)* 7.85 m/s; 24.7 m/s²
    (b) −6.28 m/s; −14.8 m/s²
33. *(a)* 0.313 Hz
    (b) 3.14 s
    (c) $x = (40\text{ cm})\cos[(2\text{ s}^{-1})t + \delta]$
35. 22.5 J
37. *(a)* 0.368 J
    (b) 3.84 cm
39. 1.38 kN/m
41. *(a)* 6.89 Hz
    (b) 0.145 s
    (c) 0.100 m
    (d) 4.33 m/s
    (e) 187 m/s²
    (f) 36.3 ms; 0
43. *(a)* 682 N/m
    (b) 0.417 s
    (c) 1.51 m/s
    (d) 22.7 m/s²
45. *(a)* 3.08 kN/m
    (b) 4.16 Hz
    (c) 0.240 s
47. *(a)* 0.438 m/s
    (b) 0.379 m/s; 120 m/s²
    (c) 95.5 ms
49. 0.262 s
51. 10.1 kJ
53. *(a)* 0.997 Hz
    (b) 0.502 s
    (c) 0.294 N
55. *(a)* 46.66 cm
    (b) 0.261 s
    (c) 0.767 m/s
57. *(a)* 0.270 J
    (b) −0.736 J
    (c) 1.01 J
    (d) 0.270 J
59. *(a)* 1.90 cm
    (b) 0.0542 J
    (c) ±0.224 J
    (d) 0.334 J
61. 12.2 s
63. 11.7 s
65. $T = 2\pi\sqrt{\dfrac{L}{g(1 - \sin\theta)}}$
67. 1.10 s
69. 0.504 kg·m²
71. *(b)* 3.17 s
73. 21.1 cm from the center of the meter stick
77. *(a)* 1.63572 m
    (b) 14.5 mm, upward
79. 13.5°

81. 3.14%

85. *(a)* 0.314

(b) -3.13×10^{-2} percent

87. *(a)* 1.57%

(c) $0.430E_0$

89. *(a)* 1.01 Hz

(b) 2.01 Hz

(c) 0.352 Hz

91. *(a)* 4.98 cm

(b) 14.1 rad/s

(c) 35.4 cm

(d) 1.00 rad/s

93. *(a)* 0

(b) 4.00 m/s

95. *(a)* 14.1 cm; 0.444 s

(b) 23.1 cm; 0.363 s

(c) $(14.1\text{ cm})\sin[(14.1\text{ s}^{-1})t]$; $(23.1\text{ cm})\sin[(17.3\text{ s}^{-1})t]$

97. *(a)* $v = -(1.2\text{ m/s})\sin\left[(3\text{ rad/s})t + \frac{\pi}{4}\right]$

(b) -0.849 m/s

(c) 1.20 m/s

(d) 1.31 s

99. *(a)* The normal force is identical to the tension in a string of length r that keeps the particle moving in a circular path and a component of mg provides, for small displacements θ_0 or s_2, the linear restoring force required for oscillatory motion.

(b) The particles meet at the bottom. Because s_1 and s_2 are both much smaller than r, the particles behave like the bobs of simple pendulums of equal length and, therefore, have the same periods.

101. 1.62 s

103. 3.86×10^{-7} N·m/rad

105. g' is closer to g than is g''. Thus the error is greater if the clock is elevated.

107. *(a)* $\mu_s = \dfrac{Ak}{(m_1 + m_2)g}$

(b) A is unchanged. E is unchanged since $E = \frac{1}{2}kA^2$. ω is reduced by increasing the total mass of the system and T is increased.

109. *(b)* 2.04 cm/s²

113. *(a)* $x = 0$

(b) $v_s = x_0\sqrt{\dfrac{k}{m_b + m_p}}$

(c) $x_f = x_0\sqrt{\dfrac{m_p}{m_b + m_p}}$

115. *(a)*

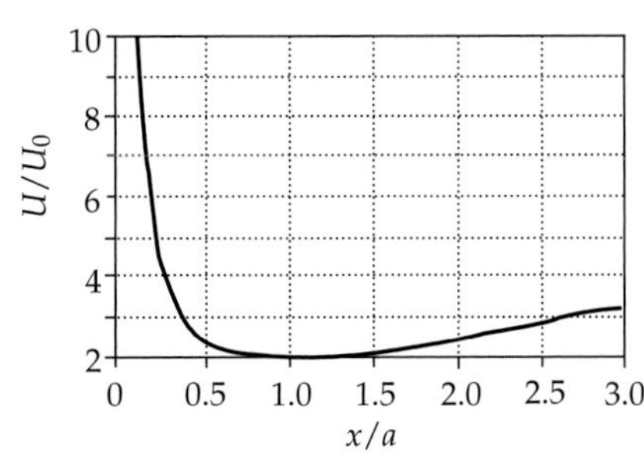

(b) $x_0 = a$ or $\alpha_0 = 1$

(c) $U(x_0 + \varepsilon) = U_0[1 + \beta + (1 + \beta)^{-1}]$

(d) $U(x_0 + \varepsilon) = \text{constant} + U_0\dfrac{\varepsilon^2}{a^2}$

119. 6.44×10^{13} rad/s

121. $7.78\sqrt{\dfrac{R}{g}}$

123. *(a)* 0.0478

(b) 0.00228

127. *(a)*

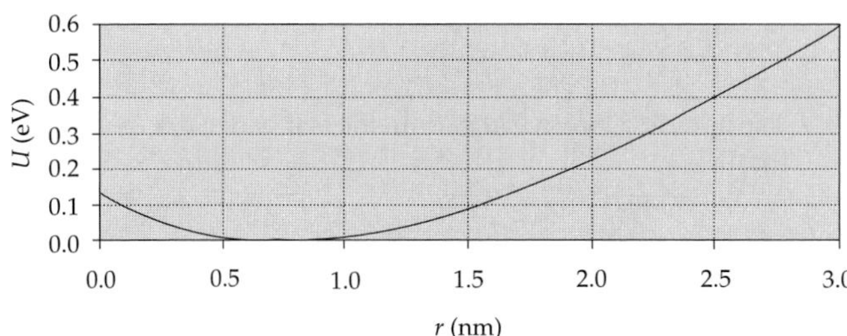

(b) $r = r_0$; $k = 2\beta^2 D$

(c) $\omega = 2\beta\sqrt{\dfrac{D}{m}}$

Chapter 15

1. The speed of a transverse wave on a rope is given by $v = \sqrt{F/\mu}$ where F is the tension in the rope and μ is its linear density. The waves on the rope move faster as they move up because the tension increases due to the weight of the rope below.

3. True

5. The speed of the wave v on the bullwhip varies with the tension F in the whip and its linear density μ according to $v = \sqrt{F/\mu}$. As the whip tapers, the wave speed in the tapered end increases due to the decrease in the mass density, so the wave travels faster.

7. No; Because the source and receiver are at rest relative to each other, there is no relative motion of the source and receiver and there will be no Doppler shift in frequency.

9. The light from the companion star will be shifted about its mean frequency periodically due to the relative approach to and recession from the earth of the companion star as it revolves about the black hole.

11. *(a)* True

(b) False

(c) False

13. There was only one explosion. Sound travels faster in water than air. Abel heard the sound wave in the water first, then, surfacing, heard the sound wave traveling through the air, which took longer to reach him.

15.

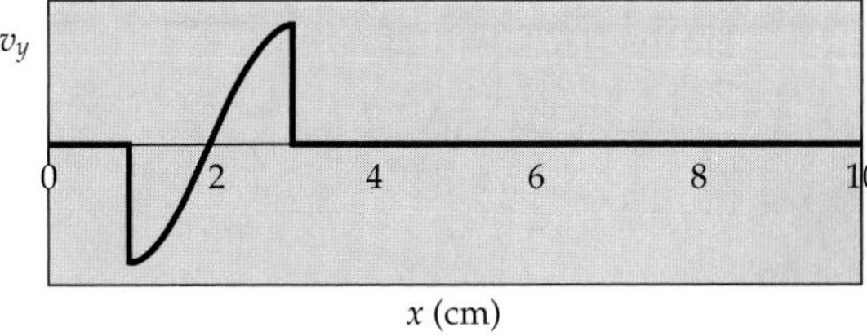

17. Path C. Because the wave speed is highest in the water, and more of path C is underwater than A or B, the sound wave will spend the least time on path C.

19. *(a)* 78.5 m
   (b) 69.7 m
   (c) 70.5 m . . . about 1% larger than our result in part (*b*) and 11% smaller than our first approximation in (*a*).

21. 270 m/s; 20.6%

23. 1.32 km/s

25. 19.6 g

27. *(a)* 265 m/s
   (b) 15.0 g

29. *(b)* 40.0 N

33. The lightning struck 680 m from the ball park, 58.4° W (or E) of north.

39. *(a)* $y(x,t) = A \sin k(x - vt)$
   (b) $y(x,t) = A \sin 2\pi\left(\frac{x}{\lambda} - ft\right)$
   (c) $y(x,t) = A \sin 2\pi\left(\frac{x}{\lambda} - \frac{1}{T}t\right)$
   (d) $y(x,t) = A \sin\frac{2\pi}{\lambda}(x - vt)$
   (e) $y(x,t) = A \sin 2\pi f\left(\frac{x}{v} - t\right)$

41. 9.87 W

43. *(a)* The wave is traveling in the $-x$ direction.; 5.00 m/s
   (b) 10.0 cm; 50.0 Hz; 0.0200 s
   (c) 0.314 m/s

45. *(a)* 6.82 J
   (b) 44.0 W

47. *(a)* 79.0 mW
   (b) Increasing f by a factor of 10 would increase P_{av} by a factor of 100. Increasing A by a factor of 10 would increase P_{av} by a factor of 100. Increasing F by a factor of 10^4 would increase v by a factor of 100 and P_{av} by a factor of 100.
   (c) Depending on the adjustability of the power source, increasing f or A would be the easiest.

49. *(a)* 0.750 Pa
   (b) 4.00 m
   (c) 85.0 Hz
   (d) 340 m/s

51. *(a)* 3.68×10^{-5} m
   (b) 8.27×10^{-2} Pa

53. *(a)* The displacement s is zero.
   (b) 3.68 μm

55. *(a)* 138 Pa
   (b) 21.7 W/m^2
   (c) 0.217 W

57. *(a)* 50.3 W
   (b) 2.00 m
   (c) 4.45×10^{-3} W/m^2

59. *(a)* 20.0 dB
   (b) 100 dB

61. 90.0 dB

65. *(a)* 100 m
   (b) 0.126 W

67. *(a)* 100 dB
   (b) 50.3 W
   (c) 2.00 m
   (d) 96.5 dB

69. *(a)* 81.1 dB
   (b) 80.0 dB; Eliminating the two least intense sources does not reduce the intensity level significantly.

71. 87.8 dB

73. 57.0 dB

75. *(a)* 260 m/s
   (b) 1.30 m
   (c) 262 Hz

77. *(a)* 1.70 m
   (b) 247 Hz

79. 153 Hz

81. 1021 Hz or a fraction increase of 2.06%; Because this fractional change in frequency is less than the 3% criterion for recognition of a change in frequency, it would be *impossible* to use your sense of pitch to estimate your running speed.

83. 349 mi/h

85. 7.78 kHz

87. 15.0 km west of P

89. *(a)* $f' = (1 - u_r/v)(1 - u_s/v)^{-1} f_0$

91. 1.33 m/s

93. *(a)* 824 Hz
   (b) 849 Hz

95. 184 m

97. -2.07×10^{-5} nm; 99 2.25×10^8 m/s

99. 2.25 × 10^8 m/s . . . where the upper arrow means the 8 is an exponent.

101. 20.8 cm

103. 3.42 m/s

105. 529 Hz; 474 Hz

107. 7.99 m

109. *(a)* 55.1 N/m^2
   (b) 3.46 W/m^2
   (c) 0.109 W

111. 77.0 kN

113. 204 m

115. 24.0 cm

117. *(b)* $v_0 = \sqrt{\frac{F}{\mu}}$
   (c) As seen by an observer at rest, the pulse remains at the same position because its speed along the chain is the same as the speed of the chain. With respect to a fixed point on the chain, the pulse travels through 360°.

119. *(b)* 2.21 s

Chapter 16

1.

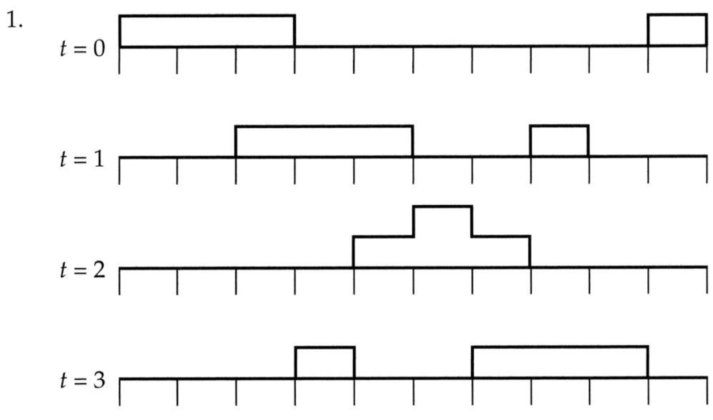

3. (c)
5. (b)
7. (a)
9. since $v \propto T$, increasing the temperature increases resonant frequencies.
11. No; the wavelength of a wave is related to its frequency and speed of propagation ($\lambda = v/f$). The frequency of the plucked string will be the same as the wave it produces in air, but the speeds of the waves depend on the media in which they are propagating. Since the velocities of propagation differ, the wavelengths will not be the same.
13. When the edges of the glass vibrate, sound waves are produced in the air in the glass. The resonance frequency of the air columns depends on the length of the air column, which depends on how much water is in the glass.
15. (b)
17. The pitch is determined mostly by the resonant cavity of the mouth, and the frequency of sounds he makes is directly proportional to their speed. Since $v_{He} > v_{air}$ (see Equation 15-5), the resonance frequency is higher if helium is the gas in the cavity.
19. Pianos are tuned by ringing the tuning fork and the piano note simultaneously and tuning the piano string until the beats are far apart (i.e., the time between beats is very long). If we assume that 2 s is the maximum detectable period for the beats, then one should be able to tune the piano string to at least 0.5 Hz.
21. 34.0 Hz; Because $v \propto T$, the frequency will be somewhat higher in the summer.
23. 7.07 cm
25. (a) 90.0°
    (b) $\sqrt{2}A$
27. (a) 0
    (b) $2I_0$
    (c) $4I_0$
29. (a) $\frac{1}{4}\lambda$
    (b) $\frac{1}{4}\lambda$
31. (a) 60.0 cm
    (b) $\dfrac{2\pi}{5}$
    (c) 24.0 m/s
33. 4726 Hz; 9452 Hz
35. (b)

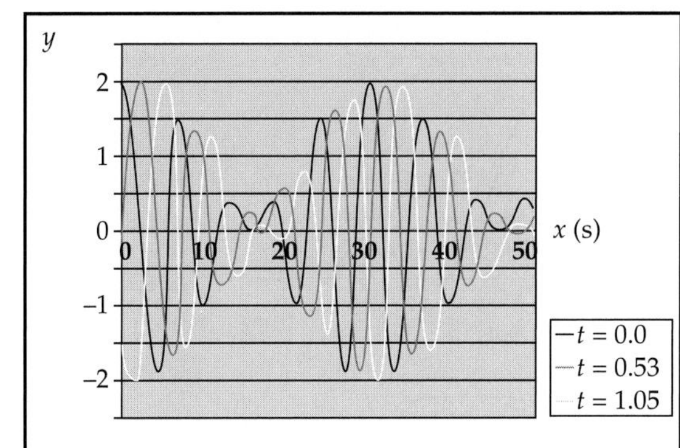

    (c) 0.500 m/s
37. 1.81; 51.5°
39. (a) 0.279 m
    (b) 1.22 kHz
    (c)

m	θ_m
	(rad)
3	0.432
4	0.592
5	0.772
6	0.992
7	1.354
8	undefined

    (d) 0.0698 rad
41. 1.98 rad or 113°
43. (a) 70.5 Hz
    (b) The person on the street hears no beat frequency as the sirens of both ambulances are Doppler shifted up by the same amount (approximately 35 Hz).
45. (a) 2.00 m; 25.0 Hz
    (b) $y_3(x,t) = (4\text{ mm})\sin kx \cos \omega t$, where $k = \pi\text{m}^{-1}$ and $\omega = 50\pi\text{s}^{-1}$.
47. (a) 521 m/s
    (b) 2.80 m; 186 Hz
    (c) 372 Hz; 558 Hz
49. 141 Hz
51. (a) 31.4 cm; 47.7 Hz
    (b) 15.0 m/s
    (c) 62.8 cm

53. (a)

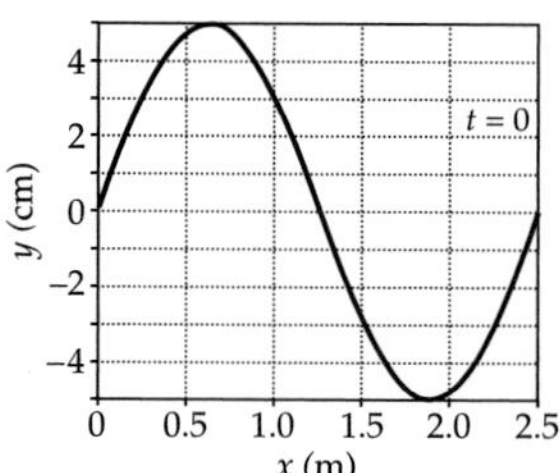

t = T/4 and 3T/4

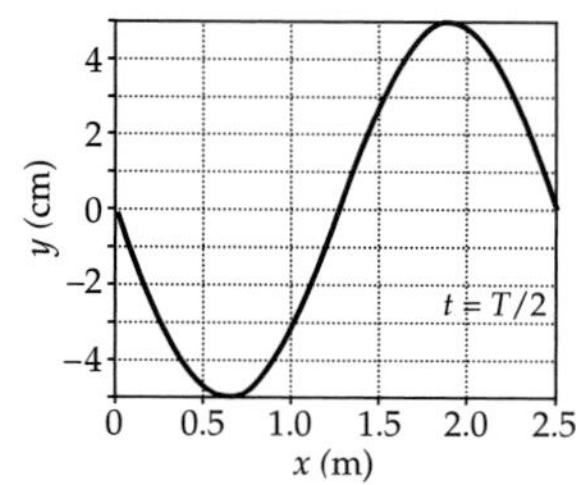

(b) 12.6 ms

(c) Since the string is moving either upward or downward when $y(x) = 0$ for all x, the energy of the wave is entirely kinetic energy.

55. (a) 70.8 Hz

(b) 4.89 Hz

(c) 35

57. 452 Hz; It would be better to have the pipe expand so that v/L, where L is the length of the pipe, is independent of temperature.

59. (a) 80 cm

(b) 480 N

(c) You should place your finger 9.23 cm from the scroll bridge.

61. (a) 75.0 Hz

(b) The harmonics are the 5th and 6th.

(c) 2.00 m

63. (a) 0.574 g/m

(b) 1.29 g/m; 2.91 g/m; 6.55 g/m

65. (a) The two sounds produce a beat because the third harmonic of the A string equals the second harmonic of the E string, and the original frequency of the E string is slightly greater than 660 Hz. If $f_E = (660 + \Delta f)$Hz, a beat of $2\Delta f$ will be heard.

(b) 661.5 Hz

(c) 79.6 N

69. 76.8 N; 19.2 N; 8.53 N

71. (a) N/f_0

(b) $\Delta x/N$

(c) $2\pi N/\Delta x$

(d) N is uncertain because the waveform dies out gradually rather than stopping abruptly at some time; hence, where the pulse starts and stops is not well defined.

73. (a) 3.40 kHz; 10.2 kHz; 17.0 kHz

(b) Frequencies near 3400 Hz will be most readily perceived.

75. $\frac{1}{3}\lambda$

77. 6.62 m

79. (a) 1.90 cm; 3.59 m/s

(b) 0; 0

(c) 1.18 cm; 2.22 m/s

(d) 0; 0

81. (a) At resonance, standing waves are set up in the tube. At a displacement antinode, the powder is moved about; at a node the powder is stationary, and so it collects at the nodes.

(b) $2fD$

(c) If we let the length L of the tube be 1.2 m and assume that $v_{\text{air}} = 344$ m/s (the speed of sound in air at 20°C), then the 10th harmonic corresponds to $D = 25.3$ cm and a driving frequency of 680 Hz.

(d) If $f = 2$ kHz and $v_{\text{He}} = 1008$ m/s (the speed of sound in helium at 20°), then D for the 10th harmonic in helium would be 25.3 cm, and D for the 10th harmonic in air would be 8.60 cm. Hence, neglecting end effects at the driven end, a tube whose length is the least common multiple of 8.60 cm and 25.3 cm (218 cm) would work well for the measurement of the speed of sound in either air or helium.

83. (a) The pipe is closed at one end.

(b) 262 Hz

(c) 32.4 cm

85. (a) $y_1(x,t) = (0.01\text{ m})\sin\left[\left(\frac{\pi}{2}\text{ m}^{-1}\right)x - (40\pi\text{ s}^{-1})t\right]$;

$y_2(x,t) = (0.01\text{ m})\sin\left[\left(\frac{\pi}{2}\text{ m}^{-1}\right)x + (40\pi\text{ s}^{-1})t\right]$;

(b) 2.00 m

(c) $v_y(1\text{ m},t) = -(2.51\text{ m/s})\sin(40\pi\text{ s}^{-1})t$

(d) $a_y(1\text{ m},t) = -(316\text{ m/s}^2)\cos(40\pi\text{ s}^{-1})t$

87. $y_{\text{res}}(x,t) = 0.1\sin(kx - \omega t)$

89. (b) 203 Hz

91. (a) What you hear is the fundamental mode of the tube and its overtones. A more physical explanation is that the echo of the finger snap moves back and forth along the tube with a characteristic time of $2L/c$, leading to a series of clicks from each echo. Since the clicks happen with a frequency of $c/2L$, the ear interprets this as a musical note of that frequency.

(b) 38.6 cm

93. (a) Since no conditions were placed on its derivation, this expression is valid for all harmonics.

(b) 1.54%

95. (a) $v_y(x,t) = -\omega_1 A_1 \sin\omega_1 t \sin k_1 x - \omega_2 A_2 \sin\omega_2 t \sin k_2 x$

(b) $dK = \frac{1}{2}\mu[\omega_1^2 A_1^2 \sin^2\omega_1 t \sin^2 k_1 x + 2\omega_1\omega_2 A_1 A_2 \sin\omega_1 t \sin k_1 x \sin\omega_2 t \sin k_2 x + \omega_2^2 A_2^2 \sin^2\omega_2 t \sin^2 k_2 x]dx$

(c) $K = \frac{1}{4}m\omega_1^2 A_1^2 \sin^2\omega_1 t + \frac{1}{4}m\omega_2^2 A_2^2 \sin^2\omega_2 t$

97. (a)

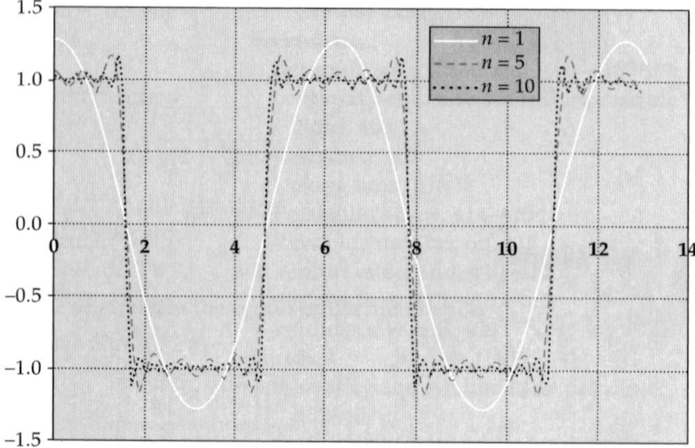

(b) $f(2\pi) = 1$ which is equivalent to the Liebnitz formula.

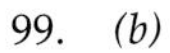

99. *(b)*

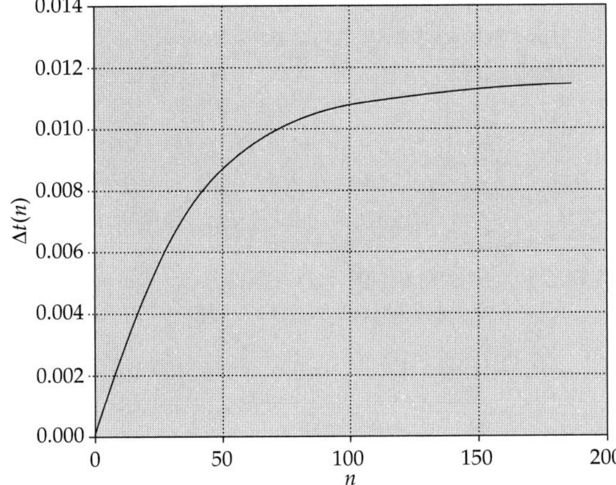

(c) The frequency heard at any time is $1/\Delta t_n$, so because Δt_n increases over time, the frequency of the culvert whistler decreases.; 7.65 kHz

Chapter 17

1. *(a)* False
   (b) False
   (c) True
   (d) False
3. Mert's room was colder.
5. From the ideal-gas law we have $P = nRT/V$. In the process depicted, both the temperature and the volume increase but the temperature increases faster than does the volume. Hence the pressure increases.
7. True
9. K_{av} increases by a factor of 2; K_{av} is reduced by a factor of $\frac{1}{2}$.
11. False
13. Since $10^7 >> 273$, it does not matter.
15. *(b)*
17. *(d)*
19. The ratio of the rms speeds is inversely proportional to the square root of the ratio of the molecular masses. The kinetic energies of the molecules are the same.
21. Because the temperature remains constant, the average speed of the molecules remains constant. When the volume decreases, the molecules travel less distance between collisions, so the pressure increases because the frequency of collisions increases.
23. The average molecular speed of He gas at 300 K is about 1.4 km/s, so a significant fraction of He molecules have speeds in excess of earth's escape velocity (11.2 km/s), and thus "leak" away into space. Over time, the He content of the atmosphere decreases to almost nothing.
25. *(a)* 3.61×10^3 K
   (b) 225 K
   (c) If $v_{rms} > \frac{1}{5}v_e$ or $T \geq 25T_{atm}$, H_2 molecules escape. Therefore, the more energetic H_2 molecules escape from the upper atmosphere.
   (d) 164 K; 10.3 K; If we assume that the temperature on the moon with an atmosphere would have been approximately 1000 K, then all O_2 and H_2 would have escaped during the time since the formation of the moon to the present.
27. *(a)* 1.24 km/s
   (b) 310 m/s
   (c) 264 m/s
   (d) O_2, CO_2, and H_2 should be found on Jupiter.
29. 1063°C
31. *(a)* 8.40 cm
   (b) 107°C
33. −319°F
35. *(a)* 54.9 torr
   (b) 3704 K
37. −40°C = −40°F
39. −183°C; −297°F
41. *(a)* $B = 3.94 \times 10^3$ K; $R_0 = 3.97 \times 10^{-3}\ \Omega$
   (b) 1.31 kΩ
   (c) −389 Ω/K; −433 Ω/K
   (d) The thermistor is more sensitive (i.e., has greater sensitivity, at lower temperatures).
43. 1.79 mol; 1.08×10^{24} molecules
45. −83.2 *glips*
47. *(a)* 3.66×10^3 mol
   (b) 60.0 mol
49. 10.0 atm
51. 1.19 kg/m³
53. 2.56 N
55. *(a)* 276 m/s
   (b) 872 m/s
57. 499 km/s; 2.07×10^{-16} J
61. $K/\Delta U = 7.95 \times 10^4$
65. *(a)* 0.142 s
   (b) 0.143 s
67. *(a)* 122 K
   (b) 244 K
   (c) 1.43 atm
69. 111 mol; 55.5 mol
71. $7m_H$
73. 400.49 K
75. *(a)* 4.10×10^{-26} m
   (b) 4.28 nm; The mean free path is larger by approximately a factor of 1000.
77. *(a)* 48.9%
   (b) 70.6%

Chapter 18

1. $\Delta T_B = 4\Delta T_A$
3. *(c)*
5. Yes, if the heat adsorbed by the system is equal to the work done by the system.
7. $W_m + Q_m = \Delta E_{int}$; For an ideal gas, ΔE_{int} is a function of T only. Since $W = 0$ and $Q = 0$ in a free expansion, $\Delta E_{int} = 0$ and T is constant. For a real gas, ΔE_{int} depends on the density of the gas because the molecules exert weak attractive forces on each other. In a free expansion, these forces reduce the average kinetic energy of the molecules and, consequently, the temperature.
9. The temperature of the gas increases. The average kinetic energy increases with increasing volume due to the repulsive interaction between the ions.

11. *(a)*

13. *(a)* False
    (b) False
    (c) False
    (d) True
    (e) True
    (f) True
    (g) True

15. *(d)*

17. If V decreases, the temperature decreases.

19. The heat capacity of a substance is proportional to the number of degrees of freedom per molecule associated with the molecule. Since there are 6 degrees of freedom per molecule in a solid, and only 3 per molecule (translational) for a monatomic liquid, you would expect the solid to have the higher heat capacity.

21. 1.63 min, an elapsed time that seems to be consistent with experience.

23. $c_p = (1.01\%)c_{water}$

25. *(a)* 10.5 MJ
    (b) 121 W

27. 7.48 kcal

29. 48.8 mg

31. 365°C

33. 20.8°C

35. 453 kg

37. *(a)* 0°C
    (b) 125 g

39. *(a)* 4.94°C
    (b) No ice is left.

41. *(a)* 2.99°C
    (b) 199.8 g
    (c) The answer would be the same.

43. 618°C

45. 2.21 kJ

47. 176°C

49. 53.7 J

51. *(a)* 6.13 W
    (b) 19.0 min

53. *(a)* 405 J
    (b) 861 J

55. *(a)* 507 J
    (b) 963 J

57. $\frac{3}{2}P_0V_0$

59. *(a)* 555 J
    (b) 555 J

61. *(a)* 55.7 g/mol
    (b) Fe

63. *(a)* 0; 6.24 kJ; 6.24 kJ
    (b) 8.73 kJ; 6.24 kJ; 2.49 kJ
    (c) 2.49 kJ

65. 59.6 L

67. $\Delta C_P = -\frac{13}{2}Nk$

69. $C_{V,water} = 5Nk$

71. *(a)* 465 K
    (b) 387 K

73. *(a)* 300 K; 7.80 L; 1.14 kJ; 1.14 kJ
    (b) 208 K; 5.41 L; 574 J; 0

75. *(a)* 263 K
    (b) 10.8 L
    (c) −1.48 kJ
    (d) 1.48 kJ

79. −142 J

81. $Q_{D\to A} = 8.98$ kJ; $Q_{A\to B} = 13.2$ kJ; $Q_{B\to C} = -8.98$ kJ; $Q_{C\to D} = -6.56$ kJ; $W_{cycle} = 6.62$ kJ

83. *(a)*

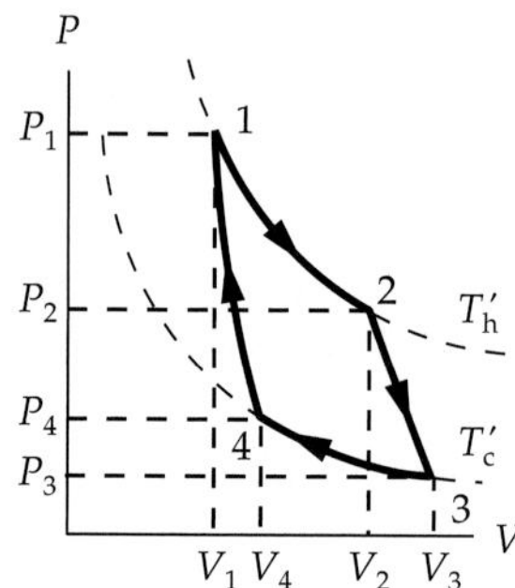

85. 180 kJ

87. *(a)* 65.2 K; 81.2 K
    (b) 1.62 kJ
    (c) 2.22 kJ

89. *(a)* 65.2 K; 81.2 K
    (b) 2.65 kJ
    (c) 3.25 kJ

91. *(a)* 9.20×10^{-2} J/kg·K
    (b) 0.0584 J/kg

93. 47.6 kPa; 51.5 K; 71.2 K; 148 kPa

95. *(a)* 2.49 kJ
    (b) 3.20 kJ

97. 171 K

99. *(a)* $W = 0$; $Q = 3.74$ kJ
    (b) $\Delta U = 3.74$ kJ; $Q = 6.24$ kJ; $W = 2.50$ kJ

101.

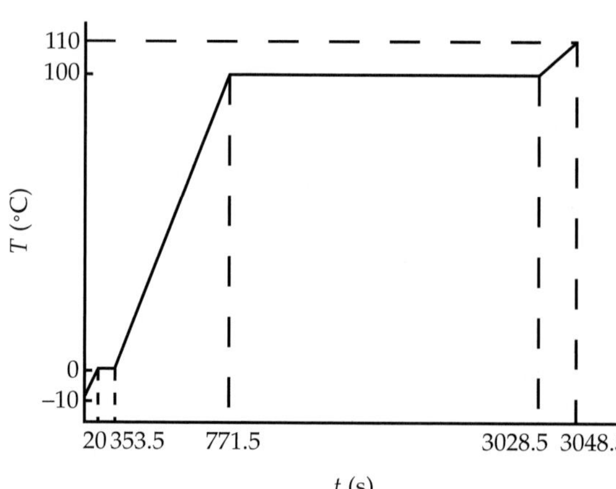

103. $4RT$

105. 396 K

107. *(a)* $\frac{1}{2}P_0$
     (b) diatomic
     (c) In the isothermal process, T is constant, and the translational kinetic energy is unchanged. In the adiabatic process, $T_3 = 1.32T_0$, and the translational kinetic energy increases by a factor of 1.32.

109. (*a*) 93.5 kPa
(*b*) 6266 K; 1.30 MPa
(*c*) 56.7 kPa

111. (*b*) $\Delta U = 4621$ J, a result in good agreement with the result of Problem 106.

Chapter 19

1. Friction reduces the efficiency of the engine.
3. Increasing the temperature of the steam increases the Carnot efficiency, and generally increases the efficiency of any heat engine.
5. (*c*)
7. (*d*)
9. Note that A→B is an adiabatic expansion. B→C is a constant volume process in which the entropy decreases; therefore heat is released. C→D is an adiabatic compression. D→A is a constant volume process that returns the gas to its original state. The cycle is that of the Otto engine (see Figure 19-3).
11.

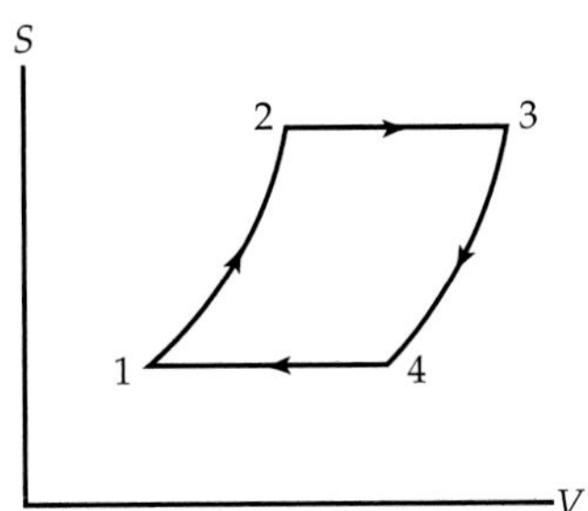

13.

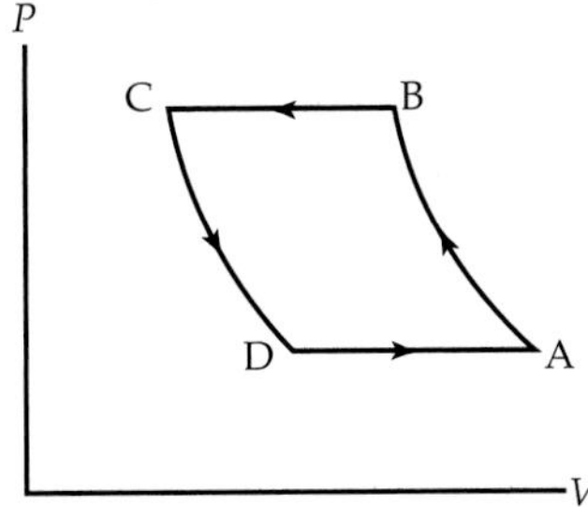

15. 56.5%
17. (*a*) 1.66×10^{17} W
(*b*) 5.66×10^{14} J/K·s
(*c*) 3.09×10^{13} J/K·s
19. 29.8 kJ/K
21. (*a*) 500 J
(*b*) 400 J
23. (*a*) 40.0%
(*b*) 80.0 W
25. (*a*)

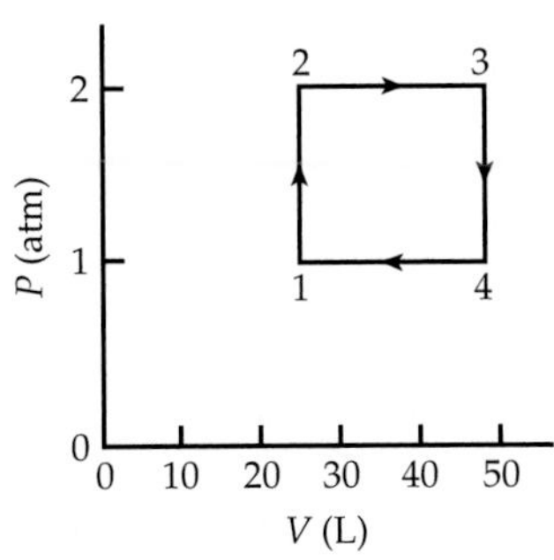

$W_{1\to2} = 0$; $Q_{1\to2} = 3.74$ kJ
$W_{2\to3} = 4.99$ kJ; $Q_{2\to2} = 12.5$ kJ
$W_{3\to4} = 0$; $Q_{3\to4} = -7.48$ kJ
$W_{4\to1} = 2.49$ kJ; $Q_{4\to1} = -6.24$ kJ

(*b*) 15.4%

27.

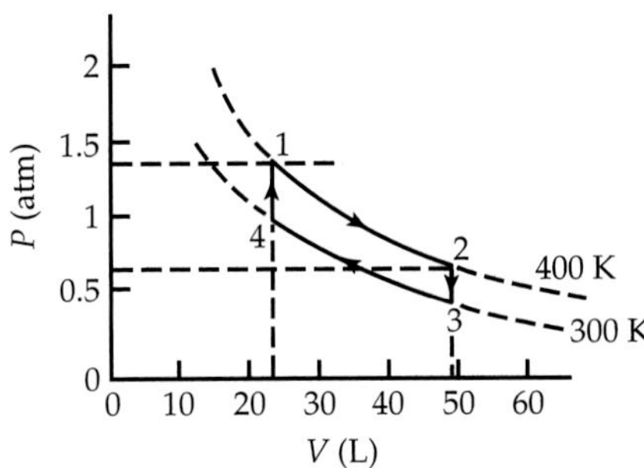

13.1%

29. (*a*) 600 K; 1800 K; 600 K
(*b*) 15.4%
31. (*a*) 5.16%; The fact that this efficiency is considerably less than the actual efficiency of a human body does not contradict the Second Law of Thermodynamics. The application of the second law to chemical reactions such as the ones that supply the body with energy have not been discussed in the text.
(*b*) Most warm-blooded animals survive under roughly the same conditions as humans. To make a heat engine work with appreciable efficiency, internal body temperatures would have to be maintained at an unreasonably high level.
35. (*a*) 33.3%
(*b*) 33.3 J
(*c*) 66.7 J
(*d*) 2.00
37. Let the first engine be run as a refrigerator. Then it will remove 140 J from the cold reservoir, deliver 200 J to the hot reservoir, and require 60 J of energy to operate. Now take the second engine and run it between the same reservoirs, and let it eject 140 J into the cold reservoir, thus replacing the heat removed by the refrigerator. If ε_2, the efficiency of this engine, is greater than 30%, then Q_{h2}, the heat removed from the hot reservoir by this engine, is $140\text{ J}/(1 - \varepsilon_2) > 200$ J, and the work done by this engine is $W = \varepsilon_2 Q_{h2} > 200$ J. The end result of all this is that the second engine can run the refrigerator, replacing the heat taken from the cold reservoir, and do additional mechanical work. The two systems working together then convert heat into mechanical energy without rejecting any heat to a cold reservoir, in violation of the second law.
39. (*a*) 33.3%
(*b*) If COP > 2, then 50 J of work will remove more than 100 J of heat from the cold reservoir and put more than 150 J of heat into the hot reservoir. So running engine (*a*) to operate the refrigerator with a COP > 2 will result in the transfer of heat from the cold to the hot reservoir without doing any net mechanical work in violation of the second law.
41. (*a*) 100°C
(*b*) $Q_{1\to2} = 3.12$ kJ; $Q_{2\to3} = 0$; $Q_{3\to1} = -2.91$ kJ
(*c*) 6.73%
(*d*) 35.5%
43. (*a*) 5.26
(*b*) 3.19 kW
(*c*) 4.81 kW

45. (a) 173 kJ
(b) 121 kJ

47. $\Delta S_u = 2.40\ \text{J/K}$

49. (a) 11.5 J/K
(b) Since the process is not quasi-static, it is nonreversible and the entropy of the universe must increase.

51. 1.22 kJ/K

53. (a) 0
(b) 267 K

55. (a) 244 kJ/K
(b) −244 kJ/K
(c) $\Delta S_u > 0$

57. (a) −117 J/K
(b) 137 J/K
(c) 20.3 J/K

59. 1.97 kJ/K

61. (a) 0.417 J/K
(b) 125 J

63. (a) 20.0 J
(b) 66.7 J; 46.7 J

65. (a) 51.0%
(b) 102 kJ
(c) 98.0 kJ

67. 113 W/K

69. (a) Process (1) is more wasteful of *mechanical* energy. Process (2) is more wasteful of *total* energy.
(b) 1.67 J/K; 0.833 J/K

71. 313 K

73. 10.0 W

75. (a) 253 kPa
(b) 462 K
(c) 6.96 kJ; 25.9%

77. (a) 253 kPa
(b) 416 K
(c) 6.58 kJ; 34.8%

79. 180 J

83. $\approx 10^{478}$

Chapter 20

1. The glass bulb warms and expands first, before the mercury warms and expands.

3. (c)

5. (a) With increasing altitude P decreases; from curve OF, T of the liquid-gas interface diminishes, so the boiling temperature decreases. Likewise, from curve OH, the melting temperature increases with increasing altitude.
(b) Boiling at a lower temperature means that the cooking time will have to be increased.

7. The thermal conductivity of metal and marble is much greater than that of wood; consequently, heat transfer from the hand is more rapid.

9. (c)

11. In the absence of matter to support conduction and convection, radiation is the only mechanism.

13. (a)

15. The temperature of an object is inversely proportional to the maximum wavelength at which the object radiates (Wein's displacement law). Since blue light has a shorter wavelength than red light, an object for which the wavelength of the peak of thermal emission is blue is hotter than one which is red.

17. 18.1 mW/(m·K)

19. 2.90 nm

21. (a) $\gamma \equiv \dfrac{\Delta A/A}{\Delta T}$
(b) $\gamma \approx 2\alpha\Delta T$

23. 217°C

25. $15.4 \times 10^{-6}\ \text{K}^{-1}$

27. 5.24 m

29. 0.255 mm

31. (a) The clock runs slow.
(b) 8.21 s

33. $3.68 \times 10^{-12}\ \text{N/m}^2$

35. (a) 90°C
(b) 82°C
(c) 170 kPa

37. (b) $\left(P_r + \dfrac{3}{V_r^2}\right)(3V_r - 1) = 8T_r$

39. 2.07 kBtu/h

41. (a) $I_{Cu} = 962\ \text{W}$; $I_{Al} = 569\ \text{W}$
(b) 1.53 kW
(c) 0.0523 K/W

43. (a) Conservation of energy requires that the thermal current through each shell be the same.

45. $I = \dfrac{2\pi kL}{\ln(r_1/r_2)}(T_2 - T_1)$

47. $9.35 \times 10^{-3}\ \text{m}^2$

49. 1598°C

51. 2.10 km

53. 5767 K

55. 1.18 cm

57. (b) $\dfrac{\beta_{exp} - \beta_{th}}{\beta_{th}} < \boxed{0.3\%}$

59. 1.26×10^{10} kW; <0.002%

61. 132 W ignoring the cylindrical insulation; 142 W taking the insulation into account.

63. $L_2 = L_1$; $\omega_2 \approx (1 - 2\alpha\Delta T)\omega_1$; $E_2 = E_1(1 - 2\alpha\Delta T)$

65. (a) 0.698 cm/h
(b) 11.9 d

67. (b) 40.5 min

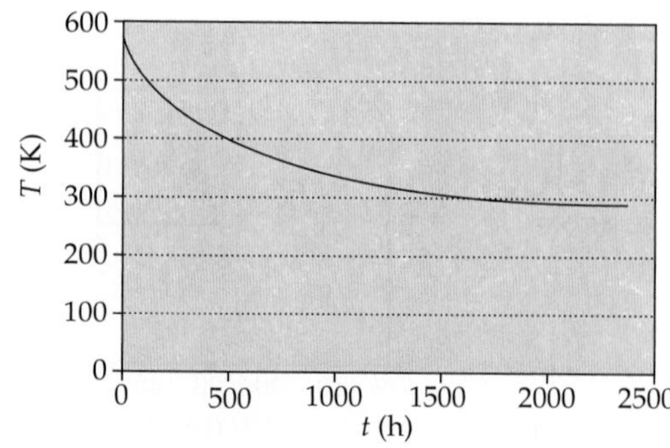

INDEX

Physical Constants†

Atomic mass constant	$m_u = \frac{1}{12}m(^{12}\text{C})$	$1\ \text{u} = 1.660\,538\,73(13) \times 10^{-27}$ kg
Avogadro's number	N_A	$6.022\,141\,99(47) \times 10^{23}$ particles/mol
Boltzmann constant	$k = R/N_A$	$1.380\,6503(24) \times 10^{-23}$ J/K $8.617\,342(15) \times 10^{-5}$ eV/K
Bohr magneton	$m_B = e\hbar/(2m_e)$	$9.274\,008\,99(37) \times 10^{-24}$ J/T = $5.788\,381\,749(43) \times 10^{-5}$ eV/T
Coulomb constant	$k = 1/(4\pi\epsilon_0)$	$8.987\,551\,788\ldots \times 10^{9}$ N·m²/C²
Compton wavelength	$\lambda_C = h/(m_e c)$	$2.426\,310\,215(18) \times 10^{-12}$ m
Fundamental charge	e	$1.602\,176\,462(63) \times 10^{-19}$ C
Gas constant	R	$8.314\,472(15)$ J/(mol·K) = $1.987\,2065(36)$ cal/(mol·K) = $8.205\,746(15) \times 10^{-2}$ L·atm/(mol·K)
Gravitational constant	G	$6.673(10) \times 10^{-11}$ N·m²/kg²
Mass of electron	m_e	$9.109\,381\,88(72) \times 10^{-31}$ kg = $0.510\,998\,902(21)$ MeV/c^2
Mass of proton	m_p	$1.672\,621\,58(13) \times 10^{-27}$ kg = $938.271\,998(38)$ MeV/c^2
Mass of neutron	m_n	$1.674\,927\,16(13) \times 10^{-27}$ kg = $939.565\,330(38)$ MeV/c^2
Permittivity of free space	ϵ_0	$8.854\,187\,817\ldots \times 10^{-12}$ C²/(N·m²)
Permeability of free space	μ_0	$4\pi \times 10^{-7}$ N/A²
Planck's constant	h	$6.626\,068\,76(52) \times 10^{-34}$ J·s = $4.135\,667\,27(16) \times 10^{-15}$ eV·s
	$\hbar = h/(2\pi)$	$1.054\,571\,596(82) \times 10^{-34}$ J·s = $6.582\,118\,89(26) \times 10^{-16}$ eV·s
Speed of light	c	$2.997\,924\,58 \times 10^{8}$ m/s
Stefan-Boltzmann constant	σ	$5.670\,400(40) \times 10^{-8}$ W/(m²·K⁴)

† The values for these and other constants can be found in Appendix B as well as on the Internet at http://physics.nist.gov/cuu/Constants/index.html. The numbers in parentheses represent the uncertainties in the last two digits. (For example, 2.044 43(13) stands for 2.044 43 ± 0.000 13.) Values without uncertainties are exact. Values with ellipses are exact (like the number $\pi = 3.1415\ldots$).

Derivatives and Definite Integrals

$$\frac{d}{dx}\sin ax = a\cos ax \qquad \int_0^\infty e^{-ax}\,dx = \frac{1}{a} \qquad \int_0^\infty x^2 e^{-ax^2}\,dx = \frac{1}{4}\sqrt{\frac{\pi}{a^3}}$$

$$\frac{d}{dx}\cos ax = -a\sin ax \qquad \int_0^\infty e^{-ax^2}\,dx = \frac{1}{2}\sqrt{\frac{\pi}{a}} \qquad \int_0^\infty x^3 e^{-ax^2}\,dx = \frac{4}{a^2}$$

$$\frac{d}{dx}e^{ax} = ae^{ax} \qquad \int_0^\infty xe^{-ax^2}\,dx = \frac{2}{a} \qquad \int_0^\infty x^4 e^{-ax^2}\,dx = \frac{3}{8}\sqrt{\frac{\pi}{a^5}}$$

The a in the six integrals is a positive constant.

Vector Products

$\vec{A}\cdot\vec{B} = AB\cos\theta$ $\qquad \vec{A}\times\vec{B} = AB\sin\theta\,\hat{n}$ ($\hat{n}$ obtained using right-hand rule)